DIET-MICROBE INTERACTIONS IN THE GUT

DIET-MICROBE INTERACTIONS IN THE GUT

Effects on Human Health and Disease

Edited by

Kieran Tuohy

Department of Food Quality and Nutrition, Research and Innovation Centre, Foundation Edmund Mach (FEM), San Michele all' Adige, Trento, Italy

Daniele Del Rio

Department of Food Science, University of Parma, Parma, Italy

AMSTERDAM • BOSTON • HEIDELBERG • LONDON
NEW YORK • OXFORD • PARIS • SAN DIEGO
SAN FRANCISCO • SINGAPORE • SYDNEY • TOKYO
Academic Press is an imprint of Elsevier

Academic Press is an imprint of Elsevier
32 Jamestown Road, London NW1 7BY, UK
225 Wyman Street, Waltham, MA 02451, USA
525 B Street, Suite 1800, San Diego, CA 92101-4495, USA

British Library Cataloguing-in-Publication Data
A catalogue record for this book is available from the British Library

Library of Congress Cataloging-in-Publication Data
A catalog record for this book is available from the Library of Congress

ISBN: 978-0-12-407825-3

For information on all Academic Press publications visit our website at elsevierdirect.com

Typeset by MPS Limited, Chennai, India
www.adi-mps.com

Printed and bound in USA

14 15 16 17 10 9 8 7 6 5 4 3 2 1

Contents

Foreword

Recent metagenomic studies are confirming what pioneers in gut microbiology have been saying for a long time, that diet:microbe interactions in the gut are critical for human health and disease. Overcoming the limitations of culture based microbiology to study the ecology of the mainly anaerobic, culture recalcitrant and fastidious microorganisms present within the human gut, the post genomics technologies of metagenomics and metabolomics are revealing the intricate relationship between microbiota species architecture and metabolic output, and diverse host physiological functions linked to health and chronic non-communicable diseases. The gut microbiota is now emerging as an important metabolic and immunological organ in its own right, rivalling (and overcoming) the liver in metabolic diversity and activity, and playing a key role in immune education and homeostasis. This newly recognised metabolic and immunologically paramount organ is closely linked to other bodily systems, including adipose tissue, muscle, and even the brain, through the flux of metabolites produced in, and absorbed from, the gut. Indeed, the gut:brain axis is currently receiving much attention not only for its role in regulating satiety and whole body energy balance, but also for its involvement in regulating brain development, neurological diseases including autism, depression and dementia, and even mood and sleep patterns. We now know that aberrant microbiota profiles and/or metabolite production are associated with a range of chronic diseases characterised by loss of metabolic homeostasis and unresolved systemic inflammation, including the diseases of obesity, diabetes (type 1 and 2), metabolic syndrome, non-alcoholic fatty liver disease, as well as some cancers, especially colon cancer, and many autoimmune diseases, where unresolved inflammation is thought to contribute to autoantibody formation and trigger disease. While the gut microbiota have been implicated in the production of toxic compounds or in the activation of pro-carginogens ingested with food (e.g. cooked food mutagens, xenobiotics and mycotoxins), many microbial metabolites including breakdown products derived by plant polyphenols, short chain fatty acids, deconjugated bile acids returning from the gut via the enterohepatic circulation, and B vitamins produced by bifidobacteria are emerging as key beneficial mediators of human cellular activity throughout the body. These metabolites, which comprise in large part the metabolic flux from the colonic microbiota, have been implicated in regulating glucose and fat uptake and metabolism, energy storage, thermogenesis, host hormone secretion, inflammatory molecule production, and epigenetic processes via histone deacetylase inhibition on the one hand and acetylation and methylation, on the other. In this framework of harmful or beneficial bio-activation of dietary components exerted by the gut microbiota, the emerging remarkable variability in its composition helps explaining the parallel variability observed in the way we react to food intake. Our ability to obtain the maximum benefit from a healthy diet derives not only from our genes, but also from the presence or absence of specific microbial species in our gut. Moreover, it appears that our nutritional anthropology plays an important role in shaping not only the relative abundance of the gut bacteria or also apparently microbiota composition. Distinct microbiome profiles in distinct populations from diverse corners of the world and accustomed to very different foods, diets and life-styles highlight the need to act quickly to preserve the natural history of our gut microbiota before the onslaught of dietary globalization, bereft of safe microbial "passengers" and denuded of fibers, prebiotics and plant polyphenols.

Indeed, processes within the gut or rather at the gut wall appear to play a critical role in the persistent low grade systemic inflammation characteristic of many chronic human diseases, especially those associated with modern diet and life-styles. Increased intestinal permeability leads to translocation of inflammatory molecules such as lipopolysaccharide across the gut wall, which then act as continuous triggers for low level systemic inflammation. This gut leakiness appears to be strongly influenced by diet, especially high fat - low fiber modern or Western style at one extreme contributing to gut wall permeability, and ancestral or traditional dietary patterns such as the Mediterranean diet, high in fermentable fiber, prebiotics, fruit and vegetables (and indeed certain probiotic or fermentative microorganisms) at the other, supporting gut barrier function. Bifidobacteria, lactobacilli, SCFA and bile acid metabolism have been shown to regulate expression of tight junction proteins between intestinal epithelial cells and reduce gut permeability. Similarly gut inflammation and oxidative damage play their part and are intricately regulated by diet:microbe interactions within the intestine. Indeed, this process strongly mirrors the gut leakiness and low grade chronic inflammation characteristic of old age, and at least in models of ageing, represents an important event horizon which heralds systemic inflammation, metabolic derangement, typically diabetes, and eventually death.

Of course the ancients had it all along - "*death sits in the bowel*" Hippocrates c. 400 BC. However, we now have direct mechanistic evidence that gut microbial activities represent a lynch-pin upon which the destructive degenerative processes of aberrant metabolic and inflammatory pathway activation are held at bay until overwhelmed by advancing age or aberrant diet. When this occurs, or how, in terms of genetic pre-disposition to chronic disease, for any individual is determined by their genes, but also their life-style. Adherence to the Mediterranean style diet has been proven to protect against chronic non-communicable diseases and improve mental well-being. This dietary pattern too is likely to support beneficial gut microbiota function, given the abundance of fermentable fibers/prebiotics, polyphenols and fermentative microorganisms. Thus life at the boundaries of human existence – our mucosal surfaces, appears critical to our optimal health (both physical and mental), to our risk of developing chronic disease, and eventually to the onset of our physiological decline in old age. Happily, diet is one parameter we can change, and is already recognised as an important modifiable risk factor in metabolic and inflammatory chronic diseases like obesity, type 2 diabetes, cardiovascular disease and cancer. We hold that, based on the most recent metagenomic, metabolomic, nutritional and clinical evidence, at least one means by which healthy diets mediate their ability to protect against non-communicable diseases and maintain mental health, is by regulating both the composition and activity of the human gut microbiota. In this volume, experts in diverse fields of research, clinicians, immunologists, microbiologists, epideamiologists, biochemists and nutritionists, present the most recent thinking underpinning research on diet:microbiota interactions in the gut, and go on to discuss how through carful consideration of our dietary anthropology and current high-resolution post genomics data we may steer a path towards better dietary control of our closely co-evolved microbial partners in the gut and improve host health.

Acknowledgements

Kieran M. Tuohy wishes to thank the Autonomous Province of Trento for funding the TrentinoGUT project, an incoming team grant in 2009, under which his contribution to this book has been supported.

List of Contributors

Philip Allsopp, PhD Northern Ireland Centre for Food and Health, Centre for Molecular Biosciences, University of Ulster, Coleraine, County Londonderry, N. Ireland, UK

Marialaura Bonaccio, PhD IRCCS Istituto Neurologico Mediterraneo Neuromed, Pozzilli, Isernia, Italy

Duncan T. Brown, PhD Department of Food and Nutritional Sciences, School of Chemistry, Food and Pharmacy, The University of Reading, Reading, UK

Emma M. Brown, BSc, PhD Northern Ireland Centre for Food and Health, Centre for Molecular Biosciences, University of Ulster, Coleraine, County Londonderry, N. Ireland, UK

Renato Bruni, PhD Bioactives and Health, Interlaboratory Group, Department of Food Sciences, University of Parma, Parma, Italy

Luca Calani, PhD Bioactives and Health, Interlaboratory Group, Department of Food Sciences, University of Parma, Parma, Italy

Duccio Cavalieri, PhD Department of Computaitonal Biology, Research and Innovation Centre, Fondazione Edmund Mach, San Michele all'Adige, Trento, Italy

Florencia Ceppa, PhD Department of Food Quality and Nutrition, Research and Innovation Centre, Fondazione Edmund Mach, San Michele all'Adige, Trento, Italy

Angela Ceribelli, MD Division of Rheumatology and Clinical Immunology, Humanitas Clinical and Research Center, Milan, Italy

Martina Cirlini, PhD Department of Food Science, University of Parma, Parma, Italy

Adele Costabile, PhD Department of Food and Nutritional Sciences, School of Chemistry, Food and Pharmacy, The University of Reading, Reading, UK

Simone Cuva, PhD University of Trento, Department of Cognitive Science and Education, Rovereto, Italy

Chiara Dall'Asta, PhD Department of Food Science, University of Parma, Parma, Italy

Carlotta De Filippo, B.Sc, PhD Department of Food Quality and Nutrition, Fondazione Edmund Mach-Research and Innovation Centre, San Michele all'Adige, Trento, Italy

Daniele Del Rio, PhD Bioactives and Health Interlab Group, Department of Food Science, University of Parma, Parma, Italy

Maria Benedetta Donati, MD, PhD IRCCS Istituto Neurologico Mediterraneo Neuromed, Pozzilli, Isernia, Italy

Francesca Fava, PhD Department of Food Quality and Nutrition, Research and Innovation Centre, Fondazione Edmund Mach, San Michele all'Adige, Trento, Italy

Cesare Furlanello, M. Sc. MPBA/Center for Information and Communication Technology, Fondazione Bruno Kessler, Trento, Italy

Giovanni de Gaetano, MD, PhD IRCCS Istituto Neurologico Mediterraneo Neuromed, Pozzilli, Isernia, Italy

Mattia Gasperotti, PhD Department of Food Quality and Nutrition, Research and Innovation Centre, Fondazione Edmund Mach, San Michele all'Adige, Trento, Italy

Chris I.R. Gill, BSc, PhD Northern Ireland Centre for Food and Health, Centre for Molecular Biosciences, University of Ulster, Coleraine, County Londonderry, N. Ireland, UK

Meredith A.J. Hullar, PhD Public Health Sciences Division, Fred Hutchinson Cancer Research Center, Seattle, WA, USA

Licia Iacoviello, MD, PhD IRCCS Istituto Neurologico Mediterraneo Neuromed, Pozzilli, Isernia, Italy

Annett Klinder, PhD Department of Food and Nutritional Sciences, School of Chemistry, Food and Pharmacy, The University of Reading, Reading, UK

Athanasios Koutsos, BSc, MSc Hugh Sinclair Unit of Human Nutrition and Institute for Cardiovascular and Metabolic Research (ICMR), Department of Food and Nutritional Sciences, University of Reading, Reading, UK; Research and Innovation Centre, Fondazione Edmund Mach, San Michele all'Adige, Trento, Italy

Johanna W. Lampe, PhD, RD Public Health Sciences Division, Fred Hutchinson Cancer Research Center, Seattle, WA, USA

Samuel M. Lancaster, BA, MS Department of Genome Sciences, University of Washington, Seattle, WA, USA

Julie A. Lovegrove, BSc, PhD Hugh Sinclair Unit of Human Nutrition and Institute for Cardiovascular and Metabolic Research (ICMR), Department of Food and Nutritional Sciences, University of Reading, Reading, UK

Vatsala Maitin, PhD School of Family and Consumer Sciences, Nutrition and Foods Program, Texas State University, San Marcos, TX, USA

Andrea Mancini, MSc, PhD Department of Food Quality and Nutrition, Research and Innovation Centre, Fondazione Edmund Mach, San Michele all'Adige, Trento, Italy

Geoff McMullan, B.Sc, PhD Northern Ireland Centre for Food and Health, Centre for Molecular Biosciences, University of Ulster, Coleraine, County Londonderry, N. Ireland, UK

Pedro Mena, PhD Bioactives and Health Interlab Group, Department of Food Science, University of Parma, Parma, Italy

Lorenzo Morelli, PhD Istituto di Microbiologia, Università Cattolica del Sacro Cuore, Piacenza, Italy

Vania Patrone, PhD Istituto di Microbiologia, Università Cattolica del Sacro Cuore, Piacenza, Italy

George Pounis, MSc IRCCS Istituto Neurologico Mediterraneo Neuromed, Pozzilli, Isernia, Italy

Ian Rowland, BSc, PhD Hugh Sinclair Unit of Human Nutrition, Department of Food and Nutritional Sciences, University of Reading, Reading, UK

Karen P. Scott, PhD The Rowett Institute of Nutrition and Health, University of Aberdeen, Greenburn Road, Bucksburn, Aberdeen, United Kingdom

Carlo Selmi, MD, PhD Division of Rheumatology and Clinical Immunology, Humanitas Clinical and Research Center, Milan, Italy; Division of Rheumatology, Allergy, and Clinical Immunology, University of California, Davis, CA, USA

Qing Shen, PhD School of Family and Consumer Sciences, Nutrition and Foods Program, Texas State University, San Marcos, TX, USA

Douwe van Sinderen, M.Sc., PhD Laboratory of Probiogenomics, Department of Life Sciences, University of Parma, Italy

Nigel G. Ternan, BSc, PhD Northern Ireland Centre for Food and Health, Centre for Molecular Biosciences, University of Ulster, Coleraine, County Londonderry, N. Ireland, UK

Kieran M. Tuohy, BSc, MSc, PhD Department of Food Quality and Nutrition, Research and Innovation Centre, Fondazione Edmund Mach, San Michele all'Adige, Trento, Italy

Francesca Turroni, PhD Alimentary Pharmabiotic Centre and Department of Microbiology, Biosciences Institute, University College Cork, Cork, Ireland

Marco Ventura, PhD Laboratory of Probiogenomics, Department of Life Sciences, University of Parma, Italy

Paola Venuti, PhD Department of Psychology and Cognitive Science, University of Trento, Rovereto, Italy

Urska Vrhovsek, PhD Department of Food Quality and Nutrition, Research and Innovation Centre, Fondazione Edmund Mach, San Michele all'Adige, Trento, Italy

Seth C. Yoder, B. Sc. Public Health Sciences Division, Fred Hutchinson Cancer Research Center, Seattle, WA, USA

CHAPTER

1

The Microbiota of the Human Gastrointestinal Tract: A Molecular View

Kieran M. Tuohy and Karen P. Scott†*

***Department of Food Quality and Nutrition, Research and Innovation Centre, Fondazione Edmund Mach, San Michele all'Adige, Trento, Italy †The Rowett Institute of Nutrition and Health, University of Aberdeen, Greenburn Road, Bucksburn, Aberdeen, AB21 9SB. United Kingdom**

INTRODUCTION

All multicellular organisms with an organized intestine carry an intestinal microbiota. The gut microbiota serves a number of important functions: providing nutrients to the host, aiding in the digestion of complex food components, providing a barrier to invading pathogenic microorganisms, and in higher animals at least, helping to maintain immune homeostasis.[1,2] This closely co-evolved partnership comes at a cost in terms of energy, with carriage of a diverse microbiota of many hundreds of different microbial species and complex populations in the colon reaching 10^{11} cells/g contents necessitating maintenance of costly host defences. The gut-associated lymphoid tissue (GALT) is the single largest immune organ in the body and is designed to recognize microbial friend from foe and to keep this complex microbial ecosystem in check, it expends considerable energy to safely accommodate the gut microbiota and fight off invading pathogens.[2] However, in health at least, this cost is greatly offset by microbial biotransformation of complex plant foods into intermediates, especially fermentation end products, which can then be absorbed and metabolized by the host. For example, early estimates put the contribution of microbial fermentation in terms of short-chain fatty acids (SCFA) made available to the host at about 10% daily human energy demand.[3,4] From the microbial point of view of course, this energy balance equation looks quite different, with the host, either directly through secretions of the gastrointestinal tract (sloughed off epithelial cells, digestive enzymes and mucin and other glycoproteins from the mucus layer covering the gut wall) or indirectly through ingested food, providing a continuous supply of energy and nutrients, and also a relatively stable ecological habitat with the potential for potentiation and transfer into new hosts once the current host "passes away."

GUT MICROBIOTA METABOLISM IN HEALTH AND DISEASE

Living with a complex gut microbiota means living at the precipice of "war and peace." As in most conflicts, energy availability and ownership, and miscommunication (e.g., inability of the host's immune system to respond appropriately to commensal microorganisms) decide the balance of power, leading to health or disease. Indeed, it is becoming clear that the human gut microbiota plays a critical role in human health, with battlegrounds revolving about metabolism and immune function.[5–7] Diet is a major external force, acting on both human disease risk and gut microbiota function.[8,9] The gut microbiota, depending on dietary intake and consequently nutrient availability within the gut, can produce either harmful metabolites linked to human disease or beneficial compounds that protect against host disease. For instance both the conversion of ingested xenobiotic compounds

Diet-Microbe Interactions in the Gut.
DOI: http://dx.doi.org/10.1016/B978-0-12-407825-3.00001-0

into carcinogens and genotoxins, and the production of trimethylamines have negative effects, while the production of SCFA, vitamins (vitamin K and B vitamins including cobalamin, folate, niacin, thiamine) and conjugated linoleic acids (CLA), have positive health benefits.[5,8–15] In some cases, distinct metabolic pathways have been associated with bacteria we consider beneficial for human health. Some bifidobacteria have been shown to produce folate, and lactobacilli and bifidobacteria species are capable of producing CLA or gamma-aminobutyric acid (GABA) within the gut.[15–18] In many cases we do not know the specific bacterial species involved in metabolite production, and for many metabolites more than one microbial species might be involved.[19] We do know that diet, especially the amount of fermentable fiber and polyphenols reaching the colon, greatly impacts on both the relative abundance of bacteria present and on their metabolic activities.[1,9,20,21] The mammalian gut microbiota is predominantly fermentative, with the fermentation of dietary carbohydrates representing the principle mode of energy conversion within the gut ecosystem, providing energy and carbon sources for fermentative species themselves and supporting a complex food web whereby the end product of one microorganism is the growth substrate for another and of course, the human host.[12] The majority of this dietary carbohydrate is in the form of dietary fiber, a broad term covering many dietary components that escape digestion in the upper gut.[21] Dietary fiber mainly comprises of resistant starch and other complex polysaccharides of plant origin for which the human genome encodes few hydrolytic enzymes, necessitating extra-genomic aid from our internal microbial symbionts.[22,23] The gut microbiota cleaves complex polysaccharides using an array of glycases, which often work in synergy even when encoded by different microorganisms, releasing smaller oligosaccharides and sugars which are then fermented into SCFA.[4,9,12] The availability of a ready supply of dietary fiber supports a stable microbial consortium mainly involved in saccharolytic fermentation, the end products of which are beneficial to host health.[1] Indeed, this fits neatly with human epidemiological data showing that diets rich in fiber are inversely related to chronic human diseases, especially diseases linked to metabolism and immune function.[24–26] On the contrary, when fermentable carbohydrate is in short supply, the resident microbiota turn to other sources of energy to support growth and changes occur, both in microbial composition and metabolic activity. Fermentation of amino acids is energetically less favorable than carbohydrate fermentation, yet some microorganisms will ferment amino acids derived from endogenous or dietary protein[12,27] releasing end products which are potentially harmful to human health and have been linked to diseases such as cardiovascular disease and colon cancer.[10,28–30] Although most dietary fat will be absorbed in the upper gut, some can reach the colon. Bacteria are poor utilizers of lipids under anaerobic conditions and derive little energy from the conversion of dietary lipids into either beneficial or harmful fatty acids.[15,31,32] However, high-fat diets, both in animal models and in humans, lead to major shifts in the microbial ecology of the gut, to the detriment of beneficial, saccharolytic bacteria.[33–37] Thus although high-fat, low-fiber, diets may provide excess energy to the human host they effectively starve the gut microbiota and contribute to the establishment of aberrant microbiota profiles shown to increase the risk of metabolic and autoimmune disease.[38]

Dietary modulation of the gut microbiota does not go unnoticed by the guardian of human health, the immune system.[33,34,38–40] Inflammation, both at the gut wall and systemically, occurs in response to aberrant microbiota within the gut and also with leakage of inflammatory molecules from the gut.[33,34,38] The barrier function of the gut wall appears to be critically determined by the presence of certain key bacterial species and by microbial metabolites like the SCFA butyrate, which serves both as the main energy supply for the intestinal mucosa and as a mediator of gut wall development and differentiation.[41–46] Similarly, microbial metabolites produced by beneficial gut bacteria including acetate, butyrate, propionate, GABA, CLA, all play a role in the optimal functioning of immune cells including first responders like dendritic cells, neutraphiles, macrophages and intestinal epithelial cells. These cells play an important communication role, signaling changes within the gut lumen and directing appropriate responses from immune mediator cells.[47–51] Moreover, dietary patterns which modulate the gut microbiota have been shown to direct energy for either metabolic purposes or inflammation in intestinal cells, highlighting the critical balance between immune function and metabolism within intestinal epithelial cells.[52–54]

It is therefore no wonder that aberrant microbiota profiles have been described for a range of metabolic and autoimmune diseases.[55,56] All of these appear to be impacted by diet, especially the polar opposites represented by low-fiber modern Western-style diets and high-fiber traditional diets based on whole plant foods from around the world, especially the Mediterranean diet, which has particularly strong evidence supporting its ability to protect against metabolic disease and cancer.[5,57–63] In the Western world, or in populations consuming the modern Western-style diet regardless of geographical location, incidences of both metabolic diseases (obesity, type 2 diabetes, cardiovascular disease, certain cancers and fatty liver disease) and autoimmune diseases (e.g., inflammatory bowel disease, type 1 diabetes, celiac disease, food allergies) are reaching epidemic-like numbers. In contrast, in populations following more traditional diets high in fiber and whole plant foods and low in processed, energy-dense food and red meat, these same diseases are rare.[64–66]

METHODOLOGIES FOR STUDYING THE HUMAN GUT MICROBIOTA

The vast majority of gut microorganisms remain uncultured or are difficult to culture under laboratory conditions. Early estimates indicated that at least 70% of microbial species within the gut microbiota were new to science and not isolated in pure culture or characterized in detail.[67] In fact, recent successes in cultivating strictly anaerobic and fastidious species like *Faecalibacterium prausnitzii*, a dominant and prevalent member of the gut microbiota, *Roseburia* spp. and *Akkermansia muciniphila*,[68,69] indicate that given appropriate nutrient availability, efficient selective agents to inhibit competing microorganisms and strict anaerobic cultivation conditions, it is probably true that most gut bacteria can be isolated in pure culture given enough patience.[1] In fact, one study recently indicated that cultured bacterial isolates are available for each of the most abundant 29 species, which represented 50% of the total gut microbial diversity.[70] Isolation in pure culture is important because it enables the physiology of a microorganism to be studied, allowing measurement of the phenotypic expression of its genetic blueprint under various growth conditions and in response to different environmental stressors. However, such studies are extremely time consuming and labor intensive, and it can take many years to fully characterize a given microbial strain. The high species richness of the gut microbiota, with perhaps 1000s of unique microbial strains, requires more direct techniques to measure both the genetic potential and metabolic kinetics of the gut microbiota as a whole. Different molecular strategies have been developed, aimed at addressing three fundamental ecological questions: which species/phylotypes are present? (their variability); how many of each are there? (their relative abundance); and what they are doing? (their metabolic activity). These different techniques each have their own advantages and disadvantages (for critical discussion of different molecular tools applied to microbial ecology, see references[71–73]), and before embarking on a journey of exploration through the gut microbiota and their interactions with human diet in other contributions to this volume, it is important to describe briefly the more commonly employed tools used to study the gut microbiota.

Measuring Species Richness and Variability

Estimates of species richness within the gut microbiota are commonly made using either rapid fingerprinting techniques such as T/DGGE, ARISA or ADRA,[71,74] or information-rich targeted metagenomics approaches based on next-generation sequencing.[75] Fingerprinting methods give limited species identification but rapid profiling of microbiota composition and have been used to facilitate quick, and cheap, measurement of species richness incorporating traditional ecological measures such as the Shannon–Weavor index.[76] Although useful for determining whether microbial diversity has been altered within a community by environmental stimuli, they tell us little about what species or phylotypes are involved. Efforts to introduce species identification into these methodologies have had limited success. For temperature or denaturing gradient gel electrophoresis (T/DGGE), bands of amplified 16S rRNA can be cut from gels and sequenced. However, the process is time consuming, limiting the number of bands which can be analyzed at one time, and may be prone to misleading results since DGGE–PCR primer bands are only about 200bp of target 16S rRNA gene sequence. Moreover, very different bacteria, with similar DNA fragment melting profiles but different 16S rDNA sequence may co-migrate as single bands on denaturing gradient gels.

Automated rRNA Intergenic Spacer Analysis (ARISA) is based on the fact that bacterial species appear to have unique lengths of interspatial regions between the 16S and 23S rRNA genes. By separating these amplicons by capillary electrophoresis, a unique fingerprint of DNA spacer fragments is generated which reflects the microbial composition of the sample. The method is rapid and because of its use of automated capillary electrophoresis techniques, is reproducible and less prone to operator error or inter-gel differences common to T/DGGE. Although putatively useful for identifying specific bacteria by the length of their 16S–23S rRNA gene spacer region within simple microbial communities, ARISA does not provide the resolution or phylogenetic information necessary to describe the species richness of complex communities such as the gut microbiota. Attempts to create databases of fragment lengths of different bacteria have not been successful as they ultimately require confirmatory reanalysis using other, more targeted, phylogenetically robust methods to confirm fragment phylotyping.[77] However, it is useful to measure changes within microbial communities without phylogenetic classification.

The third fingerprinting technique, amplified ribosomal DNA restriction analysis (ARDRA), employs multiple restriction enzymes to cut 16S rRNA amplicons amplified directly from metagenomic DNA extracted from an environmental sample. ARDRA therefore, is another rapid inexpensive tool to assess microbiota dynamics but again does not give information on the phylogenetic make-up of a microbial community. Although databases of 16S rRNA restriction fragment profiles may be useful for studies in pure culture, it is not possible to assign phylogenetic identity with any certainty to the presence or absence of specific gel bands or profiles in samples of a complex microbiota. Additionally, both ARISA and ARDRA suffer from the fact that the lengths of interspatial regions and positions of restriction enzyme cleavage sites, and thus fragment lengths, are not under tight evolutionary control and different bacteria even from distantly related groups of bacteria may on occasion give similar results, necessitating confirmation of phylogenetic identity and sample reanalysis.

Next generation sequencing methods can permit the actual identification of bacterial species present, and recent advances have opened up the entire metagenome of the human gut microbiota to investigation. Metagenomics targeted at community 16S rRNA gene content, sometimes called metataxonomics, constitutes an increasingly cost effective and rapid means of describing microbial community species richness and gleaning semi-quantitative, relative abundance data from metagenomic samples. Using barcoded primer sets, metataxonomics seeks to amplify and subsequently sequence hypervariable regions within the 16S rRNA gene of all bacteria present within an environmental sample. This approach has become widely used following the reduction in cost of sequencing and increased availability of user-friendly bioinformatic tools allowing data analysis by non-expert bioinformaticists. Similarly, 16S rRNA targeted metagenomics has potentially universal coverage within the kingdoms Bacteria and Archaea, but since it is based on PCR, it too suffers from PCR bias.[78,79] Initially this approach was restricted by relatively short sequence reads of the 454-pyrosequencing platform (early methodologies generated about 200nt of usable 16S rRNA target gene sequence) and the lack of reliable and suitable tools for analyzing even shorter Ilumina datasets (50nt). However, new protocols for 454-pyrosequencing generate up to 700nt of quality sequence data, and Ilumina data now allow the bioinformatic reconstruction of the whole 16S rRNA gene, both giving increased reliability of phylogenetic classifications down to the species level.

The above methodologies are based on DNA amplification and are subject to PCR bias. This is particularly true when bacteria with very different GC content are amplified in competitive reactions such as those used to target 16S rRNA (T/DGGE or ARDRA) or interspatial rRNA gene regions (ARISA) from the diverse bacteria within the gut microbiota. DNA fragments containing high proportions of GC will melt more slowly than AT-rich DNA, rendering them less able to compete for NTPs, Taq and primer. Combined with the large variation in relative abundance of bacteria present, and the differences in cell wall disruption kinetics, differences in relative competitiveness within the PCR over many cycles distorts the true proportions of GC- vs AT-rich DNA. This can have a dramatic impact on the data output, particularly in relation to estimating the relative abundance by metataxonomics of low GC and high GC species.[80] For example bifidobacteria, with a GC content of about 60% and relative abundance lower than 2% in most fecal samples, are often underreported in metagenomic datasets because they lose out to lower GC species present at higher relative abundance (e.g., *Bacteroides* about 40%, and *Clostridium* species 25–55%) during competitive PCR using universal 16S rRNA bacterial primers.

Estimating Microbial Relative Abundance within the Gut Microbiota using Culture-Independent Methods

Quantitative culture-independent methods for estimating relative abundance of gut bacteria at various taxonomic levels (usually genus, sub-genus or species level) can be divided into direct methods, based on microscopy where microorganisms are enumerated directly in environmental samples, and indirect methods, which estimate microbial relative abundance from DNA extracted from environmental samples following microbial disruption using quantitative PCR, for example.

Fluorescent in situ hybridization (FISH) using labeled oligonucleotide probes targeting 16S rRNA allows direct and accurate quantification at various phylogenetic levels depending on probe design, from phylum to bacterial species.[81] FISH is considered the gold standard for microbial enumeration, being precise (to given phylogenetic targets), accurate and reproducible. Automated microscopy and computer-assisted image analysis may be employed to improve reproducibility and flow cytometry procedures have been established to reduce analysis time.[82,83]

Quantitative PCR (qPCR) commonly using either Taqman or Syber Green technologies is used for enumeration of gut bacteria indirectly from DNA extracted from fecal or gastrointestinal samples.[84,85] It is generally considered less accurate than FISH, given the variable extraction and purification of DNA from diverse gut bacteria with very different disruption potentials, and the need to estimate 16S rRNA gene copy number per cell when calculating numbers of cells present in a sample from standard curves of target DNA. Bacterial cells often carry more than one copy of the 16S rRNA gene, ranging from one to 15 copies, and the actual number can vary considerably even between closely related bacterial species. Therefore, assigning a 16S rRNA gene copy number for previously uncultured bacteria based on the copy number in its closest cultivated relative can introduce errors in exact quantification. Data generated can however be expressed as gene copy number, without any attempt to translate it into actual bacterial cell counts. Operator error can be reduced and reproducibility increased using robotic assistance for DNA extraction and PCR.

Measuring Microbial Activity

Advances in the "omics" technologies have truly revolutionized the way we measure the activity of the human gut microbiota. Metagenomics,[86] metatranscriptomics,[87] metabolomics[88] and proteomics or (meta)proteomics[89] are at varying degrees of development and applicability in ecological studies of the gut microbiota, but all have huge potential both in diagnostic profiling and mining novel mechanisms of microbial function. They can be divided according to methods which measure metabolic potential of expressed genes, or the metabolic potential of the genes encoded by the nucleic acid blueprint of the gut metagenome, and methods which measure metabolic kinetics, comprising the final functional proteins and metabolites produced by the gut microbiota. Metatranscriptomics and proteomics, or metaproteomics are still being developed, requiring advances in both bioinformatic tools and methodological breakthroughs to improve reproducibility. Conversely, metagenomics and metabolomics/metabonomics are comparatively much more developed and metabolomics is already being widely applied to study the gut microbiota and its impact on metabolite profiles in various biofluids including urine, blood and fecal water.[90,91] Both methods are heavily reliant on bioinformatics, computational power and multi-variate statistics, but now appear to have overcome some early technical hurdles in automated annotation of gene function or chemical structure, respectively.[92–94]

Metagenomics employs next-generation sequencing methods and bioinformatic tools for high-throughput gene assembly, alignment and annotation. It potentially describes all genes present within a given metagenome, using databases such as the Kyoto Encyclopedia of Genes and Genomes (KEGG), and Carbohydrate-Active enZYmes database (CAZY) for enzymes related to carbohydrate metabolism, to aid identification of gene function and presentation of data into manageable pathways or modules. Current limitations include the uneven coverage of metagenomic databases; the lack of complete, annotated bacterial genome sequences; restrictions of bioinformatic assumptions and cut off values necessary for ease of interpretation (e.g., KEGG modules); and the management of information-rich data readouts (e.g., assumption of 98% identity for species similarity or significance levels set at 10% or 20% relative abundance for gene functions). These are being overcome as more metagenomes and sequenced microorganisms are continually added to well-curated, publically available databases. Recent giant steps include timely publically funded projects around the world such as the Human Microbiome Project and MetaHIT.[95,96] Metabolomics and metabonomics use chemometrics and multivariate statistics to profile metabolites and fluxes in metabolites from biofluids using either/or NMR and hyphenated MS strategies.[97,98] While metabolomics aims to identify all the metabolites present in a complex biological sample, usually in practice at the single cell type level, metabonomics, aims to track global changes in metabolite profiles of the whole biological system comprising many different cell types from different tissues and organs, in response to external stimuli. Metabonomics in human terms, therefore, seeks to track changes in metabolite profiles of biofluids like blood, urine and fecal water, over time and in response to disease, drugs or diet for example and includes metabolites derived from all the different cell types present, mammalian and microbial, and as well as diet and drug/xenobiotic metabolites. However, the terms, metabonomics and metabolomics, are often used synonymously.[99] Metabonomics could be identified as a disruptive technology in terms of long-held views of the relative importance of the gut microbiota within the human system. However, of the omics technologies, it was metabonomics that first advanced the notion of humans as "super-organisms," with the human genome working in close co-evolved metabolic cooperation with the myriad of microorganisms inhabiting the gastrointestinal tract. Metabolomics/metabonomics can either be applied in a targeted manner, against a specific class of compounds of interest, or may be untargeted, aiming to profile all metabolites present within a biofluid but only identifying those compounds which appear to track with a particular biological function of interest, e.g., response to a drug or dietary intervention. Targeted metabolomics offers the advantage of accurately quantifying a restricted number of key metabolites, and is useful in measuring subtle changes in chosen metabolic pathways or metabolic end products (e.g., SCFA, phenolic acids derived from polyphenol metabolism). Untargeted metabolomics allows broad coverage of the entire profile of metabolites present within a biofluid, both known and unknown.

Recent advances in automated metabolite annotation from either NMR or MS datasets, and the establishment of open source, extensive and well curated metabolite databases are helping to harness the power of untargeted metabonomics and increase the value to the wider scientific community.

SPATIAL DISTRIBUTION OF THE GUT MICROBIOTA AND INTERACTIONS WITH DIET

The physiochemical environment encountered by microorganisms changes drastically between different anatomical regions of the human gastrointestinal tract (Figure 1.1). The flow rate of digesta, mucus layer composition and thickness, presence and activity of immune cells, host enzymatic secretions, bile acid concentrations, pH and redox potential, all change along the intestinal tract and help shape the microbiota of different gastrointestinal habitats.[100] The gut microbiota therefore changes in species richness and composition along the length of the gastrointestinal tract. The gut microbiota also changes along the transverse axis of the gut, with microbial populations close to the gut wall differing from those which reside within the gut lumen, often bound to food particles.[101] Despite differences in both species diversity and relative abundance of important members of the gut microbiota between feces and the different anatomical regions of the gut, the fecal microbiota is at least accepted as a fair approximation of colonic microbial ecology.[102]

The Stomach

In the stomach, gastric acid secretion leads to drastic swings in pH from as low as 2 to neutral depending on meal times and dietary composition. The traditional view is that few microorganisms can survive this pH range to call the stomach home, with numbers of viable cells limited to between 10^2 and 10^4 CFU/mL.[103] The low acid conditions of the stomach provide an important barrier to pathogens ingested with food and water. For example, *Vibrio cholerae* is acid sensitive, necessitating ingestion of large numbers of viable cells to cause disease (*Vibrio cholerae* has in infective dose of 10^6 CFU in drinking water). Other microorganisms can however tolerate the acidic environment and survive passage through the stomach. For example, certain *Salmonella* and *Shigella* species as well as *E. coli* O157 : H7 have an estimated infective dose of about 10 CFU, suggesting they readily survive the low pH of the stomach, moving onto their sites of infection lower down the intestine.[104]

Helicobacter pylori is probably the most famous of gastric residents because of its associations with peptic ulcers and gastric cancer. An early high throughput sequencing project targeting community 16S rRNA gene diversity within the stomach of individuals positive (n = 3) or negative (n = 3) for *H. pylori* carriage, found striking differences in phylotype richness of biopsies between *H. pylori* positive (Hp+) and *H. pylori* negative (Hp−) individuals.[105] 93–97% of reads from the Hp+ subjects were identified as *H. pylori* phylotype, while in the Hp− subjects,

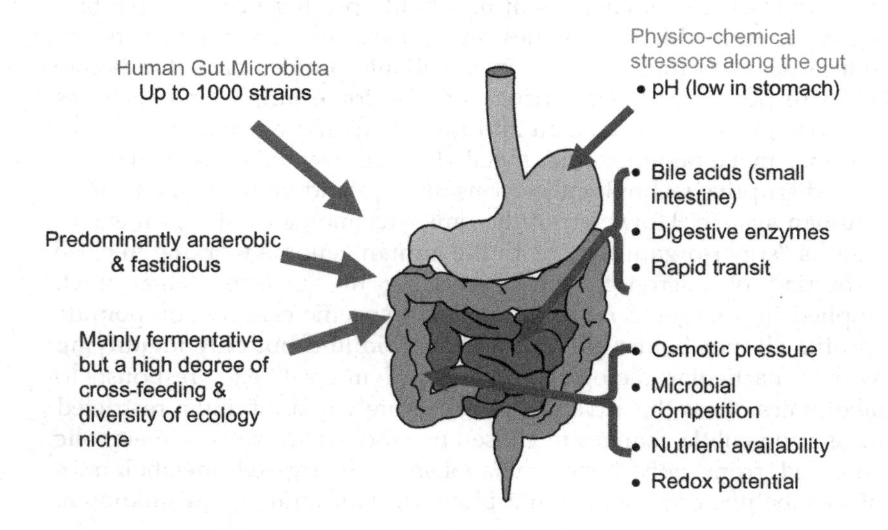

FIGURE 1.1 A schematic representing the physical and chemical challenges encountered by the gut microbiota along the length of the gastrointestinal canal.

262 phylotypes from 13 different phyla were observed. Of the 33 phylotypes common to all three Hp− subjects, the most dominant were *Streptococcus, Actinomyces, Prevotella* and *Gemella*. However, the authors admitted it was difficult to distinguish bacteria which were resident within the stomach and those transiently entering from microbial habitats in the esophagus and oropharangial cavity. Other studies, also using 16S rRNA gene fragment metagenomics, did not report differences in species richness or composition between the stomach microbiota of Hp+ and Hp− individuals. Bik *et al.*[106] used 16S rRNA targeted metagenomics to characterize the microbiota of 23 gastric biopsies and found a diverse community of some 128 phylotypes with the majority of species falling within the phyla Proteobacteria, Firmicutes, Actinobacteria, Bacteroidetes and Fusobacteria. Interestingly, 10% of phylotypes were new to science, including bacteria related to the extremophile *Deinococcus*, famously resistant to radiation, which had not previously been reported to reside within the human gut. Although 12 of the subjects tested positive for Hp, carriage of this pathogen did not appear to influence the composition of the gastric microbiota. More recently, Delgado et al.[103] used a combination of 16S rRNA targeted pyrosequencing and traditional culture based microbiology to characterize the microbial ecology of stomach mucosa biopsies and gastric juice from 12 healthy human subjects. From the sequencing data, a distinctive microbiota comprising of 59 bacterial families and 69 genera was observed at the mucosa. Bacteria belonging to the genera *Propionibacterium, Lactobacillus, Streptococcus* and *Staphylococcus* were the most commonly cultured microorganisms, and in agreement with earlier studies, total viable counts were in the region of 10^2–10^4 CFU/g or mL mucosa or gastric juice and viable microorganisms could only be isolated from five of the 12 gastric juice samples. The most abundant phylotypes present within the gastric mucosal biopsies belonged to the genera *Streptococcus, Propionibacterium* and *Lactobacillus*, in good agreement with the culture based data.

The gastric microbiota appears to be distinct from the microbiota of other habitats further up the alimentary canal, and considerable inter-individual variation exists. Considering the importance of the stomach, especially as a first line of defence against invading acid-sensitive pathogens and the medical significance of diseases linked to *H. pylori* carriage, the microbial ecology of the human stomach warrants further study.

The Small Intestine (Jejunum and Ileum)

Life for microorganisms in the small intestine too is challenging, with a fairly rapid flow of digesta limiting substrate/energy availability and increasing dilution rates, and host secretions, especially bile acids and immune activities, having direct antimicrobial actions. The immune cells in the small intestine are concentrated in Peyers patches, and there is a high secretion of immunoglobulins, especially IgA which appears to bind intestinal bacteria in a targeted manner for elimination in feces. The immune system continually "samples" the luminal contents of the small intestine mounting, at least in the absence of autoimmune or other immunological diseases, appropriate immune responses to microorganisms it encounters based on history of exposure and recognition of pathogen associated characteristics. Dendritic cells, important antigen presenting cells involved in identifying new microbial exposures, identify novel microbial and food-related antigens either via dendrites extending across the mucosal epithelium into the gut lumen or after translocation into the intestinal lumen.[107–109] The microbiota also encounters the direct antimicrobial activities of peptides like defensins and other bactericidal molecules secreted by Paneth cells.[110–112] Paneth cells reside within crypts of the small intestinal mucosa on the lumenal side, and play an active inhibitory role producing defensins, lysozyme and phospolipases. They control microbial growth within crypts of mucosal villi, an extremely sensitive location in terms of mucosal physiology, since stem cells within the villus crypt are progenitors for all epithelial cells which migrate from intestinal crypt to the villus tip before being sloughed off into the intestinal lumen.[113,114]

Another important modulator of microbial populations in the small intestine, and continuing further down the gastrointestinal tract into the colon, are bile acids secreted by the host into the small intestine to aid fat absorption. Bile acids have a strong inhibitory activity against certain bacteria, especially species lacking specific detoxification mechanisms such as bile salt hydrolases. This may be part of the reason why high fat diets which stimulate bile acid secretion into the intestine radically modulate the composition of the gut microbiota.[5,38,115,116] Although most studies are on colonic or fecal microbial populations, bile salts are also likely to play an important role in controlling microbial colonization or density within the distal ileum.

Recent molecular studies have confirmed earlier culture-based work that describes the microbiota of the small intestine as comprising between 10^7 and 10^8 microorganisms per mL ileal fluid.[117] Samples of ileal fluid collected from ileostomy patients analyzed using the HITCHIP gut microbiota 16S rRNA targeted microarray revealed a microbiota dominated by lactobacilli, streptococci, clostridia with a lower abundance of strict anaerobes including species of *Ruminococcus* and *Bacteroides*.[117] This microbiota fluctuated not only between individuals but also over time, both within study days (morning compared to afternoon) and on different days of sampling. Within an individual, the average similarity in microbiota composition of the small intestine over a 9-day period was as low as 44%, indicating high variability over time in stark contrast to other gastrointestinal sites, especially the colon. Zoetendal et al.[118] confirmed that the microbiota of the ileum is dominated by facultative anaerobes like the streptococci and lactobacilli, although they also identified clostridia and a high relative abundance of high-GC microorganisms including the Actinobacteria. Again, the species composition of the ileal microbiota was subject to considerable fluctuation over time and this fluctuation was also reflected in the ileal concentrations of SCFA, which differed greatly between morning and afternoon sampling.[118] The mean molar ratios of 20:1:4 acetate:propionate:butyrate in the ileum compares to a ratio of 3:1:1 normally recorded in human feces. Thus there are dynamic changes both in species composition and metabolic activity of the small intestine microbiota over short time frames, most likely reflecting availability of readily fermentable substrate from the host's diet.[1,118] The ileal metagenome was found to be enriched in genes involved in the rapid uptake and fermentation of simple sugars.[118]

These studies were conducted in ileostomy patients, who have often had the last few centimeters of the distal ileum, and the entire colon, removed. Other studies using high-throughput sequencing suggest that the microbiota of the terminal ileum may more closely resemble that of the colon and that oxygen tension in the ileum of ileostomy patients may influence the resident microbiota unrealistically increasing numbers of facultative anaerobes and decreasing strictly anaerobic species similar to those dominating the colonic microbiota.[119,120] In a powerful experiment in which healthy human subjects underwent gastroduodenoscopy and duodenal biopsy collection, van Baarlen et al.[121] demonstrated that the mucosa of the human small intestine reacts differently to even closely related probiotic microorganisms, in this case *Lactobacillus acidophilus* Lafti-L10, *L. casei* CRL-431 and *L. rhamnosus* GG. Transcriptome analysis showed that the probiotic strains impact on the expression of human genes involved in immune response, cell cycle, blood pressure and osmoregulation. The responses of the same individual to different probiotic strains were different, showing that closely related microorganisms can elicit different physiological responses, even from the same host.[121] Importantly, there were also considerable inter-individual differences in transcriptome response, confirming that individual humans differ in their ability to recognize and respond to given probiotic strains. This fits with studies showing that host genotype, specifically genes involved in the immune bacterial recognition processes, can elicit very different immune responses, e.g. inflammation or tolerance to the same bacterium. However, these authors previously showed that the mucosal transcriptome also responds differently to the same bacterial strain at different stages in its growth cycle (log phase, and stationary phase[122]). The small intestinal response to probiotic intereventions, in terms of altered gene expression occurred rapidly.[121] It is significant that bacterial populations within the ileum are typically in the region of 10^7-10^8 cells/mL, and ingested probiotic microorganisms, selected to survive passage through the acidic conditions of the stomach, are often at single daily doses of between 10^9 and 10^{11} viable cells. Thus, the ingested probiotics become the dominant microorganism within the small intestine in the hours following ingestion. Given the rapid and dynamic response of key human genes involved in immune and metabolic function to microbial stimuli in the gut lumen, the small intestine may be a highly active site of inter-kingdom communication between the human host and ingested microorganisms, especially probiotics. Traditionally fermented foods like artisanal cheese, which can have microbial loads of up to 10^9 CFU per 100-g serving, may have similar effects.

The Colon (Large Intestine)

The colon contains the most species-rich and densely populated human-associated microbial communities, and the most widely studied. The colonic microbiota comprises many hundreds of bacterial species, 1000s of strains and populations of about 10^{11} cells/g contents. The physiology of the colon renders it a particularly suitable bacterial habitat, as the flow of intestinal digesta slows, allowing bacteria ready access to dietary growth substrates. The pH is moderately acidic in the proximal colon, mainly due to bacterial fermentation itself, and increases to a more neutral pH in the transverse and distal colon with secretions by the host and water absorption. Indeed, using *in vitro* models of the human gut microbiota, pH has been shown to be a critical modulator of both microbiota composition and metabolic activity under the physicochemical conditions of the human colon.[123]

Although, there are studies describing the microbial ecology of colonic contents and the colonic mucosa during surgery or from samples collected during colonoscopy, the vast majority of studies have employed feces as a surrogate of the colonic microbiota because of its ease of access and collection. Studies describing the composition of the human fecal/colonic microbiota using high-throughput sequencing show that it is dominated by a few bacterial phyla, the Firmicutes, Bacteroidetes and Actinobacteria, with the Proteobacteria and Verrucomicrobia present but at lower levels.[124–126] Studying the fecal microbial ecology of six overweight men, Walker et al.[70] found 29 phylotypes which were present at more than 1% abundance. All of these had previously cultured representatives suggesting that of the dominant bacteria at least, most are culturable and closely related strains exist in culture collections. However, phylotypes which were less abundant, were also less likely to have been previously cultivated, suggesting that the commonly held belief that the majority of the gut microbiota are somehow "un-culturable" may not necessarily be the case, but may reflect the fact that sub-dominant populations within the fecal microbiota are difficult to separate from more dominant phylotypes, especially in the absence of specifically tailored selective media and growth conditions.

Early studies with culture-independent techniques highlighted that the gut microbiota profile of any given individual appears at once extremely stable over time and to be unique to that individual.[127] Moreover, even identical twins had unique profiles of microorganisms within their gut microbiota. Twins showed the closest degree of relatedness in terms of microbiota composition, followed by siblings living in the same house while parents had the greatest degree of difference in microbiota composition.[128] In a similar study, but combining NMR-based metabolite profiling with microbial fingerprinting techniques, Li et al.[129] profiled both the gut microbiota and biofluid metabolite profiles of three generations of a single Chinese family, and found that even subtle differences in gut microbiota profile may impact on biofluid metabolite profiles. These studies demonstrate clearly that although environmental factors (presumably including diet) contribute to differences in microbiota composition, genetics also plays a role. As with host phenotypes, which emerge as a combination of genetic predisposition and environmental pressure, the composition and metabolic activity of the gut microbiota also relies on the combination of genes and environment.

The fact that closely related individuals still possess a unique collection of microorganisms suggests that either a core microbiota exists, consisting of bacterial species encoding the most characteristic or essential of metabolic functions, or that there is a high degree of metabolic redundancy within the gut microbiota, with even distantly related bacteria sharing similar metabolic functions. Indeed, it appears that both these hypotheses may be true to some extent. It seems that a small core collection of microorganisms are shared amongst most humans, with other auxiliary bacteria completing the metabolic repertoire of the gut microbiota, and also that bacteria distantly related at the phylogenetic level share certain key metabolic functions. This includes habitat-specific metabolic activities which characterize bacteria well adapted to life in the gut, such as fiber fermentation or bile salt hydrolase activity. The acquisition of the gut microbiota by the host, both core and auxiliary, depends largely on extrinsic factors: exposure and colonization of the gut by microorganisms early in life from both mother and environment; microorganisms ingested with food, including fermented foods, and microorganisms associated with raw food such as fruit and vegetables; host geographical location; and overall diet, with particular dietary factors having a dramatic impact on gut microbiota composition. Host genetics, as mentioned above is also likely to play a role. However, few human genes have been identified which appear to impact on gut microbiota successional development.[130,131]

Several studies, involving large numbers of individuals, have alluded to the existence of a "core microbiota." Tap et al.[132] found that most operating taxonomic units (OTU's) identified from fecal 16S rRNA high throughput sequencing were subject unique and that only 2.1% of the total number of OTU's were present in more than 50% of the 17 subjects studied. This 2.1% accounted for 35.8% of total sequences, and was comprised of 66 different OTUs from the genera *Faecalibacterium, Ruminococcus, Eubacterium, Dorea, Bacteroides, Alistipes* and *Bifidobacterium*. Using the HITchip microbiota 16S rRNA microarray, Salonen et al.[133] demonstrated that the size of the core microbiota in 100 individuals, depended on the depth of phylogenetic coverage – the deeper the analysis, the more bacteria joined the core microbiota. In this study, the health status of the host had a greater impact on the composition of the core microbiota than the number of subjects analyzed, indicating that host health – or rather disease state – can have a pronounced impact on the composition of the gut microbiota and its dominant core members. Disease status, in this case ulcerative colitis, dramatically reduced the size of the core phylogenetic group. This of course fits well with ecological theory linking species richness of microbial ecosystems to stability and "fitness," and also the observation of reduced diversity or species richness within the fecal microbiota of patients with diverse disease states including metabolic, obesity and autoimmune diseases, compared to healthy control subjects.[134] Turnbaugh et al.[135] and later Sekelja et al.[136] described core microbiota based around *Lachnospiraceae*, within the *Clostridiale* and likely to contain high butyrate-producing species and appear to be related to *Eubacterium eligens, Dorea* and the ruminococci. These core groups appeared also to have ancient origins and an evolutionary record in the gut metagenomes of invertebrates, vertebrates and higher animals including primates. Analyzing 22 whole human fecal metagenomes and comparing with publically available bacterial single genome sequences, Arumugam et al.[137] described the existence of three dominant and prevalent groupings of gut bacteria, at the subphylum level, based around *Bacteroides*, *Prevotella* and *Ruminococcus* (also comprising bacteria related to *Blautia* and unclassified *Lachnospiraceae*) as "defining" members. These three "enterotypes" as they were dubbed, were mainly driven by species composition and co-occurrence, but were also reflected in other functional genes, or abundance of assigned orthologous groups, within the metagenomic datasets analyzed. However, the authors were careful to point out that these enterotypes were not as sharply defined as, say, human blood groups for example, but should be viewed as "densely populated areas in a multidimensional space of community composition," i.e. hot-spots within a cloud of phylotypes with "blurry" edges, and greatly influenced by the limitations of current 16S rRNA gene-based microbial classification and the often fairly arbitrary cut-off values used for different phylogenetic units. Such groupings are also determined by the limits and thresholds required for various bioinformatic and statistical analyses. Huse et al.[138] assessed two different hypervariable regions of the 16S rRNA gene (V1–3 and V3–5) of 200 individual metagenome samples from the Human Microbiome Project, and confirmed that most individuals fit the "enterotype" profile of fecal microbiota at the genus level but these distinctions fall apart at lower taxonomic levels, with little or no apparent segregations at the OTU level.

The significance of these enterotypes may therefore be limited at the biological level, as differences between microorganisms and their relation to host health and disease appears to reside at the strain level rather than higher phylogenetic levels like Phylum or Genus. A good example of this is the dichotomous behavior of *E. coli*, at once a safe and apparently effective probiotic strain, *E. coli* Nissle 1917 and concomitantly, a potentially life-threatening gastrointestinal pathogen *E. coli* O157:H7. These authors also showed that there was a considerable difference in species richness in data from the V1–3 regions compared to data from the V3–5 region, highlighting the need for care when drawing conclusions from metagenomic study data relying on partial 16S rRNA gene fragments. Our own work, and that of others, suggests that primer choice can have a dramatic impact on relative abundance and species richness within 16S rRNA based metagenomic datasets, with the variability probably due to PCR bias against high GC-rich DNA sequences. Similarly, there are suggestions of bias against GC-rich DNA in Illumina sequencing, even with protocols avoiding PCR amplification. In addition, Koren et al.,[139] using combined data from the HMP, METAHit and other human microbiome projects, concluded that for most body sites, a continuous gradient of bacterial genera may be present. For the gut, community structure appeared to be bimodal, with two discernible groupings. However, these groupings were greatly affected by the statistical methods used to identify the clusters and recommended that multiple approaches be employed when defining microbiome enterotypes or core communities. Similarly, a major limitation of current "omics" techniques, is their overreliance on statistical probability and it is fair to say that at the moment, the most advanced omics technologies, metabolomics and metagenomics, are terrific at generating theories, but these theories must then by tested in specifically tailored mechanistic studies using combinations of more traditional methodologies, model systems and finally suitably powered human intervention studies.

MODELS TO STUDY MICROBIAL ECOLOGY

Model systems are essential first steps in studying the impact of exogenous factors on the composition and/or activity of the gut microbiota. The best model is of course the human gastrointestinal tract (GIT) itself, but there is clear value in performing preliminary experiments in model systems where more variables can be compared. The most commonly used are small animal models and *in vitro* simulations, both of which have generated valuable data, yet both of which are anomalous to the human GIT. There are considerable differences between both the anatomy and gut microbiota composition of laboratory rodents and humans and even in microbiota composition within the same species but different strains of laboratory animals.[140–142] Indeed, significant laboratory and cage effects have been recognized both in the composition of the gut microbiota, their metabolic output with knock-on effects for disease susceptibility.[141,143] Additionally gnotobiotic animals, born and raised in the absence of normal microbial successional development, possess a very different physiology to conventionally reared animals. To overcome some of these limitations, germ-free animals colonized with human intestinal microbiota (Human Flora Associated, HFA, animals) have been used to study the impact of diet on various aspects of gut microbial ecology, including colon cancer risk, fermentation, xenobiotic transformation, equol production, DNA transfer, and most recently, obesity and metabolic disease.[144–148] *In vitro* systems which aim to culture the gut microbiota under laboratory conditions in mixed anaerobic culture in either batch or continuous feed modes, are limited by the absence of key metabolic (e.g., enterohepatic circulation of bile acids) and immune interactions with host cells and more fastidious members of the gut microbiota are frequently eliminated. However, although all these different strategies for studying the microbial ecology of the gut microbiota have their limitations, it is important to emphasize that with careful choice of model system, it is possible to attain biologically relevant data which can then be cross-validated with more complex experimental systems and finally in humans themselves.[142]

CONCLUSIONS

In summary, it is fair to say that the human gut microbiota is unique to each individual. A small number of prevalent and dominant microbial species are present in most individuals, but they vary greatly in relative abundance from one individual to another. It is likely that certain critical metabolic functions are conserved across many, even distantly related, bacteria within the gut microbiota. This includes the machinery to ferment complex carbohydrates, transform/ferment amino acids, produce vitamins, deconjugate bile acids and modulate or educate the immune system. Indeed, we already know from studying bacteria isolated from the gut that many of these functions are shared between distantly related bacteria that have co-evolved within the gastrointestinal habitat. It is also clear that sub-dominant species exist that play critical roles in the optimal functioning of the gut microbial community. Good examples are the methanogens and sulfate-reducing bacteria which aid H_2 disposal and alleviate microbial fermentation from end-product inhibition and reduce colonic gas volume. There is a high degree of cross-feeding whereby the end product of one microorganism serves as substrate for another, and also collaborations, where different bacteria work together in the breakdown of highly complex and recalcitrant plant polysaccharides. There is a strong interaction between the gut microbiota and the hosts' immune system. The human immune system relies on a series of microbial recognition molecules such as cell wall fragments (lipopolysaccharide, LPS) from Gram-negative bacteria, and peptidoglycan components from Gram-negative and Gram-positive bacteria; fimbrea proteins typical of gastrointestinal pathogens within the *Enterobacteriaceae*; and DNA sequences rich in CpG motifs, which can be found in different bacteria. The identifying molecules are "remembered" by our evolutionarily advanced immune system and which, when encountered in the "wrong" place, trigger appropriate inflammatory responses in the homeostatic immune system in healthy individuals.

In terms of the interactions between gut bacteria and human health, it appears that what you do, and where you do it, may be more important rather than simply who you are. Although microbial profiles and phylogenetic patterns are useful tools for tracking changes within the microbial ecology of the gut, the real challenge is to track modulations in key metabolic and immunological functions within the gut microbiota in response to environmental stressors. Diet, age and disease are key environmental stressors currently under study in terms of gut microbial ecology. Other important interactions would be those between different microbes; inter-kingdom relations between bacteria, fungi and the mammalian host; between bacteriophage and microbiota; and between human viral pathogens and the microbiota. The field of human gut microbiology has advanced tremendously over the past 20 years, both in terms of efficacious tool development and in our appreciation of how important our microbial friends really are to human health. The challenge over the next 20 years will be to gain a better understanding of the mechanisms linking microbiota composition and activity with diverse core mammalian physiological processes and how these interactions may be modified by diet to support the health of the human host.

References

1. Flint HJ, Scott KP, Louis P, Duncan SH. The role of the gut microbiota in nutrition and health. *Nat Rev Gastroenterol Hepatol.* 2012;9:577–589.
2. Nicholson JK, Holmes E, Kinross J, et al. Host–gut microbiota metabolic interactions. *Science.* 2012;336:1262–1267.
3. Cummings JH, Pomare EW, Branch WJ, Naylor CP, Macfarlane GT. Short chain fatty acids in human large intestine, portal, hepatic and venous blood. *Gut.* 1987;28:1221–1227.
4. Cummings JH, Macfarlane GT. The control and consequences of bacterial fermentation in the human colon. *J Appl Bacteriol.* 1991;70:443–459.
5. Conterno L, Fava F, Viola R, Tuohy KM. Obesity and the gut microbiota: does up-regulating colonic fermentation protect against obesity and metabolic disease? *Genes Nutr.* 2011;6:241–260.

6. Guinane CM, Cotter PD. Role of the gut microbiota in health and chronic gastrointestinal disease: understanding a hidden metabolic organ. *Therap Adv Gastroenterol*. 2013;6:295–308.
7. Magrone T, Jirillo E. The interplay between the gut immune system and microbiota in health and disease: nutraceutical intervention for restoring intestinal homeostasis. *Curr Pharm Des*. 2013;19:1329–1342.
8. Rowland IR, Mallett AK, Wise A. The effect of diet on the mammalian gut flora and its metabolic activities. *Crit Rev Toxicol*. 1985;16:31–103.
9. Louis P, Scott KP, Duncan SH, Flint HJ. Understanding the effects of diet on bacterial metabolism in the large intestine. *J Appl Microbiol*. 2007;102:1197–1208.
10. Koeth RA, Wang Z, Levison BS, et al. Intestinal microbiota metabolism of L-carnitine, a nutrient in red meat, promotes atherosclerosis. *Nat Med*. 2013;19:576–585.
11. Johnson CH, Patterson AD, Idle JR, Gonzalez FJ. Xenobiotic metabolomics: major impact on the metabolome. *Annu Rev Pharmacol Toxicol*. 2012;52:37–56.
12. Macfarlane GT, Macfarlane S. Bacteria, colonic fermentation, and gastrointestinal health. *J AOAC Int*. 2012;95:50–60.
13. LeBlanc JG, Milani C, de Giori GS, Sesma F, van Sinderen D, Ventura M. Bacteria as vitamin suppliers to their host: a gut microbiota perspective. *Curr Opin Biotechnol*. 2013;24:160–168.
14. Cummings JH, Macfarlane GT. Role of intestinal bacteria in nutrient metabolism. *J Parenter Enteral Nutr*. 1997;21:357–365.
15. Wall R, Ross RP, Shanahan F, et al. Metabolic activity of the enteric microbiota influences the fatty acid composition of murine and porcine liver and adipose tissues. *Am J Clin Nutr*. 2009;89:1393–1401.
16. Coakley M, Ross RP, Nordgren M, Fitzgerald G, Devery R, Stanton C. Conjugated linoleic acid biosynthesis by human-derived Bifidobacterium species. *J Appl Microbiol*. 2003;94:138–145.
17. D'Aimmo MR, Mattarelli P, Biavati B, Carlsson NG, Andlid T. The potential of bifidobacteria as a source of natural folate. *J Appl Microbiol*. 2012;112:975–978.
18. Barrett E, Ross RP, O'Toole PW, Fitzgerald GF, Stanton C. γ-Aminobutyric acid production by culturable bacteria from the human intestine. *J Appl Microbiol*. 2012;113:411–417.
19. Bolca S, Verstraete W. Microbial equol production attenuates colonic methanogenesis and sulphidogenesis in vitro. *Anaerobe*. 2010;16:247–252.
20. Fava F, Lovegrove JA, Gitau R, Jackson KG, Tuohy KM. The gut microbiota and lipid metabolism: implications for human health and coronary heart disease. *Curr Med Chem*. 2006;13(25):3005–3021.
21. Tuohy KM, Conterno L, Gasperotti M, Viola R. Up-regulating the human intestinal microbiome using whole plant foods, polyphenols, and/or fiber. *J Agric Food Chem*. 2012;60(36):8776–8782.
22. Tuohy KM, Duncan T, Brown DT, Klinder A, Costabile A. Shaping the human microbiome with prebiotic foods–current perspectives for continued development. *Food Sci Technol Bull*. 2010;7:49–64.
23. Englyst KN, Liu S, Englyst HN. Nutritional characterization and measurement of dietary carbohydrates. *Eur J Clin Nutr*. 2007;61(suppl 1): S19–S39.
24. Wolever TM. Workshop report. Fiber and CHD management. *Adv Exp Med Biol*. 1997;427:315–317.
25. Ludwig DS, Pereira MA, Kroenke CH, et al. Dietary fiber, weight gain, and cardiovascular disease risk factors in young adults. *J Am Med Assoc*. 1999;82:1539–1546.
26. Schoenaker DA, Toeller M, Chaturvedi N, Fuller JH, Soedamah-Muthu SS. EURODIAB prospective complications study group. Dietary saturated fat and fibre and risk of cardiovascular disease and all-cause mortality among type 1 diabetic patients: the EURODIAB prospective complications study. *Diabetologia*. 2012;55:2132–2141.
27. Scott KP, Gratz SW, Sheridan PO, Flint HJ, Duncan SH. The influence of diet on the gut microbiota. *Pharmacol Res*. 2013;69:52–60.
28. Smith EA, Macfarlane GT. Dissimilatory amino acid metabolism in human colonic bacteria. *Anaerobe*. 1997;3:327–337.
29. Hughes R, Magee EA, Bingham S. Protein degradation in the large intestine: relevance to colorectal cancer. *Curr Issues Intest Microbiol*. 2000;1:51–58.
30. Nyangale EP, Mottram DS, Gibson GR. Gut microbial activity, implications for health and disease: the potential role of metabolite analysis. *J Proteome Res*. 2012;11:5573–5585.
31. Morotomi M, Guillem JG, LoGerfo P, Weinstein IB. Production of diacylglycerol, an activator of protein kinase C, by human intestinal microflora. *Cancer Res*. 1990;50:3595–3599.
32. Vulevic J, McCartney AL, Gee JM, Johnson IT, Gibson GR. Microbial species involved in production of 1,2-sn-diacylglycerol and effects of phosphatidylcholine on human fecal microbiota. *Appl Environ Microbiol*. 2004;70:5659–5666.
33. Cani PD, Amar J, Iglesias MA, et al. Metabolic endotoxemia initiates obesity and insulin resistance. *Diabetes*. 2007;56:1761–1772.
34. Cani PD, Neyrinck AM, Fava F, et al. Selective increases of bifidobacteria in gut microflora improve high-fat-diet-induced diabetes in mice through a mechanism associated with endotoxaemia. *Diabetologia*. 2007;50:2374–2383.
35. Fava F, Gitau R, Griffin BA, Gibson GR, Tuohy KM, Lovegrove JA. The type and quantity of dietary fat and carbohydrate alter faecal microbiome and short-chain fatty acid excretion in a metabolic syndrome 'at-risk' population. *Int J Obes*. 2013;37:216–223.
36. Zhang C, Zhang M, Wang S, et al. Interactions between gut microbiota, host genetics and diet relevant to development of metabolic syndromes in mice. *ISME J*. 2010;4:232–241.
37. Zhang C, Zhang M, Pang X, Zhao Y, Wang L, Zhao L. Structural resilience of the gut microbiota in adult mice under high-fat dietary perturbations. *ISME J*. 2012;6:1848–1857.
38. Devkota S, Wang Y, Musch MW, et al. Dietary-fat-induced taurocholic acid promotes pathobiont expansion and colitis in Il10−/− mice. *Nature*. 2012;487:104–108.
39. Alcock J, Franklin ML, Kuzawa CW. Nutrient signaling: evolutionary origins of the immune-modulating effects of dietary fat. *Q Rev Biol*. 2012;87:187–223.
40. Vulevic J, Juric A, Tzortzis G, Gibson GR. A mixture of trans-galactooligosaccharides reduces markers of metabolic syndrome and modulates the fecal microbiota and immune function of overweight adults. *J Nutr*. 2013;143:324–331.

41. Suzuki T, Yoshida S, Hara H. Physiological concentrations of short-chain fatty acids immediately suppress colonic epithelial permeability. *Br J Nutr*. 2008;100:297–305.
42. Peng L, Li ZR, Green RS, Holzman IR, Lin J. Butyrate enhances the intestinal barrier by facilitating tight junction assembly via activation of AMP-activated protein kinase in Caco-2 cell monolayers. *J Nutr*. 2009;139:1619–1625.
43. Ferreira TM, Leonel AJ, Melo MA, et al. Oral supplementation of butyrate reduces mucositis and intestinal permeability associated with 5-Fluorouracil administration. *Lipids*. 2012;47:669–678.
44. Jones C, Badger SA, Regan M, et al. Modulation of gut barrier function in patients with obstructive jaundice using probiotic LP299v. *Eur J Gastroenterol Hepatol*. 2013;25:1424–1430.
45. Everard A, Belzer C, Geurts L, et al. Cross-talk between *Akkermansia muciniphila* and intestinal epithelium controls diet-induced obesity. *Proc Natl Acad Sci USA*. 2013;110:9066–9071.
46. Ewaschuk JB, Diaz H, Meddings L, et al. Secreted bioactive factors from *Bifidobacterium infantis* enhance epithelial cell barrier function. *Am J Physiol Gastrointest Liver Physiol*. 2008;295(5):G1025–G1034.
47. Smith PM, Howitt MR, Panikov N, et al. The microbial metabolites, short-chain fatty acids, regulate colonic treg cell homeostasis. *Science*. 2013;341:569–573.
48. Meijer K, de Vos P, Priebe MG. Butyrate and other short-chain fatty acids as modulators of immunity: what relevance for health? *Curr Opin Clin Nutr Metab Care*. 2010;13:715–721.
49. Jin Z, Mendu SK, Birnir B. GABA is an effective immunomodulatory molecule. *Amino Acids*. 2013;45:87–94.
50. Tricon S, Burdge GC, Williams CM, Calder PC, Yaqoob P. The effects of conjugated linoleic acid on human health-related outcomes. *Proc Nutr Soc*. 2005;64:171–182.
51. Dilzer A, Park Y. Implication of conjugated linoleic acid (CLA) in human health. *Crit Rev Food Sci Nutr*. 2012;52:488–513.
52. Shulzhenko N, Morgun A, Hsiao W, et al. Crosstalk between B lymphocytes, microbiota and the intestinal epithelium governs immunity versus metabolism in the gut. *Nat Med*. 2011;17:1585–1593.
53. Tamrakar AK, Schertzer JD, Chiu TT, et al. NOD2 activation induces muscle cell-autonomous innate immune responses and insulin resistance. *Endocrinology*. 2010;151:5624–5637.
54. Jin C, Flavell RA. Innate sensors of pathogen and stress: Linking inflammation to obesity. *J Allergy Clin Immunol*. 2013;132:287–294.
55. de Goffau MC, Luopajärvi K, Knip M, et al. Fecal microbiota composition differs between children with β-cell autoimmunity and those without. *Diabetes*. 2013;62:1238–1244.
56. Brown K, DeCoffe D, Molcan E, Gibson DL. Diet-induced dysbiosis of the intestinal microbiota and the effects on immunity and disease. *Nutrients*. 2012;4:1095–1119.
57. Franz MJ, Powers MA, Leontos C, et al. The evidence for medical nutrition therapy for type 1 and type 2 diabetes in adults. *J Am Diet Assoc*. 2010;110:1852–1889.
58. Cosnes J. Smoking, physical activity, nutrition and lifestyle: environmental factors and their impact on IBD. *Dig Dis*. 2010;28:411–417.
59. D'Aversa F, Tortora A, Ianiro G, Ponziani FR, Annicchiarico BE, Gasbarrini A. Gut microbiota and metabolic syndrome. *Intern Emerg Med*. 2013;8(Suppl. 1):S11–115.
60. Arvaniti F, Priftis KN, Papadimitriou A, et al. Adherence to the Mediterranean type of diet is associated with lower prevalence of asthma symptoms, among 10–12 year old children: the PANACEA study. *Pediatr Allergy Immunol*. 2011;22:283–289.
61. Bonaccio M, Lacoviello L, de Gaetano G, Moli–Sani Investigators. The mediterranean diet: the reasons for a success. *Thromb Res*. 2012;129:401–404.
62. Willcox DC, Willcox BJ, Todoriki H, Suzuki M. The Okinawan diet: health implications of a low-calorie, nutrient-dense, antioxidant-rich dietary pattern low in glycemic load. *J Am Coll Nutr*. 2009;28(suppl):500S–516S.
63. Frassetto LA, Schloetter M, Mietus-Synder M, Morris Jr RC, Sebastian A. Metabolic and physiologic improvements from consuming a paleolithic, hunter–gatherer type diet. *Eur J Clin Nutr*. 2009;63:947–955.
64. Prescott SL. Early-life environmental determinants of allergic diseases and the wider pandemic of inflammatory noncommunicable diseases. *J Allergy Clin Immunol*. 2013;131:23–30.
65. Kong AP, Xu G, Brown N, So WY, Ma RC, Chan JC. Diabetes and its comorbidities—where East meets West. *Nat Rev Endocrinol*. 2013;9:537–547.
66. Selmi C, Tsuneyama K. Nutrition, geoepidemiology, and autoimmunity. *Autoimmun Rev*. 2010;9:A267–A270.
67. Suau A, Bonnet R, Sutren M, et al. Direct analysis of genes encoding 16S rRNA from complex communities reveals many novel molecular species within the human gut. *Appl Environ Microbiol*. 1999;65:4799–4807.
68. Duncan SH, Barcenilla A, Stewart CS, Pryde SE, Flint HJ. Acetate utilization and butyryl coenzyme A (CoA):acetate-CoA transferase in butyrate-producing bacteria from the human large intestine. *Appl Environ Microbiol*. 2002;68:5186–5190.
69. Derrien M, Vaughan EE, Plugge CM, de Vos WM. *Akkermansia muciniphila* gen. nov., sp. nov., a human intestinal mucin-degrading bacterium. *Int J Syst Evol Microbiol*. 2004;54:1469–1476.
70. Walker AW, Ince J, Duncan SH, et al. Dominant and diet-responsive groups of bacteria within the human colonic microbiota. *ISME J*. 2011;5:220–230.
71. Inglis GD, Thomas MC, Thomas DK, Kalmokoff ML, Brooks SP, Selinger LB. Molecular methods to measure intestinal bacteria: a review. *J AOAC Int*. 2012;95:5–23.
72. de Bruijn FJ. In: de Bruijn FJ, ed. *Handbook of Molecular Microbial Ecology I: Metagenomics and Complementary Approaches*. Oxford, UK: Wiley–Blackwell; 2011.
73. de Bruijn FJ. In: de Bruijn FJ, ed. *Handbook of Molecular Microbial Ecology II: Metagenomics in Different Habitats*. Oxford, UK: Wiley–Blackwell; 2011.
74. McCartney AL. Application of molecular biological methods for studying probiotics and the gut flora. *Br J Nutr*. 2002;88(Suppl. 1): S29–S37.
75. Ventura M, Turroni F, Canchaya C, Vaughan EE, O'Toole PW, van Sinderen D. Microbial diversity in the human intestine and novel insights from metagenomics. *Front Biosci (Landmark Ed)*. 2009;14:3214–3221.

76. Gafan GP, Lucas VS, Roberts GJ, Petrie A, Wilson M, Spratt DA. Statistical analyses of complex denaturing gradient gel electrophoresis profiles. *J Clin Microbiol.* 2005;43:3971–3978.
77. Popa R, Popa R, Mashall MJ, Nguyen H, Tebo BM, Brauer S. Limitations and benefits of ARISA intra-genomic diversity fingerprinting. *J Microbiol Methods.* 2009;78:111–118.
78. Sim K, Cox MJ, Wopereis H, et al. Improved detection of bifidobacteria with optimised 16S rRNA-gene based pyrosequencing. *PLoS One.* 2012;7(3):e32543.
79. Berry D, Ben Mahfoudh K, Wagner M, Loy A. Barcoded primers used in multiplex amplicon pyrosequencing bias amplification. *Appl Environ Microbiol.* 2011;77:7846–7849.
80. Rajendhran J, Gunasekaran P. Microbial phylogeny and diversity: small subunit ribosomal RNA sequence analysis and beyond. *Microbiol Res.* 2011;166:99–110.
81. Amann R, Fuchs BM, Behrens S. The identification of microorganisms by fluorescence in situ hybridisation. *Curr Opin Biotechnol.* 2001;12:231–236.
82. Jansen GJ, Wildeboer-Veloo AC, Tonk RH, Franks AH, Welling GW. Development and validation of an automated, microscopy-based method for enumeration of groups of intestinal bacteria. *J Microbiol Methods.* 1999;37:215–221.
83. Rochet V, Rigottier-Gois L, Rabot S, Doré J. Validation of fluorescent in situ hybridization combined with flow cytometry for assessing interindividual variation in the composition of human fecal microflora during long-term storage of samples. *J Microbiol Methods.* 2004;59:263–270.
84. Furet JP, Firmesse O, Gourmelon M, et al. Comparative assessment of human and farm animal faecal microbiota using real-time quantitative PCR. *FEMS Microbiol Ecol.* 2009;68:351–362.
85. Ahmed S, Macfarlane GT, Fite A, McBain AJ, Gilbert P, Macfarlane S. Mucosa-associated bacterial diversity in relation to human terminal ileum and colonic biopsy samples. *Appl Environ Microbiol.* 2007;73:7435–7442.
86. Furrie E. A molecular revolution in the study of intestinal microflora. *Gut.* 2006;55:141–143.
87. Berry D, Schwab C, Milinovich G, et al. Phylotype-level 16S rRNA analysis reveals new bacterial indicators of health state in acute murine colitis. *ISME J.* 2012;6:2091–2106.
88. Yap IK, Li JV, Saric J, et al. Metabonomic and microbiological analysis of the dynamic effect of vancomycin-induced gut microbiota modification in the mouse. *J Proteome Res.* 2008;7:3718–3728.
89. Erickson AR, Cantarel BL, Lamendella R, et al. Integrated metagenomics/metaproteomics reveals human host–microbiota signatures of Crohn's disease. *PLoS One.* 2012;7(11):e49138.
90. Martin FP, Collino S, Rezzi S, Kochhar S. Metabolomic applications to decipher gut microbial metabolic influence in health and disease. *Front Physiol.* 2012;3:113.
91. Blottière HM, de Vos WM, Ehrlich SD, Doré J. Human intestinal metagenomics: state of the art and future. *Curr Opin Microbiol.* 2013;16:232–239.
92. Delmont TO, Simonet P, Vogel TM. Mastering methodological pitfalls for surviving the metagenomic jungle. *Bioessays.* 2013;35:744–754.
93. De Filippo C, Ramazzotti M, Fontana P, Cavalieri D. Bioinformatic approaches for functional annotation and pathway inference in metagenomics data. *Brief Bioinform.* 2012;13:696–710.
94. Stanstrup J, Gerlich M, Dragsted LO, Neumann S. Metabolite profiling and beyond: approaches for the rapid processing and annotation of human blood serum mass spectrometry data. *Anal Bioanal Chem.* 2013;405:5037–5048.
95. Human Microbiome Project Consortium. Structure, function and diversity of the healthy human microbiome. *Nature.* 2012;486:207–214.
96. Qin J, Li R, Raes J, et al. A human gut microbial gene catalogue established by metagenomic sequencing. *Nature.* 2010;464:59–65.
97. Brindle JT, Nicholson JK, Schofield PM, Grainger DJ, Holmes E. Application of chemometrics to 1H NMR spectroscopic data to investigate a relationship between human serum metabolic profiles and hypertension. *Analyst.* 2003;128:32–36.
98. Xie G, Zhang S, Zheng X, Jia W. Metabolomics approaches for characterizing metabolic interactions between host and its commensal microbes. *Electrophoresis.* 2013;34:2787–2798.
99. Nicholson JK, Lindon JC. Systems biology: Metabonomics. *Nature.* 2008;455(7216):1054–1056.
100. Savage DC. Microbial ecology of the gastrointestinal tract. *Annu Rev Microbiol.* 1977;31:107–133.
101. Macfarlane S, McBain AJ, Macfarlane GT. Consequences of biofilm and sessile growth in the large intestine. *Adv Dent Res.* 1997;11:59–68.
102. Harrell L, Wang Y, Antonopoulos D, et al. Standard colonic lavage alters the natural state of mucosal-associated microbiota in the human colon. *PLoS One.* 2012;7:e32545.
103. Delgado S, Cabrera-Rubio R, Mira A, Suárez A, Mayo B. Microbiological survey of the human gastric ecosystem using culturing and pyrosequencing methods. *Microb Ecol.* 2013;65:763–772.
104. Kothary M, Babu U. Infective dose of foodborne pathogens in volunteers: a review. *J Food Saf.* 2001;21:49–68.
105. Andersson AF, Lindberg M, Jakobsson H, Bäckhed F, Nyrén P, Engstrand L. Comparative analysis of human gut microbiota by barcoded pyrosequencing. *PLoS One.* 2008;3(7):e2836.
106. Bik EM, Eckburg PB, Gill SR, et al. Molecular analysis of the bacterial microbiota in the human stomach. *Proc Natl Acad Sci USA.* 2006;17:732–737.
107. Rescigno M, Urbano M, Valzasina B, et al. Dendritic cells express tight junction proteins and penetrate gut epithelial monolayers to sample bacteria. *Nat Immunol.* 2001;2:361–367.
108. Nicoletti C, Regoli M, Bertelli E. Dendritic cells in the gut: to sample and to exclude? *Mucosal Immunol.* 2009;2:462.
109. Farache J, Koren I, Milo I, et al. Luminal bacteria recruit CD103 + dendritic cells into the intestinal epithelium to sample bacterial antigens for presentation. *Immunity.* 2013;38:581–595.
110. Schenk M, Mueller C. The mucosal immune system at the gastrointestinal barrier. *Best Pract Res Clin Gastroenterol.* 2008;22:391–409.
111. Fava F, Danese S. Intestinal microbiota in inflammatory bowel disease: friend of foe? *World J Gastroenterol.* 2011;17:557–566.
112. Sorini C, Falcone M. Shaping the (auto)immune response in the gut: the role of intestinal immune regulation in the prevention of type 1 diabetes. *Am J Clin Exp Immunol.* 2013;2:156–171.
113. Salzman NH. Paneth cell defensins and the regulation of the microbiome: détente at mucosal surfaces. *Gut Microbes.* 2010;1:401–406.

114. Lehrer RI, Lu W. α-Defensins in human innate immunity. *Immunol Rev*. 2012;245:84–112.
115. Duncan SH, Belenguer A, Holtrop G, Johnstone AM, Flint HJ, Lobley GE. Reduced dietary intake of carbohydrates by obese subjects results in decreased concentrations of butyrate and butyrate-producing bacteria in feces. *Appl Environ Microbiol*. 2007;73:1073–1078.
116. Islam KB, Fukiya S, Hagio M, et al. Bile acid is a host factor that regulates the composition of the cecal microbiota in rats. *Gastroenterology*. 2011;141:1773–1781.
117. Booijink CC, El-Aidy S, Rajilić-Stojanović M, et al. High temporal and inter-individual variation detected in the human ileal microbiota. *Environ Microbiol*. 2010;12:3213–3227.
118. Zoetendal EG, Raes J, van den Bogert B, et al. The human small intestinal microbiota is driven by rapid uptake and conversion of simple carbohydrates. *ISME J*. 2012;6:1415–1426.
119. Hartman AL, Lough DM, Barupal DK, et al. Human gut microbiome adopts an alternative state following small bowel transplantation. *Proc Natl Acad Sci USA*. 2009;106:17187–17192.
120. Wang M, Ahrné S, Jeppsson B, Molin G. Comparison of bacterial diversity along the human intestinal tract by direct cloning and sequencing of 16S rRNA genes. *FEMS Microbiol Ecol*. 2005;54:219–231.
121. van Baarlen P, Troost F, van der Meer C, et al. Human mucosal in vivo transcriptome responses to three lactobacilli indicate how probiotics may modulate human cellular pathways. *Proc Natl Acad Sci USA*. 2011;108(suppl 1):4562–4569.
122. van Baarlen P, Troost FJ, van Hemert S, et al. Differential NF-kappaB pathways induction by *Lactobacillus plantarum* in the duodenum of healthy humans correlating with immune tolerance. *Proc Natl Acad Sci USA*. 2009;106:2371–2376.
123. Walker AW, Duncan SH, McWilliam Leitch EC, Child MW, Flint HJ. pH and peptide supply can radically alter bacterial populations and short-chain fatty acid ratios within microbial communities from the human colon. *Appl Environ Microbiol*. 2005;71:3692–3700.
124. Eckburg PB, Bik EM, Bernstein CN, et al. Diversity of the human intestinal microbial flora. *Science*. 2005;308:1635–1638.
125. Durbán A, Abellán JJ, Jiménez-Hernández N, et al. Assessing gut microbial diversity from feces and rectal mucosa. *Microb Ecol*. 2011;61:123–133.
126. Codling C, O'Mahony L, Shanahan F, Quigley EM, Marchesi JR. A molecular analysis of fecal and mucosal bacterial communities in irritable bowel syndrome. *Dig Dis Sci*. 2010;55:392–397.
127. Zoetendal EG, Akkermans AD, De Vos WM. Temperature gradient gel electrophoresis analysis of 16S rRNA from human fecal samples reveals stable and host-specific communities of active bacteria. *Appl Environ Microbiol*. 1998;64:3854–3859.
128. Zoetendal E, Akkermans A, Akkermans-Van Vleit W, De Visser J, De Vos W. The host genotype affects the bacterial community in the human gastrointestinal tract. *Microbial Ecol Health Dis*. 2001;13:129–134.
129. Li M, Wang B, Zhang M, et al. Symbiotic gut microbes modulate human metabolic phenotypes. *Proc Natl Acad Sci USA*. 2008;105:2117–2122.
130. De Palma G, Capilla A, Nadal I, et al. Interplay between human leukocyte antigen genes and the microbial colonization process of the newborn intestine. *Curr Issues Mol Biol*. 2010;12:1–10.
131. Lee SM, Donaldson GP, Mikulski Z, Boyajian S, Ley K, Mazmanian SK. Bacterial colonization factors control specificity and stability of the gut microbiota. *Nature*. 2013;501:426–429.
132. Tap J, Mondot S, Levenez F, et al. Towards the human intestinal microbiota phylogenetic core. *Environ Microbiol*. 2009;11:2574–2584.
133. Salonen A, Salojärvi J, Lahti L, de Vos WM. The adult intestinal core microbiota is determined by analysis depth and health status. *Clin Microbiol Infect*. 2012;18(suppl 4):16–20.
134. Mondot S, de Wouters T, Doré J, Lepage P. The human gut microbiome and its dysfunctions. *Dig Dis*. 2013;31:278–285.
135. Turnbaugh PJ, Quince C, Faith JJ, et al. Organismal, genetic, and transcriptional variation in the deeply sequenced gut microbiomes of identical twins. *Proc Natl Acad Sci USA*. 2010;107:7503–7508.
136. Sekelja M, Berget I, Næs T, Rudi K. Unveiling an abundant core microbiota in the human adult colon by a phylogroup-independent searching approach. *ISME J*. 2011;5:519–531.
137. Arumugam M, Raes J, Pelletier E, et al. Enterotypes of the human gut microbiome. *Nature*. **473**: 174–180.
138. Huse SM, Ye Y, Zhou Y, Fodor AA. A core human microbiome as viewed through 16S rRNA sequence clusters. *PLoS One*. 2012;7(6): e34242.
139. Koren O, Knights D, Gonzalez A, et al. A guide to enterotypes across the human body: meta-analysis of microbial community structures in human microbiome datasets. *PLoS Comput Biol*. 2013;9(1):e1002863.
140. Andersson KE, Axling U, Xu J, et al. Diverse effects of oats on cholesterol metabolism in C57BL/6 mice correlate with expression of hepatic bile acid-producing enzymes. *Eur J Nutr*. 2013;52:1755–1769.
141. Gulati AS, Shanahan MT, Arthur JC, et al. Mouse background strain profoundly influences Paneth cell function and intestinal microbial composition. *PLoS One*. 2012;7(2):e32403.
142. Rumney CJ, Rowland IR. *In vivo* and *in vitro* models of the human colonic flora. *Crit Rev Food Sci Nutr*. 1992;31:299–331.
143. Swann JR, Tuohy KM, Lindfors P, et al. Variation in antibiotic-induced microbial recolonization impacts on the host metabolic phenotypes of rats. *J Proteome Res*. 2011;10:3590–3603.
144. Rumney CJ, Rowland IR, Coutts TM, et al. Effects of risk-associated human dietary macrocomponents on processes related to carcinogenesis in human-flora-associated (HFA) rats. *Carcinogenesis*. 1993;14:79–84.
145. Silvi S, Rumney CJ, Cresci A, Rowland IR. Resistant starch modifies gut microflora and microbial metabolism in human flora-associated rats inoculated with faeces from Italian and UK donors. *J Appl Microbiol*. 1999;86:521–530.
146. Bowey E, Adlercreutz H, Rowland I. Metabolism of isoflavones and lignans by the gut microflora: a study in germ-free and human flora associated rats. *Food Chem Toxicol*. 2003;41(5):631–636.
147. Tuohy K, Davies M, Rumsby P, Rumney C, Adams MR, Rowland IR. Monitoring transfer of recombinant and nonrecombinant plasmids between *Lactococcus lactis* strains and members of the human gastrointestinal microbiota *in vivo*—impact of donor cell number and diet. *J Appl Microbiol*. 2002;93:954–964.
148. Ridaura VK, Faith JJ, Rey FE, et al. Gut microbiota from twins discordant for obesity modulate metabolism in mice. *Science*. 2013;341 (6150):1241214.

C H A P T E R

2

A Nutritional Anthropology of the Human Gut Microbiota

Carlotta De Filippo and Kieran M. Tuohy

Department of Food Quality and Nutrition, Research and Innovation Centre, Fondazione Edmund Mach, San Michele all'Adige, Trento, Italy

HUMAN DIET OR MICROBIOTA, WHICH CAME FIRST?

The feeding strategy of *Homo sapiens* appears to be characterized by an extraordinary omnivorism, which has no equal among mammals with the exception to some extent of the Suidae and the brown bear. This strategy allows him to have a diet that is able to capture all substances and nutrients necessary for its energy and structural needs, according to the best sources, such as foods, available in the ecosystem of origin and from a certain point in its evolution, adapted to remote ecosystems. We can therefore say that diet is one of the main factors that differentiates and drives evolution of human populations. Dietary differences originated from cultural evolution and geographic differences in availability of crops and cultivation and animal husbandry. It is widely recognized that a varied and balanced diet is essential to an individual's health. The adverse effects of nutrient deficiency are numerous and well documented.[1–4] Because nutritionally related problems continue to be the cause behind many diseases that hinder progress towards universally adequate health, all countries should be actively pursuing the improvement of their people's nutritional status. Recently we witnessed an explosion of food consumption studies in both urban and rural areas of developing countries.[5–7] These types of studies are vital to our understanding of more "transitional" and/or "traditional" diets vs. the modern-day Western-style diet. Furthermore, food-consumption in rural communities in particular generally involves a large proportion of the food coming from home-production or gathering or, at the very least, having been grown, produced and purchased locally. Therefore, diets are usually monotonous and simple because they are dependent on the availability of foods in the home or local markets as well as the prices of those foods. However, the foods themselves, often consumed with little processing or using traditional fermentation technologies, represent complex mixtures of non-digestible carbohydrates and fibers, polyphenols and live fermentative microorganisms, thereby representing both complex nutritional support for the gut microbiota and an important source of passenger microorganisms with immune-modulatory and metabolic potential. The relative invariability of these traditional diets may potentially be reflected in gut colonization by relatively homogeneous and characteristic microbiomes. Recent discoveries highlighting the importance of gut microbiota have demonstrated how the availability of the nutrients present in the foods comprising everyone's diet is highly dependent on the human gut microbiota. The question then becomes, to what extent is the human gut microbiota dependent on changes in diet and how robust is the human microbiota from birth to death? To propose potential answers to these questions first of all we have to understand what is the human microbiota.

Diet-Microbe Interactions in the Gut.
DOI: http://dx.doi.org/10.1016/B978-0-12-407825-3.00002-2

METAGENOMICS AND CULTIVATION-INDEPENDENT ASSESSMENT OF HUMAN GUT MICROBIOTA

The human gut microbiota is composed of commensal microorganisms inherited largely from our mothers at birth, passengers' microorganisms, mainly environmental, with which we come into continuous contact via the food we eat, and potential pathogens, exogenous invaders which try to overcome the body's defenses and cause disease. In the 20th century our knowledge of the human microbiota was constrained by the ability to describe and study the biological functions of less than a hundred cultivable bacteria. The species we described until the year 2000 were also the most easily cultivated, and given the special attention of funding agencies towards pathogens, we fundamentally ignored the genome to function relation for the vast majority of our commensal organisms which do not cause disease and a handful of bacterial species used in food production and shown to dominate the gut microbiota of breast-fed infants, the lactobacilli and bifidobacteria, respectively.

Furthermore, for a century the study of microorganisms has been limited by the ability to cultivate them. The established view is that only a subset of the microbial species which make up our microbiome can be easily cultivated. Recently, the scientific revolution driven by high-throughput sequencing techniques (Next Generation Sequencing, NGS), has made possible the unraveling of the evolutionary history of human gut microbiome. Key to this endeavor has been the emergence of bioinformatics tools necessary to describe the microbial ecology-encoded high-resolution NGS data derived from diverse microbiomes.

Large-scale projects such as the European Metagenomics of the Human Intestinal Tract MetaHIT[8] and the US Human Microbiome Project, HMP[9,10] have made substantial progress towards this goal and the amount of metagenomic information is exponentially increasing, especially that obtained for individuals living in industrialized countries. The first EU-funded MetaHIT consortium produced Illumina sequences of fecal samples of 124 European individuals, including healthy, overweight and obese adults, as well as patients with inflammatory bowel disease (IBD).[8] When extended to Japanese and American populations, MetaHIT also established that the worldwide population could be classified into three distinct enterotypes.[11] The NIH-funded Human Microbiome Project, HMP Consortium, is also developing and indexing another fundamental reference set of microbial genome sequences from a population of 242 healthy adults, sampled at different body sites, generating 5177 microbial taxonomic profiles from 16S ribosomal RNA genes and over 3.5 terabases of metagenomic sequence so far.[9,10] In parallel, they have sequenced approximately 800 reference strains isolated from the human body, generating data that represent the largest resource describing the abundance and variety of the human microbiome. The project encountered an estimated 81–99% of the genera, enzyme families and community configurations occupied by the healthy Western microbiome.[9,10] The information deposited in these resources promises to be a goldmine for pathway and network inference, reconstructing the super-meta-pathway subtending the interaction between humans and their microbiomes.

MICROBIOME AND HUMAN NUTRITIONAL PHENOTYPE

The role of the gut microbiota in provision of nutritionally relevant molecules for human health and nutrition is still largely unknown, but indeed these organisms do contribute metabolic and digestive functions absent from the human genome.[12] A glimpse of the metabolic pathway complexity contained in metagenomics datasets first emerged from the study of Gill et al.[13]: the human genome lacks most of the enzymes required for degradation of plant polysaccharides and they are supplied by the human gut microbiome which can metabolize cellulose, starch and unusual sugars such as arabinose, mannose, and xylose, thanks to at least 81 different glycoside hydrolase families. With the aim of understanding the dietary modulation of gut microbiota, Zhu et al.[14] undertook a large-scale analysis of 16S rRNA gene sequences to profile the microbiota inhabiting the digestive system of giant pandas using a metagenomic approach. They performed predicted gene functional classification, finding the presence of putative cellulose-metabolizing symbionts in this little-studied microbial environment, explaining how giant pandas are able to partially digest bamboo fiber, despite a genome lacking enzymes that can degrade cellulose. This study showed the paramount importance of pathway analysis in reconstructing microbial metabolic networks. The co-occurrence of microbial species in similar abundance could be seen as an indication of their being part of an integrated network, providing a set of mutually complementary functions integrated in a multi-organismal pathway.

The size of this super-network can be estimated integrating the HMP and the MetaHIT studies, indicating that gene content in gut microbiota is at least 150-fold higher than the human genome. The microbiome identifies over 19,000 different functions among which at least 5000 have never been seen before, at least 6000 are shared by all individuals, the so-called "minimal metagenome," and at least 1200 are required for any bacterium to survive in the human gut, the "minimal gut microbiome."[8–13] The focus of these studies, performed in Western Europe and North American industrialized countries, suggests that this meta-network still represents just the tip of the iceberg of the overall human microbial metabolic diversity, that could vary significantly in regions with different diets and different environmental conditions, such those in African developing countries. Interestingly, many of these enzymes are necessary for the digestion of vegetables and whole plant foods. The composition of gut microbiota in animals and humans varies according to the type of food consumed; vegetarian animals and humans have more microbial diversity than carnivores and omnivores. This microbial diversity is also associated with higher enzymatic diversity, which is likely necessary for plant food digestion.

Once the composition and functional state of the healthy gut microbiota are understood, features associated with disease status can be determined. However, the complexity of the microbiota, and the variation between and within individuals and populations complicates the definition of what this ideal state may be. Ecological principles can aid in understanding the host–microbe interactions and specific functions of the gut microbiota. Other advanced "omics" level technologies such as proteomics and metabolomics, coupled with metabolic network modeling,[15,16] show how host and environmental factors can affect gut microbial ecology over a lifetime. The challenge remains to combine information from multi-omics technologies describing the basic structure and function of the human microbiome in order to derive knowledge of real relevance to human health and disease.

THE GUT MICROBIOTA IN HUMAN EVOLUTION

The nutritional requirements of humans were established by natural selection during millions of years in which humans and their hominid ancestors consumed foods exclusively from a "menu" of wild animals and uncultivated plants.[17] During evolution, changes in the length and compartmentalization of the digestive tract have enabled vertebrates to occupy diverse habitats and exploit different feeding strategies. Of course many of these innovations in gut physiology were driven by the need to optimize two basic biological functions, nutrient absorption and microbial fermentation of slowly digestible plant foods. The co-evolution of gut anatomy, microbes and diet has been first proposed by studies on mammals[18] in which the authors have made a comparative metagenomic study of the fecal microbiota of human beings and 60 other mammalian species, living in zoos and in the wild, to see how taxonomic position and diet affect the composition of the commensal microbiota and to understand how these relationships have co-evolved. They found that both diet and phylogeny influence the increase in bacterial diversity from carnivore to omnivore to herbivore. Although there is a general trend for herbivores harboring the most diverse communities and carnivores the least, the overall relationship between microbiota and its host appears to be the species: baboons in the St Louis Zoo have much the same gut microbiota as wild baboons in Namibia.

Decades of anthropological research have been devoted to elucidating dietary history, in part because dietary shifts were likely associated with major anatomical and cultural changes (e.g., the increase in relative brain size and the advent of modern civilization via agriculture).

Dietary habits are considered one of the main factors contributing to the diversity of human gut microbiota.[19] Profound changes in diet and lifestyle conditions began with the so-called "Neolithic revolution" with the introduction of agriculture and animal husbandry approximately 10,000 years ago.[2] The origin and spread of agriculture and animal husbandry over the past 10,000 years, with centers of domestication in Asia, Europe, South America, and Africa, represent the first major shift in human diets. After that time, food resources became more abundant and constant. The food production and storage technologies associated with this dietary shift led to population densities that are orders of magnitude greater than what is possible under hunter–gatherer subsistence economies. Furthermore, with the advent of agriculture, novel foods were introduced as staples for which the hominid genome had little evolutionary experience. More importantly, food-processing procedures were developed, following the industrial revolution and the more recent Green Revolution of the post-World War II era, which allowed for quantitative and qualitative food and nutrient combinations that had not previously been encountered over the course of hominid evolution and above all, the emergence of energy-dense, nutrient-poor largely sterile foods as the major dietary components in developed economies. A major food macro-component which differentiates ancient and hunter–gatherer diets from our modern diet is the quantity and quality of dietary fiber. It has been estimated that the Paleolithic diet delivered more than 100 g/day of fiber, whereas current recommended daily allowances in the US are 20–30 g/day and average American intake is between 10 and 20 g/day.[20] The quality or chemical make-up of fibers forming the larger proportion of the current dietary intake is also likely to be different. A high proportion of dietary fiber in the current diet is derived from cereal grains compared to a much more diverse collection of fibers from a wide range of fruit, vegetables, roots, legumes and nuts which would have been consumed in hunter gatherers and in Paleolithic times.[2,21] Such foods contain higher levels of fermentable carbohydrates such as resistant starch compared to cereal-bran-derived fiber. Resistant starch is a major contributor towards fermentable carbohydrate in the diet. Fermentation drives the energy economy of the colon, impacting on the relative prevalence of the resident saccharolytic microbiota and production of short-chain fatty acids (SCFAs), such as acetate, propionate and butyrate. Up to 95% of these SCFAs produced in the colon may be taken up and utilized by the host. Butyrate derived directly from fiber carbohydrate fermentation or upon cross-feeding of acetate and lactate by butyrigenic bacteria, forms a principal energy currency within the colonic mucosa, playing an important role in mucosal architecture and barrier function. Acetate and propionate, in particular, enter the portal blood stream to the liver. Here, acetate is metabolized to acetyl-CoA and plays a role in de novo lipogenesis, while propionate acts as an inhibitor of lipid synthesis and cholesterogenesis by interacting with 3-hydroxy-3-methylglutaryl-CoA (HMG CoA) reductase. Acetate also acts as an important energy source for muscle and brain tissue. The production of high levels of propionate through colonic fermentation has been proposed as a possible mechanism responsible for lowering serum and hepatic cholesterol levels in animal feeding studies with various non-digestible carbohydrates and foods with a low glycemic index, but high fiber content. Similarly, SCFAs produced in the colon may be involved in the regulation of gastrointestinal peptides and hormones involved in satiety control and play an important role in fortification of the intestinal barrier and reducing mucosal permeability, which may impact on systemic inflammation (for a review, see[22]).

However, overall, the spread of agriculture was associated with an astounding relative reduction in the diversity of nutrient intake, including polyphenols, which are commonly much lower in domesticated plant foods compared to wild varieties. For example, 50–70% of the calories in the agricultural diet are from starch alone. In addition to a reduction in nutritional diversity, agricultural diets may also have been associated with a caloric availability that exceeds growth and energetic requirements, as observed among the most developed contemporary agricultural economies. These modern diets too are relatively rich in fat, and high-fat feeding both in animal models of metabolic syndrome and in humans, has been shown to impact not only on energy metabolism, but also on gut function (mucosal permeability), gut microbiota composition as discussed above, and immune function. High-fat feeding acts as a trigger for inflammatory process linked to insulin resistance, non-alcoholic fatty liver disease and obesity (i.e., metabolic endotoxemia).[23]

Another noteworthy effect of the Neolithic revolution, the concentration of large human populations in limited areas, created selective pressure that favored pathogens specialized in colonizing human hosts and probably produced the first wave of pandemic human diseases.[24] It has been hypothesized that bacteria specialized in human-associated niches, including our gut commensal microbiota, underwent intense transformation during the social and demographic changes that took place with the first Neolithic settlements.[25,26] Western developed countries successfully controlled infectious diseases during the second half of the last century, by improving sanitation and using antibiotics and vaccines. At the same time, a rise in new diseases such as allergic, metabolic disease and autoimmune disorders like IBD both in adults and in children have been observed.[24] The fact that the incidence of IBD and allergies is greater in industrialized, Western, societies than in developing countries organized on communities living as farmers or still as hunter gatherers[27] leads to the hypothesis that improvements in hygiene, together with decreased microbial exposure, in childhood could be responsible for immune malfunctioning.[28] The interplay between diet and microbiota probably shaped the immune system itself. It is possible to hypothesize that our immune system changed profoundly with the transition to agriculture and husbandry into villages and cities. In particular the consumption of dairy foods could have selected alleles for tolerance to milk and at the same time favored the recognition of lactic bacteria and bifidobacteria or bread yeasts as "friends." In this perspective we can speculate that studies on the evolution of host–microbe interaction, a field currently in its infancy, will witness rapid progress in the coming years. Such comparisons could shed light on how attributes that are specific to modern human biology and nutrition, such as the availability of foods that are more diverse, abundant and often heavily processed, affect our microbial partners.

POPULATION METAGENOMIC VARIATION WITHIN THE HUMAN MICROBIOTA

We do not yet completely understand how the different environments and wide range of diets which modern humans around the world experience has affected the microbial ecology of the human gut. Gut microbial composition depends on different dietary habits just as health depends on microbial metabolism, but the association of microbiota with different diets in human populations has not yet been shown. The relative importance of genetic make-up and uptake of environmental microorganisms to provide humans with these fundamental microbial functions is still under debate.

Twins and mother–daughter pairs have more similar microbiota compositions than unrelated individuals, suggesting that there could be a genetic influence over the microbiota.[29,30] However, monozygotic and dizygotic adult twins have equally similar microbiota, suggesting that environment rather than genetics may drive familial similarities.[30]

Through cultural innovation and changes in habitat and ecology, there have been a number of major dietary shifts in human evolution, including meat eating, cooking, and those associated with plant and animal domestication. The analysis performed on the gut microbiota of humans living a modern lifestyle revealed that the microbial community is typical of omnivorous primates.[18]

Contemporary human beings are genetically adapted to the environment in which their ancestors survived, which is known to have conditioned their genetic make-up. Whilst humans range from being almost completely carnivorous, such as the Eskimo groups in northern temperate and arctic regions, to largely vegetarian, such as the hunter–gatherers in southern Africa, the majority of societies effectively balance their diet with an omnivorous mix of meat and vegetables.[31]

Initial studies indicate that different aspects of the gut microbiota can distinguish human populations according to their histories and lifestyles, including diet.

The stomach-associated bacterium *Helicobacter pylori* exemplifies co-evolution between microorganisms and humans. Patterns of *H. sapiens* migration from Africa across the globe can be traced from the strain diversity of this bacterial species.[32] Comparing the human gut microbiota and microbiome worldwide is the only way, in principle, to reveal if there is a core set of gut microbial genes and organismal lineages that are shared by most, if not all, humans.

Populations can be Separated by Characteristic Differences in the Gut Microbiota

In our recent paper[33] investigating diet–gut microbiota interaction, we discuss how the ability of modern *H. sapiens* to live in different environments and to follow a wide range of different diets has affected our gut microbial ecology. We characterized the fecal microbiota of 14 African children (BF) living in a rural village (Boulpon, Burkina Faso, Africa) and of 15 European children (EU) living in an urban area (Florence, Italy), by sequencing the 16S ribosomal RNA gene, with the aim of elucidating the effects of different diets on gut microbiota. The experimental design included a "natural" control set, breast-fed children from Burkina Faso and Italy. There are five major advancements. The first key finding is that Burkina children cluster separately from Florentine children and breast-fed toddlers form a third cluster in between (Figure 2.1A). This can be solely the result of diet. Contributions from sanitation could play a role, but this is indeed minor; in fact, if sanitation or hygiene did cause the difference then one would expect that breast-fed children would also cluster differently. While we clearly observed that mothers' milk significantly reduces the differences and that is why the breast-fed children cluster separately, the cluster contains both Florence and Burkina toddlers, and is apart from the other Burkina and Tuscan children. This fits well with existing knowledge of how breast milk shapes the gut microbiota and illustrates the universality of bifidobacteria as a closely co-evolved microbial partner within the gut microbiota at this early stage of development. The second key finding is that the BF microbiota was significantly enriched in *Bacteroidetes* and depleted in *Firmicutes* compared to their EU counterparts, suggesting a co-evolution of intestinal bacteria with their diet, rich in plant polysaccharides (Figure 2.1B). The third key finding is that the fecal samples from children in Burkina Faso especially differed from the Italian subjects because of the presence of *Prevotella* and *Xylanibacter* (*Bacteroidetes*), *Treponema* (*Spirochaetes*) and *Butyrivibrio* (*Firmicutes*); all appeared in the African but were not found in the Italian samples. This peculiar microbiota is at risk of being lost in urbanized Africans. We hypothesize that these distinctive bacterial genera might help to extract energy from the polysaccharides in the children's heavier fiber diet. These bacteria

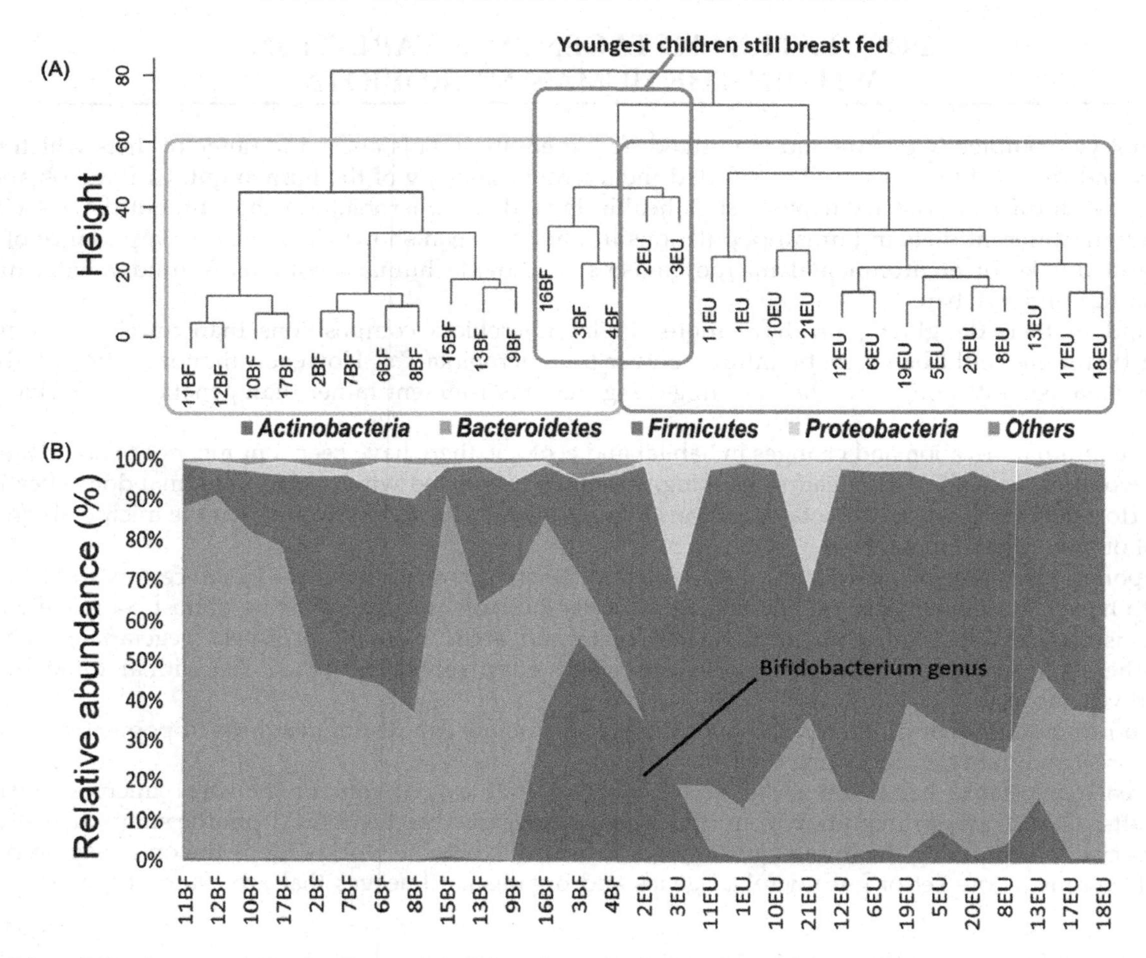

FIGURE 2.1 (A) Dendrogram obtained with complete linkage hierarchical clustering of the samples from BF and EU populations based on their genera. The subcluster located in the middle of the tree contains samples taken from the five youngest children (1–2 years old) still breast fed (three from the BF group and two from the EU group). (B) Relative abundances (percentage of sequences) of the four most abundant bacterial phyla in each individual among the BF and EU children. Firmicutes (red) and Bacteroidetes (green) abundances significantly differentiate the BF from the EU children. Blue area in middle shows abundance of Actinobacteria, mainly represented by *Bifidobacterium* genus, in the five youngest EU and BF children.

are capable of fermenting cellulose and xylan through a number of carbohydrate-active enzymes, producing anti-inflammatory effects at the same time. The speculation is that there is a link between these microbes that colonize the human gut and termites as part of the Burkinabé diet. The fourth key finding is that short-chain fatty acids levels are statistically much higher in Burkina children with respect to the European ones, and the four Burkina-specific species (*Prevotella, Xylanibacter, Treponema and Butyrivibrio*) have the enzymes needed to digest fibers and produce these beneficial molecules. The presence of these species would allow Burkina Faso children to maximize the energy intake from indigestible components by producing high levels of SCFAs that supply the host with an additional amount of energy. Normal colonic epithelia derive 60–70% of their energy supply from SCFAs, particularly butyrate. Propionate is largely taken up by the liver and is a good precursor for gluconeogenesis, liponeogenesis and protein synthesis. Acetate enters the peripheral circulation to be metabolized by peripheral tissues and is a substrate for cholesterol synthesis. Together, they account for 10% of calories extracted from a Western diet each day and probably an amount of calories more abundant than a typical rural Village of Burkina Faso diet. In addition, SCFAs have several functions: they can reduce inflammation in colitis, promoting tissue renewal, increase the absorption of sodium and water in diarrhea, regulate enteric neurons and control of gastrointestinal motility. The last key finding is that biodiversity is significantly reduced in European children with respect to Burkina Faso children. Basically, a diverse gut microbiota is a stable gut microbiota, and homeostasis within the microbiota-derived functions linked to host metabolism and immune function in particular are likely to be key determinants of host health or disease. The different bacterial compositions are likely to have profound influences on other organs as well as the immune system, possibly explaining the low incidence of inflammatory bowel diseases in African children and adults. Exposure to the large variety of environmental microbes associated with a high-fiber diet could increase the potentially beneficial bacterial genomes, enriching the microbiome. Reduction in microbial richness is possibly one of the undesirable effects of globalization and of eating generic, energy-dense, nutrient-poor, low fiber, uncontaminated foods.

Similar results were recently reported by Yatsunenko et al.[13] showing that both children and adults from Malawi and the Amazonas state of Venezuela had an increase in *Prevotella* while the microbiota of those in the US had more *Bacteroides*. Although genetically different, the populations investigated also differ in other factors that could affect the microbiota, such as environmental exposures, provision of adequate sanitation and levels of cleanliness, diet and antibiotic use. The relative homogeneity of the microbial composition of Europeans and Americans, with respect to Africans, likely reflects this globalization, rather than a transcendent tendency of humans to select the same bacteria worldwide.[9] Cultural factors, especially diet, may be crucial in shaping the gut microbiota, since the Burkina Faso, Malawi and Amazonas diet is dominated by fiber-rich plant-derived polysaccharide foods,[13,33] and very different from the western-style diet rich in simple sugars, proteins and fat. The ratio of these two main genera of gut bacteria, *Prevotella* and *Bacteroides*, correlates well with the overall pattern of diversity across healthy adults[12] and with differences in long-term diet between western-world individuals. *Bacteroides* was associated with a long-term diet rich in animal protein, amino acids and saturated fats, and *Prevotella* was associated with a diet rich in carbohydrates.[34]

Interestingly, gene transfer between bacteria in the gut has recently been shown to be influenced by the interactions between food and the gut microbiota. Our early studies showed that a high-fat, Western-style diet can increase conjugation rates and plasmid transfer between ingested microorganisms and the resident microbiota of human-flora-associated rats compared to HFA-rats on standard rodent chow.[35] In Japanese individuals, whose diet includes regular consumption of sushi, Hehemann and coworkers,[36] working with *Zobellia galactanivorans*, a member of the marine *Bacteroidetes*, discovered an enzyme (porphyranase) responsible for breaking down porphyran, an abundant polysaccharide in the red algae species *Porphyra*, on which *Z. galactanivorans* is often found. While searching gene-sequence databases, investigators came across predicted porphyranase sequences in metagenomes derived from human feces and in the genome of the resident human gut bacterium *Bacteroides plebeius*, suggesting that *B. plebeius* acquired the genes laterally from marine bacteria. Because it turned out that these sequences were present in Japanese individuals but not in residents of the US, the authors concluded that *Z. galactanivorans* were introduced via *Porphyra* ("nori"), the traditional seaweed used to wrap sushi and a common component of the Japanese diet. The researchers hypothesize that by constantly consuming seaweeds, Japanese communities produced a selective force that led to retaining the beneficial porphyranase genes in their gut microbiomes.[34]

Recently metabolism by intestinal microbiota of dietary L-carnitine, abundant in red meat to trimethylamine (TMA), which is then transformed into trimethylamine-N-oxide (TMAO) in the liver has been shown to accelerate atherosclerosis.[37] Chronic dietary L-carnitine supplementation in mice altered cecal microbial composition, markedly enhanced synthesis of TMA and TMAO, and increased atherosclerosis, but this did not occur if intestinal microbiota was concurrently suppressed. The study comparing vegans with omnivorous human subjects, showed that the latter produced more TMAO than did vegans or vegetarians following ingestion of L-carnitine through a microbiota-dependent mechanism. The presence of specific bacterial taxa in human feces was associated with both plasma TMAO concentration and dietary status. The metagenomics results from the study on vegans interestingly paralleled the results from our study on African children[33] demonstrating a predominance of fermentative bacteria, a lower relative abundance of disease-causing enterobacteria, and also much higher fecal concentrations of fiber fermentation end products, SCFAs. The Burkina metagenome and other recent studies are highlighting the importance of microorganisms in liberating from foods molecules of paramount importance in gut health and immune training, as well as metabolic features possibly lost from Western microbiomes. The results from the Burkina and the vegans study also indicated that intestinal microbiota may contribute to the well-established link between high levels of red meat consumption and cardiovascular disease risk, showing how meat assumption alters the vegetarian diet microbiome and increases health risks as a whole.

Recently David et al.[38] showed convincingly with a very elegant conclusive experiment that diet has a dominant role in determining the microbiome. The authors demonstrated that the short-term consumption of diets composed entirely of animal or plant products alters microbial community structure and overwhelms inter-individual differences in microbial gene expression. They did this by recruiting 10 volunteers who were willing to collect daily fecal samples. They each ate two different diets for five straight days – a plant-based one that was rich in grains, legumes, fruit and vegetables, and an animal-based one composed of meat, eggs and cheese. David and Turnbaugh's team also found that the altered gut communities did different things. During the plant diet, they became better at breaking down carbohydrates; during the animal diet, protein digestion was their forte. On the meat-heavy days, they activated more genes for breaking down harmful chemicals found in charred meat, and for making vitamins. The point that the paper makes can be summarized as follows: (i) dietary switch from carnivorous to vegetarian diets is associated to drastic changes in microbiota; (ii) these changes do not require centuries to happen but days; (iii) microbial composition changes correspond to function composition changes; (iv) every change is reversible. In addition, the microorganisms required for these changes are present in our foods, and can expand to dominate the gut environment based on the food sources. The results obtained made use of a novel integrated approach that follows microorganisms from foods to the gut. Also, technically, for the first time the authors merged classical DNA-based metagenomics with metatranscriptomics and metabolomics. Overall, the study shows the importance of preserving the microbiomes present in our foods and indicates these as a primary source for metabolic functions. This suggests that preserving microorganisms present in foods is a requirement and is essential for metabolic extraction of nutrients. This likely reflects on the health effects of foods seen also as inducing the presence of a precise microbiota, whose role could potentially extend to immune training.

THE WESTERN DIET METAGENOME IS OBESITY PRONE

Human nutritional evolution arose through a multi-million-year process in which genetic change reflected the environmental challenges. Thus our bodies are designed for a nutritional environment very different to the present day diet. Nutritional homeostatic processes and in-built checks and controls regulating dietary intake are based on foods which formed the mainstay of these ancient diets. Moreover, recent metagenomic investigations have confirmed that genes involved in carbohydrate transport and metabolism are strongly over-represented within the human gut microbiota compared to the genetic make-up of other bacterial communities, reaffirming the importance of carbohydrate fermentation as the chief force driving the energy economy of the large bowel and establishing carbohydrate fermentation as an important functional activity of the gut microbiota selected for by host evolutionary and dietary strategies over the millennia.[14] The consequences of recent divergence from this dietary symbiosis with our fermentative gut microbiota may be nowhere more evident than in the epidemic of obesity sweeping populations in both the developed and developing world and the sharp increase in diseases of affluence (type 2 diabetes, cardiovascular disease, non-alcoholic fatty liver disease and certain cancers). Obesity is not a state our hunter–gatherer predecessors were accustomed to and is one for which our bodies are ill suited considering the chronic associated pathologies. This obese xenobiologic state also impacts on our intestinal microbiota. Ley et al.[39] were the first to show that the gut microbiota of genetically obese laboratory animals is different to that of non-obese, wild-type counterparts on the same diet. The major difference being reduced prevalence of *Bacteroidetes* and a higher prevalence of *Firmicutes* in the obese. A similar microbiota profile is observable in obese humans and Ley et al.[40] showed that upon weight loss, prevalence of the *Bacteroidetes* returns to that observed in lean individuals. These authors went on to suggest that the gut microbiota of the obese is more energetically efficient compared to lean counterparts.[41] Concomitantly, by using the metabonomics ^{1}H-NMR approach to characterize changes in metabolite profiles present in biofluids at the whole organism, Dumas et al.[42] found that in mice (genetically predisposed to impaired glucose homeostasis and non-alcoholic fatty liver disease, NAFLD) on a high-fat diet, microbial activities within the intestinal tract lead to reduced choline bioavailability, mimicking choline-deficient diets. It has already been shown that such diet favors the onset NAFLD and insulin resistance. Specifically, ^{1}H-NMR spectral profiling using PCA was able to identify low circulating plasma phosphatidylcholine and high urinary excretion levels of methylamines, co-metabolic products of the host animal and gut microbiota, as differentiative of high-fat fed animals with NAFLD and impaired glucose metabolism. More recently, Holmes et al.,[43] using similar metabolite profiling of urine samples collected from 4600 individuals in Japan, China, the UK and USA, correlated reduced blood pressure and heart disease risk with metabolites of host–microbiota co-metabolic origin, including formate and hippurate, and dietary intake of fiber. These observations suggest that dietary disruption of the intestinal microbiota leads to altered microbial–host co-metabolic processing and altered risk of metabolic disease. It also raises the possibility that such diet–microbiota interactions may be modified through the fermentable carbohydrates. Observations in extant hunter–gatherer populations confirm the theory linking modern diet and lifestyle with increased risk of metabolic disease; in particular in Australia, Aboriginal populations who adopt the Western-style diet also experience an increased prevalence of type 2 diabetes. Surprisingly, this situation is reversed when these populations drop the Western-style diet and take up their more traditional diets, high in fermentable carbohydrates.[44] Similarly, the Prima Indians in Arizona, who follow a traditional desert-type diet rich in whole plant foods and particularly elevated in fermentable carbohydrates, are at a 2.5-times lower risk of developing diabetes in adjusted risk models compared to the same ethnic group following a modern "Anglo" type diet common in America which is low in fermentable fibers.[45] Traditional peoples, for example those in the Americas and the Indian subcontinent, do have a higher genetic predisposition to diabetes and the diseases of obesity, with the "thrifty-gene" theory suggesting that genetic selection for energy harvesting and storage from diets low in available energy contribute to metabolic disease risk in these populations once they adopt high-energy, Western-style diets.[46,47] Few genes, initially suspected of playing a role in increased risk of obesity or metabolic disease, are proving important at the population level and so far, true "thrifty" genes remain elusive.[48] However, recent metagenomic studies highlighting the importance of the gut microbiota in mammalian energy metabolism and ability to harvest energy from non-digestible food components, raises the intriguing possibility that some of these thrifty genes are in fact of microbial origin and contribute to increase metabolic disease risk upon transition to low-fiber, energy-dense, Western-style diets. However, no studies so far have characterized the microbiology of this transition between ancient and modern dietary selective pressures, but such a study may shed new light on the fundamental co-metabolic processes defining the human condition as a "super-organismal" metabolic network of human encoded gene functions and the metagenome of the thousands of microbial species living in and passing through the human gut. A deep insight into gut microbiota–host co-metabolism will establish the scientific basis for rational dietary modulation of the gut microbiota for improved host health, a strategy which can be applied at the population level to tackle modern-diet-associated diseases linked to aberrant metabolic and immune function.

Understanding the compositional and functional differences in the gut microbiota can lay the foundation to relate these differences to human health. Differences in the microbiota and the microbiome could help to explain the variation in the gut metabolic processes of individuals, including the metabolism of drugs and food.[49] Many of these metabolic pathways are outside the common functional core, so they can underlie host-specific responses.

CONCLUSIONS

Understanding how cultural traditions affect the microbiota will highlight the factors that result in the marked differences in the incidence of diseases associated with microbiota. Studying these associations will require expanded studies that sample a greater number of populations and control the confounding factors. We strongly believe that there is a need to sample our human microbiome as thoroughly and as rapidly as possible, particularly in societies that are undergoing dramatic cultural, socioeconomic and technological transformations not only to preserve the natural history of our gut microbiota but also to help design new dietary strategies aimed at reducing the burden of modern chronic diseases of affluence.

References

1. Cordain L. Cereal grains: humanity's double-edged sword. *World Rev Nutr Diet*. 1999;84:19–73.
2. Cordain L, et al. Origins and evolution of the Western diet: Health implications for the 21st century. *Am J Clin Nutr*. 2005;81:341–354.
3. Ramakrishnan U, Yip R. Experiences and challenges in industrialized countries: control of iron deficiency in industrialized countries. *J Nutr*. 2002;132(4 Suppl):820S–824S.
4. Müller O, Krawinkel M. Malnutrition and health in developing countries. *CMAJ*. 2005;173:279–286.
5. Savy M, Martin-Prével Y, Traissac P, et al. Dietary diversity scores and nutritional status of women change during the seasonal food shortage in rural Burkina Faso. *J Nutr*. 2006;136:2625–2632.
6. Arimond M, Wiesmann D, Becquey E, et al. Simple food group diversity indicators predict micronutrient adequacy of women's diets in 5 diverse, resource-poor settings. *J Nutr*. 2010;140:2059S–2069S.
7. Kennedy G, Fanou-Fogny N, Seghieri C, et al. Food groups associated with a composite measure of probability of adequate intake of 11 micronutrients in the diets of women in urban Mali. *J Nutr*. 2010;140:2070S–2078S.
8. Qin J, Li R, Raes J, et al. A human gut microbial gene catalogue established by metagenomic sequencing. *Nature*. 2010;464:59–65.
9. Human Microbiome Project Consortium. A framework for human microbiome research. *Nature*. 2012; 486: 215–221.
10. Human Microbiome Project Consortium. Structure, function and diversity of the healthy human microbiome. *Nature* 2012; 486: 207–214.
11. Arumugam M, Raes J, Pelletier E, et al. Enterotypes of the human gut microbiome. *Nature*. 2011;473:174–180.
12. Yatsunenko T, Rey FE, Manary MJ, et al. Human gut microbiome viewed across age and geography. *Nature*. 2012;486:222–227.
13. Gill SR, Pop M, Deboy RT, et al. Metagenomic analysis of the human distal gut microbiome. *Science*. 2006;312:1355–1359.

14. Zhu L, Zhu L, Wu Q, Dai J, Zhang S, Wei F. Evidence of cellulose metabolism by the giant panda gut microbiome. *Proc Natl Acad Sci USA*. 2011;108:17714–17719.
15. Borenstein E, Kupiec M, Feldman MW, Ruppin E. Large-scale reconstruction and phylogenetic analysis of metabolic environments. *Proc Natl Acad Sci USA*. 2008;105:14482–14487.
16. Freilich S, Kreimer A, Borenstein E, et al. Metabolic-network-driven analysis of bacterial ecological strategies. *Genome Biol*. 2009;10:R61.
17. Eaton SB, Konner M. Paleolithic nutrition. A consideration of its nature and current implications. *N Engl J Med*. 1985;312:283–289.
18. Ley RE, Hamady M, Lozupone C, et al. Evolution of mammals and their gut microbes. *Science*. 2008;320:1647–1651.
19. Bäckhed F, Ley RE, Sonnenburg JL, Peterson DA, Gordon JI. Host–bacterial mutualism in the human intestine. *Science*. 2005;307:1915–1920.
20. Eaton SB, Cordain L. Evolutionary aspects of diet: old genes, new fuels. Nutritional changes since agriculture. *World Rev Nutr Diet*. 1997;81:26–37.
21. Eaton SB. The ancestral human diet: what was it and should it be a paradigm for contemporary nutrition? *Proc Nutr Soc*. 2006;65:1–6.
22. Conterno L, Fava F, Viola R, Tuohy KM. Obesity and the gut microbiota: does up-regulating colonic fermentation protect against obesity and metabolic disease? *Genes Nutr*. 2011;6(3):241–260.
23. Cani PD, Amar J, Iglesias MA, et al. Metabolic endotoxemia initiates obesity and insulin resistance. *Diabetes*. 2007;56:1761–1772.
24. Blaser MJ. Who are we? Indigenous microbes and the ecology of human diseases. *EMBO Rep*. 2006;7:956–960.
25. Mira A, Mira A, Pushker R, Rodríguez-Valera F. The Neolithic revolution of bacterial genomes. *Trends Microbiol*. 2006;14:200–206.
26. Strachan DP. Hay fever, hygiene, and household size. *BMJ*. 1989;299:1259–1260.
27. Loftus EV. Clinical epidemiology of inflammatory bowel disease: incidence, prevalence, and environmental influences. *Gastroenterology*. 2004;126:1504–1517.
28. Rook GAW, Brunet LR. Microbes, immunoregulation and the gut. *Gut*. 2005;54:317–320.
29. Dicksved J, Halfvarson J, Rosenquist M, et al. Molecular analysis of the gut microbiota of identical twins with Crohn's disease. *ISME J*. 2008;2:716–727.
30. Turnbaugh PJ, Hamady M, Yatsunenko T, et al. A core gut microbiome in obese and lean twins. *Nature*. 2009;457:480–484.
31. Luca F, Perry GH, Di Rienzo A. Evolutionary adaptations to dietary changes. *Annu Rev Nutr*. 2010;30:291–314.
32. Falush D, Wirth T, Linz B, et al. Traces of human migrations in *Helicobacter pylori* populations. *Science*. 2003;299:1582–1585.
33. De Filippo C, Cavalieri D, Di Paola M, et al. Impact of diet in shaping gut microbiota revealed by a comparative study in children from Europe and rural Africa. *Proc Natl Acad Sci USA*. 2010;107:14691–14696.
34. Wu GD, Chen J, Hoffmann C, et al. Linking long-term dietary patterns with gut microbial enterotypes. *Science*. 2011;334:105–108.
35. Tuohy K, Davies M, Rumsby P, Rumney C, Adams MR, Rowland IR. Monitoring transfer of recombinant and nonrecombinant plasmids between *Lactococcus lactis* strains and members of the human gastrointestinal microbiota in vivo—impact of donor cell number and diet. *J Appl Microbiol*. 2002;93(6):954–964.
36. Hehemann JH, Correc G, Barbeyron T, et al. Transfer of carbohydrate-active enzymes from marine bacteria to Japanese gut microbiota. *Nature*. 2010;464:908–912.
37. Koeth RA, Wang Z, Levison BS, et al. Intestinal microbiota metabolism of L-carnitine, a nutrient in red meat, promotes atherosclerosis. *Nat Med*. 2013;19(5):576–585.
38. David LA, Maurice CF, Carmody RN, et al. Diet rapidly and reproducibly alters the human gut microbiome. *Nature*. 2014;505 (7484):559–563.
39. Ley RE, Bäckhed F, Turnbaugh P, et al. Obesity alters gut microbial ecology. *Proc Natl Acad Sci USA*. 2005;102:11070–11075.
40. Ley RE, Turnbaugh PJ, Klein S, Gordon JI. Microbial ecology: human gut microbes associated with obesity. *Nature*. 2006;444:1022–1023.
41. Turnbaugh PJ, Ley RE, Mahowald MA, et al. An obesity-associated gut microbiome with increased capacity for energy harvest. *Nature*. 2006;444:1027–1031.
42. Dumas ME, Barton RH, Toye A, et al. Metabolic profiling reveals a contribution of gut microbiota to fatty liver phenotype in insulin-resistant mice. *Proc Natl Acad Sci USA*. 2006;103:12511–12516.
43. Holmes E, Loo RL, Stamler J, et al. Human metabolic phenotype diversity and its association with diet and blood pressure. *Nature*. 2008;453:396–400.
44. O'Dea K. Westernisation, insulin resistance and diabetes in Australian aborigines. *Med J Aust*. 1991;155:258–264.
45. Williams DE, Knowler WC, Smith CJ, et al. The effect of Indian or Anglo dietary preference on the incidence of diabetes in Pima Indians. *Diabetes Care*. 2001;24:811–816.
46. Hegele RA. Genes and environment in type 2 diabetes and atherosclerosis in aboriginal Canadians. *Curr Atheroscler Rep*. 2001;3:216–221.
47. Rey D, Fernandez-Honrado M, Areces C, et al. Amerindians show no association of PC-1 gene Gln121 allele and obesity: a thrifty gene population genetics. *Mol Biol Rep*. 2012;39:7687–7693.
48. Zabaneh D, Balding DJ. A genome-wide association study of the metabolic syndrome in Indian Asian men. *PLoS One*. 2010;5:e11961.
49. Clayton TA, Baker D, Lindon JC, Everett JR, Nicholson JK. Pharmacometabonomic identification of a significant host–microbiome metabolic interaction affecting human drug metabolism. *Proc Natl Acad Sci USA*. 2009;106:14728–14733.

CHAPTER

3

Probiotic Microorganisms for Shaping the Human Gut Microbiota – Mechanisms and Efficacy into the Future

Lorenzo Morelli and Vania Patrone

Istituto di Microbiologia, Università Cattolica del Sacro Cuore, Piacenza, Italy

INTRODUCTION

Scientific literature retrievable from bibliographic databases clearly highlights the booming use of the term "probiotic(s)" as a keyword; its occurrence in the PubMed database was nearly negligible at the beginning of the 1990s, climbed to hundreds per year at the beginning of the 2000s and is nowadays at the level of a thousand times per year (data retrieved from www.gopubmed.org).

The word "probiotic" is a strange blend of Latin (*pro* = for, in favor of) and Greek (*bios* = life) and it was first coined to define substances able to support the growth of microorganisms,[1] a function nowadays attributed to "prebiotics." The definition of probiotics was then reshaped to identify "A live microbial feed supplement which beneficially affects the host animal by improving its intestinal balance,"[2] with a clear limitation to the area of animal nutrition; at that time there was a stringent need to develop feed additives able to replace the use of antibiotics as growth promoters. However, the definition, from the very beginning of the 1980s, was used to also identify the use of beneficial bacteria for humans.[3] Then, in the last 30 years, dozens of bacteria have been isolated, characterized and assessed *in vivo* in order to be used as food ingredients or food supplements and to exert their beneficial actions on human beings. Nowadays there are two major lines of research: one dealing with the use of these bacteria to treat or cure pathological conditions, while the second is related to the use as food or a food supplement, and the intended use is reserved to healthy people willing to reduce the risk of entering into a pathological situation or to improve so-called "physiological functions."

In this chapter we will focus our attention only on the second line of research; the reason for this choice will be clarified in the following section, devoted to the history and the regulatory aspects of the definition of probiotics.

In detail, this chapter will deal with the five words characterizing its title, namely: *probiotics, microbiota, efficacy* and *mechanisms*, concluding with some views for the *future*.

LET'S START WITH THE DEFINITION OF PROBIOTICS

Two sequential FAO/WHO documents[4,5] have defined "Probiotics" as: "live microorganisms which, when consumed in adequate amounts, confer a health benefit on the host." This definition was provided by an expert committee which dealt with issues specifically related to probiotics in food; the FAO/WHO experts stated emphatically that in their intentions the term probiotic was used exclusively to indicate beneficial bacteria consumed as a part of food, thus excluding any reference to bio-therapeutic agents. As a consequence, the 2001 document was directed to evaluate and to provide guidance on the administration of beneficial bacteria to

Diet-Microbe Interactions in the Gut.
DOI: http://dx.doi.org/10.1016/B978-0-12-407825-3.00003-4

"otherwise healthy people," with the scope to retain their health and well-being, and potentially reduce their long-term risk of illness, and it was not designed to cover their use to treat any disease.

Such an assumption has very important regulatory implications: beneficial bacteria can be utilized as therapy for pathological conditions, but in this case they must be considered drugs and the pharmaceutical rules apply. Several key aspects of the probiotic definition emerge from this document and the specifications outlined by the FAO/WHO Working Group for the Evaluation of Probiotics in Food.[5]

First, a probiotic must be alive when administered. Viability of the probiotic strains is essential for their persistence in the gut: only live bacteria may be able to pass the gastric and ileal environments and then reproduce themselves in the large intestine. Probiotics that remain viable upon reaching the target body site are more likely to be metabolically active and exercise functions with the potential to influence human physiology and confer a health benefit.

Second, probiotics must be delivered at an effective dose. In the gut, the resident microbiota clearly outnumbers administered probiotic microorganisms and this may reduce the chances of probiotics having a major impact on the microbial ecology of the gut environment. A number of viable cells are required to be able to survive gastric transit, to resist the action of bile salts and then reproduce within the gut to ensure a colonization rate sufficient to provide the expected beneficial effects.

The amount of potentially probiotic bacteria which is to be ingested by consumers to receive the beneficial effect (the "effective dose" in contrast with the "infectious dose" of pathogenic bacteria) is not easily defined, as there is strong evidence that this could be a strain-specific feature, and that it also depends on the intended target effect,[6] as well as host-related factors and those of the vector food.

The French Agency for Food Safety gave rough and broad guidelines in its dossier devoted to Probiotics and Prebiotics,[7] suggesting that the amount of viable cells to be present in the gut after consuming such products must be greater than or equal to 10^6 CFU/mL in the small intestine (ileum) and 10^8 CFU/g in the colon, even if the same AFSSA document admits that: "the scientific basis for these statements is relatively weak."

Third, a probiotic must be a taxonomically defined microbe or combination of microbes. The correct identification of the probiotic microorganism at the strain level is of paramount relevance, since it is fully acknowledged that the beneficial properties are strain-specific, and also to enable accurate post-marketing surveillance and epidemiological studies for safety/efficacy purposes.

Fourth, a probiotic must be safe for its intended use. The large majority of bacterial species used as probiotics have a long history of safe use in food; this means that the whole body of knowledge available substantiates their safety as food products or dietary supplements and no toxicological studies are needed. However, it must be pointed out that, on the basis of long history of safe use, only the use in "healthy" people and via the oral route is warranted; this implies that these bacteria are not necessarily safe whatever the conditions of the treated subjects or if administered, for example, in pathological settings.

Fifth, a probiotic must undergo controlled evaluation to document health benefits in the target host. The most common health claims related to probiotics are structure–function claims, i.e., claims describing the role of dietary supplements in maintaining normal body structure or function; these claims can not imply that a dietary component affects a disease state in terms of treatment/cure.[8]

In the last decade, the FAO/WHO documents have been taken into account by industrial, scientific and regulatory communities. While several countries adopted the definition suggested by the Consultation in their regulatory documents,[9–11] the word "probiotic" is often used by scientists in a frame of clinical trials and pathological conditions without a clear knowledge of the "food area" in which the documents were allocated. The consequences of this misinterpretation are that bacteria considered to be safe on the basis of their historical presence in foods have been used to treat patients affected by severe pathologies, sometimes causing threats to their health.[12,13] In addition, the incorrect interpretation of the probiotic definition has led to the use of the word to indicate preparations of dead microbial cells or crude microbial cell fractions[14] or bacteria with no documented benefits for human health.[15]

In this chapter we will avoid any reference to the use of probiotics in patients and we will focus on the impact of these bacteria on the gut microbiota of healthy people or at least people not affected by gastrointestinal pathologies.

SHAPING THE MICROBIOTA

Why could it be of relevance for human well-being to try to "shape" the composition of the bacterial population inhabiting the intestine and why not retain the composition sorted out by the naturally occurring mechanisms of selection of ingested bacteria?

It should be taken into consideration that the community structure and function of the gut microbiota contribute considerably to the phenotype of the host, influencing metabolism, immune system activity, and resistance to infection.[16–18] Consistent with this, alterations to the gut microbiota and consequent dysregulation of host–microbe interactions have been implicated in a number of intestinal and extra-intestinal disorders and illnesses.[19,20] In later decades, probiotics have received huge attention as dietary interventions aimed at modulating the microbial ecology of the gastrointestinal tract to garner human health benefits. The rationale for consumption of living microorganisms to maintain or improve health is to harness the beneficial effects of the commensal microbiota or introduce beneficial function by providing additional biological activities. Establishing relevant links between the composition and activities of intestinal microbial populations and the host healthy phenotypes is a prerequisite of developing an effective probiotic intervention. However, the complexity and variability of the human microbiota complicate the definition of what the ideal gut microbial community profile may be at individual or population level.[21] Substantial progress in acquiring knowledge about the microbial communities in the human gut has been recently enabled by culture-independent genomic studies based on microarray technologies and high-throughput sequencing. The study of the composition of the human colonic microbiota has long relied on culture-based analysis of fecal samples. This approach has the advantage of allowing phenotypic investigation of living strains, but the assessment of gut microbiota diversity by culture-based techniques presents major inherent limitations since it is estimated that only a small fraction of human intestinal bacteria is cultivable. The application of culture-independent, molecular methods, such as FISH, quantitative PCR, DGGE and microarrays, has provided a more comprehensive picture of the community composition both in qualitative and quantitative terms. DGGE has been widely applied as a fingerprinting technique to investigate variations in the gut dominant bacteria profiles due to exogenous factors such as antibiotic treatment or probiotic administration. Sanger sequencing of the 16S ribosomal RNA gene (rDNA) has rapidly become the standard strategy for bacterial phylogenetic analysis allowing the creation of large databases of publicly available 16S rDNA sequences from a great many bacterial taxa. The development of next-generation DNA sequencing technologies has revolutionized microbial ecology studies in recent years by allowing high-throughput sequence analysis of DNA. In particular, gut microbiota diversity investigations have applied either pyrosequencing of 16S rDNA amplicons method or analysis of 16S rDNA fragments retrieved from shotgun metagenomics. A major advantage of these approaches is the possibility to examine an enormous number of sequences per sample, enabling a better coverage of each sample and thereby increasing the chances to detect the least abundant bacterial species. In addition, a greater number of samples may be analyzed in parallel, thus providing more robust results for comparisons. However, it should be pointed out that reads generated by next-generation sequencing technologies are relatively short in length and as such they provide poor phylogenetic information at low taxonomic levels as species or strains in comparison to higher hierarchical ranks (i.e., from phyla to genera). These studies have confirmed that the composition of the human colonic microbiota could be depicted as a timeline with three following major sections:

- the neonatal period, from birth up to 2 years of life;
- then the following *adult* period;
- and finally the *golden age* (using the Metchnikoff definition) period.

We will shortly review the state-of-the-art regarding the microbiota composition at three specific moments of life, as well as regarding the potential to beneficially influence the composition of the gut microbiota by means of probiotic administration.

THE NEONATAL PERIOD

The assembly of a resident community of microorganisms in the gut represents one of the most fascinating examples of ecosystem establishment: it is when the host lays the foundations for entering into a life-long relationship with his microbial partners. The early colonization pattern of gut microbiota is heavily influenced by mode of delivery, as this determines microbial exposure at the time of birth.[22] Naturally delivered infants are firstly exposed to maternal vaginal and fecal bacteria, so *Lactobacillus, Prevotella, Atopobium* are prevalent in their gut, whereas the microbiota of cesarean section babies are more similar to the skin communities of the mothers, with an abundance of *Staphylococcus* spp.[23] Microbial colonization of the gut in infants delivered by caesarean section is delayed compared to naturally delivered infants. Infants born through cesarean section have lower numbers of *Bifidobacterium* (Figure 3.1) and *Bacteroides*, whereas they are more colonized by *Clostridium* species,

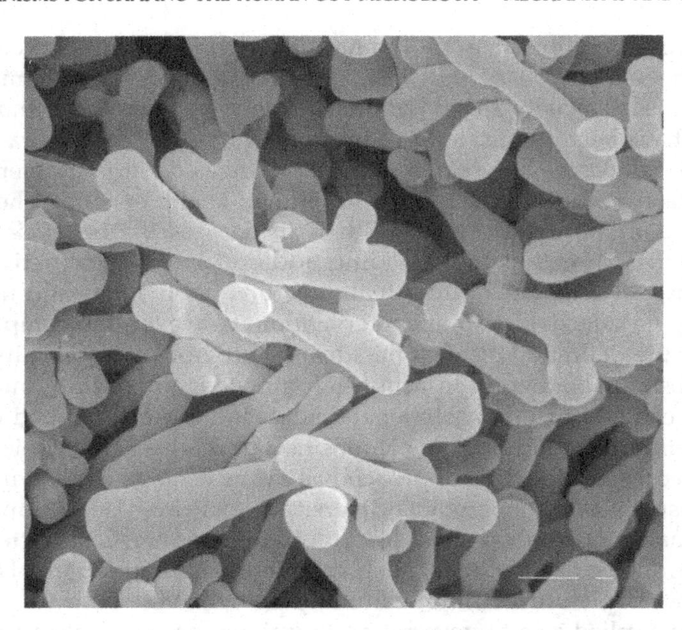

FIGURE 3.1 Scanning electron micrograph (SEM) of *Bifidobacterium* spp. isolated from fecal samples. *Courtesy of Centro di Microscopia Elettronica, Facoltà di Agraria, Università Cattolica del Sacro Cuore.*

in comparison with vaginally born infants.[24,25] Since at birth the gut contains oxygen, the initial colonizers are aerotolerant bacteria like Enterobacteriaceae and *Streptococcus* that contribute to lower the redox potential through their metabolic activity, thereby creating a more reduced environment in the gut that allows proliferation of strict anaerobic bacteria like *Bifidobacterium, Bacteroides, Clostridium.*[26] However, recent molecular-based studies detected the presence of strict anaerobic genera at relatively early stages of microbiota development.[27–29] During the first weeks, the development of the microbial community in infant gut is influenced mostly by type of feeding, which provides sources of colonizing bacteria[30] as well as nutrients for their growth.[31] Thus, exclusively breast-fed babies acquire a microbiota dominated by beneficial *Bifidobacterium* species within the first week, which may represent up to 90% of the total fecal bacteria whereas the intestines of formula-fed infants harbor a more diverse variety of bacterial genera, including Enterobacteriaceae, *Streptococcus, Bacteroides, Clostridium, Bifidobacterium.*[24,26,32,33] Anyway, recent results indicate that the microbiota of breast-fed infant might be more complex than previously thought: *Ruminococcus* genus was found in a high percentage of the infants fed breast milk suggesting a possible major role played by ruminococci as dominant bacteria, at the same level as bifidobacteria.[34] The introduction of solid food, by providing new substrates that may promote the survival and dominance of different species, evens out the differences between formula- and breast-fed babies. The phylogenetic diversity increases gradually over time, with *Clostridium coccoides* group and Bacteroidetes becoming predominant[35,36] so that by the second year of life the microbiota profile resembles that of the adult.

Microbial stimulation plays a pivotal role in the development and maturation of the intestinal immune system at birth[37,38] and a growing body of evidence suggests that the early development of a gut microbiota dominated by bifidobacteria is associated with improved infant health.[39] Naturally delivered, exclusively breast-fed infants are characterized by a low incidence of morbidity and mortality, and immune-related disorders compared to babies who have experienced other modes of delivery and feeding. Since type of feeding is the major factor driving the configuration of the gut microbiota early in life, one can expect to positively affect the health status during infancy by favoring the most "efficient" microbiota in terms of health protection through dietary intervention.[40] Moreover, the high instability of the infant-associated microbial community, characterized by low species richness and diversity, is likely to elicit reduced resistance to colonization and offer more niche opportunities for a potential new settler as a probiotic microorganism. A number of studies indicate that the composition of infant gut microbiota can be modified to resemble that of healthy breast-fed infants by probiotic administration.[41–43] A probiotic approach based on modulation of the gut microbiota might carry outstanding potential health benefits in particularly vulnerable subsets of infant populations, such as preterm babies. Preterm infants have delayed colonization of the gut with bifidobacteria by several weeks: this is believed to result in an abnormal intestinal immune system development and a higher risk of gastrointestinal diseases such as neonatal necrotizing enterocolitis (NEC).[44] It has been suggested that introducing probiotics into the dietary regimen of preterm infants might be beneficial to avoid overgrowth of pathogenic organisms and thereby prevent nosocomial infections.

ADULT LIFE AND THE PROPOSED ENTEROTYPE CLASSIFICATION

By the age of about 2 years, the community structure of the gut microbiota is fully developed and remains relatively stable in healthy individuals throughout adulthood.[45,46] The gut microbiota of healthy, adult humans is dominated by strictly anaerobic bacteria, and its members belong to four major bacterial phyla: mostly Firmicutes and *Bacteroidetes*, which account for about 90%, followed by Proteobacteria and Actinobacteria.[47,48] The next-generation sequencing approach has revealed that, despite this distribution at the phylum level, the microbial colonization pattern varies remarkably among healthy individuals at the species level[49] and is strongly modulated by host genetic and environmental factors.[50,51] Given the high diversity of the microbial communities in the gut, it is not surprising that researchers are attempting to reduce this complexity, and find microbial patterns that can be linked to host susceptibility to disease. The concept that all humans are populated by a "core microbiota" has recently been addressed: some studies on the fecal microbiota of healthy humans have detected a limited number of bacterial species that seem to be shared among individuals.[52,53] An alternative line of evidence supports the hypothesis that the reservoir of bacterial phylotypes varies substantially among individuals, but metabolic genes and pathways are conserved to constitute a "core microbiome."[48,54] From this perspective, it is likely that multiple microbial community configurations with different compositions may be able to sustain health by performing nearly equivalent functions.[55] In addition, perturbations of core microbiological functions, rather than imbalances in composition and numbers of indigenous microbial populations, might play a central role in the progression of dysfunctions and disease states. A very interesting paper by Arumugam and colleagues[56] suggests that the human gut microbiome may occur in three host–microbial symbiotic states, defined by the variation in the levels of one of three genera: *Bacteroides*, *Prevotella* and *Ruminococcus*. By analyzing parallel fecal metagenomes of individuals of different nationality, age, gender, the authors showed that the enterotypes do not seem to correlate with any of these factors. Although fascinating, such categorization of the gut microbiota has given rise to much debate within the scientific community. Huse and co-workers, by analyzing the 16S rRNA sequence of the microbiota of 200 individuals, found no enterotypes, showing community gradients rather than community clusters with a continuous ratio of *Prevotella* to *Bacteroides*.[57] In this line of evidence, recent results by a large meta-analysis of enterotypes in the human gut suggest that communities of gut bacteria exhibit a smooth abundance gradient of key genera rather than falling into discrete clusters.[58] Whether the individual gut microbiota may be described in terms of discrete enterotypes may be of crucial importance insofar as these enterotypes are relevant for health, for example affecting host susceptibility to infection or disease. Interestingly, it has been showed that in healthy subjects enterotypes are strongly associated with diet, particularly protein and animal fat (*Bacteroides* enterotype) and carbohydrates (*Prevotella* enterotype).[59] Very recently, it has been suggested that children with type 1 diabetes tend to have the *Bacteroides* enterotype, while their healthy counterparts have the *Prevotella* enterotype.[60] On the contrary, a recent examination of the microbiota composition and associated health status in a large cohort of healthy but frail elderly individuals found that individual microbiota was characterized by dominance of two or three bacteria co-abundance groups rather than one of three enterotypes.[61] Much work is still required to gain insight into the relationship between patterns of gut microbiota composition and health-related host features.

Data generated by new sequencing technologies has allowed us to gain insight into ecological dynamics within the human gut: as stated before, the gut microbiota of adults is now acknowledged to be much more diverse than previously thought and more stable over time. Such temporal constancy suggests that the adult gut microbial community possesses a certain amount of *resilience*, i.e., the ability to resist disturbances and the tendency to return to the previous state if change results from such disturbance.[55] Species richness and functional diversity may be further important features contributing to resilience and driving the performance of the gut microbial community: consistent with this, many chronic inflammatory diseases associated with adult gut microbiome dysbiosis exhibit an overall reduction of bacterial diversity.[21] While it is well known that the gut microbial ecology can be disturbed by antibiotics use or diet, the actual, long-term effects of these perturbations are still being elucidated. Research suggests that the human gut microbiota of generally healthy adults changes detectably after short-term dietary changes, but that enterotype identity remains stable.[59] The concept of resilience appears fundamental when considering the opportunities for probiotic approaches of modulating the microbial ecology of the gut in adult populations. Resident, well-established microbial communities in the gut of the adult host are likely to display significant resistance to alterations by allochthonous probiotic strains. The development of probiotic interventions targeting the gut microbiota to positively regulate human physiology thus requires a deeper understanding of the resilience-related dynamics of the gut microbial ecosystem. Studies conducted in healthy volunteers have shown that administration of probiotic strains does not influence the total composition of the intestinal microbiota and the only alterations observed are linked to the presence in fecal flora of the species supplemented.[62–64] Such results were confirmed by a very recent study, in which a high-throughput sequencing-based analysis of fecal samples from healthy individuals subjected to intervention with six commercially available probiotics showed no significant changes in the overall structure of gut microbiota regardless of types of the probiotics used.[65] Nevertheless, an administered probiotic bacterium, if selected on the basis of appropriate criteria, could replace some of the strains already present in the intestine and display functional activities resulting in beneficial consequences for the host. Within this scenario, positive effects may result from the biological activities of the probiotic bacterium itself that physiologically benefit the host regardless of a direct action on the resident microbial community of the intestinal tract.

THE AGED PERIOD

As in the earliest stages of life, old age is characterized by a number of complex transformations in the structure and functions of the human gut microbiota. These shifts reflect the dramatic physiological changes related to age, which are undoubtedly linked to exogenous and endogenous factors including nutrition, mobility, intestinal functionality, infection, and medication, whose relevance has not been fully elucidated. It must be taken into account that, while the threshold for the transition from infancy to adulthood can ultimately be set at weaning, there is no clear indication of when an individual, and by extension, his microbiota, can be defined as "old." This probably accounts, at least in part, for the inconsistent and often contradictory results provided by studies performed on the gut microbiota of elderly people. It has widely been considered for a long time that old age was associated with a decrease in the abundance and species diversity of bifidobacteria in the intestine bacterial community.[66–69] However, in recent culture-independent studies no differences in bifidobacteria levels were observed between young adults and healthy elderly, except for a reduction in centenarians.[45,70] Analogously, *Bacteroides* have been reported to decrease in a number of studies,[45,67–69] and increase in others.[71,72] Conversely, many investigators concur with the statement that facultative anaerobes, including enterobacteria, enterococci, streptococci, staphylococci, generally increase during the old age.[45,66,68,69,72] Since many opportunistic pathogens belong to these bacterial genera, their increase may represent a very critical factor for the development of disease from infection in elderly people. Lower levels of *Clostridium* cluster XIVa and *Faecalibacterium prausnitzii* have been found in elderly compared to younger adults.[69,72] Pyrosequencing characterization of the fecal microbiota in 161 subjects and nine younger control subjects in the context of the Irish ELDERMET project showed a clear shift to a more *Clostridium* cluster IV- and *Bacteroides*-dominated community in the elderly.[73] In addition, an extreme variability was reported between individuals and reduced diversity of the overall gut bacterial community. In their later paper[61] the same researchers reported that the composition of the intestinal microbiota in old people was related to the places where the elderly people lived, with the bacterial patterns of the older people in institutional care displaying a higher proportion of Bacteroidetes and a lower proportion of Firmicutes. The same groups were obtained by clustering subjects by dietary patterns. Moreover, every specific bacterial profile correlated significantly with measures of frailty, comorbidity and with markers of inflammation, besides nutritional status. Despite the possibility that other factors may account for the associations found in this study, as a whole these data support a role of dietary intervention relying on modulation of the gut microbiota to promote health in aging subjects. Within this scenario, augmented levels of health-promoting bacteria in the fecal microbiota of healthy elderly have been reported following supplementation of several probiotic *Bifidobacterium* strains.[74,75] Although further investigations are needed to establish cause and effect relationships between microbiota composition and health indices, the shifts of the gut microbiota in elderly subjects might be linked to immunosenescence, that is the age-dependent decrease in immunological competence and the associated cronic, low-grade inflammatory status.[76] It is worth mentioning the work by Ouwehand et al.[77] who revealed that probiotic intervention augmented levels of certain *Bifidobacterium* species in instituzionalized elderly, and observed significant negative correlations between *Bifidobacterium* populations and cytokine levels. Microbiota modulation by probiotics may provide opportunities to sustain health in older people in several common age-related conditions such as intestinal constipation, increased susceptibility to infection or severity of infection, and decreased response to vaccination.

MECHANISMS AND EFFICACY

In the previous sections it was strongly suggested that the beneficial actions of probiotic bacteria are strain-specific, therefore there is the need to achieve a strain-specific characterization.

While for species identification there is a general consensus about the use of similarities among the 16S rRNA-encoding DNA, it seems to be more debated how it could be possible to identify a bacterial population at the strain level. From the epidemiological point of view, a practical approach is to define two genetically "indistinguishable" bacterial isolates which share at 100% their pulsed-field gel electrophoresis (PFGE) profiles; strains are defined as "closely related" if they show PFGE profiles with at least 85%, i.e., showing between two to three band differences, probably due to a single genetic event.[78]

However, this method of differentiating strains does not seem to be applicable to probiotic strains, as several authors have shown that very subtle genetic differences could totally impair or totally change features relevant for the probiotic action. A spontaneously occurring mutant *Lactobacillus crispatus* strain was indistinguishable using PFGE (three enzymes), ribotyping and randomly amplified polymorphic DNA (RAPD) from the wild parent strain but the co-aggregation phenotype of the wild type was missing in the mutant.[79] This subtle phenotypic difference has a major impact on such features as the adhesion to mucus,[80] survival and persistence in mice[81] and in the human gut[79] and, what is even more relevant, only the wild-type strain has been shown to be able to protect mice from DSS-induced colitis, while the mutant strain has lost any protective action.[81]

In a similar way it was shown in *L. acidophilus* strain NCFM that a single gene variation has a dramatic impact on its probiotic efficacy.[82] In this case a mutant was obtained by means of the gene-knocking out technique, causing the deletion of the gene encoding a surface protein of the studied strain. The mutant was therefore different from the type in a single gene but it exhibited lower growth rates, increased sensitivity to sodium dodecyl sulfate and greater resistance to bile; with a similar genetic approach, the same strain has been characterized for its mechanism of immune modulation.

A knockout mutant of the same *L. acidophilus* NCFM strain, lacking the S-layer A protein (SlpA) was significantly impaired in its binding to dendritic cells as *L. acidophilus* NCFM attaches to dendritic cells and induces changes in interleukin production.[83] This mutant incurred a chromosomal rearrangement, which caused the expression of a second S-layer protein, SlpB. In the SlpB-expressing strain, the impact on the immune system was the opposite of the wild type, as proinflammatory cytokines were produced by the action of this strain with dendritic cells, with an opposite action detected in the wild type.

These results clearly and strongly suggest that a single gene mutation can induce major changes in the probiotic activity of a given strain; if we then have to deal with the efficacy and mechanisms of probiotics it seems mandatory to focus the attention on single, well characterized strains.

EFFICACY IN HEALTHY PEOPLE

Metchnikoff is believed to be the "father" of the probiotic concept as his area of research covered both "immunity" (he was awarded with the Nobel prize for his pioneering studies on macrophages) and the more broad "resistance" concept; in one of his books he wrote on this latter subject, describing the "disharmonies" in all living beings, from insects to humans.[84]

In a further book he focused his attention on a very special "disharmony", stating: "The dependence of the intestinal microbes on the food makes it possible to adopt measures to modify the flora in our bodies and to replace the harmful microbes by useful microbes."[85]

These original suggestions by Metchnikoff were also supported by the studies of Tissier, a French pediatrician, who observed that children affected by diarrhea had a low number of Y-shaped bacteria in their stools, whereas stools of healthy children had an abundance of bacteria with this strange morphology.[86] The practical suggestion was therefore that the administration of bifidobacteria could be advisable for patients with diarrhea, in order to restore the "healthy" status of the intestinal flora.

After these studies a large number of products based on beneficial bacteria have been developed on the basis of the ecological concept proposed by these two authors.

They have developed the concept of "rebalance" of the gut microbiota, a rationale which was generated by their comparison of the composition of the gut microbiota between healthy and unhealthy people. While different patterns of microbial colonization associated with disease states have been largely confirmed,[87] specific features of a healthy gut microbiome remain to be defined so that it is still impossible to establish a picture of the ideal bacterial biota of the human gut.

Even if the focus of this section is on the gut microbiota, it could be worthwhile to note that a different scenario is provided by the microbiota of the vaginal ecosystem, whereas the presence of a large majority of lactobacilli does ensure the well-being of women (for a review, see Ref. 88).

As regards the gut microbiota, however, the ecological use of probiotics holds when the microbiota composition has been altered by external or internal factors, such as antibiotic treatments or inflammatory diseases. For this kind of application there is strong and convincing evidence of the usefulness of the probiotics.[89]

On the contrary, it seems extremely difficult to demonstrate the beneficial effect of probiotics administration to healthy adults, which are supposed to have a quite stable intestinal microbiota. We have to keep in mind that ecological niches such as the gut try to resist changes when they have reached homeostasis; this attribute has been defined before as the "resilience" of the intestinal microbiota.

While the impact of diet in shaping the composition of the gut microbiota is well supported by analytical data (for a review, see Ref. 90), it seems more difficult that a single bacterial strain and even a complex mixture of strains could significantly alter the overall pattern of the intestinal bacterial content, thus restricting the ecological use of probiotic to specific conditions of altered microbiota.

A solid evidence of efficacy of probiotics in shaping the gut microbiota composition has been provided by their administration to infants delivered pre-term, which in turn has been shown to be positively related[91] to the reduction of mortality due to necrotizing NEC; however, we will not address here these peculiar applications, as pre-terms are hard to be allocated into a "normal," "healthy" population group.

However, probiotic strains can replace other indigenous strains even when administered to healthy individuals with a stable, resilient, microbiota, without altering the overall microbiota composition in terms of genera and phyla and causing a beneficial modulation, i.e., diverting the overall action towards a specific physiological function, in the functionality of the microbiota. This replacement effect was shown in 2006 for the first time by our research group by means of genetic identification.[92] Seven healthy subjects (Table 3.1) were dosed with a *L. paracasei* strain (Figure 3.2); after 15 days of administration there was not a statistically significant difference in the total amount of lactobacilli but the probiotic strain represented 66.6% of the total vancomycin-insensitive CFUs isolated from fecal samples, having replaced the lactobacilli originally colonizing the gut of the treated individuals.

The replacement effect could allow planning the use of carefully selected strains in order to colonize the gut of healthy individuals with bacteria able to modulate specific beneficial actions, i.e., towards the immune system or against specific pathogens in the absence of a major effect on the global composition of the gut microbiota. The rationale to replace some strains with other strains closely related to the indigenous ones is that the mechanisms, which are in place in our gut to select bacteria before they reach the colon, have evolved, during the millennia of human evolution, in order to avoid the presence of pathogens, and probably not necessarily to select commensal bacteria on the basis of their ability to modulate the immune system to reduce inflammation or reduce the risk of allergic reactions.

In the following sections we will review some information on this replacement effect in healthy individuals.

It has been demonstrated *in vitro* that *L. acidophilus* and *L. delbrueckii* are able to bind ferric hydroxide at their cell surface, rendering it unavailable to pathogenic microorganisms;[95] this mechanism, if confirmed, could

TABLE 3.1 Some Examples of the Replacement Studies

Strains Used	Treatment	Replaced Bacterial Population	Percentage of the Total Count of the Administered Strain(s) During Treatment	Refs
Lactobacillus paracasei B21060	7 subjects three times a day for 15 days a preparation containing 5×10^9 CFU	Lactobacilli	From 66% to 74% of lactobacilli according to the subject	[92]
Streptococcus salivarius strains 20P3 and 5	219 children treated for 2 days or 9 days with 6×10^9 CFU/day	Oral streptococci	11% of the treated children with a reduction of oral *S. pyogenes* and increase in *S. salivarius*	[93]
Lactobacillus rhamnosus hct 70	40 children treated for 3 weeks with 6×10^9 CFU/day	Oral streptococci	The difference in *mutans* streptococci counts was statistically highly significant ($p < 0.001$).	[94]

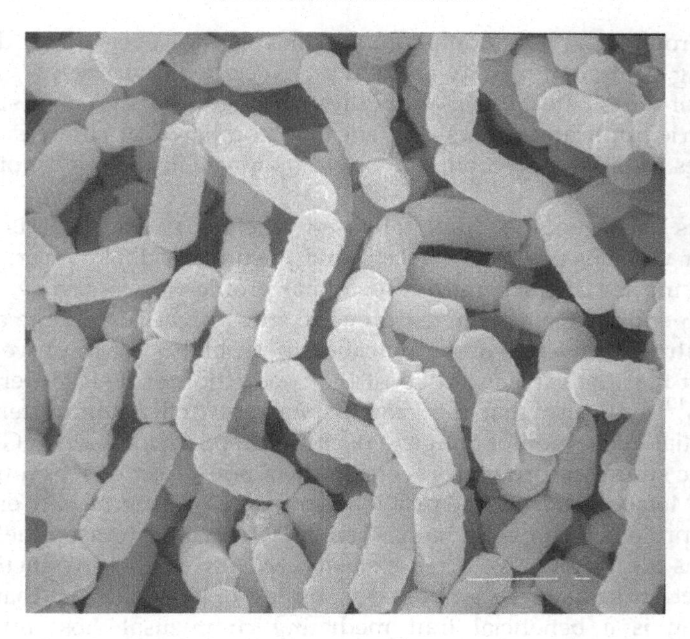

FIGURE 3.2 Scanning electron micrograph (SEM) of *Lactobacillus paracasei* isolated from fecal samples. *Courtesy of Centro di Microscopia Elettronica, Facoltà di Agraria, Università Cattolica del Sacro Cuore.*

explain the clinically observed action of several probiotics against pathogenic bacteria but could also be exploited to reduce the risk of infection by iron-requiring pathogens such as the *Enterobacteriaceae*.

For the same purpose of resistance towards pathogens, the use of bacteriocin-producing probiotic strains (a bacteriocin is an antibiotic-like compound, generally a peptide, produced by bacteria in order to outcompete other bacteria inhabiting the same environment) could be of interest. It has been demonstrated that bacteriocin Abp118 produced by *L. salivarius* strain UCC118 is able to protect mice against infection with the invasive food-borne pathogen *Listeria monocytogenes*.[96] It has also been shown that the same *L salivarius* UCC118 strain is responsible for substantial alteration of the composition of the gut microbiota of mice *in vivo*.[97]

The replacement effect has also been studied in the oral microbiota.[98] In this case bacteriocin-producing probiotics have been shown to have positive effects replacing the cariogenic bacteria.

Dierksen and coworkers[93] have treated a large number of subjects with a salivaricin A (SalA)-producing *Streptococcus salivarius* strain to replace strains of the same species naturally occurring in the mouth and tongue of 219 children. The replacement effect was successful (Table 3.1). It has to be pointed out that SalA bacteriocin is known to be active against *Streptococcus pyogenes*, and that it could be assumed that the consumption of SalA-producing probiotic *S. salivarius* could potentially reduce the *S. pyogenes* oral infections.

This was not the only example of the use of probiotic bacteria for reducing the risk of caries; in 2012, Juneja and Kakade successfully used, in a 3-week dosage period, a milk containing the probiotic *L. rhamnosus* hct 70 strain to feed Indian children at high risk of caries, with a control group which was drinking plain milk.[94]

Saliva samples were collected to estimate the amount of salivary *mutans* streptococci counts; the difference in the post follow up *mutans* streptococci count of the treated group, compared to the reference group, was highly significant with a p value <0.001.

To provide further support to the use of specific strains of probiotics in reducing the risk of infections by pathogenic bacteria their effect on the efficacy of vaccinations should be mentioned. As a curiosity it could be noted that, if the first modulation of the immune system deliberately induced by humans was vaccination and that the first industrial application of the ability of lactic acid bacteria to enhance an immune response was about their adjuvant activity in vaccinations.[99] Studies in humans have been published for probiotics as adjuvants for vaccines of both attenuated pathogenic bacteria and viruses. Live-attenuated influenza vaccines (LAIV) have been used to elucidate the adjuvant effects of probiotics in the elderly[100,101] and adults.[102,103] Clinical protocols used were quite similar for the four studies: a consumption of the probiotic product for a prolonged period (from 1 month to 13 weeks) with the vaccination step generally more or less in the middle of the administration period. The daily dose of the probiotic strain used was always between 10^9 and 10^{10} CFU/d. Similar but not identical positive effects have been obtained by using four different lactobacilli and are related to higher post-vaccination specific antibody titers in probiotic groups compared to controls; however, some differences in the adjuvant action were noticed, including a different activity according to specific flu viruses.

An increased production of virus-neutralizing antibodies has also been shown in subjects treated with a polio vaccine;[104] in another pediatric application it was shown that, compared to controls, probiotic supplementation improved antibody responses in subjects receiving a single monovalent dose but not those who received three monovalent doses.[105]

As regards adjuvant effects for bacterial or bacterial-based vaccines, positive effects of a range of probiotic lactobacilli have been shown for vaccines against *Salmonella* and pneumococci. However, while it is well known that different *Lactobacillus* strains induce distinct mucosal cytokine profiles (for a review, see Ref. 106), data dealing with strain-specific actions in supporting vaccine efficacy are still scarce, despite the above-cited research works and their relevance for the future of research and application of probiotics for improved vaccine efficacy.

The available data suggest that the capacity to boost the production of IL-10 differs considerably in response to different probiotic strains[107] and also that the growth phase of orally administered individual *Lactobacillus* strains is linked to a differential expression of antigen-specific antibody subclasses IgG1 and IgG2a.[108]

Another activity of specific strains of probiotics seems to be a non-direct effect on the gut microbiota by interaction with the host tissues. Metabolites from the probiotic strains *B. infantis* Y1 may enhance gut barrier function by altering tight junction protein expression and thereby increasing transepithelial resistance[109] and live *Escherichia coli* Nissle 1917 has been shown to increase *zonula occludens* (ZO)-1 production.[110] Improvement in gut barrier integrity might impact on invasion of pathogens: Fanning et al.[111] showed that *B. breve* UCC2003 surface exopolysaccharide production is a beneficial trait mediating commensal–host interaction through immune modulation and reducing colonization levels of the mouse gut pathogen *Citrobacter rodentium*, which is used as a laboratory model for attaching and effacing *Escherichia coli* infections (EHEC and EPEC strains). In addition, recent findings suggest that acetate produced as final catabolite by bifidobacteria could act *in vivo* to promote the defense functions of host epithelial cells in germ-free mice and protect them against lethal infection by *E. coli* O157:H7.[112]

After this long section devoted to the role of probiotics in enhancing the barrier against oral and intestinal infections, we will move to another effect of probiotics administration to healthy subjects, the impact of probiotics on intestinal transit, an increasingly important measure of well-being given the high incidence of constipation amongst the elderly and pregnant women, where efficacious and safe treatments, without side-effects are scarce. In a study in which 20 pregnant women aged ≥ 18 years with functional constipation were dosed with a mix of three bifidobacteria and three lactobacilli for four weeks, the change in defecation frequency, which was the primary outcome of the study,[113] significantly increased from 3.1 at baseline to 6.7 in week four ($p < 0.01$). Also secondary outcomes (stool consistency, sensation of incomplete evacuation, sensation of anorectal obstruction, manual manoeuvres to facilitate defecation, abdominal pain) showed a good degree of improvement, with no side effects reported.

Immediately after delivery, another problem could affect the well-being of the mother: infantile colic, which is painful for the infant and distressing for the parents. The cause of infantile colic remains unclear; from the epidemiologic point of view it could be noted that colic affects 10–30% of infants worldwide and that this condition is encountered in male and female neonates and infants with equal frequency. The colic syndrome is commonly observed in neonates and infants aged 2 weeks to 4 months and the incidence of colic in breast-fed and bottle-fed infants is similar. *Lactobacillus reuteri* has been suggested to be able to relieve colic symptoms in breast-fed infants within 1 week of treatment[114] with a 50% reduction in crying time from baseline when compared with the placebo group. In another study the efficacy of *L. reuteri* in reducing the colic problems was compared to the use of simethicone:[115] 83 breast-fed, colic infants were randomly assigned to receive either *L. reuteri* (41 infants) or simethicone (42 infants). After 28 days of treatment, 39 patients (95%) in the probiotic group had a decrease in daily average crying time of at least 50% compared with three patients (7%) in the group treated with simethicone.

Even if some authors[116,117] state that there is insufficient evidence to recommend for or against the use of probiotics in the management of colic, this application is of interest for its potential to alleviate a problem of well-being in a healthy, very young individual and, moreover, to reduce the stress of the parents. Moreover, recent data, obtained by a range of research groups in the most recent years, clearly support the efficacy of this treatment.[114]

Another field of application of probiotics in healthy people is their potential to reduce the risk of the so-called winter diseases. As an example, the common cold is responsible for the largest proportion of school and work absenteeism and causes a huge economic burden. In a recent meta-analysis[118] the authors identified 10 studies grouping a total of 2894 participants, 1588 in the probiotics group and 1306 used as control group. The effect of probiotics on the prevention of the common cold had a relative risk (RR) of 0.92 (95% CI, 0.85 to 1.00, I2 = 26%), while the RR of administration of probiotics without any active intervention (vitamin and mineral) was 0.87 (95% CI, 0.78 to 0.97). While the authors conclude that probiotics have a marginal effect on the prevention of the common cold, it seems significant that this modest effect can be marked as a positive action of probiotics; it is important to remember that we are dealing with food or food supplements and not with a pharmaceutical product or drug!

CONCLUSIONS

The above-described scenarios imply that for the future, the selection of probiotics intended to be used in healthy people will require, as a mandatory step, the deep knowledge of the mechanism(s) of the desired beneficial action. To date, the selection criteria used (i.e., resistance to acid, to bile salts, adhesion to epithelial cells, etc.) will represent a pre-requisite in order to guarantee the survival of the bacterial cells into the gut and then to exert their beneficial action in that site.

After the primary selection, it will then be necessary to further evaluate the strains on the basis of functional parameters such as the potential to inhibit pathogens or the potential to modulate some specific immune functions, etc. Then, by means of very accurate processes of characterization, also based on genomic characterization, it will be possible to identify strains able to colonize the gut, replace the presence of strains belonging to their own genus or species and then exert the specific beneficial action required by the consumer.

Thus, the replacement effect causes a new shape of the microbiota composition, not as a different ratio among the different bacterial groups but as a different composition in strains within the genera already characterizing the microbiota. This could be considered as the ultimate consequence of the so-called strain specificity of the probiotic bacteria: the selection of a specific strain for a specific action.

References

1. Fuller R. Probiotics in man and animals. *J Appl Bacteriol*. 1989;66:365–378.
2. Wadström T. Streptococcus faecium M 74 in control of diarrhoea induced by a human enterotoxigenic *Escherichia coli* strain in an infant rabbit model. *Zentralbl Bakteriol Mikrobiol Hyg A*. 1984;257:357–363.
3. Dorofeĭchuk VG, Volkov AI, Kulik NN, Karaseva GN, Zilmina VS. Antacid bifilact and its effectiveness in the treatment of chronic gastroduodenitis and ulcer disease in children. *Vopr Pitan*. 1983;6:30–33.
4. *Food and Agriculture Organization of the United Nations*. Health and Nutritional Properties of Probiotics in Food including Powder Milk with Live Lactic Acid Bacteria. <http://www.who.int/foodsafety/publications/fs_management/en/probiotics.pdf>; 2001.
5. *Food and Agriculture Organization of the United Nations*. Guidelines for the evaluation of probiotics in food. <ftp://ftp.fao.org/es/esn/food/wgreport2.pdf>; 2002.
6. Aureli P, Capurso L, Castellazzi AM, et al. Probiotics and health: an evidence-based review. *Pharmacol Res*. 2011;63:366–376.
7. *Agence Française de Sécurité Sanitaire des Aliments*. Effets des probiotiques et prébiotiques sur la flore et l'immunité de l'homme adulte. <http://www.isapp.net/docs/AFFSAprobioticprebioticfloraimmunity05.pdf>; 2005.
8. *World Gastroenterology Organization*. WGO Practice Guideline – Probiotics and Prebiotics. <http://www.worldgastroenterology.org/assets/export/userfiles/Probiotics_FINAL_20110116.pdf>; 2011.
9. *Ministero della Salute Italiano*. Linee guida probiotici e prebiotici. <http://www.salute.gov.it/imgs/C_17_pubblicazioni_1016_allegato.pdf>; 2005.
10. *Food Directorate Health Products and Food Branch Health Canada*. Guidance Document –The Use of Probiotic Microorganisms in Food. <http://www.hc-sc.gc.ca/fn-an/legislation/guide-ld/probiotics_guidance-orientation_probiotiques-eng.php>; 2009.
11. Indian Council of Medical Research. *Department of Health Research of the Ministry of Health & Family Welfare and Department of Biotechnology*. New Delhi: Ministry of Science and Technology; 2011:Guidelines for evaluation of probiotics in food. <icmr.nic.in/guide/PROBIOTICS_GUIDELINES.pdf
12. Vahabnezhad E, Mochon AB, Wozniak LJ, Ziring DA. *Lactobacillus bacteremia* associated with probiotic use in a pediatric patient with ulcerative colitis. *J Clin Gastroenterol*. 2013;47:437–439.
13. Land MH, Rouster-Stevens K, Woods CR, Cannon ML, Cnota J, Shetty AK. *Lactobacillus* sepsis associated with probiotic therapy. *Pediatrics*. 2005;115:178–181.
14. Sanders ME, Hamilton J, Reid G, Gibson GR. A nonviable preparation of *Lactobacillus acidophilus* is not a probiotic. *Clin Infect Dis*. 2007;44:886.
15. Reid G. Probiotics and prebiotics – progress and challenges. *Int Dairy J*. 2008;18:969–975.
16. Nicholson JK, Holmes E, Kinross J, et al. Host–gut microbiota metabolic interactions. *Science*. 2012;336:1262–1267.
17. Hooper LV, Littman DR, Macpherson AJ. Interactions between the microbiota and the immune system. *Science*. 2012;336:1268–1273.
18. Wardwell LH, Huttenhower C, Garrett WS. Current concepts of the intestinal microbiota and the pathogenesis of infection. *Curr Infect Dis Rep*. 2011;13:28–34.

19. DuPont AW, DuPont HL. The intestinal microbiota and chronic disorders of the gut. *Nat Rev Gastroenterol Hepatol*. 2011;8:523–531.
20. Clemente JC, Ursell LK, Parfrey LW, Knight R. The impact of the gut microbiota on human health: an integrative view. *Cell*. 2012;148:1258–1270.
21. Lozupone CA, Stombaugh JI, Gordon JI, Jansson JK, Knight R. Diversity, stability and resilience of the human gut microbiota. *Nature*. 2012;489:220–230.
22. Costello EK, Stagaman K, Dethlefsen L, Bohannan BJ, Relman DA. The application of ecological theory toward an understanding of the human microbiome. *Science*. 2012;336:1255–1262.
23. Dominguez-Bello MG, Costello EK, Contreras M, et al. Delivery mode shapes the acquisition and structure of the initial microbiota across multiple body habitats in newborns. *Proc Natl Acad Sci USA*. 2010;107:11971–11975.
24. Penders J, Thijs C, Vink C, et al. Factors influencing the composition of the intestinal microbiota in early infancy. *Pediatrics*. 2006;118:511–521.
25. Adlerberth I, Wold AE. Establishment of the gut microbiota in western infants. *Acta Paediatr*. 2009;98:229–238.
26. Palmer C, Bik EM, DiGiulio DB, Relman DA, Brown PO. Development of the human infant intestinal microbiota. *PLoS Biol*. 2007;5:e177.
27. Wang M, Ahrné S, Antonsson M, Molin G. T-RFLP combined with principal component analysis and 16SrRNA gene sequencing: an effective strategy for comparison of fecal microbiota in infants of different ages. *J Microbiol Methods*. 2004;59:53–69.
28. Hopkins MJ, Macfarlane GT, Furrie E, Fite A, Macfarlane S. Characterisation of intestinal bacteria in infant stools using real-time PCR and northern hybridisation analyses. *FEMS Microbiol Ecol*. 2005;54:77–85.
29. Jost T, Lacroix C, Braegger CP, Chassard C. New insights in gut microbiota establishment in healthy breast fed neonates. *PLoS ONE*. 2012;7:e44595.
30. Martín R, Langa S, Reviriego C, et al. Human milk is a source of lactic acid bacteria for the infant gut. *J Pediatr*. 2003;143:754–758.
31. Le Huërou-Luron I, Blat S, Boudry G. Breast- v. formula-feeding: impacts on the digestive tract and immediate and long-term health effects. *Nutr Res Rev*. 2010;23:23–36.
32. Favier CF, Vaughan EE, De Vos WM, Akkermans AD. Molecular monitoring of succession of bacterial communities in human neonates. *Appl Environ Microbiol*. 2002;68:219–226.
33. Fanaro S, Chierici R, Guerrini P, Vigi V. Intestinal microflora in early infancy: composition and development. *Acta Paediatr*. 2003;91:48–55.
34. Coppa GV, Gabrielli O, Zampini L, et al. Oligosaccharides in 4 different milk groups, Bifidobacteria, and Ruminococcus obeum. *J Pediatr Gastroenterol Nutr*. 2011;53:80–87.
35. Koenig JE, Spor A, Scalfone N, et al. Succession of microbial consortia in the developing infant gut microbiome. *Proc Natl Acad Sci USA*. 2011;108(Suppl. 1):4578–4585.
36. Fallani M, Amarri S, Uusijarvi A, et al. Determinants of the human infant intestinal microbiota after the introduction of first complementary foods in infant samples from five European centres. *Microbiology*. 2011;157(Pt 5):1385–1392.
37. Mackie RI, Sghir A, Gaskins HR. Developmental microbial ecology of the neonatal gastrointestinal tract. *Am J Clin Nutr*. 1999;69:1035–1045.
38. Gronlund MM, Arvilommi H, Kero P, Lehtonen OP, Isolauri E. Importance of intestinal colonisation in the maturation of humoral immunity in early infancy: a prospective follow up study of healthy infants aged 0–6 months. *Arch Dis Child Fetal Neonatal Ed*. 2000;83:186–192.
39. Rautava S, Luoto R, Salminen S, Isolauri E. Microbial contact during pregnancy, intestinal colonization and human disease. *Nat Rev Gastroenterol Hepatol*. 2012;9:565–576.
40. Morelli L. Postnatal development of intestinal microflora as influenced by infant nutrition. *J Nutr*. 2008;138:1791S–1795S.
41. Rinne M, Kalliomaki M, Arvilommi H, Salminen S, Isolauri E. Effect of probiotics and breastfeeding on the *Bifidobacterium* and *Lactobacillus/Enterococcus* microbiota and humoral immune responses. *J Pediatr*. 2005;147:186–191.
42. Mohan R, Koebnick C, Schildt J, et al. Effects of Bifidobacterium lactis Bb12 supplementation on intestinal microbiota of preterm infants: a double-blind, placebo-controlled, randomized study. *J Clin Microbiol*. 2006;44:4025–4031.
43. Underwood MA, Salzman NH, Bennett SH, et al. A randomized placebo-controlled comparison of 2 prebiotic/probiotic combinations in preterm infants: impact on weight gain, intestinal microbiota, and fecal short-chain fatty acids. *J Pediatr Gastroenterol Nutr*. 2009;48:216–225.
44. Butel MJ, Suau A, Campeotto F, et al. Conditions of bifidobacterial colonization in preterm infants: a prospective analysis. *J Pediatr Gastroenterol Nutr*. 2007;44:577–582.
45. Rajilić-Stojanović M, Heilig HG, Molenaar D, et al. Development and application of the human intestinal tract chip, a phylogenetic microarray: analysis of universally conserved phylotypes in the abundant microbiota of young and elderly adults. *Environ Microbiol*. 2009;11:1736–1751.
46. Costello EK, Lauber CL, Hamady M, Fierer N, Gordon JI, Knight R. Bacterial community variation in human body habitats across space and time. *Science*. 2009;326:1694–1697.
47. Eckburg PB, Bik EM, Bernstein CN, et al. Diversity of the human intestinal microbial flora. *Science*. 2005;308:1635–1638.
48. The Human Microbiome Project Consortium. Structure, function and diversity of the healthy human microbiome. *Nature*. 2012;486:207–214.
49. Zoetendal EG, Rajilić-Stojanović M, de Vos WM. High-throughput diversity and functionality analysis of the gastrointestinal tract microbiota. *Gut*. 2008;57:1605–1615.
50. Benson AK, Kelly SA, Legge R, et al. Individuality in gut microbiota composition is a complex polygenic trait shaped by multiple environmental and host genetic factors. *Proc Natl Acad Sci USA*. 2010;107:18933–18938.
51. Yatsunenko T, Rey FE, Manary MJ, et al. Human gut microbiome viewed across age and geography. *Nature*. 2012;486:222–227.
52. Tap J, Mondot S, Levenez F, et al. Towards the human intestinal microbiota phylogenetic core. *Environ Microbiol*. 2009;11:2574–2584.
53. Qin J, Li R, Raes J, et al. A human gut microbial gene catalogue established by metagenomic sequencing. *Nature*. 2010;464:59–65.
54. Turnbaugh PJ, Hamady M, Yatsunenko T, et al. A core gut microbiome in obese and lean twins. *Nature*. 2009;457:480–484.
55. Bäckhed F, Fraser CM, Ringel Y, et al. Defining a healthy human gut microbiome: current concepts, future directions, and clinical applications. *Cell Host Microbe*. 2012;12:611–622.

56. Arumugam M, Raes J, Pelletier E, et al. Enterotypes of the human gut microbiome. *Nature*. 2011;473:174–180.
57. Huse SM, Ye Y, Zhou Y, Fodor AA. A core human microbiome as viewed through 16S rRNA sequence clusters. *PLoS ONE*. 2012;7:e34242.
58. Koren O, Knights D, Gonzalez A, et al. A guide to enterotypes across the human body: meta-analysis of microbial community structures in human microbiome datasets. *PLoS Comput Biol*. 2013;9:e1002863.
59. Wu GD, Chen J, Hoffmann C, et al. Linking long-term dietary patterns with gut microbial enterotypes. *Science*. 2011;334:105–108.
60. Murri M, Leiva I, Gomez-Zumaquero JM, et al. Gut microbiota in children with type 1 diabetes differs from that in healthy children: a case-control study. *BMC Med*. 2013;11:46.
61. Claesson MJ, Jeffery IB, Conde S, et al. Gut microbiota composition correlates with diet and health in the elderly. *Nature*. 2012;488:178–184.
62. Satokari RM, Vaughan EE, Akkermans AD, Saarela M, De Vos WM. Polymerase chain reaction and denaturing gradient gel electrophoresis monitoring of fecal bifidobacterium populations in a prebiotic and probiotic feeding trial. *Syst Appl Microbiol*. 2001;24:227–231.
63. Rochet V, Rigottier-Gois L, Ledaire A, et al. Survival of *Bifidobacterium animalis* DN-173 010 in the faecal microbiota after administration in lyophilized form or in fermented product—a randomised study in healthy adults. *J Mol Microbiol Biotechnol*. 2008;14:128–136.
64. Firmesse O, Mogenet A, Bresson JL, Corthier G, Furet JP. *Lactobacillus rhamnosus* R11 consumed in a food supplement survived human digestive transit without modifying microbiota equilibrium as assessed by real-time polymerase chain reaction. *J Mol Microbiol Biotechnol*. 2008;14:90–99.
65. Kim SW, Suda W, Kim S, et al. Robustness of gut microbiota of healthy adults in response to probiotic intervention revealed by high-throughput pyrosequencing. *DNA Res*. 2013;:1–13:Epub ahed of print Apr 9.
66. Gavini F, Cayuela C, Antoine JM, et al. Differences in distribution of bifidobacterial and enterobacterial species in human fecal microflora of three different (children, adults, elderly) age groups. *Microb Ecol Health Dis*. 2001;13:40–45.
67. Hopkins MJ, Macfarlane GT. Changes in predominant bacterial populations in human feces with age and with *Clostridium difficile* infection. *J Med Microbiol*. 2002;51:448–454.
68. Woodmansey EJ, McMurdo MET, Macfarlane CT, Macfarlane S. Comparison of compositions and metabolic activities of fecal microbiotas in young adults and in antibiotic-treated and non-antibiotic treated elderly subjects. *Appl Environ Microbiol*. 2004;70:6113–6122.
69. Mueller S, Saunier K, Hanisch C, et al. Differences in fecal microbiota in different european study populations in relation to age, gender, and country: a cross-sectional study. *Appl Environ Microbiol*. 2006;72:1027–1033.
70. Biagi E, Nylund L, Candela M, et al. Through ageing, and beyond: gut microbiota and inflammatory status in seniors and centenarians. *PLoS ONE*. 2010;5:e10667.
71. Zwielehner J, Liszt K, Handschur M, Lassl C, Lapin A, Haslberger AG. Combined PCR-DGGE fingerprinting and quantitative-PCR indicates shifts in fecal population sizes and diversity of *Bacteroides*, bifidobacteria and Clostridium cluster IV in institutionalized elderly. *Experimen Gerontol*. 2009;44:440–446.
72. Mäkivuokko H, Tiihonen K, Tynkkynen S, Paulin L, Rautonen N. The effect of age and non-steroidal anti-inflammatory drugs on human intestinal microbiota composition. *Br J Nutr*. 2010;103:227–234.
73. Claesson M,J, Cusack S, O'Sullivan O, et al. Composition, variability, and temporal stability of the intestinal microbiota of the elderly. *Proc Natl Acad Sci USA*. 2011;108(Suppl 1):4586–4591.
74. Ahmed M, Prasad J, Gill H, Stevenson L, Gopal P. Impact of consumption of different levels of Bifidobacterium lactis HN019 on the intestinal microflora of elderly human subjects. *J Nutr Health Aging*. 2007;11:26–31.
75. Lahtinen SJ, Tammela L, Korpela J, et al. Probiotics modulate the *Bifidobacterium* microbiota of elderly nursing home residents. *Age (Dordr)*. 2009;31:59–66.
76. Biagi E, Candela M, Fairweather-Tait S, Franceschi C, Brigidi P. Aging of the human metaorganism: the microbial counterpart. *Age (Dordr)*. 2012;34:247–267.
77. Ouwehand AC, Bergsma N, Parhiala R, et al. *Bifidobacterium* microbiota and parameters of immune function in elderly subjects. *FEMS Immunol Med Microbiol*. 2008;53:18–25.
78. Tenover FC, Arbeit RD, Goering RV, et al. Interpreting chromosomal DNA restriction patterns produced by pulsed-field gel electrophoresis: criteria for bacterial strain typing. *J Clin Microbiol*. 1995;33:2233–2239.
79. Cesena C, Morelli L, Alander M, et al. Lactobacillus crispatus and its nonaggregating mutant in human colonization trials. *J Dairy Sci*. 2001;84:1001–1010.
80. Kirjavainen PV, Ouwehand AC, Isolauri E, Salminen SJ. The ability of probiotic bacteria to bind to human intestinal mucus. *FEMS Microbiol Lett*. 1998;167:185–189.
81. Castagliuolo I, Galeazzi F, Ferrari S, et al. Beneficial effect of auto-aggregating *Lactobacillus crispatus* on experimentally induced colitis in mice. *FEMS Immunol Med Microbiol*. 2005;43:197–204.
82. Goh YJ, Azcárate-Peril MA, O'Flaherty S, et al. Development and application of a upp-based counterselective gene replacement system for the study of the S-layer protein SlpX of *Lactobacillus acidophilus* NCFM. *Appl Environ Microbiol*. 2009;75:3093–3105.
83. Konstantinov SR, Smidt H, de Vos WM, et al. S layer protein A of *Lactobacillus acidophilus* NCFM regulates immature dendritic cell and T cell functions. *Proc Natl Acad Sci USA*. 2008;105:19474–19479.
84. Metchnikoff E. *Nature of Man*. 3rd ed. New York and London: Putnam & Sons; 1903.
85. Metchnikoff E. *The prolongation of life: optimistic studies*. New York: Putnam and Sons; 1908.
86. Tissier H. Traitement des infections intestinales par la méthode de la flore bactérienne de l'intestin. *C R Soc Biol*. 1906;60:359–361.
87. Gerritsen J, Smidt H, Rijkers GT, de Vos WM. Intestinal microbiota in human health and disease: the impact of probiotics. *Genes Nutr*. 2011;6:209–240.
88. Reid G. Probiotic Lactobacilli for urogenital health in women. *J Clin Gastroenterol*. 2008;42(Suppl 3 Pt 2):S234–S236.
89. Sanders ME, Guarner F, Guerrant R, et al. An update on the use and investigation of probiotics in health and disease. *Gut*. 2013;62:787–796.
90. Scott KP, Gratz SW, Sheridan PO, Flint HJ, Duncan SH. The influence of diet on the gut microbiota. *Pharmacol Res*. 2013;69:52–60.
91. Deshpande G, Rao S, Patole S, Bulsara M. Updated meta-analysis of probiotics for preventing necrotizing enterocolitis in preterm neonates. *Pediatrics*. 2010;125:921–930.

92. Morelli L, Garbagna N, Rizzello F, Zonenschain D, Grossi E. In vivo association to human colon of *Lactobacillus paracasei* B21060: map from biopsies. *Dig Liver Dis.* 2006;38:894–898.
93. Dierksen KP, Moore CJ, Inglis M, Wescombe PA, Tagg JR. The effect of ingestion of milk supplemented with salivaricin A-producing *Streptococcus salivarius* on the bacteriocin-like inhibitory activity of streptococcal populations on the tongue. *FEMS Microbiol Ecol.* 2007;59:584–591.
94. Juneja A, Kakade A. Evaluating the effect of probiotic containing milk on salivary mutans streptococci levels. *J Clin Pediatr Dent.* 2012;37:9–14.
95. Elli M, Zink R, Rytz A, Reniero R, Morelli L. Iron requirement of *Lactobacillus* spp. in completely chemically defined growth media. *J Appl Microbiol.* 2000;88:695–703.
96. Corr SC, Li Y, Riedel CU, O'Toole PW, Hill C, Gahan CGM. Bacteriocin production as a mechanism for the antiinfective activity of *Lactobacillus salivarius* UCC118. *Proc Natl Acad Sci USA.* 2007;104:7617–7621.
97. Murphy EF, Cotter PD, Hogan A, et al. Divergent metabolic outcomes arising from targeted manipulation of the gut microbiota in diet-induced obesity. *Gut.* 2013;62:220–226.
98. Tagg JR, Dierksen KP. Bacterial replacement therapy: adapting 'germ warfare' to infection prevention. *Trends Biotechnol.* 2003;21:217–223.
99. Schiffrin EJ, Rochat F, Link-Amster H, Aeschlimann JM, Donnet-Huighes A. Immunomodulation of human blood cells following the ingestion of lactic acid bacteria. *J Dairy Sci.* 1995;78:491–497.
100. Bunout D, Barrera G, Hirsch S, et al. Effects of a nutritional supplement on the immune response and cytokine production in free-living Chilean elderly. *JPEN J Parenter Enteral Nutr.* 2004;28:348–354.
101. Boge T, Rémigy M, Vaudaine S, Tanguy J, Bourdet-Sicard R, van der Werf S. A probiotic fermented dairy drink improves antibody response to influenza vaccination in the elderly in two randomised controlled trials. *Vaccine.* 2009;27:5677–5684.
102. Olivares M, Díaz-Ropero MP, Sierra S, et al. Oral intake of *Lactobacillus fermentum* CECT5716 enhances the effects of influenza vaccination. *Nutrition.* 2007;23:254–260.
103. Davidson LE, Fiorino AM, Snydman DR, Hibberd PL. *Lactobacillus* GG as an immune adjuvant for live-attenuated influenza vaccine in healthy adults: a randomized double-blind placebo-controlled trial. *Eur J Clin Nutr.* 2011;65:501–507.
104. de Vrese M, Rautenberg P, Laue C, Koopmans M, Herremans T, Schrezenmeir J. Probiotic bacteria stimulate virus-specific neutralizing antibodies following a booster polio vaccination. *Eur J Nutr.* 2005;44:406–413.
105. Soh SE, Ong DQ, Gerez I, et al. Effect of probiotic supplementation in the first 6 months of life on specific antibody responses to infant Hepatitis B vaccination. *Vaccine.* 2010;28:2577–2579.
106. Sanz Y, De Palma G. Gut microbiota and probiotics in modulation of epithelium and gut-associated lymphoid tissue function. *Int Rev Immunol.* 2009;28:397–413.
107. Van Overtvelt L, Moussu H, Horiot S, et al. Lactic acid bacteria as adjuvants for sublingual allergy vaccines. *Vaccine.* 2010;28:2986–2992.
108. Maassen CB, Boersma WJ, van Holten-Neelen C, Claassen E, Laman JD. Growth phase of orally administered Lactobacillus strains differentially affects IgG1/IgG2a ratio for soluble antigens: implications for vaccine development. *Vaccine.* 2003;21:2751–2757.
109. Ewaschuk JB, Diaz H, Meddings L, et al. Secreted bioactive factors from *Bifidobacterium infantis* enhance epithelial cell barrier function. *Am J Physiol Gastrointest Liver Physiol.* 2008;295:G1025–G1034.
110. Zyrek AA, Cichon C, Helms S, Enders C, Sonnenborn U, Schmidt MA. Molecular mechanisms underlying the probiotic effects of *Escherichia coli* Nissle 1917 involve ZO-2 and PKCzeta redistribution resulting in tight junction and epithelial barrier repair. *Cell Microbiol.* 2007;9:804–816.
111. Fanning S, Hall LJ, Cronin M, et al. Bifidobacterial surface-exopolysaccharide facilitates commensal–host interaction through immune modulation and pathogen protection. *Proc Natl Acad Sci USA.* 2012;109:2108–2013.
112. Fukuda S, Toh H, Hase K, et al. Bifidobacteria can protect from enteropathogenic infection through production of acetate. *Nature.* 469:543–547.
113. de Milliano I, Tabbers MM, van der Post JA, Benninga MA. Is a multispecies probiotic mixture effective in constipation during pregnancy? 'A pilot study'. *Nutr J.* 2012;11:80.
114. Szajewska H, Gyrczuk E, Horvath A. *Lactobacillus reuteri* DSM 17938 for the management of infantile colic in breastfed infants: a randomized, double-blind, placebo-controlled trial. *J Pediatr.* 2013;162:257–262.
115. Savino F, Pelle E, Palumeri E, Oggero R, Miniero R. *Lactobacillus reuteri* (American Type Culture Collection Strain 55730) versus simethicone in the treatment of infantile colic: a prospective randomized study. *Pediatrics.* 2007;119:e124–e130.
116. Critch J. Infantile colic: is there a role for dietary interventions? *Paediatr Child Health.* 2011;16:47–49.
117. Mommaerts JL, Devroey D. It's too soon to recommend probiotics for colic. *J Fam Pract.* 2011;60:251–252.
118. Kang EJ, Kim SY, Hwang IH, Ji YJ. The effect of probiotics on prevention of common cold: a meta-analysis of randomized controlled trial studies. *Korean J Fam Med.* 2013;34:2–10.

CHAPTER

4

Bifidobacteria of the Human Gut: Our Special Friends

Marco Ventura, Francesca Turroni† and Douwe van Sinderen†*

***Laboratory of Probiogenomics, Department of Life Sciences, University of Parma, Italy**
†Alimentary Pharmabiotic Centre and Department of Microbiology, Biosciences Institute, University College Cork, Cork, Ireland

TAXONOMY OF BIFIDOBACTERIA

The genus *Bifidobacterium* belongs to the phylum *Actinobacteria*, one of the dominant phyla of the kingdom Bacteria, with five subclasses, six orders, and 14 suborders. This phylum includes Gram-positive microorganisms, whose genomic DNA contains a high G + C content, ranging from 51% (*Corynebacterium*) to 70% (*Streptomyces* and *Frankia*). An exception to this is the genome of the obligate pathogen *Tropheryma whipplei*, with less than 50% G + C.[1]

The Actinobacteria phylum includes a large variety of species exhibiting different cell morphologies (from coccoid to fragmenting hyphal), possessing various physiological metabolic properties and occupying diverse ecological niches. With respect to the latter, the phylum Actinobacteria includes genera originating from a wide variety of ecological environments, and includes, among others, soil inhabitants (*Streptomyces*), plant commensals (*Leifsonia*), nitrogen-fixing symbionts (*Frankia*) and gastrointestinal commensals (*Bifidobacterium*). Actinobacterial members also include microorganisms that interact with the human host, some of which are disease-causing, examples of which are certain members of the genera *Corynebacterium*, *Mycobacterium*, *Nocardia*, *Tropheryma* and *Propionibacterium*, while others represent bacteria exerting health-promoting activities on their hosts such as particular members of the genus *Bifidobacterium*.

In 1899 Henri Tissier isolated for the first time bifidobacteria from feces of breast-fed infants; these microorganisms were initially named Bacillus bifidus communis,[2] because of their particular y-shaped morphology and prevalence in infant feces. In fact, the term *bifidus* in Latin means "forked in two parts."

The possible formation of a connection between lactic acid bacteria (LAB) and the propionic acid bacteria led Orla-Jensen in 1924 to propose a new classification that considered this *Bacillus bifidus* as a new separated genus with the name *Bifidobacterium*.[3]

However, bifidobacteria for much of the last century were classified as members of other taxonomic groups and only with the 8th edition of Bergey's Manual, bifidobacteria were clustered in a separate genus, designated *Bifidobacterium*, which originally consisted of only 11 species.[4] The work carried out by Scardovi and Sgorbati[5] led to the identification of 15 novel bifidobacterial species isolated from a variety of distinctly different ecological niches (e.g., the human gut and insect gut).

At the time of writing, the genus *Bifidobacterium* encompasses 48 species, including four taxa (*Bifidobacterium longum*, *Bifidobacterium pseudolongum*, *Bifidobacterium animalis*, *Bifidobacterium thermacidophilum*) that are further divided into subspecies, all of which share more than 93% identity in their 16S rRNA sequences.[6] Among the 48 currently recognized bifidobacterial species are four species that were isolated from the digestive tract of a bumble bee, i.e., *Bifidobacterium actinocoloniiforme*, *Bifidobacterium bohemicum*, *Bifidobacterium coagulans* and *Bifidobacterium*

Diet-Microbe Interactions in the Gut.
DOI: http://dx.doi.org/10.1016/B978-0-12-407825-3.00004-6

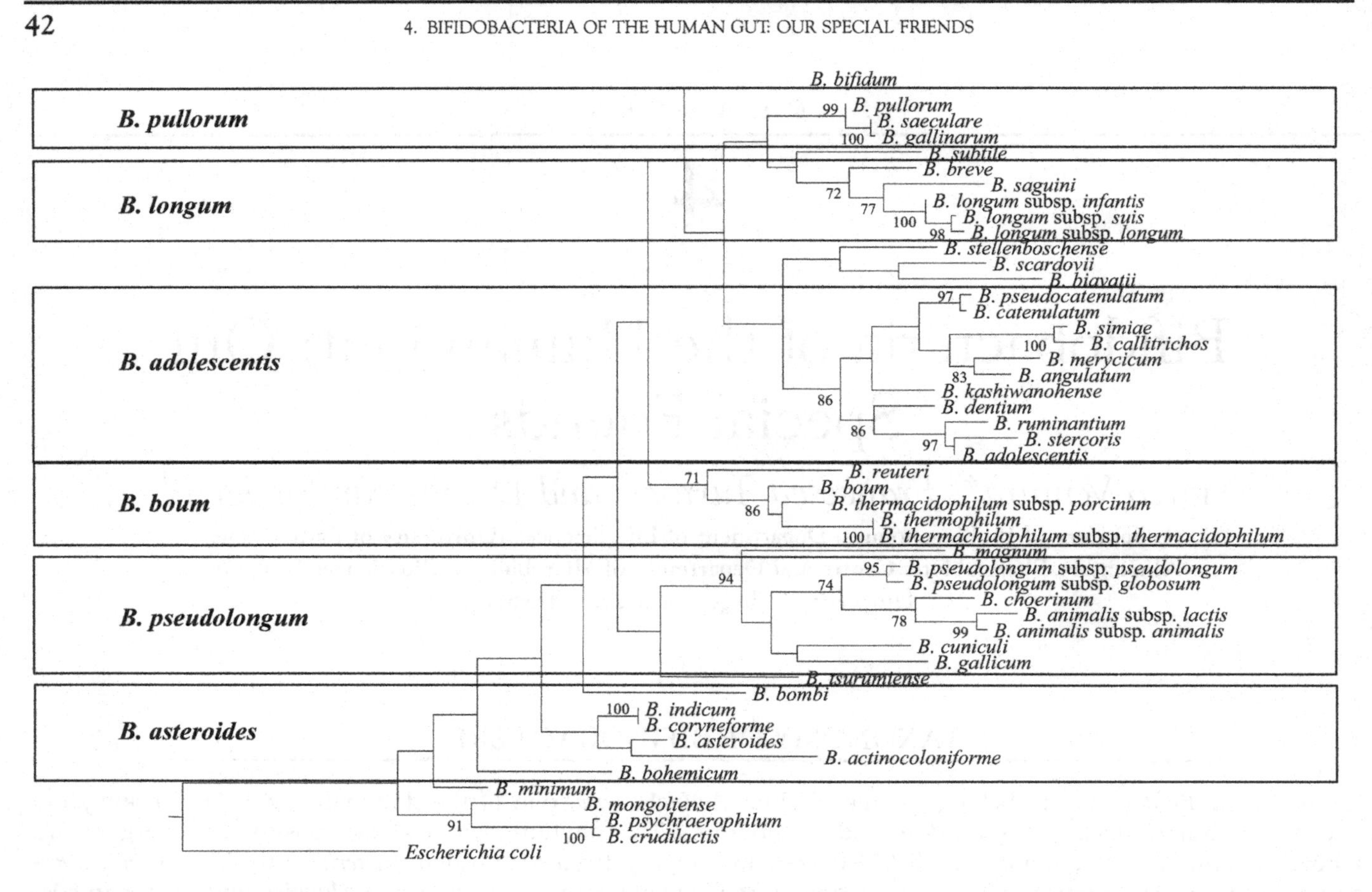

FIGURE 4.1 Phylogenetic tree of the genus *Bifidobacterium* based on 16S rRNA gene sequences. The six phylogenetic groups are highlighted.

bombi,[7–9] as well as five bifidobacterial taxa identified from feces of common marmoset (Bifidobacterium *reuteri*, *Bifidobacterium callitrichos*, *Bifidobacterium saguini* and *Bifidobacterium stellenboschense*) and one species that was isolated from feces of a red-handed tamarin (*Bifidobacterium biavatii*).[10]

In recent years, a relatively large number of publicly available bifidobacterial genomes has allowed the definition of a phylogenetic tree obtained using 16S rRNA genes (Figure 4.1). However, the investigation of phylogeny within a specific microbial taxon using a single molecular marker (i.e., the 16S rRNA gene sequence) is sometimes questionable. In fact, factors such as Horizontal Gene Transfer (HGT) events, different rates of recombination or incongruous mutation rates are responsible for atypical evolutionary trends.[11] In this context, and with the recent advent of the genomic era, the availability of gene sequences of many possible alternative molecular markers aside from the 16S rRNA gene, allowed bacterial taxonomists to build so-called phylogenetic supertrees, which provide a more robust image of the evolutionary development of a specific bacterial taxonomic unit.[12] When this approach was applied to the *Bifidobacterium* genus by employing 16S rRNA gene sequences and six housekeeping genes, i.e., clpC, dnaJ, xfp, dnaB, rpoC and purF, six distinct phylogenetic groups were identified: *Bifidobacterium longum, Bifidobacterium adolescentis, Bifidobacterium pullorum, Bifidobacterium pseudolongum, Bifidobacterium boum* and *Bifidobacterium asteroides*.[13] Notably, the achieved supertree allowed the identification of the putative ancestor of the genus *Bifidobacterium*, which was determined to be most closely related to representatives of the *B. asteroides* phylogenetic group (*Bifidobacterium asteroides, Bifidobacterium bombi, Bifidobacterium indicum, Bifidobacterium coryneforme, Bifidobacterium actinocoloniiforme, Bifidobacterium bohemicum*). It therefore seems that bifidobacteria took up residence in the insect gut and subsequently specialized to occupy the gastrointestinal tract of mammals, reptiles and birds. From an evolutionary perspective it is interesting to note that insects represent one of the first terrestial, multicellular creatures that developed a specialized digestive organ.[14]

BIFIDOBACTERIAL ECOLOGY

Bifidobacteria have been isolated from six different ecological niches, of which three are directly linked to the human and animal intestinal environment: e.g., the human gut, animal intestine (bovine, rabbit, murine, chicken and insect) and oral cavity, while others (sewage, blood and food) are probably the consequence of contamination from the gastro-intestinal tract (GIT)[11] (see below). Bifidobacteria are widely distributed among living organisms that provide their offspring with parental care such as mammals, birds and social insects. No bifidobacteria have been isolated so far from other animals such as reptiles and fish. Therefore, an important reason of their ecological distribution may be due to direct transmission of bifidobacterial cells from parent/carer to offspring.

Lamendella et al.[15] described that *Bifidobacterium cuniculi, Bifidobacterium angulatum* and *Bifidobacterium gallinarum* show a strict ecological adaptation to a particular animal gut, in this case rabbit, human and chicken, respectively. On the other hand, bifidobacteria belonging to the species *Bifidobacterium animalis, Bifidobacterium adolescentis, Bifidobacterium dentium* and *Bifidobacterium catenulatum* appear to display a more cosmopolitan lifestyle.[11]

Bifidobacteria are common inhabitants of the mammalian gut, but are also found in three other ecological niches: human blood (*Bifidobacterium scardovii*), sewage (e.g., *Bifidobacterium minimum* and *Bifidobacterium thermacidophilum*) and food products (e.g., *Bifidobacterium. animalis* subsp. *lactis*). These three apparently atypical ecological origins are completely different from that of the gut. However, it is plausible that the identification of bifidobacteria in these environments may have been a consequence of accidental contaminations during the sampling procedures and/or from "natural" contaminations from human/animal gut origins.

BIFIDOBACTERIAL POPULATIONS IN THE HUMAN GUT

The microbiota of a baby from birth until the time of weaning is characterized by high levels of bifidobacteria as judged on the basis of microbiota analyses of infant feces.[16] Bifidobacteria rapidly colonize the intestine of infants during the first weeks of life due to selection by a diet that is exclusively based on breast milk and/or formula milk, as confirmed by fluorescence in situ hybridization (FISH), quantitative real-time PCR (qPCR), and microarray analyses.[17]

The relative abundance of bifidobacteria within the human gut decreases with age, even if ecological analyses based on FISH and metagenomic studies estimated that their presence in the adult colon is around $4.3 \pm 4.4\%$ of fecal microbes.[18,19] In adult feces, *B. adolescentis* and *B. catenulatum* species were commonly detected, followed by *B. longum* and *B. bifidum*.

The HITChip is a phylogenetic microarray containing over 4800 oligonucleotides that target the V1 or the V6 region of the 16S rRNA gene from 1132 microbial phylotypes present in the human gastrointestinal tract.[20] Microbiota analyses using the HITChip showed that proportions and compositions of bifidobacteria and *Bacteroidetes* were consistently stable within adult individuals, whereas the *Firmicutes* showed large inter-individual variation.

Recently, a polyphasic approach, consisting of bifidobacterial isolation from human intestinal mucosal samples and fecal samples on selective media (culture-dependent method) followed by sequencing of corresponding 16S rRNA internal transcribed spacer (ITS) sequences, allowed an investigation into the biodiversity of the bifidobacterial population present in the human gut.[21] This study identified the most abundant bifidobacterial species present in this environment, which was shown to consist of *B. longum, B. pseudolongum, B. animalis* subsp. *lactis, B. adolescentis, B. bifidum, B. pseudocatenulatum* and *B. breve*. The analysis of the corresponding Internal 16S–23S transcribed spacer region (ITS) sequences also demonstrated that the bifidobacterial distribution in various human subjects revealed not only an inter-subject but also an intra-subject variability. Although fecal samples from infants are dominated by a relatively small number of different bifidobacterial species and strains, they do reveal a high inter-individual variation in taxonomic composition. In contrast, the bifidobacterial populations isolated from adults seemed more complex despite the fact that they exhibited a remarkable conservation of bifidobacterial species and strains.[22] Furthermore, the latter authors also evaluated the bifidobacterial composition of the human intestine by a microbiomic approach, through the analyses of five colonic mucosal samples from healthy adults.[21] The obtained data from this latter work showed that each individual possesses a specific population of colonic bifidobacteria, a finding that is consistent with the previously described considerable inter-variability of the overall intestinal microbiota.[18,23] These samples were dominated by 16S rDNA sequences that are closely related to the *B. pseudolongum* phylogenetic groups, followed by members of the *B. longum* (17%) and

B. adolescentis (8.5%) phylogenetic groups. Moreover, this culture-independent approach led to the identification of many novel bifidobacterial 16S rRNA gene sequences, which are presumed to represent as yet unidentified and thus novel bifidobacterial species that have escaped isolation due to their strictly anaerobic as well as very peculiar growth requirements.

BIFIDOBACTERIA AS PROBIOTICS

Probiotics are described as "live microorganisms, which when administered in adequate amounts confer a health benefit on the host" (FAO/WHO, 2002). Bifidobacteria are largely used as probiotics in many food products such as yoghurt, milk, infant formula, cheese, and dietary supplements.[24] The probiotic concept was recognized for the first time by Metchnikoff in 1908 when he observed that the introduction of some fermented foods in the diet had beneficial properties on human health.[25] Since Metchnikoff's original observations, various researches confirmed the ability of probiotic strains to exert positive effects on the host's health status, such as immunostimulation, modulation of the intestinal microbiota, cholesterol reduction, alleviation of acute gastro-enteritis, short-chain fatty acid (SCFAs) production, alleviation of constipation and the reduction of allergic disease symptoms, lactose intolerance and intestinal inflammation.[26]

Probiotic bacteria have been extensively exploited by food industries for several decades. In fact, food industries developed the so called "probiotic products" where a food substrate is added with probiotic bacteria such as bifidobacteria. Probiotic products represent a growing business area and still attract significant economic interest from various commercial sectors, such as the food, feed and pharmaceutical industries.[27]

However, the probiotic concept should be interpreted very carefully, particularly when one considers how the numerous different members of the gut microbiota and the highly complex interactions that occur within this vast microbial consortium may affect the metabolism and the immune-system of the host. The intestinal microbiota and its host enjoy a "fluctuating" relationship that depends on the type of bacterial species present. Symbiotic components may be present where both partners derive benefit from this relationship, but this may turn into commensalism where only a single partner enjoys a positive effect from the interaction.[28]

A healthy host exists in a state of equilibrium with the gut microbiota and in this way this microbiota contributes to a balanced host immunity, to homeostasis at the intestinal mucosa and to metabolism, while the host provides it with a constant and convenient habitat, and at the same time keeps microbial numbers in check. A disorder in the host–microbiota equilibrium may happen where a pathogen enters the host intestine and interacts with both the host and the microbiota. If both these two components (host and indigenous microbiota) are resilient, the balance is restored with a consequent recovery of the intestinal mucosa. A disruption of the intestinal homeostasis is the result of a disturbance of the host–microbiota balance that in extreme cases may lead to the death of the host. If the pathogen acquires a niche for itself among the indigenous bacteria of the gut, it will eventually create a new equilibrium between the host and the microbiota with consequent chronic presence of the pathogen.[29]

The claimed health-promoting activities exerted by bifidobacteria are numerous, and include establishment of a healthy microbiota in preterm infants,[30] cholesterol reduction, lactose intolerance, prevention of infectious diarrhea, prevention of cancer, protection against infectious diseases, modulation of mucosal barrier function, amino acid and vitamin production, inhibition of nitrate reduction, stimulation of calcium uptake by enterocytes,[26] short-chain fatty acid production,[31] stimulation of intestinal epithelia through induction of anti-inflammatory cytokine interleukin (IL)-10 and junctional adhesion molecules.[32]

The interaction of different *Bifidobacterium* species and the host may have distinct effects on the immune system, e.g., supplementation of *B. longum* showed some decrease in the expression of genes encoding human pro-inflammatory cytokines,[33] while ingestion of *B. animalis* subsp. *lactis* has been reported to cause an increase in the anti-inflammatory cytokine IFN-α and in phagocytic activity.[34]

In another study supplementation of live cells of *B. breve* NCIMB 702258 significantly increased c9, t11 Conjugate Linoleic Acid (CLA) in the liver of the human host, leading to reduced production of the inflammatory metabolites interferon (IFN)-γ, tumor necrosis factor (TNF)-α, interleukin (IL)-8, and IL-12, combined with an increase in the regulatory cytokine IL-10. Mice supplemented with *B. longum* promoted dendritic cell maturation, increased IFN-γ gene expression and the interferon-γ/IL-4 ratio in intestinal mucosa, up-regulated IL-10, IL-12, IFN-γ and raised immunoglobulin secretion in cultured peripheral blood mononuclear cells.

From the above-mentioned publications it is concluded that bifidobacteria, sometimes in combination with other bacterial species, may have important immunodulatory effects on the host organism. However, the precise mechanisms of the immunodulatory activity of bifidobacteria and other intestinal bacteria are poorly understood and are currently the target of ongoing scientific investigations.

A recent study has highlighted that the induction of Regulatory T cells (T_{reg}) was promoted by gut commensal bacteria; this subpopulation of T cells is responsible for maintaining the Th1/Th2 balance.[35] In this context, it has been shown that *Bifidobacterium bifidum*-derived membrane vesicles induce dendritic cell-mediated T_{reg} differentiation.[36]

BIFIDOBACTERIAL GENOMICS

The first bacterial genome sequence of a bacterial pathogen was published in 1995, and within 15 years, more than 1000 microorganisms were fully sequenced and deposited in the NCBI database. This large amount of genomic data has shed light on the genetic basis of bacterial pathogenesis and has lead to the formulation of a new genomics-based discipline called pathogenomics.[37] Only relatively recently, genomic efforts have focused on the decoding of genome sequences from human gut commensals, including health-promoting bacteria, such as bifidobacteria. In recent years, 27 bifidobacterial genomes have been completely sequenced of the 48 species so far recognized,[14,31,38–43] with another 25 strains whose genome sequences are still unfinished (NCBI source). However, despite the apparently large number of available bifidobacterial genomes, which have been shown to range from 1.9 Mb for *B. animalis* subsp. *lactis* DSM10140 to 2.8 Mb for *B. longum* subsp. *infantis* ATCC15697, they represent just nine of the 48 currently recognized bifidobacterial species. In this context, more than one genome sequence is publicly available for the species *B. longum* subsp. *longum*[39,40] and *B. animalis* subsp. *lactis*.[38,44,45]

All this genomic data provides intriguing information about specific characteristics of bifidobacteria (e.g., metabolic capabilities, genetics and phylogeny) and provides scientific support to understand the molecular mechanisms responsible for their adaptation to a specific ecological niche (see below).

In this context, probiogenomics, a recently coined discipline that investigates genomic information of health-promoting bacteria, has also provided some compelling evidence on how probiotic bacteria have adapted to the GIT environment and how they interact with their hosts.[46] Bifidobacterial genome sequences can notably contribute to the understanding of genetic adaptation to specific niches such as to the infant gut in the case of *B. longum* subsp. *infantis* ATCC15697[31] or to the oral cavity such as the case of *B. dentium* Bd1,[41] or to the human gut in the case of *B. bifidum* PRL2010.[42] Also, in silico analysis of the recently sequenced genome of *B. asteroides* PRL2011 reveals the genetic adaptation of this bifidobacterial strain to honeybee intestine.[14]

Probiogenomics has led efforts to identify and understand genetic adaptations of bifidobacteria to the infant gut, for example through in silico analyses of the genome sequence of *B. longum* subsp. *infantis* ATCC 15697, which revealed the presence of a large genomic locus whose encoded protein products are involved in the metabolism of Human Milk Oligosaccharides (HMO).[31] This cluster consists of a 43-kb DNA region, encoding specific intracellular glycoside hydrolases, such as fucosidases, sialidases, β-hexosaminidase and β-galactosidase, extracellular solute binding proteins and permeases, predicted to hydrolyze and internalize human milk-derived oligosaccharides thereby generating monosaccharides that enter the fructose-6-phosphate phosphoketolase central metabolic pathway.[31] Moreover, the genome of this organism contains a complete urease operon, predicted to be involved in the utilization of urea that may represent an important supply of nitrogen in milk.[31]

Another dominant infant gut bifidobacterial species is *B. bifidum*.[21] Various physiological investigations revealed that members of the *B. bifidum* species are able to metabolize host-derived glycans, such as mucin.[47] Mucin is the principal component of the mucus gel that covers the GIT epithelium and it represents the first barrier between host and intestinal bacteria, as well as between host and nutrients present in the gut. Also other human gut microorganisms (*Bacteroides* spp., *Ruminococcus* spp., *Clostridium* spp. and *Akkermansia muciniphila*) have been described to possess an important role in mucin degradation,[48–50] although the genetic requirements for this phenotype in these latter bacteria are still largely unknown. The complete genome sequence of *B. bifidum* PRL2010 was published in 2010[42] and revealed intriguing metabolic strategies followed by *B. bifidum* to metabolize mucin-derived carbohydrates. In silico analyses coupled with functional genomic investigations have revealed the existence of specific *B. bifidum* enzymatic pathways involved in the utilization of host-derived glycans. For example, an extracellular endo-α-N-acetylgalactosaminidase was identified that releases galacto-N-biose (GNB) from core 1-type O-glycans in mucin glycoproteins.[51] The generated carbohydrates may therefore undergo further degradation by other extracellular enzymes, such as a β-galactosidase and β-N-acetylhexosaminidase, and the resulting products are believed to be translocated across the cell membrane for further hydrolysis, phosphorylation, isomerization and/or deacetylation.[51]

The availability of the complete genome sequences of the oral opportunistic pathogen *B. dentium* Bd1 represents another key example of the importance of genomics in understanding the biology of bifidobacteria.[41] Recent studies of oral bifidobacteria related to dental caries in adults and children revealed that *B. dentium* represents about 8% of the culturable bacteria isolated from carious lesions.[41] Genome analysis of *B. dentium* Bd1 displayed that this strain possesses the necessary metabolic capabilities to utilize a large variety of carbohydrates, including both simple sugars as well as complex carbohydrates.[41] This may reflect the ecological origin of this microorganism, in fact the human oral cavity contains a large amount of simple carbohydrates that are preferentially utilized by the microbiota encountered in this GIT compartment. At the same time Bd1 is able to survive in the human fecal material, where the only carbon source is represented by complex carbohydrates. Moreover, the Bd1 genome sequence revealed the presence of seven loci that encode pilus-like structures and these appendages may be involved in adhesion to the dental plaque as seen for other oral pathogens, e.g., *Actinomyces naeslundii*.[52]

COMPARATIVE GENOMICS AND BIFIDOBACTERIA

Comparative genomics is a particular activity within the discipline of genomics and is aimed at comparing genome sequences from a wide variety of organisms (from bacteria to human). Comparison of whole genome sequences generates a complete and detailed picture of how organisms are related to each other at the genetic level. Comparative genomic analyses also provide information aimed at understanding the function and evolutionary processes that act on genomes in light of adaptation to their specific ecological niches.

In the case of bifidobacteria whole genome comparisons have revealed a higher extent of conservation and synteny across entire genomes with limited phylogenetic diversity compared to other gut inhabitants such as *Bacteroides*.[11] A recent publication[53] has compared the genome sequences from nine bifidobacterial strains and identified a presumed bifidobacterial pan-genome encompassing more than 5000 genes.[53] Furthermore, a set of genes was identified that was shared by all nine sequenced *Bifidobacterium* species, thus representing the presumed core genome, represented by 967 genes. The sequences within this core genome were shown to specify various housekeeping functions, i.e., those involved in replication, transcription, translation, cell envelope biogenesis and signal transduction.[53] Furthermore, the above-mentioned comparative genome analysis allowed the identification of the so called Truly Unique Genes (TUG), which involve genes present only in a reference bifidobacterial genome and are absent in any other genome. Most of these unique genes have unknown functions but are presumed to be important for bifidobaterial persistence in the GIT such as the ability of these bacteria to interact with their natural environment, host physiology and other microorganisms.[53]

An alternative strategy to whole genome sequencing is represented by performing Comparative Genome Hybridization (CGH) experiments, which is based on DNA microarray hybridization, also known as genomotyping. This technique has been used to compare inter- and intra-species genome variability in bacteria.[54] It allowed the comparison of genome sequences and highlighted DNA regions of genetic variability that are indicative of physiological plasticity. Recently, a small number of *B. longum* strains were analyzed[55] using CGH to explore the genome variability, employing the genome sequence of *B. longum* subsp. *infantis* ATCC15697 as a reference.[55] This study revealed the presence of large genomic regions, which appear to be highly variable between the strains investigated, and which correspond to specific ecological niche adaptation islands encompassing genes encoding enzymes involved in HMO metabolism, such as fucosidases, a sialidase, ABC-transporters and type 1 Solute Binding Proteins (SBP).[55] CGH experiments were also successfully applied to investigate the genome variability within the *B. dentium* species,[41] as well as within the *B. bifidum* taxon[42] and the *B. adolescentis* phylogenetic group.[56]

In *B. dentium* and *B. bifidum* genomes, the observed variable genomic regions encompass genes and/or gene clusters predicted to encode R/M systems, Exo-Polysaccharide Structures (EPS), pili and other mobile genetic elements (e.g., IS elements, transposons, phages and integrated plasmids).[57,58]

INTERACTION BETWEEN BIFIDOBACTERIA AND THEIR HOSTS

Among the structures and molecules encoded by gut bacteria known to play a pivotal role in the interaction between the human intestinal microbiota and its host, the following are worth mentioning: (i) exopolysaccharides; (ii) pili; (iii) the serpin-like protease inhibitor; and (iv) bacteriocins.

Exopolysaccharides (EPS)

These complex multilayered glycan structures constitute a crucial interface between the particular bacterial producer and its environment. They have been shown to play a key role in anchoring proteins, and consequently play an important role in promoting bacterial adhesion and colonization. Most of these studies were focused on microbial pathogens,[59] but they still may provide interesting hypotheses regarding the molecular mechanisms/strategies used by other bacteria such as commensals for intestinal adhesion, colonization and persistence.

One of the structures implicated in bifidobacterial adhesion is an extracellular surface-associated polysaccharide structure, and it has been claimed that such capsular polysaccharides possess a crucial role for bifidobacterial adherence to host cells, while they may also provide resistance to stomach acids and bile salts.[60]

Recently, the genome sequence of B. dentium and B. bifidum genomes UCC2003 has been fully decoded[43] and revealed the presence of a gene cluster predicted to be involved in the production of two different cell surface-associated EPSs.[61] Surface EPS produced by UCC2003 influenced *in vivo* persistence of bifidobacterial cells, while EPS-producing B. dentium and B. bifidum genomes UCC2003 cells stimulate only a weak adaptive immune response compared with EPS-deficient mutants lacking this cell envelope-associated structure. Specifically, EPS production was shown to be linked to the evasion of adaptive B-cell responses. In addition, the presence of surface EPS-expressing UCC2003 cells in a murine model reduced colonization levels of the murine pathogen *Citrobacter rodentium*. Notably, mouse *C. rodentium* infection is used as a model for human attaching and effacing intestinal pathogens like *E. coli* EHEC and EPEC strains. These findings indicate a crucial and beneficial role for bifidobacterial surface EPS in modulating various aspects of host–microbe interaction, including host-mediated immune tolerance of the commensal, while providing protection against a pathogen in an as yet unknown manner.[61]

Pilus-Like Structure

Another key structure produced by gut bacteria, which are implicated in host–microbe interactions is represented by pili. Pili are proteinaceous appendages present on the surface of both Gram-negative and Gram-positive bacteria. In Gram-negative pili, subunits are linked by non-covalent bonds, while in Gram-positive bacteria the pilin subunits can be covalently polymerized by specific transpeptidase enzymes named sortases.[62]

The pathogen *Corynebacterium diphtheria* was one of the first Gram-positive microorganisms in which the presence of pili was identified. Further knowledge on the role of pili in Gram-positive bacteria was obtained by the identification of these structures in the oral pathogen *Actinomyces naeslundii*.[52,63] In these bacteria, the genes required for the composition and assembly of pili are organized in pathogenic islets that encode one major pilin (known as SpaA, SpaD, and SpaH subunits in *C. diphtheriae*, or FimA and FimP subunits in *A. naeslundii*), one or two minor pilins (called SpaB, SpaC, SpaE, SpaF, SpaG and SpaI subunits in *C. diphtheriae*, or FimB and FimQ subunits in *A. naeslundii*), and a pilus-specific sortase.[62,64] The sortase plays a key role in the covalent polymerization of the major pilins; indeed, this enzyme generates the intermolecular amide bonds between the C terminus of one subunit and a lysine residue on the next.[62]

In *Corynebacterium*, SpaA and, sometimes SpaB subunits, form the pilus shaft. The major subunits are commonly located at the base of the shaft, while SpaC subunits are present at the tip.[62,65] Genomes of *Actinomyces* have been shown to harbor two different loci (*fim*QP and the *fim*AB) encoding for two different types of fimbriae (called type 1 and type 2). These fimbriae are constituted of the major fimbrial protein, i.e., FimA or FimP, forming the shaft and the minor fimbrial protein, i.e., FimB or FimQ, localized at the tip and the cell surface.[66]

Genome analyses of various Gram-positive bacteria revealed the presence of pilus gene clusters containing sortases in pathogens such as *Clostridium perfringens*, enterococci, many streptococcal species, *Actinomyces* taxon[64] and, recently, also in *Lactobacillus rhamnosus* GG a human gut commensal and well-studied probiotic strain.[67,68]

Many studies focusing on Gram-negative pathogens identified pili as important players for the attachment/colonization of microbial cells to host tissues and consequently as key bacterial structures in the establishment of bacterial infection.[59] The involvement of pili structures in the establishment of pathogenesis has also been established in Gram positive bacteria like *Corynebacterium*, where SpaA-type pilus is required for specific adherence to human pharyngeal epithelial cells.[69] Furthermore, in the oral pathogens *Actinomyces* the two encoced pilus structures displayed different functions, type I fimbria mediated the interaction of *Actinomyces* to tooth enamel,[66] while the other pili structures (type II fimbria) interacted with oral streptococci and with host cells, causing dental plaque formation.[66]

On the other hand, pili were recently also identified in the genome sequences of human gut commensals such as *Bifidobacterium* and *Lactobacillus*. In *Lactobacillus rhamnosus* GG, two sortase-dependent pili, SpaCBA and SpaEF, have been shown to be involved in bacterial attachment to human cells, and bacterial establishment within the human gut.[67]

A recent study involving *B. breve* UCC2003 describes that under *in vivo* conditions this microorganism expresses pili structures belonging to a type IVb or Tad (tight-adherence) pilus family,[43] which is encoded by the so-called *tad* locus on its genome. A mutant of UCC2003 *tad* locus revealed that these structures are necessary for *in vivo* colonization and immunogold transmission electron microscopy showed the existence of Tad pili at the poles of the UCC2003 cells recovered from the murine gut.

Notably, the *tad* locus is highly conserved within bifidobacterial genomes, which might imply a common pilus-mediated mechanism for bifidobacterial gut colonization and persistence.[70]

Serine Protease Inhibitor

Another key molecule effector for bifidobacterial–host interaction is represented by a serpin-like protease inhibitor, which was originally identified in the genome sequence of *B. longum* subsp. *longum* NCC2705.[71] Ivanov et al.[71] demonstrated that the serpin encoded by NCC2705 inhibited host proteases like the human neutrophil and pancreatic elastases, which may be found at the sites of intestinal inflammation.[40,71] Analyses of bifidobacterial genome sequences have highlighted the presence of genes encoding serpin-like proteins among enteric bifidobacteria including *B. breve*, *B. longum* subsp. *longum* and *B. longum* subsp. *infantis*.[72]

Bacteriocins

Bacteriocins are ribosomally synthesized antimicrobial peptides produced by bacteria that have bactericidal or bacteriostatic effects on similar or closely related bacterial strains.[73] Bacteriocins show an antagonistic activity against various pathogens such as *Listeria monocytogenes*.[74] Another known function of bacteriocins is their role as signaling peptides, communicating to other bacteria through quorum sensing and crosstalk mechanisms within bacterial population, and to cells of the host immune system. In the case of bifidobacteria, just a few strains have been shown to produce bacteriocins: *B. bifidum* NCFB1454 produces bifidocin B.[75,76] Recently, a bifidobacterial strain, *B. thermophilum RBL67*, has been described, which is capable of producing a bacteriocin-like inhibitory substance or "lantibiotic,"[77] while genome analyses of *B. longum* subsp. *longum* DJ010A revealed a complete lantibiotic gene cluster.[78]

CONCLUSIONS

For millennia mammals have evolved with their commensal partners, and adaptive co-evolution has formed very complex links between the gut microbiota and their host. Imbalances in the gut microbiota may contribute to certain human diseases. Furthermore, it is generally accepted that gut bacteria such as bifidobacteria clearly affect the regulatory network of the immune system, which compounds the intricate connection between gut microbiota composition and host health. Here, we have provided a number of telling examples of molecules, such as pili and capsular or surface polysaccharides, that mediate bifidobacterial host–microbe interaction. The first decade of genomic exploration of the biology of gut commensals, such as bifidobacteria, has afforded unprecedented insights into the genetic adaptation of these microorganisms to the human gut through the decoding of their genome sequences (probiogenomics). The next decade holds the promise of being even more rewarding as new discoveries on the molecular mechanisms underpinning host–microbe interactions are generated by means of functional genomics efforts.

References

1. Ventura M, Canchaya C, Tauch A, et al. Genomics of actinobacteria: tracing the evolutionary history of an ancient phylum. *Microbiol Mol Biol Rev*. 2007;71(3):495–548.
2. Tissier H. *Recherchers sur la flora intestinale normale et pathologique du nourisson*. Paris, France: University of Paris; 1900.
3. Ventura M, Canchaya C, Zhang Z, Bernini V, Fitzgerald GF, van Sinderen D. How high G + C Gram–positive bacteria and in particular bifidobacteria cope with heat stress: protein players and regulators. *FEMS Microbiol Rev*. 2006;30(5):734–759.
4. R. E. Buchanan (Editor) NEGE. Bergey's manual of systematic bacteriology; 2009.
5. Scardovi V, Sgorbati B. Electrophoretic types of transaldolase, transketolase, and other enzymes in bifidobacteria. *Antonie Van Leeuwenhoek*. 1974;40(3):427–440.
6. Miyake T, Watanabe K, Watanabe T, Oyaizu H. Phylogenetic analysis of the genus *Bifidobacterium* and related genera based on 16S rDNA sequences. *Microbiol Immunol*. 1998;42(10):661–667.
7. Killer J, Kopecny J, Mrazek J, et al. *Bifidobacterium actinocoloniiforme* sp. nov. and *Bifidobacterium bohemicum* sp. nov., from the bumblebee digestive tract. *Int J Syst Evol Microbiol*. 2011;61(Pt 6):1315–1321.
8. Killer J, Kopecny J, Mrazek J, et al. *Bombiscardovia coagulans* gen. nov., sp. nov., a new member of the family Bifidobacteriaceae isolated from the digestive tract of bumblebees. *Syst Appl Microbiol*. 2010;33(7):359–366.
9. Killer J, Kopecny J, Mrazek J, et al. *Bifidobacterium bombi* sp. nov., from the bumblebee digestive tract. *Int J Syst Evol Microbiol*. 2009;59(Pt 8):2020–2024.
10. Endo A, Futagawa-Endo Y, Schumann P, Pukall R, Dicks LM. *Bifidobacterium reuteri* sp. nov., *Bifidobacterium callitrichos* sp. nov., *Bifidobacterium saguini* sp. nov., *Bifidobacterium stellenboschense* sp. nov. and *Bifidobacterium biavatii* sp. nov. isolated from faeces of common marmoset (*Callithrix jacchus*) and red-handed tamarin (*Saguinus midas*). *Syst Appl Microbiol*. 2012;35(2):92–97.
11. Ventura M, Canchaya C, Fitzgerald GF, Gupta RS, van Sinderen D. Genomics as a means to understand bacterial phylogeny and ecological adaptation: the case of bifidobacteria. *Antonie Van Leeuwenhoek*. 2007;91(4):351–372.
12. Bininda-Emonds OR. The evolution of supertrees. *Trends Ecol Evol*. 2004;19(6):315–322.
13. Ventura M, Canchaya C, Del Casale A, et al. Analysis of bifidobacterial evolution using a multilocus approach. *Int J Syst Evol Microbiol*. 2006;56(Pt 12):2783–2792.
14. Bottacini F, Milani C, Turroni F, et al. *Bifidobacterium asteroides* PRL2011 genome analysis reveals clues for colonization of the insect gut. *PLoS One*. 2012;7(9):e44229.
15. Lamendella R, Santo Domingo JW, Kelty C, Oerther DB. Bifidobacteria in feces and environmental waters. *Appl Environ Microbiol*. 2008;74 (3):575–584.
16. Harmsen HJ, Raangs GC, He T, Degener JE, Welling GW. Extensive set of 16S rRNA-based probes for detection of bacteria in human feces. *Appl Environ Microbiol*. 2002;68(6):2982–2990.
17. Kleerebezem M, Vaughan EE. Probiotic and gut lactobacilli and bifidobacteria: molecular approaches to study diversity and activity. *Annu Rev Microbiol*. 2009;63:269–290.
18. Eckburg PB, Bik EM, Bernstein CN, et al. Diversity of the human intestinal microbial flora. *Science*. 2005;308(5728):1635–1638.
19. Mueller S, Saunier K, Hanisch C, et al. Differences in fecal microbiota in different European study populations in relation to age, gender, and country: a cross-sectional study. *Appl Environ Microbiol*. 2006;72(2):1027–1033.
20. Salonen A, Nikkila J, Jalanka-Tuovinen J, et al. Comparative analysis of fecal DNA extraction methods with phylogenetic microarray: effective recovery of bacterial and archaeal DNA using mechanical cell lysis. *J Microbiol Methods*. 2010;81(2):127–134.
21. Turroni F, Foroni E, Pizzetti P, et al. Exploring the diversity of the bifidobacterial population in the human intestinal tract. *Appl Environ Microbiol*. 2009;75(6):1534–1545.
22. Turroni F, Marchesi JR, Foroni E, et al. Microbiomic analysis of the bifidobacterial population in the human distal gut. *ISME J*. 2009;3 (6):745–751.
23. Palmer C, Bik EM, DiGiulio DB, Relman DA, Brown PO. Development of the human infant intestinal microbiota. *PLoS Biol*. 2007;5(7): e177.
24. Russell DA, Ross RP, Fitzgerald GF, Stanton C. Metabolic activities and probiotic potential of bifidobacteria. *Int J Food Microbiol*. 2011;149 (1):88–105.
25. Metchnikoff E. The Prolongation of Life. 1908.
26. Lee JH, O'Sullivan DJ. Genomic insights into bifidobacteria. *Microbiol Mol Biol Rev*. 2010;74(3):378–416.
27. Stanton C, Ross RP, Fitzgerald GF, Van Sinderen D. Fermented functional foods based on probiotics and their biogenic metabolites. *Curr Opin Biotechnol*. 2005;16(2):198–203.
28. Marco ML, Pavan S, Kleerebezem M. Towards understanding molecular modes of probiotic action. *Curr Opin Biotechnol*. 2006;17 (2):204–210.
29. Sekirov I, Finlay BB. The role of the intestinal microbiota in enteric infection. *J Physiol*. 2009;587(Pt 17):4159–4167.
30. Wang C, Shoji H, Sato H, et al. Effects of oral administration of *Bifidobacterium breve* on fecal lactic acid and short-chain fatty acids in low birth weight infants. *J Pediatr Gastroenterol Nutr*. 2007;44(2):252–257.
31. Sela DA, Chapman J, Adeuya A, et al. The genome sequence of *Bifidobacterium longum* subsp. *infantis* reveals adaptations for milk utilization within the infant microbiome. *Proc Natl Acad Sci USA*. 2008;105(48):18964–18969.
32. Chichlowski M, De Lartigue G, German JB, Raybould HE, Mills DA. Bifidobacteria isolated from infants and cultured on human milk oligosaccharides affect intestinal epithelial function. *J Pediatr Gastroenterol Nutr*. 2012;55(3):321–327.
33. Furrie E. A molecular revolution in the study of intestinal microflora. *Gut*. 2006;55(2):141–143.
34. Arunachalam K, Gill HS, Chandra RK. Enhancement of natural immune function by dietary consumption of *Bifidobacterium lactis* (HN019). *Eur J Clin Nutr*. 2000;54(3):263–267.
35. Vael C, Desager K. The importance of the development of the intestinal microbiota in infancy. *Curr Opin Pediatr*. 2009;21(6):794–800.
36. Lopez P, Gonzalez-Rodriguez I, Gueimonde M, Margolles A, Suarez A. Immune response to *Bifidobacterium bifidum* strains support Treg/ Th17 plasticity. *PLoS One*. 2011;6(9):e24776.

37. Pallen MJ, Wren BW. Bacterial pathogenomics. *Nature*. 2007;449(7164):835–842.
38. Kim JF, Jeong H, Yu DS, et al. Genome sequence of the probiotic bacterium *Bifidobacterium animalis* subsp. *lactis* AD011. *J Bacteriol*. 2009;191(2):678–679.
39. Lee JH, Karamychev VN, Kozyavkin SA, et al. Comparative genomic analysis of the gut bacterium *Bifidobacterium longum* reveals loci susceptible to deletion during pure culture growth. *BMC Genomics*. 2008;9:247.
40. Schell MA, Karmirantzou M, Snel B, et al. The genome sequence of *Bifidobacterium longum* reflects its adaptation to the human gastrointestinal tract. *Proc Natl Acad Sci USA*. 2002;99(22):14422–14427.
41. Ventura M, Turroni F, Zomer A, et al. The *Bifidobacterium dentium* Bd1 genome sequence reflects its genetic adaptation to the human oral cavity. *PLoS Genet*. 2009;5(12):e1000785.
42. Turroni F, Bottacini F, Foroni E, et al. Genome analysis of *Bifidobacterium bifidum* PRL2010 reveals metabolic pathways for host-derived glycan foraging. *Proc Natl Acad Sci USA*. 2010;107(45):19514–19519.
43. O'Connell Motherway M, Zomer A, Leahy SC, et al. Functional genome analysis of *Bifidobacterium breve* UCC2003 reveals type IVb tight adherence (Tad) pili as an essential and conserved host-colonization factor. *Proc Natl Acad Sci USA*. 2011;108(27):11217–11222.
44. Barrangou R, Briczinski EP, Traeger LL, et al. Comparison of the complete genome sequences of *Bifidobacterium animalis* subsp. *lactis* DSM 10140 and Bl–04. *J Bacteriol*. 2009;191(13):4144–4151.
45. Bottacini F, Dal Bello F, Turroni F, et al. Complete genome sequence of *Bifidobacterium animalis* subsp. *lactis* BLC1. *J Bacteriol*. 2011;193 (22):6387–6388.
46. Ventura M, O'Flaherty S, Claesson MJ, et al. Genome-scale analyses of health-promoting bacteria: probiogenomics. *Nat Rev Microbiol*. 2009;7(1):61–71.
47. Ruas-Madiedo P, Gueimonde M, Fernandez-Garcia M, de los Reyes-Gavilan CG, Margolles A. Mucin degradation by *Bifidobacterium* strains isolated from the human intestinal microbiota. *Appl Environ Microbiol*. 2008;74(6):1936–1940.
48. Collado MC, Derrien M, Isolauri E, de Vos WM, Salminen S. Intestinal integrity and *Akkermansia muciniphila*, a mucin–degrading member of the intestinal microbiota present in infants, adults, and the elderly. *Appl Environ Microbiol*. 2007;73(23):7767–7770.
49. Derrien M, van Passel MW, van de Bovenkamp JH, Schipper RG, de Vos WM, Dekker J. Mucin-bacterial interactions in the human oral cavity and digestive tract. *Gut Microbes*. 2010;1(4):254–268.
50. Derrien M, Collado MC, Ben-Amor K, Salminen S, de Vos WM. The Mucin degrader *Akkermansia muciniphila* is an abundant resident of the human intestinal tract. *Appl Environ Microbiol*. 2008;74(5):1646–1648.
51. Wada J, Ando T, Kiyohara M, et al. *Bifidobacterium bifidum* lacto-N-biosidase, a critical enzyme for the degradation of human milk oligosaccharides with a type 1 structure. *Appl Environ Microbiol*. 2008;74(13):3996–4004.
52. Ton-That H, Marraffini LA, Schneewind O. Sortases and pilin elements involved in pilus assembly of *Corynebacterium diphtheriae*. *Mol Microbiol*. 2004;53(1):251–261.
53. Bottacini F, Medini D, Pavesi A, et al. Comparative genomics of the genus *Bifidobacterium*. *Microbiology*. 2010;156(Pt 11):3243–3254.
54. Clewley JP. Genomotyping: comparative bacterial genomics using arrays. *Commun Dis Public Health*. 2002;5(3):258–259.
55. LoCascio RG, Desai P, Sela DA, Weimer B, Mills DA. Broad conservation of milk utilization genes in *Bifidobacterium longum* subsp. *infantis* as revealed by comparative genomic hybridization. *Appl Environ Microbiol*. 2010;76(22):7373–7381.
56. Duranti S, Turroni F, Milani C, et al. Exploration of the genomic diversity and core genome of the *Bifidobacterium adolescentis* phylogenetic group by means of a polyphasic approach. *Appl Environ Microbiol*. 2013;79(1):336–346.
57. Ventura M, Turroni F, Lima-Mendez G, et al. Comparative analyses of prophage-like elements present in bifidobacterial genomes. *Appl Environ Microbiol*. 2009;75(21):6929–6936.
58. Ventura M, Turroni F, Foroni E, et al. Analyses of bifidobacterial prophage-like sequences. *Antonie Van Leeuwenhoek*. 2010;98(1):39–50.
59. Pizarro-Cerda J, Cossart P. Bacterial adhesion and entry into host cells. *Cell*. 2006;124(4):715–727.
60. Perez PF, Minnaard Y, Disalvo EA, De Antoni GL. Surface properties of bifidobacterial strains of human origin. *Appl Environ Microbiol*. 1998;64(1):21–26.
61. Fanning S, Hall LJ, Cronin M, et al. Bifidobacterial surface-exopolysaccharide facilitates commensal–host interaction through immune modulation and pathogen protection. *Proc Natl Acad Sci USA*. 2012;109(6):2108–2113.
62. Ton-That H, Schneewind O. Assembly of pili in Gram-positive bacteria. *Trends Microbiol*. 2004;12(5):228–234.
63. Mishra A, Wu C, Yang J, Cisar JO, Das A, Ton-That H. The *Actinomyces oris* type 2 fimbrial shaft FimA mediates co-aggregation with oral streptococci, adherence to red blood cells and biofilm development. *Mol Microbiol*. 2010.
64. Telford JL, Barocchi MA, Margarit I, Rappuoli R, Grandi G. Pili in gram-positive pathogens. *Nat Rev Microbiol*. 2006;4(7):509–519.
65. Mandlik A, Swierczynski A, Das A, Ton-That H. Pili in Gram-positive bacteria: assembly, involvement in colonization and biofilm development. *Trends Microbiol*. 2008;16(1):33–40.
66. Yeung MK. *Actinomyces:surface macromolecules and bacteria–host interactions. Gram-positive pathogens*. Washington DC: American Society for Microbiology; 2000.
67. Kankainen M, Paulin L, Tynkkynen S, et al. Comparative genomic analysis of *Lactobacillus rhamnosus* GG reveals pili containing a human-mucus binding protein. *Proc Natl Acad Sci USA*. 2009;106(40):17193–17198.
68. von Ossowski I, Reunanen J, Satokari R, et al. Mucosal adhesion properties of the probiotic *Lactobacillus rhamnosus* GG SpaCBA and SpaFED pilin subunits. *Appl Environ Microbiol*. 2010;76(7):2049–2057.
69. Mandlik A, Swierczynski A, Das A, Ton-That H. *Corynebacterium diphtheriae* employs specific minor pilins to target human pharyngeal epithelial cells. *Mol Microbiol*. 2007;64(1):111–124.
70. Ventura M, Turroni F, Motherway MO, MacSharry J, van Sinderen D. Host–microbe interactions that facilitate gut colonization by commensal bifidobacteria. *Trends Microbiol*. 2012;20(10):467–476.
71. Ivanov D, Emonet C, Foata F, et al. A serpin from the gut bacterium *Bifidobacterium longum* inhibits eukaryotic elastase-like serine proteases. *J Biol Chem*. 2006;281(25):17246–17252.
72. Turroni F, Foroni E, O'Connell Motherway M, et al. Characterization of the serpin-encoding gene of *Bifidobacterium breve* 210B. *Appl Environ Microbiol*. 2010;76(10):3206–3219.

73. Klaenhammer TR. Genetics of bacteriocins produced by lactic acid bacteria. *FEMS Microbiol Rev*. 1993;12(1–3):39–85.
74. Corr SC, Li Y, Riedel CU, O'Toole PW, Hill C, Gahan CG. Bacteriocin production as a mechanism for the antiinfective activity of *Lactobacillus salivarius* UCC118. *Proc Natl Acad Sci USA*. 2007;104(18):7617–7621.
75. Yildirim Z, Winters DK, Johnson MG. Purification, amino acid sequence and mode of action of bifidocin B produced by *Bifidobacterium bifidum* NCFB 1454. *J Appl Microbiol*. 1999;86(1):45–54.
76. Yildirim Z, Johnson MG. Characterization and antimicrobial spectrum of bifidocin B, a bacteriocin produced by *Bifidobacterium bifidum* NCFB 1454. *J Food Prot*. 1998;61(1):47–51.
77. Mathys S, Meile L, Lacroix C. Co-cultivation of a bacteriocin-producing mixed culture of *Bifidobacterium thermophilum* RBL67 and *Pediococcus acidilactici* UVA1 isolated from baby faeces. *J Appl Microbiol*. 2009;107(1):36–46.
78. Lee JH, Li X, O'Sullivan DJ. Transcription analysis of a lantibiotic gene cluster from *Bifidobacterium longum* DJO10A. *Appl Environ Microbiol*. 2011;77(17):5879–5887.

CHAPTER

5

Shaping the Human Microbiome with Prebiotic Foods – Current Perspectives for Continued Development*

Kieran M. Tuohy[†,‡], *Duncan T. Brown*[‡], *Annett Klinder*[‡], *Adele Costabile*[‡] *and Francesca Fava*[†]

[†]**Department of Food Quality and Nutrition, Research and Innovation Centre, Fondazione Edmund Mach, San Michele all'Adige, Trento, Italy** [‡]**Department of Food and Nutritional Sciences, School of Chemistry, Food and Pharmacy, The University of Reading, Reading, UK**

INTRODUCTION

The human intestinal tract is colonized by a complex, metabolically diverse and ecologically dynamic community of microorganisms, dubbed the intestinal microbiome. It is estimated that between 400 and 1000 different species of microorganism comprise the human gut microbiota with climax populations reaching densities of up to 10^{12} cells per mL colonic contents.[1,2] Comparative 16 S rRNA gene sequencing studies have shown that the majority of the intestinal bacteria come from only a few bacterial phyla; the Firmicutes, Bacteroidetes, Actinobacteria, Proteobacteria, Fusobacteria and Verrucomicrobia.[1–3] This microbiome has long been associated with particular disease states such as gastroenteritis, antibiotic associated diarrhoea, colon cancer and inflammatory bowel disease (IBD).[4,5] However, more recently the gut microbiome has also been shown to play important roles in a number of innate physiological functions, including the entero-hepatic circulation of bile acids,[6] nitrogen/ammonia cycling,[7] maintenance of a tolerogenic immune system,[8,9] bioconversion of plant polyphenolic secondary metabolites[10] and fermentation of non-digestible food components such as dietary fiber.[11,12] Indeed, recent studies employing post-genomics approaches have shown that distinct profiles of microorganisms inhabit the human intestine in health and in disease states such as ulcerative colitis, Crohn's disease,[13,14] colon cancer,[15] obesity and diabetes.[16,17] The fecal microbiota of individuals with these chronic diseases commonly display alterations in species richness, an increased relative abundance of bacterial groups which harbor many pathogens, such as the *Enterobacteriaceae* and certain clostridial groups, and reduced populations of beneficial bacteria, especially the bifidobacteria and often other important saccharolytic or butyrate producing bacteria such as *Faecalibacterium prausnitzii*.[14,18,19] Such disease-associated microbial profiles hint at possible roles for the gut microbiota in the etiology or maintenance of these chronic diseases. The gut microbiota along the intestinal tract is shaped by a number of endogenous and exogenous forces, including host physiology (production of gastric acid and digestive enzymes, transit times), the immune system, and the host's diet. Indeed, interactions between aberrant intestinal microbiota and diet may be responsible for the emergence of modern diseases of affluence, closely linked to Western lifestyles (e.g., obesity, type 2 diabetes, certain cancers) or autoimmune diseases like IBD and allergy.[12,14–16] The proposition that

*This is an update of: "Shaping the human microbiome with prebiotic foods – current perspectives for continued development." *Food Science and Technology Bulletin* 2010; 7(4): 49–64. Available from: http://dx.doi.org/10.1616/1476-2137.15989 handle: http://hdl.handle.net/10449/19776. Re-published with the permission of International Food Information Service (IFIS Publishing).

Diet-Microbe Interactions in the Gut.
DOI: http://dx.doi.org/10.1016/B978-0-12-407825-3.00005-8

bifidobacteria in particular are health promoting, stems from the fact that the gut microbiota of breast-fed infants is dominated by bifidobacteria, and that breast feeding is seen as the optimal nutrition for human infants impacting on health not only in infancy but throughout life.[20–22] Feeding infants *Bifidobacterium lactis* Bb12 has been shown to improve body weight, reduce fecal pH, increase fecal acetate and lactate, and improve immune biomarkers (calprotectin and fecal IgA concentrations) compared to placebo treatment.[23] Moreover, evidence from *in vitro* and animal studies has shown how these bacteria interact with mammalian cells to affect particular physiological processes. Bifidobacteria produce biologically active compounds, including folate and conjugated linoleic acid which are seen as protective against coronary vascular disease (CVD). Bifidobacteria have been shown to produce both these molecules *in vivo*, contributing to systemic folate levels and to improved lipid profiles (elevated polyunsaturated fatty acids) in host tissues leading to reduced inflammatory molecule production by adipocytes.[24–26] Bifidobacteria also produce a range of short-chain fatty acids (SCFA) upon carbohydrate fermentation in the colon, including acetate, lactate and formate, which are discussed in more detail below. Certain bifidobacteria have also been shown to produce soluble factors which have direct immune modulatory ability, particularly playing a role in regulating maturation of mucosal dendritic cells, contributing to immune homeostasis in the gut.[27] They have also been shown to impact on mucosal architecture, increasing transepithelial electrical resistance and inducing tight junction proteins, key components of improved mucosal barrier function.[28–31] In addition, different bifidobacterial strains appear to possess strain-dependent abilities to inhibit important gastrointestinal pathogens and mollify the inflammatory response to other bacteria, again contributing to stable tolerogenic relations between the intestinal immune system and the diverse collection of microorganisms which comprise the intestinal microbiome.[32,33] Reflecting these health-promoting activities, the PASSCLAIM expert panel on gut health and immunity concluded that "a healthy, or balanced, flora (microbiota) is, therefore, one that is predominantly saccharolytic and comprises significant numbers of bifidobacteria and lactobacilli."[34] Morever, the European Food Safety Authority may also recognize a reduction in harmful bacteria in the gut as potentially beneficial to human health (http://www.efsa.europa.eu/en/scdocs/doc/1233.pdf).

LINKING MICROBIOME STRUCTURE AND FUNCTION

With the exception of recognized pathogenic strains and certain groups of beneficial bacteria like the bifidobacteria and lactobacilli, we know very little about the ecology of the vast majority of intestinal bacteria let alone their impact on human health and disease. Therefore, only monitoring changes in bacterial populations within the gut microbiota by itself gives limited insight into their biochemical and ecological roles.[12] Recently, the power of high throughput DNA sequencing has shed new light on the ecological/physiological function of the mammalian intestinal microbiome. Gill et al.[35] characterized the entire genetic make up of the gut microbiota present in fecal samples collected from two healthy individuals and found that roughly one-third of the genes present within the gut microbiota were shared between the two fecal samples. The remainder of the putative genes were unique to each individual, highlighting the individual nature of the gut microbiota present in each human in terms of species present and in the genetic potential of these microorganisms. This also highlights the fact that our unique gut microbiome is likely to interact with its human host in different ways at the individual level. They also found that the gut microbiomes of the two individuals studied were enriched for genes involved in utilization of complex plant polysaccharides, complementing enzymatic capabilities of the human genome, which is, in general, restricted in its ability to break down and derive energy from these major dietary components. They suggested that the major function of the gut microbiome, therefore, was to extract energy from complex plant polysaccharides, many of which are non-digestible by the host and commonly thought of as dietary fiber. Until very recently, human diets included large quantities of these fiber compounds, hinting at a possible link between aberrant gut microbiota composition/activity associated with certain chronic human diseases and modern diets deficient in complex non-digestible plant polysaccharides.[12] SCFA are the main end-products of bacterial fermentation in the colon, with acetate, butyrate and propionate being produced in the largest amounts. These SCFA, the majority of which are absorbed by the host, are involved in a range of different host physiological functions. SCFA produced in the colon are important sources of energy for the cells of the gut wall, with butyrate providing 50% of the daily energy requirements of colonic mucosa. They are also involved in mucosal transport processes, proliferation, differentiation and programed cell death (apoptosis) of mucosal cells,[36] maturation of mucosal immune cells,[37] regulation of cholesterol biosynthesis and lipogenesis in the liver,[36,38] and supply energy for tissues outside the gastrointestinal tract including kidney, heart, brain and mussle tissues.[39,40] They are also involved in regulating expression of gut hormones, some involved in governing intestinal permeability and barrier function, and in controlling satiety, thus impacting on body weight and metabolic disease.[16]

Fermentation of dietary carbohydrates is therefore seen as an important and health-promoting activity of the human gut microbiota, with many of the dominant anaerobic species being involved. Many different bacteria belonging to the *Clostridium letpum* and *C. coccoides* groups, the *Bacteroides* and bifidobacteria appear to be important in SCFA production, either as direct polysaccharide and oligosaccharide degraders, or upon cross-feeding off simpler sugars and fermentation end products produced by other bacteria.[41]

Metabonomics has also recently been applied to studies examining ecological functions of the gut microbiome. Metabonomics is defined as "the quantitive measurement of the multiparametric metabolic response of living systems to pathophysiological stimuli or genetic modification"[42] and utilizes analytical chemistry techniques to assay for the thousands of low-molecular-weight metabolites typically present in biofluids such as urine, blood and fecal water. Most commonly metabonomics employs proton nuclear magnetic resonance (^{1}H-NMR), liquid chromatography coupled mass spectrometry (LC-MS), gas chromatography coupled mass spectrometry (GC-MS), and variants such as high-performance liquid chromatography coupled mass spectrometry (HPLC-MS) and two-dimensional gas chromatography mass spectrometry (GC × GC-MS) to generate profiles of metabolites and monitor how these profiles change in response to external stimuli such as drugs, diet or disease.[43] Rather than attempt to identify every metabolite signal recorded using these analytical techniques, metabonomics uses a range of bioinformatic and chemometric methods to compare metabolite profiles,[42] identifying differential signals corresponding to metabolites which change in response to test stimuli. This approach to studying metabolism is relatively new, and to date, few studies have been carried out involving the gut microbiome. However, it is becoming clear that many of the metabolites which differentiate individual responses to dietary components, drugs and disease are derived from gut microbiotal activities.[44] Holmes et al.[45] used metabonomics to profile metabolites in urine from 4630 individuals across China, Japan, the USA and UK, and were able to identify discriminatory metabolites in 24-h urine samples. The main differentiating metabolites were of dietary origin and many derived from microbiotal and host co-metabolism of dietary components, while epidemiologically these populations exhibit contrasting diets and rates of stroke and coronary heart disease. Formate was inversely related to blood pressure among participants, while other discriminatory metabolites, hippurate and methylamines, may also be derived from intestinal microbial activity. Li et al.[46] analyzed the urinary NMR profile and compared it to microbiota profiles generated by denaturing gradient gel electrophoresis (DGGE), a culture-independent molecular microbiology profiling method in four generations of a single Chinese family. They were able to link particular bacterial species present within the gut microbiota with unique urinary metabolites. They found that well-known gut microbiotal metabolites – including dimethylamine, a product of microbial catabolism of choline, were statistically linked with the presence of *Faecalibacterium prausnitzii*. The combination of metagenomics or culture-independent microbiology techniques and metabonomics is proving highly effective in linking microbiome structure to ecological function, and may be a useful tool for proving cause and effect of functional foods targeting the human gut microbiota. Recent advances in metagenomics and metabonomics, and how they are revolutionizing our understanding of diet: microbe interactions within the gut are discussed in more detail in other chapters of this edition (Chapters 1,2).

PROBIOTICS

Various methods have been investigated to increase the relative abundance of microorganisms seen as beneficial within the human gastrointestinal tract. Metchnikov was one of the first to propose that ingestion of milk fermented with *Lactobaccillus* species could improve host health. This idea later developed into the probiotic concept. Probiotics are "live microorganisms which when administered in adequate amount confer a health benefit on the host" (FAO).[47] Species commonly used as probiotics must survive passage through the stomach and small intestine, and exert some beneficial effect on host health, and commonly include species of *Lactobacillus* and *Bifidobacterium*. Incidents of human infection by both *Bifidobacterium* species and *Lactobacillus* species are very rare and after many years of use in the dairy industry they have achieved GRAS (Generally Regarded As Safe) status. Studies on the ability of probiotics to protect or treat certain diseases in humans have not always shown positive results, however, but there is growing acceptance for the use of probiotics in lactose-intolerant individuals, and those undergoing antibiotic therapy.[48] Findings of definitive reductions in pre-cancerous lesions in experimental animals,[49] and reduced fecal water genotoxicity and DNA damage in humans[50] suggest that probiotics may also have potential in reducing the risk of colon cancer as part of a long-term lifestyle strategy. Meanwhile other studies have proved less successful, causing only minor effects of short duration.[51,52] An inherent limitation of the probiotic approach is that many probiotic traits appear to be strain specific. For example, different species of *Lactobacillus* and *Bifidobacterium* induce human immune cells to produce very different cytokine profiles.[53,54] The ability of bifidobacteria and lactobacilli to inhibit intestinal pathogens, including *Clostridium difficile*, pathogenic *Escherichia coli* strains and *Campylobacter jejuni* has also been shown to be strain specific, with different strains of the same bacterial species showing very different anti-pathogen activities.[32] Bacteria often display a high degree of genetic plasticity, acquiring and losing particular genetic traits in order to increase their fitness to colonize particular ecological niches. *Bifidobacterium longum*, an important commensal microorganism and candidate probiotic strain, has recently been shown to lose genetic determinants for traits useful in colonizing the intestinal tract (e.g., lantibiotic encoding region responsible for inhibition of *Clostridium difficile*) upon repeated passage on growth media in pure culture.[55] This illustrates the importance of identifying probiotic traits at the genetic level and of following the genetic stability of these traits during probiotic testing and production. The strain-specific nature of probiotics is further illustrated by the various faces of *Escherichia coli*. Most humans harbor stable populations of *E. coli* within their intestinal tract in the absence of any disease symptomology, despite the fact that many *E. coli* strains cause gastroenteritis. Moreover, another strain of *E. coli*, the Nissle 1917 strain, has a long and successful history of use as a probiotic supplement indicating that commensal, pathogenic and probiotic strains of the same bacterial species exist and that differences in probiotic or health-promoting traits as well as virulence factors occur at the strain level.[56] Another limitation to the probiotic approach is that studies are often carried out on ill-defined target populations or disease syndromes where more than one pathological mechanism may be at work. A good example is Traveller's diarrhoea, which can be mediated by many different pathogenic agents, bacterial as well as viral, and it may be unrealistic to expect a single probiotic strain to inhibit such a wide range of gastrointestinal pathogens. Similarly, our immune system recognizes bacteria encountered at mucosal surfaces using pattern recognition receptors (PRR) such as toll-like receptors (TLRs) and nucleotide binding and oligomerization domain (NOD) proteins. These molecules carried on antigen-presenting and epithelial cells recognize particular microbial-associated molecular patterns on and in microorganisms, and govern how our immune systems differentiate friend from foe, directing the appropriate immune response, tolerance or inflammation. However, mutations in these PRR are common, with certain mutations leading to increased risk of diseases such as IBD. Mutations in NOD2 and TLR2 have recently been shown to alter cytokine profiles produced by immune cells in response to bacteria, including lactobacilli, bifidobacteria and the pathogen *Salmonella* Typhimurium.[57,58] A final and apparently critical limitation of probiotic studies to date is the inability to prove convincingly a cause and effect relationship between probiotic ingestion and various health effects in healthy human volunteers. Changes in health biomarkers such as various gastrointestinal symptoms (e.g,. abdominal discomfort, pain, bloating, etc.) or immune function, are often highly subjective, show large variability between apparently healthy subjects or are not always accepted as appropriate biomarkers of particular physiological processes or health parameters by regulatory authorities. Therefore it is not surprising that not all human intervention studies with particular probiotic strains have been successful, given the strain-specific nature of probiotics, the multi-factorial nature of the diseases, and heterogeneity in immune responses to different probiotics. Indeed, these considerations are apparently reflected in the fact that no single probiotic food product for human use has been awarded a general health claim by the European Food Standards Agency under article 13 or 14 of the Regulation (EC) No 1924/2006 on nutrition and health claims made on foods. Similarly, no health claims have been forthcoming for probiotics from the Food and Drug Administration in the United States. On the other hand, considering the wealth of animal data, long history of use and potential benefit in the fight against chronic disease, there is an onus on such regulatory bodies to provide expert consensus positions on the strength of various physiological biomarkers considering the recent growth in awareness of the importance of the gut in maintaining human health. This is particularly true in Europe, where the EU has funded much of the supportive science and expert position statements such as the PASSCLAIM documents.[34]

PREBIOTICS

An alternative or complementary approach used to modulate the human gut microbiota towards one with increased levels of beneficial bacteria is the prebiotic approach. Prebiotics are dietary fibers and usually non-digestible plant polysaccharides or non-digestible oligosaccharides. The most recent definition of prebiotics, states that, "a dietary prebiotic is a selectively fermented ingredient that results in specific changes, in the composition and/or activity of the gastrointestinal microbiota, thus conferring benefit(s) upon host health."[59] This definition and it's previous versions, imply that prebiotics mediate their health-promoting activities upon fermentation by the gut microbiota, necessitating that prebiotics must reach the colon and in effect be non-digestible in the upper gut. A defining characteristic of prebiotics, as opposed to dietary fibers, as defined recently by the European Commission (see Table 5.1), is that they have a very particular effect on the gut microbiota, increasing numbers or activities of beneficial microorganisms, and that it is through this microbiome modulation that improvements in human health are mediated. Of the putative beneficial microorganisms within the gut microbiota, only the lactobacilli and bifidobacteria have been studied in much detail in terms of their health-promoting activities. In healthy humans, bifidobacteria usually outnumber lactobacilli by a factor of about 100 and appear to be more efficient utilizers of prebiotic oligosaccharides. Figure 5.1 describes some of the putative health effects ascribed to prebiotics. The vast majority of the scientific support for these health effects still comes from animal studies, where mechanisms of effect have also been identified, most linked either to the activities of SCFA produced upon prebiotic fermentation or to direct immunological activities resulting from microbiota modulation.[61–63] However, notable examples of human intervention studies illustrate that dietary supplementation with prebiotics does have the potential to improve host health, for example, improving calcium absorption in adolescents,[64,65] reducing fecal genotoxicity in patients at risk of colon cancer,[50] improving blood lipid profiles in patients with dyslipidemia[66–68] and potentially reducing energy intake and other contributors to the metabolic syndrome in overweight individuals

TABLE 5.1 Current Definitions of Prebiotics and Dietary Fibers

Year	Definition	Reference
2009	A dietary prebiotic is a selectively fermented ingredient that results in specific changes, in the composition and/or activity of the gastrointestinal microbiota, thus conferring benefit(s) upon host health.	Gibson et al.[59]
2008	Fiber definition: for the purposes of this Directive "fiber" means carbohydrate polymers with three or more monomeric units, which are neither digested nor absorbed in the human small intestine and belong to the following categories: • edible carbohydrate polymers naturally occurring in the food as consumed; • edible carbohydrate polymers which have been obtained from food raw material by physical, enzymatic or chemical means and which have a beneficial physiological effect demonstrated by generally accepted scientific evidence; • edible synthetic carbohydrate polymers which have a beneficial physiological effect demonstrated by generally accepted scientific evidence.	COMMISSION DIRECTIVE 2008/100/EC of 28 October 2008[60]

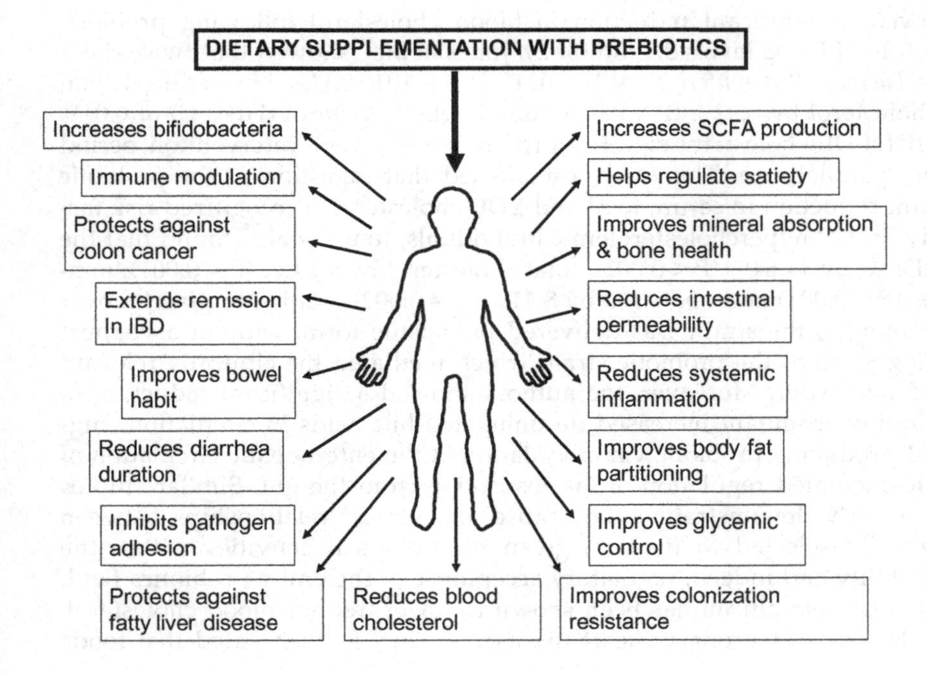

FIGURE 5.1 Health effects associated with dietary supplementation with prebiotic functional foods.

through improved satiety control.[69–71] Despite these intervention studies, human data is still scarce and in particular, there are few studies linking microbiota modulation with the observed physiological benefits; a key component in bridging cause and effect and a critical step in providing health claims quality data. Since their emergence in the early 1990s the prebiotic functional food market has grown into a multi-million Euro industry, worth an estimated Euro 295M in 2008 and expected to grow to Euro 676M by 2015 (http://www.frost.com/prod/servlet/press-release.pag?docid = 168776171). This great commercial and emerging socioeconomic impact of prebiotics in Europe largely grew out of enlightened public–private co-operations under EU research initiatives Frameworks 5 and 6, such as, *SYNCAN* – Synbiotics and cancer prevention in humans (760064) and the ProEuHealth cluster of projects (http://proeuhealth.vtt.fi/). However, further studies, particularly in humans measuring microbiota modulation and recognized biomarkers of physiological effects are critically required to support the field. This is evident from the fact that, in common with probiotics, prebiotics have not achieved health claims status in the EU or US. This situation still holds in 2014.

The fact that no probiotics or prebiotic products have so far attained health claims status in the EU or US has come as a surprise to much of the food industry. However, the challenge of meeting requirements of regulators such as the EFSA and USFDA may herald a paradigm shift in the functional foods market.[72] To date, most added value to functional food products has come from marketing, with product success largely based on marketing creativity or budgets.[72] The future of probiotics and prebiotics lies in precise product characterization (e.g., correct bacterial characterization to strain level) and in providing the scientific support for health claims. It is interesting to note that as many as 30 different probiotic feeds for production animal use have been approved by EU authorities. This is despite similar requirements for full product characterization to human probiotics and the fact that three *in vivo* studies in the target animal are usually included in evaluation portfolios (EFSA has stated that a single human study showing a statistically significant effect may be adequate for human probiotic demonstration of efficacy).[72] This shift in industry away from marketing and towards applied research, will place probiotics and prebiotics on sound scientific footing and, together with a strong regulatory authority to enforce a level playing field and protect consumers from ill-defined products or exaggerated health claims, will ensure the emergence of efficacious products with real potential to improve human health at the population level. Industry should take encouragement from the fact that similar approaches have worked in the past, with probiotic feeds for animals, where product success is much more determined by demonstrable efficacy – farmers make decisions on supplements based on immediate impact on production quality and yields, while the end users themselves are less likely to be swayed by creative marketing. Similarly, products likely to share some mechanisms of effect with prebiotics, e.g., soluble fibers and whole-grain cereals, have well-established and specific health claims for heart health from the FDA in the United States and recently, EFSA.[73]

In 2012, two well-powered human dietary interventions with a probiotic microorganism capable of hydrolyzing bile acids, were conducted to establish the efficacy of this strain to lower cholesterol levels in the hypercholesterolemic. Both studies showed a significant reduction in blood cholesterol following probiotic ingestion. In 114 hypercholesterolemic adults (LDL-C levels 3.4 mmol/L), Jones et al.[74] showed that twice-daily doses of bile salt hydrolase (BSH)-active *Lactobacillus reuteri* NCIMB 30242 at 5×1010 CFU/day, reduced total cholesterol by 4·81% ($P = 0.031$), LDL-cholesterol by 8.92% ($P = 0.016$), non-HDL-cholesterol (HDL-C) of 6.01% ($P = 0.029$) and improved the ratio of HDL:LDL-cholesterol significantly, over a six-week intervention period compared to the placebo (yoghurt alone) parallel arm. This study established that ingestion of the probiotic containing yoghurt resulted in a significant reduction in serum total and LDL-cholesterol, a recognized risk factor for CVD. Moreover, in a second study, in 127 hypercholesterolemic individuals, Jones et al.[75] found that the same BSH active strain reduced blood LDL-C by 11.64% ($P < 0.001$), total cholesterol by 9.14%, ($P < 0.001$), non-HDL-cholesterol (non-HDL-C) by 11.30% ($P < 0.001$) and apoB-100 by 8.41% ($P = 0.002$) relative to placebo over a 9-week intervention period. This time though, the strain was delivered in capsule form, without a yoghurt vehicle, clearly establishing that it was ingestion of the probiotic strain which mediated the clinically relevant reduction in serum cholesterol, not the food carrier. Moreover, the authors showed a significant reduction in absorption of sterols from the intestine and concomitant increased deconjugated bile acids in circulation, suggesting a mechanism of action of the BSH-producing probiotic via modulation of the enterhepatic circulation of bile acids and farnesoid X receptor (FXR)-mediated regulation of fat absorption from the gut. Similar studies along the same vein with other strains clearly demonstrating the "cause and effect" relationships between ingestion of particular probiotics specifically selected to mediate given physiological activities within the human body will no doubt boost the credibility and indeed, regulatory acceptance of the entire probiotics field. Similarly, in 2010, EFSA recognized that "Oat beta-glucan has been shown to lower/reduce blood cholesterol. Blood cholesterol lowering may reduce the risk of (coronary) heart disease."[73] They further stated that foods should provide at least 3 g oat beta-glucan per day to bear the claim. Interestingly, we and others have shown that beta-glucan mediates a similar modulation of the gut microbiota to inulin in laboratory animals[76] and have recently also shown that oat porridge (100% oats) mediates both a prebiotic modulation of the gut microbiota and a significant reduction in blood cholesterol in hyperlipidemic individuals.[77]

TESTING PREBIOTICS

A wide range of complex, non-digestible carbohydrates are currently being commercialized for a range of functional properties and finding applications as food stabilizers, fat and sugar substitutes and texture agents/hydrocolloids.[78,79] There is also intense industrial interest in establishing the potential health-promoting activities of these non-digestible carbohydrates. All these compounds with a degree of polymerization greater than 3 are likely to fulfill the criteria laid out under the current definition of dietary fiber and are commercialized on this basis (see Table 5.1). However, prebiotics, are recognized by industry and consumers alike as a class of dietary fibers particularly associated with improved health and the prebiotic label adds value to traditional dietary fibers. Therefore, testing the prebiotic potential of different dietary fibers is of considerable commercial and scientific interest. The main criterion which separates prebiotics from other dietary fibers is their potential to selectively modulate the human gut microbiota.[80] This modulation, usually seen as a relative increase in numbers of bifidobacteria within the gut microbiota, is based on the ability of bifidobacteria to outcompete other bacteria present within the gut microbiota and is not based solely on the ability of an organism to grow on the test prebiotic in pure culture. Bifidobacteria appear to be specialized oligosaccharide degrading bacteria, possessing a range of glycolytic enzymes and carbohydrate uptake channels with high affinity for prebiotic oligosaccharides and their composite sugars.[81–83] Putative prebiotics, therefore, must first be tested for their ability to selectively stimulate relative population levels of bifidobacteria within the gut microbiota.[84] This involves the use of mixed culture microbiology techniques where the selective enrichment of bifidobacteria over other microorganisms present can be assessed using a test prebiotic as sole growth substrate. Several *in vitro* culture systems mimicking the colonic microbiota have been developed and used to test the microbiota modulatory abilities of different non-digestible carbohydrates.[85] Batch culture systems, where test carbohydrates can be fermented using fecal inocula under anaerobic and pH-controlled conditions offer an initial screening process to establish the bifidogenic nature of putative prebiotics. Recently, such systems have been employed to establish the ability of Agave fructans, bifidobacterial extracellular polysaccharide, cereal fractions, glucomannans, pecticoligosaccharides, and galactooligosaccharides to selectively stimulate the growth of bifidobacteria.[86–92] However, such observations need confirmation in human feeding studies in order to confirm the microbiota modulatory capabilities of putative prebiotics. Table 5.2 was compiled to illustrate both the diversity of carbohydrates which have shown prebiotic potential *in vivo* and the fact that the well-established prebiotics, inulin oligofructose/fructooligosacchatides, galactooligosaccharides and lactulose, have repeatedly been proven to increase relative numbers of fecal bifidobacteria, using both traditional culture-based microbiology methods and direct culture-independent microbial enumeration (using FISH or qPCR). With the reduced cost and increased accessibility of next-generation sequencing over the last few years, our ability to study the compositon and ecological flux of the human gut microbiota has much improved. However, few studies to date have employed high-throughput sequencing to track changes in the relative abundance of gut bacteria in response to prebiotics, but those which have do indeed support the bifidogenic potential of prebiotics and their dose response in terms of Actinobacterial and bifidobacterial relative abundance.[135,148,149] Advances in molecular microbiology have also brought other gut commensal microorganisms to the fore in terms of importance both within the gut and possibly in terms of host health. This is particularly true of butyrate-producing bacteria which constitute some of the most dominant and prevalent species within the human gut microbiota.[150] The fructans, due both to the number of studies showing a significant increase in fecal bifidobacteria and the weight of evidence supporting their ability to improve host health warrant particular attention.

The most widely studied prebiotics are fructans; more specifically inulin, inulin-derived oligofructose and fructooligoasccharides synthesized from sucrose. Prebiotic fructans are usually polymers of β(2-1)-bound D-fructose terminating in an α(1-2)-linked D-glucose monomer, with a degree of polymerization (DP) varying from 2 to 60. Inulin has a DP of greater than 20 and is found in more than 35,000 plant species and a wide variety of foods[151,152] while oligofructose (DP between 2 and 20) is derived from longer chains by enzymatic hydrolysis or synthesized from sucrose (fructooligosaccharides, FOS). Due to their wide availability in natural plant foods

TABLE 5.2 Human Dietary Interventions with Prebotic Functional Foods

Prebiotic	Microbiological Methods	Dose	Design	Results	References
Inulin	Culture	15 g/day 15 days	Placebo-controlled, cross-over study	Bifidobacteria↑ Gram-positive cocci↓	Gibson et al.[93]
Inulin	Culture	20–40 g/day 19 days	Placebo-controlled, cross-over study	Bifidobacteria↑ Enterococci↓ Enterobacteria↓	Kleesen et al.[94]
Inulin	Culture	9 g/day 28 days	Placebo-controlled, cross-over study	Bifidobacteria↑ Total facultative anaerobes↓	Brighenti et al.[95]
Inulin	FISH	Up to 34 g/day 64 days	Placebo-controlled, parallel study	Bifidobacteria↑	Kruse et al.[96]
Inulin (long chain)	FISH	8 g/day 14 days	Placebo-controlled, cross-over study	Bifidobacteria↑	Tuohy et al.[97]
Inulin	FISH	5 g/day and 8 g/day, 14 days	Placebo-controlled, cross-over study	Bifidobacteria↑	Kolida et al.[98]
Jerusalem artichoke inulin (JA) or chicory inulin (CH)	FISH	7.7 g/day 7 days	Placebo-controlled, parallel study	Bifidobacteria↑ Bacteroides↓ Clostridia↓	Kleessen et al.[99]
Inulin	Culture	0.75, 1.00 or 1.25 g/day 35 days	Placebo-controlled, parallel study	Bifidobacteria↑ Clostridia↓ Gram-positive cocci↓	Yap et al.[100]
Inulin	qPCR	10 g/day, 16 days, 12 healthy adults	Treatment compared to no treatment period	Bifidobacteria↑ *Faecalibacterium prausnitzii*↑	Ramirez-Farias et al.[101]
Inulin		~20 g/day 28 days	Placebo-controlled, cross-over	Bifidobacteria↑	Petry et al.[102]
Very long chain inulin (VLCI)	FISH	10 g/day	Double-blind, crossover, placebo controlled, 21 days	VLCI: Bifidobacteria↑, Lactobacilli↑, Atopobium↑& Bacteroides-Prevotella↓	Costabile et al.[103]
FOS	Culture	15 g/day 15 days	Placebo-controlled, cross-over study	Bifidobacteria↑ Bacteroides↓, clostridia↓, fusobacteria↓	Gibson et al.[93]
FOS	Culture	0–20 g/day 7 days	Placebo-controlled, parallel study	Bifidobacteria↑	Bouhnik et al.[104]
FOS	FISH	5.1 g/L enteral feed	Double-blind, randomized controlled (fiber 8.9 g/L enteral feed) study	Bifidobacteria↑, Clostridia↓	Whelan et al.[105]
FOS	Culture	10 g/day, 28 days		Bifidobacteria↑	Yen et al.[106]
scFOS (Actilight™)	Culture	Daily dose of 2.5, 5.0, 7.5 and 10 g/day 7 days	Placebo-controlled, cross-over study	Bifidobacteria↑	Bouhnik et al.[107]
FOS + PHGG	FISH	6.6 g/day FOS 3.4 g/day PHGG 21 days	Placebo-controlled, cross-over study	Bifidobacteria↑	Tuohy et al.[108]
FOS + GOS	Culture	4 g/L FOS 8 g/L GOS 28 days	Placebo-controlled, parallel study	Lactobacilli↑Bifidobacteria↑	Moro et al.[109]
FOS + GOS	Culture	10 g/L 28 days	Placebo-controlled, parallel study	Bifidobacteria↑	Boehm et al.[110]
FOS + GOS	Culture	8 g/L 42 days	Placebo-controlled, parallel study	Percentage of bifidobacteria↑	Knol et al.[111]

(*Continued*)

TABLE 5.2 (Continued)

Prebiotic	Microbiological Methods	Dose	Design	Results	References
Synergy1 inulin; GOS:FOS; Standard infant formula	FISH	Synergy 1 (0.4 g/dL or 0.8 g/dL), GOS:FOS (90:10) 0.8 g/dL	Randomized, controlled, parallel study	Bifidobacteria ↑ in all prebiotic formula, increased with dose	Veereman-Wauters et al.[112]
Mixed scGOS, lcFOS, +/− pectic-acidic oligosaccharides	FISH (flow cytometry)		Double-blind, placebo controlled, parallel study	Bifidobacteria ↑ *Bacteroides* ↓ *Clostridium coccoides* ↓	Magne et al.[113]
GOS (Transgalacto-OS, TOS)	Culture	0–10 g/d 56 days	Placebo-controlled, parallel study	Bifidobacteria ↑ Lactobacilli ↑	Ito et al.[114]
GOS (Transgalacto-OS, TOS)	Culture	2.5 g/d 21 days	Feeding study	Bifidobacteria ↑	Ito et al.[115]
GOS (Transgalacto-OS, TOS)	Culture	10 g/day 21 days	Feeding study	Bifidobacteria ↑	Bouhnik et al.[116]
GOS (Transgalacto-OS, TOS)	Culture	8.5 g/day and 14.4 g/day 21 days	Placebo-controlled, parallel study	Bifidobacteria ↔	Alles et al.[117]
B-GOS	FISH	7 g/d 7 days	Placebo-controlled, cross-over study	Bifidobacteria ↑	Depeint et al.[118]
B-GOS	FISH	5.5 g/d 10 week	Placebo-controlled, cross-over study	Bifidobacteria ↑	Vulevic et al.[119]
B-GOS	FISH		Double-blind, placebo controlled, crossover study	Bifidobacteria ↑, Bacteroides ↓, Clostridria ↓	Vulevic et al.[120]
GOS (Transgalacto-OS, TOS)	FISH	3.5 g/d and 7 g/d 12-week	Placebo-controlled, parallel study	Bifidobacteria ↑	Silk et al.[121]
Lactulose	Culture	3 g/d 14 days	Feeding study	Bifidobacteria ↑ Lecithinase-positive clostridia ↓	Terada et al.[122]
Lactulose	Culture	5 g/L & 10 g/L 21 days	Feeding study	Bifidobacteria ↑ Clostridia ↓	Nagendra et al.[123]
Lactulose	Culture	20 g/d 4 weeks	Placebo	Bifidobacteria ↑ Lactobacilli ↑	Ballongue et al.[124]
Lactulose	FISH + Culture	10 g/d 26 days	Placebo-controlled, parallel study	Bifidobacteria ↑	Tuohy et al.[125]
Lactulose	Culture	10 g/d 6 weeks	Placebo-controlled, parallel study	Bifidobacteria ↑	Bouhnik et al.[126]
IMO	Culture	13.5 g/d 14 days	Feeding study	Bifidobacteria ↑	Kohmoto et al.[127]
IMO	Culture	5–20 g/d Variable dose 12 days	Feeding study	Bifidobacteria ↑	Kaneko et al.[128]
IMO	Culture	15 g/d 7 days	Feeding study	Bifidobacteria ↑ Lactobacilli ↑ *C. perfringens* ↓	Gu et al.[129]
SOS	Culture	3–5 g/day 15 g/day	Placebo-controlled, cross-over study	Bifidobacteria ↑ Clostridia ↓ Bacteroides ↓	Benno et al.[130]
SOS	Culture	10 g/day 21 days	Placebo-controlled, cross-over study	Bifidobacteria ↑ Clostridia ↓	Hayakawa et al.[131]

(*Continued*)

TABLE 5.2 (Continued)

Prebiotic	Microbiological Methods	Dose	Design	Results	References
Raffinose	FISH	2 g/day 4 weeks	Placebo-controlled, cross-over study	Bifidobacteria ↑	Dinoto et al.[132]
Resistant starch	Culture	10 g/day 7 days	Placebo-controlled, parallel study	Bifidobacteria ↑	Bouhnik et al.[133]
Manitol, Manitol + PDX, Manitol + RS, in chocolate	FISH	22.8 g/day to 45.6 g/day over 42 days	Placebo controlled, parallel, double blinded	All 3 test prebiotics Bifidobacteria ↑, PDX also Lactobacilli ↑	Beards et al.[134]
Resistant starch (RS2 and RS4)	454-pyrosequencing with 16 S rRNA tagged primers	~30 g RS	Double-blind, crossover, controlled (native wheat starch)	RS4 (chemically modified resistant starch): Actinobacteria (bifidobacteria) ↑, Bacteroidetes (Parabacteroides distonis) ↑, RS2 (native granular resistant starch), *Ruminococcus bromii* ↑ and *Eubacterale rectale* ↑	Martinez et al.[135]
XOS Inulin + XOS	qPCR	5 g XOS, or inulin (3 g) + XOS (1 g), 28 days	Randomized, placebo controlled, parallel	XOS alone: Bifidobacteria ↑ Inulin + XOS: Bifidobacteria ↑	Lecerf et al.[136]
Acacia gum	Culture	10 g/day and 15 g/day 10 days	Placebo-controlled, parallel study	Bifidobacteria ↑	Cherbut et al.[137]
Red wine (RW), dealcoholized red wine (DRW), or Gin	qPCR	RW, DRW, 272 mL/d, Gin 100 mL/day 20 days	Randomized, crossover, controlled study	RW and DRW: Enterococcus ↑, Bacteroides ↑, Bifidobacterium ↑, Bacteroides uiformis ↑, Eggerthella lenta ↑, Blautia coccoides-eubacterium rectale group ↑ GIN; clostridia ↑	Queipo-Ortuno et al.[138]
Red wine (RW), dealcoholized red wine (DRW), or Gin	qPCR	RW, DRW, 272 mL/d, Gin 100 mL/day 20 days	Randomized, crossover, controlled study	RW: Bifidobacterium ↑, Prevotella ↑	Clemente-Postigo et al.[139]
Wild blueberry powder drink (WBPD)	RT-PCR	25 g WBPD/day, 6 weeks	Randomized, crossover, placebo controlled	WBPD: Bifidobacterium ↑, Lactobacilli ↑ Placebo: Lactobacilli ↑	
Cocoa-derived flavanols (high or low dose)	FISH	494 mg or 23 mg cocoa flavanols/day, 28 days	Randomized, placebo controlled, cross-over study	High dose flavanols: Bifidobacterium ↑, Lactobacilli ↑	Tzounis et al.[140]
Refined psyllium fiber	Culture	7.0 g/day	Feeding study (before/after)	No change in bifidobacteria overall, increase in 6 or 11 subjects with low starting levels	Elli et al.[141]
Whole-grain wheat	FISH	48 g/day 21 days	Placebo-controlled, cross-over study	Bifidobacteria ↑ Lactobacilli ↑	Costabile et al.[142]
Arabinoxylan-oligosaccharides (AXOS)	FISH	2.2 g/day	Double-blind, placebo controlled	Bifidobacteria ↑ & Lactobacilli ↑ in both treatment and placebo AXOS butyrate ↑	Walton et al.[143]
Wheat bran extract containing arabino-xylan-oligosaccharides	FISH	0, 2.2 or 4.8 g/day for 21 days	Double-blind, controlled, crossover	Bifidobacteria ↑ at highest dose AXOS	Maki et al.[144]

(Continued)

TABLE 5.2 (Continued)

Prebiotic	Microbiological Methods	Dose	Design	Results	References
Wheat bran extract containing arabino-xylan-oligosaccharides		0, 3, 10 g/day, 21 days	Double-blind, controlled, crossover	Bifidobacteria ↑ at highest dose	Francois et al.[145]
Gum Arabic	qPCR	Dose response (5–40 g/day), 4 weeks	Parallel, different doses, negative control (water), positive control 10 g/day inulin	Bifidobacteria ↑ Lactobacilli ↑ Bacteroides ↑ At 10 g/day	Calame et al.[146]
Arabinoxylan – oligosaccharides	qPCR	10 g/day for 21 days	20 healthy adults, placebo controlled, cross-over study	Bifidobacteria ↑ (Bifidobacteria ↑ with placebo)	Cloetens et al.[147]

The microbiological methods are listed (culture = traditional culture-based microbiological techniques, FISH = fluorescent in situ hybridization, qPCR = quantitative PCR), together with the dose of prebiotic, the study design and the main microbiological results in terms of fecal microbiota modulation.

they have always been a part of the human diet although today, on average, Europeans consume 2–10 g/day, while in the USA average consumption is only 1–4 g/day.[153] Fructans β(2-1) linkages make them resistant to digestion by human enzymes in the upper gut and so they reach the colon largely unchanged as demonstrated *in vivo* with ileostomy patients.[154,155] Once in the large intestine, fructans promote the growth bifidobacteria in particular, which express a β(2-1)-hydrolyzing β-fructanfuranosidase enzyme, allowing them to make use of the fructose. Fructan-induced bifidogenesis has been demonstrated repeatedly *in vitro* using pH controlled batch cultures,[86,156] in more complicated continuous flow three-stage gut models[157,158]) and *in vivo* in animal models[159–161] and human dietary interventions (see Table 5.2). Roberfroid et al.[162] suggested that a dose of between 4 and 15 g/day of inulin and oligofructose was sufficient to induce a significant rise in fecal bifidobacteria in adult humans.

The final criterion for a prebiotic is that its fermentation in the colon has some beneficial impact on host health. *In vitro* studies using models of the colonic microbiota inoculated with human feces and studies in animals, have shown that fermentation of prebiotic fructans leads to accumulation of acetate and butyrate in intestinal/gut model contents.[163,164] Fermentation of other prebiotics and certain dietary fibers has also been shown to increase propionate production in these systems.[89,165] Small amounts of lactate and succinate can also be observed using *in vitro* models, but *in vivo*, these SCFA are rapidly converted into butyrate and propionate by the gut microbiota. Bifidobacteria and lactobacilli ferment carbohydrates mainly to acetate and lactate, but do not themselves produce butyrate. Recent studies have shown that dominant members of the Firmicutes, *Eubacterium halli*, *Roseburia*, *Faecalibacterium prausnitzii* and *Anaerostipes caccae* are able to cross-feed off acetate and lactate within the colonic milieu converting them into butyrate, providing a mechanism by which prebiotic modulation of acetate-producing bifidobacteria can lead to elevated butyrate concentrations within the gut.[166,167] SCFA have been implicated in many different health effects, and are likely to constitute and important mechanism of action for prebiotics and dietary fiber in general. As discussed earlier, SCFA have been shown to interact at many levels with mammalian physiology. Butyrate has long been known to regulate cellular proliferation, differentiation, apoptosis and gene expression in mammalian intestinal mucosa through its inhibition of histone deacylase, an important epigenetic determinant and a critical player in cellular differentiation and potential anti-cancer mediator.[168,169] More recently, butyrate has been shown to play a role in regulating dendritic cell maturation, an important process in immune tolerogenesis, also through inhibition of histone deacylase.[170] Butyrate also modulates production of inflammatory cytokines in response to bacterial lipopolysaccharide (LPS) and induces neutrophil apoptosis, again contributing to immune tolerance of the commensal microbiota[171] and to fortify mucosal integrity through regulation of tight junction protein assembly thus reducing bacterial translocation.[172] Acetate is recognized as substrate for cholesterol production in the liver, but also to play an important role in regulating satiety through induction of gut hormones which regulate food intake.[173–175] SCFA have also been shown to act as chemoattractants for neutrophils, and in contrast to propionate and butyrate, acetate has been shown to increase reactive oxygen species of neutrophils, increasing their ability to kill invading bacterial pathogens.[171] Propionate too has received much recent attention for its ability to inhibit histone deacylatase and also for its ability to induce gut hormones involved in controlling food intake and in inhibiting hepatic lipogenesis.[176] Propionate has also been shown to increase production of leptin (the obesity hormone) and to reduce inflammatory molecule production by human adipocytes and to play a role in adipogenesis, all important targets for modulating the risk of the metabolic syndrome and cardiovascular disease risk.[177,178] SCFA are ligands for the G-protein-coupled receptors (GPR41, GPR43), important cellular signaling molecules known to regulate gut hormone production, and peroxisome proliferator activated receptors (PPARs), nuclear receptors which act as transcription factors for many important genes, including genes involved in inflammation, antioxidant defenses, G-coupled protein receptors, insulin sensitivity and vascular reactivity, an important regulator of blood pressure. These pathways are recognized targets for the discovery of novel pharmaceuticals for treating cancer, heart disease, obesity and dementia. Although we still know little about the interactions between colonically derived SCFA, individually or in mixtures, and these pathways, initial studies do raise the tantalizing possibility of elucidating the molecular basis of the health benefits of these fermentation end products, and probiotics and prebiotics by extension, seen in animal feeding studies and human epidemiological data linking high-fiber diets with protection from these diseases, especially diseases linked to aberrant metabolic or inflammatory processes.[179]

The ability of prebiotics to protect against or treat a range of human diseases has been tested to varying degrees using animal models and human dietary interventions.[61] Figure 5.1, lists some of the putative health effects, although most still require further placebo-controlled human intervention studies in suitable target populations, measuring accepted biomarkers of disease for confirmation of efficacy. Brighenti[66] recently reviewed the evidence in support of using prebiotic fructans to lower blood cholesterol levels and found that although there were a limited number of studies, the fructans could be shown to significantly reduce triacylglycerides across studies. Dietary intervention with fructans also appeared to be successful in lowering total cholesterol in hyperlipidemic individuals. Possible mechanisms linked to SCFA have been proposed for this improvement in blood lipid profiles, including modulation of acetate:propionate ratios impacting on *de novo* lipogenesis in the liver,[180] or SCFA-mediated alteration in energy intake via upregulation of satiating gut hormones.[175] Prebiotic fructans have also been shown to reduce the chronic systemic inflammation characteristics of obesity and diabetes in animal models of diet-induced metabolic disease through a mechanism involving fortification of the intestinal mucosa, reduced gut permeability and lower translocation of bacterial lipopolysaccharide across the mucosal barrier, lowering systemic markers of inflammation.[181] Animal studies have routinely shown that prebiotic fructans can reduce the number and size of chemically induced aberrant crypt foci (precancerous lesions in the gut wall), intestinal tumors and metastasis of tumor cells.[182] However, studies in humans are rare, with only one study on fructans alone showing reduced cellular proliferation in the rectal crypt of individuals fed FOS.[183] A synbiotic (a functional food comprised of both a probiotic and prebiotic moiety) combination of the fructan Synergy1 (a mixture of short-chain oligofructose and long-chain inulin) and the probiotics *Lactobacillus rhamnosus* GG and *Bifidobacterium lactis* Bb12 have recently been shown to reduce DNA damage in colonic biopsies taken from subjects with colonic polyps or who previously had colon cancer (51). Dietary supplementation with prebiotics has also been shown to improve mineral absorption, especially calcium, impacting on bone health. Griffin et al.[184] found a significant increase in calcium absorption in adolescent girls given FOS and inulin (4 g/day) and a daily supplementation of calcium (1.5 g/day) for 3 weeks. More recently, long-term (one year) fructan intake delivered in a calcium-fortified orange juice drink increased calcium absorption and improved bone mineralization compared to a maltodextrin control delivered in the same fruit drink in adolescents.[185] Although the mechanism by which prebiotic intervention improved mineral absorption and bone health remains to be determined, the same group later showed that prebiotic-induced improvement in calcium absorption derives from improved colonic absorption, that responders to the prebiotic intervention had greater calcium accretion in their skeleton and that the prebiotic intervention group had a lower body mass index and improved body composition compared to the control group after 1 year of intervention[64,65,185]

Recent studies, have expanded the range of substances which have been shown to mediate this type of modulation wihtin the gut microbiota. Costabile et al.[142] found that ingestion of 100% whole-grain wheat can increase both bifidobacteria and lactobacilli within the fecal microbiota of helathy human subjects. The same team later went on to state that whole-grain maize cereal induced a small increase in fecal bifidobacteria again in healthy volunteers. Connolly et al.[77] later showed that a 100% oats breakfast cereal, in a double-blind placebo-controlled feeding study in hypercholesterolemic subjects both increased fecal lactobacilli and bifidobacteria, and reduced blood cholesterol and LDL-cholesterol levels. No other significant alterations within the gut microbiota was observed in these feeding studies. Tzounis et al.[140] showed that high-cocoa flavanol (HCF) chocolate drink (494 mg cocoa flavanols/d) compared to a low-cocoa flavanol (LCF) chocolate flavored drink (23 mg cocoa flavanols/d) significantly increased fecal bifidobacteria and lactobacilli and reduced clostridia, after 4 weeks of daily ingestion, and concomitantly a significant reduction in plasma triacylglycerol ($P < 0.05$) and C-reactive protein ($P < 0.05$) concentrations was observed. Further, Queipo-Ortuño et al.[138] showed that red wine and de-alcoholized red wine may mediate similar prebiotic effects both on the gut microbiota and blood markers of CVD risk. In this case the authors also observed increased numbers of fecal butyrate-producing bacteria and the Bacteroides/Prevotella group of carbohydrate degraders. This effect was not observed with gin, the alcohol control in the study, which instead gave a significant increase in fecal clostridia and a reduction in most other gut bacteria enumerated including bifidobacteria and lactobacilli. Such studies clearly show that a broad range of whole plant foods and plant-derived polyphenols may mediate a prebiotic type modulation of the gut microbiota and, moreover, concomitantly show a clear "cause and effect" relationship between ingestion of these foods and lowering of blood cholesterol. Indeed, this duality of beneficial health effect, may go some way to explaining the well-recognized health effects of whole plant foods or diets rich in whole plant foods, especially against CVD.[186] Further clarity on how these functional activities can be ascribed to given foods and what dose will be needed to allow acceptable product description for regulatory agencies, but they do hold much promise for the ongoing development of the prebiotics field or research and commercial market given their apparent efficacy and the increasing recognition of the important role played by gut bacteria in maintaining host health.

CONCLUSION

Post-genomics studies are providing new insights into the composition and functioning of the human gut microbiota. It is becoming apparent that the collection of microorganisms inhabiting our intestine constitutes a closely co-evolved partner to our own metabolic potential encoded by the human genome.[187] The high relative abundance of bacterial genes involved in carbohydrate utilization and energy metabolism suggests that the colonic microbiota has evolved to recover energy from complex plant polysaccharides, common in traditional human diets, but which by and large are in short supply in modern Western-style diets.[12,35] Fermentation of non-digestible carbohydrates by bacteria in the colon produces SCFA which can be utilized by the host and which have been shown to impact on many human physiological processes.[179] The importance of these microbiome activities is illustrated by the fact that both epidemiological and direct intervention studies are showing that upregulation of carbohydrate fermentation in the colon leads to improvements in a wide range of human disease states. Traditionally, prebiotics have been viewed as a specific group of non-digestible carbohydrates which appear to be particularly effective at up-regulating this native function of the gut microbiota and moreover, in bringing about an increase in the relative abundance of bacteria seen as beneficial for human health, specifically the bifidobacteria. Thus prebiotics can impact on host health both locally within the gut and systemically. Bifidobacteria have been shown to beneficially modulate the immune system (playing a role in mucosal immune tolerance and homeostasis), induce expression of tight junction proteins thereby fortifying the mucosal barrier, inhibit intestinal pathogens and produce biologically active compounds (e.g., folate and conjugated linoleic acid). This combination of increased saccharolytic fermentation and relative abundance of beneficial gut bacteria may thus represent a useful and measurable biomarker of gut health. Indeed, an independent panel of experts commissioned by the EU in 2004 concluded that "a healthy, or balanced, flora (microbiota) is, therefore, one that is predominantly saccharolytic and comprises significant numbers of bifidobacteria and lactobacilli."[34] Recent studies have expanded this view of prebiotics to include certain whole plant foods (especially the cereal grains) and plant polyphenolic compounds. Studies showing both beneficial microbiota modulation and reduction in markers of CVD are suggesting that a prebiotic activity may be one way by which these foods, which have long been associated with protection from chronic disease, may mediate their health effects.[186] There is also increasing evidence from animal studies and some few human interventions, that prebiotic dietary supplementation can bring about improved host health, improving lipid profiles in the blood, reducing systemic inflammation, improving systemic glycemic control, improving bowel habits and inhibiting intestinal pathogens, and enhancing mineral absorption and improving bone strength. Through these processes prebiotics may have the potential to reduce the risk of developing coronary vascular disease, type 2 diabetes, liver disease, intestinal cancer and autoimmune diseases. However, because of the dearth of human studies, particularly those measuring multiple parameters, e.g., blood lipid profiles and biomarkers of prebiotic modulation of the gut microbiota (e.g., bifidobacterial levels, SCFA concentrations, gut hormones, etc.) prebiotics have yet to be fully accepted as alternative or adjunct therapies for the prophylaxis and possibly treatment of chronic human diseases.

References

1. Suau A, Bonnet R, Sutren M, et al. Direct analysis of genes encoding 16S rRNA from complex communities reveals many novel molecular species within the human gut. *Appl Environ Microbiol.* 1999;65:4799–4807.
2. Bäckhed F, Ley RE, Sonnenburg JL, Peterson DA, Gordon JI. Host–bacterial mutualism in the human intestine. *Science.* 2005;25 (307):1915–1920.
3. Wilson KH, Blitchington RB. Human colonic biota studied by ribosomal DNA sequence analysis. *Appl Environ Microbiol.* 1996;62:2273–2278.
4. Neish AS. Microbes in gastrointestinal health and disease. *Gastroenterology.* 2009;136:65–80.
5. Saulnier DM, Kolida S, Gibson GR. Microbiology of the human intestinal tract and approaches for its dietary modulation. *Curr Pharm Des.* 2009;15:1403–1414.
6. Claus SP, Tsang TM, Wang Y, et al. Systemic multicompartmental effects of the gut microbiome on mouse metabolic phenotypes. *Mol Syst Biol.* 2008;4:219.
7. Bergen WG, Wu G. Intestinal nitrogen recycling and utilization in health and disease. *J Nutr.* 2009;139:821–825.
8. Cerutti A, Rescigno M. The biology of intestinal immunoglobulin A responses. *Immunity.* 2008;28:740–750.
9. Abt MC, Artis D. The intestinal microbiota in health and disease: the influence of microbial products on immune cell homeostasis. *Curr Opin Gastroenterol.* 2009;25:496–502.
10. Del Rio D, Costa LG, Lean ME, Crozier A. Polyphenols and health: what compounds are involved? *Nutr Metab Cardiovasc Dis.* 2009;20:1–6.
11. Jacobs DM, Gaudier E, van Duynhoven J, Vaughan EE. Non-digestible food ingredients, colonic microbiota and the impact on gut health and immunity: a role for metabolomics. *Curr Drug Metab.* 2009;10:41–54.
12. Tuohy KM, Gougoulias C, Shen Q, Walton G, Fava F, Ramnani P. Studying the human gut microbiota in the trans-omics era—focus on metagenomics and metabonomics. *Curr Pharm Des.* 2009;15:1415–1427.
13. Martinez C, Antolin M, Santos J, et al. Unstable composition of the fecal microbiota in ulcerative colitis during clinical remission. *Am J Gastroenterol.* 2008;103:643–648.
14. Macfarlane GT, Blackett KL, Nakayama T, Steed H, Macfarlane S. The gut microbiota in inflammatory bowel disease. *Curr Pharm Des.* 2009;15:1528–1536.
15. Scanlan PD, Shanahan F, Clune Y, et al. Culture-independent analysis of the gut microbiota in colorectal cancer and polyposis. *Environ Microbiol.* 2008;10:789–798.
16. Cani PD, Delzenne NM. Interplay between obesity and associated metabolic disorders: new insights into the gut microbiota. *Curr Opin Pharmacol.* 2009;9:737–743.
17. Vrieze A, Holleman F, Zoetendal EG, de Vos WM, Hoekstra JB, Nieuwdorp M. The environment within: how gut microbiota may influence metabolism and body composition. *Diabetologia.* 2010;53:606–613.
18. Cucchiara S, Iebba V, Conte MP, Schippa S. The microbiota in inflammatory bowel disease in different age groups. *Dig Dis.* 2009;27:252–258.
19. Sokol H, Seksik P, Furet JP, et al. Low counts of *Faecalibacterium prausnitzii* in colitis microbiota. *Inflamm Bowel Dis.* 2009;15:1183–1189.
20. Fallani M, Young D, Scott J, et al. Other Members of the INFABIO Team Intestinal microbiota of 6-week-old infants across Europe: geographic influence beyond delivery mode, breast-feeding, and antibiotics. *J Pediatr Gastroenterol Nutr.* 2010;51:77–84.
21. Le Huërou-Luron I, Blat S, Boudry G. Breast- v. formula-feeding: impacts on the digestive tract and immediate and long-term health effects. *Nutr Res Rev.* 2010;23:23–36.
22. Zivkovic AM, German JB, Lebrilla CB, Mills DA. Microbes and Health Sackler Colloquium: Human milk glycobiome and its impact on the infant gastrointestinal microbiota. *Proc Natl Acad Sci USA.* 2010;108(suppl 1):4653–4658.
23. Mohan R, Koebnick C, Schildt J, Mueller M, Radke M, Blaut M. Effects of *Bifidobacterium lactis* Bb12 supplementation on body weight, fecal pH, acetate, lactate, calprotectin, and IgA in preterm infants. *Pediatr Res.* 2008;64:418–422.
24. Pompei A, Cordisco L, Amaretti A, et al. Administration of folate-producing bifidobacteria enhances folate status in Wistar rats. *J Nutr.* 2007;137:2742–2746.
25. Strozzi GP, Mogna L. Quantification of folic acid in human feces after administration of *Bifidobacterium* probiotic strains. *J Clin Gastroenterol.* 2008;42(suppl 3):S179–S184.
26. Wall R, Ross RP, Shanahan F, et al. Metabolic activity of the enteric microbiota influences the fatty acid composition of murine and porcine liver and adipose tissues. *Am J Clin Nutr.* 2009;89:1393–1401.
27. Heuvelin E, Lebreton C, Grangette C, Pot B, Cerf-Bensussan N, Heyman M. Mechanisms involved in alleviation of intestinal inflammation by bifidobacterium breve soluble factors. *PLoS One.* 2009;4:e5184. [Epub 2009 April 17].
28. Commane DM, Shortt CT, Silvi S, Cresci A, Hughes RM, Rowland IR. Effects of fermentation products of pro- and prebiotics on trans-epithelial electrical resistance in an *in vitro* model of the colon. *Nutr Cancer.* 2005;51:102–109.
29. Putaala H, Salusjärvi T, Nordström M, et al. Effect of four probiotic strains and *Escherichia coli* O157:H7 on tight junction integrity and cyclo-oxygenase expression. *Res Microbiol.* 2008;159:692–698.
30. Takeda Y, Nakase H, Namba K, et al. Upregulation of T-bet and tight junction molecules by *Bifidobactrium longum* improves colonic inflammation of ulcerative colitis. *Inflamm Bowel Dis.* 2009;15:1617–1618.
31. Khailova L, Dvorak K, Arganbright KM, et al. *Bifidobacterium bifidum* improves intestinal integrity in a rat model of necrotizing enterocolitis. *Am J Physiol Gastrointest Liver Physiol.* 2009;297:G940–G949.
32. Likotrafiti E, Manderson KS, Fava F, Tuohy KM, Gibson GR, Rastall RA. Molecular identification and antipathogenic activities of putative probiotic bacteria isolated from faeces of healthy elderly individuals. *Microb Ecol Health Dis.* 2004;16:105–112.
33. Weiss G, Rasmussen S, Nielsen Fink L, Jarmer H, Nøhr Nielsen B, Frøkiaer H. *Bifidobacterium bifidum* actively changes the gene expression profile induced by *Lactobacillus acidophilus* in murine dendritic cells. *PLoS One.* 2010;5:e11065.
34. Cummings JH, Antoine JM, Azpiroz F, et al. PASSCLAIM–gut health and immunity. *Eur J Nutr.* 2004;43(suppl 2):II118–II173.

35. Gill SR, Pop M, Deboy RT, et al. Metagenomic analysis of the human distal gut microbiome. *Science*. 2006;312:1355–1359.
36. Vanhoutvin SA, Troost FJ, Hamer HM, et al. Butyrate-induced transcriptional changes in human colonic mucosa. *PLoS One*. 2009;4(8):e6759.
37. Millard AL, Mertes PM, Ittelet D, Villard F, Jeannesson P, Bernard J. Butyrate affects differentiation, maturation and function of human monocyte-derived dendritic cells and macrophages. *Clin Exp Immunol*. 2002;130:245–255.
38. Beylot M. Effects of inulin-type fructans on lipid metabolism in man and in animal models. *Br J Nutr*. 2005;93(suppl 1):S163–S168.
39. Cummings JH, Macfarlane GT. Role of intestinal bacteria in nutrient metabolism. *JPEN J Parenter Enteral Nutr*. 1997;21:357–365.
40. Macfarlane S, Macfarlane GT. Regulation of short-chain fatty acid production. *Proc Nutr Soc*. 2003;62:67–72.
41. Louis P, Scott KP, Duncan SH, Flint HJ. Understanding the effects of diet on bacterial metabolism in the large intestine. *J Appl Microbiol*. 2007;102:1197–1208.
42. Lindon JC, Nicholson JK, Holmes E, Everett JR. Metabonomics: metabolic processes studied by NMR spectroscopy of biofluids. *Concepts Magn Reson*. 2000;12:289–320.
43. Goodacre R, Vaidyanathan S, Dunn WB, Harrigan GG, Kell DB. Metabolomics by numbers: acquiring and understanding global metabolite data. *Trends Biotechnol*. 2004;22:245–252.
44. van Duynhoven J, Vaughan EE, Jacobs DM, et al. Microbes and Health Sackler Colloquium: metabolic fate of polyphenols in the human superorganism. *Proc Natl Acad Sci USA*. 2010;108(suppl.1):4531–4538.
45. Holmes E, Loo RL, Stamler J, et al. Human metabolic phenotype diversity and its association with diet and blood pressure. *Nature*. 2008;453:396–400.
46. Li M, Wang B, Zhang M, et al. Symbiotic gut microbes modulate human metabolic phenotypes. *Proc Natl Acad Sci USA*. 2007;105:2117–2122.
47. FAO. Report of a Joint FAO/WHO Expert Consultation on Evaluation of Health and Nutritional Properties of Probiotics in Food Including Powder Milk with Live Lactic Acid Bacteria, "Health and Nutritional Properties of Probiotics in Food including Powder Milk with Live Lactic Acid Bacteria", Amerian Córdoba Park Hotel, Córdoba, 2001. Argentina 1–4 October. <http://www.who.int/foodsafety/publications/fs_management/en/probiotics.pdf>.
48. Marteau PR, de Vrese M, Cellier CJ, Schrezenmeir J. Protection from gastrointestinal diseases with the use of probiotics. *Am J Clin Nutr*. 2001;73(suppl):430S–436S.
49. Reddy BS. Possible mechanisms by which pro- and prebiotics influence colon carcinogenesis and tumor growth. *J Nutr*. 1999;129:1478S–1482S.
50. Rafter J, Bennett M, Caderni G, et al. Dietary synbiotics reduce cancer risk factors in polypectomized and colon cancer patients. *Am J Clin Nutr*. 2007;85:488–496.
51. Montesi A, García-Albiach R, Pozuelo MJ, Pintado C, Goni I, Rotger R. Molecular and microbiological analysis of caecal microbiota in rats fed with diets supplemented either with prebiotics or probiotics. *Int J Food Microbiol*. 2005;98:281–289.
52. Prantera C, Scribano ML. Probiotics and Crohn's disease. *Dig Liver Dis*. 2002;34(suppl 2):S66–S67.
53. Fink LN, Zeuthen LH, Christensen HR, Morandi B, Frøkiaer H, Ferlazzo G. Distinct gut-derived lactic acid bacteria elicit divergent dendritic cell-mediated NK cell responses. *Int Immunol*. 2007;19:1319–1327.
54. Zeuthen LH, Fink LN, Frøkiaer H. Toll-like receptor 2 and nucleotide-binding oligomerization domain-2 play divergent roles in the recognition of gut-derived lactobacilli and bifidobacteria in dendritic cells. *Immunology*. 2008;124:489–502.
55. Lee JH, Karamychev VN, Kozyavkin SA, et al. Comparative genomic analysis of the gut bacterium *Bifidobacterium longum* reveals loci susceptible to deletion during pure culture growth. *BMC Genomics*. 2008;9:247.
56. de Vrese M, Schrezenmeir J. Probiotics, prebiotics and synbiotics. *Adv Biochem Eng Biotechnol*. 2008;111:1–66.
57. Salucci V, Rimoldi M, Penati C, et al. Monocyte-derived dendritic cells from Crohn patients show differential NOD2/CARD15-dependent immune responses to bacteria. *Inflamm Bowel Dis*. 2008;14:812–818.
58. Mileti E, Matteoli G, Iliev ID, Rescigno M. Comparison of the immunomodulatory properties of three probiotic strains of Lactobacilli using complex culture systems: prediction for in vivo efficacy. *PLoS One*. 2009;4(9):e7056.
59. Gibson GR, Scott KP, Rastall RA, et al. Dietary prebiotics: Current status and new definition. *Food Sci Technol Bull*. 2010;7:1–19.
60. COMMISSION DIRECTIVE 2008/100/EC of 28 October 2008 amending Council Directive 90/496/EEC on nutrition labelling for foodstuffs as regards recommended daily allowances, energy conversion factors and definitions. <http://eur-lex.europa.eu/LexUriServ/LexUriServ.do?uri=OJ:L:2008:285:0009:0012:EN:PDF>.
61. Macfarlane GT, Steed H, Macfarlane S. Bacterial metabolism and health-related effects of galacto-oligosaccharides and other prebiotics. *J Appl Microbiol*. 2008;104:305–344.
62. Saulnier DM, Spinler JK, Gibson GR, Versalovic J. Mechanisms of probiosis and prebiosis: considerations for enhanced functional foods. *Curr Opin Biotechnol*. 2009;20:135–141.
63. Cani PD, Delzenne NM. Involvement of the gut microbiota in the development of low grade inflammation associated with obesity: focus on this neglected partner. *Acta Gastroenterol Belg*. 2010;73:267–269.
64. Abrams SA, Hawthorne KM, Aliu O, Hicks PD, Chen Z, Griffin IJ. An inulin-type fructan enhances calcium absorption primarily via an effect on colonic absorption in humans. *J Nutr*. 2007;137:2208–2212.
65. Abrams SA, Griffin IJ, Hawthorne KM, Ellis KJ. Effect of prebiotic supplementation and calcium intake on body mass index. *J Pediatr*. 2007;151:293–298.
66. Brighenti F. Dietary fructans and serum triacylglycerols: a meta-analysis of randomized controlled trials. *J Nutr*. 2007;137 (11 suppl):2552S–2556S.
67. Russo F, Chimienti G, Riezzo G, et al. Inulin-enriched pasta affects lipid profile and Lp(a) concentrations in Italian young healthy male volunteers. *Eur J Nutr*. 2008;47:453–459.
68. Jackson KG, Lovegrove JA. Impact of probiotics, prebiotics and synbiotics on lipid metabolism in humans. *Nutr Aging*. 2012;1:181–200.
69. Cani PD, Joly E, Horsmans Y, Delzenne NM. Oligofructose promotes satiety in healthy human: a pilot study. *Eur J Clin Nutr*. 2006;60:567–572.
70. Parnell JA, Reimer RA. Weight loss during oligofructose supplementation is associated with decreased ghrelin and increased peptide YY in overweight and obese adults. *Am J Clin Nutr*. 2009;89:1751–1759.

71. Russo F, Riezzo G, Chiloiro M, et al. Metabolic effects of a diet with inulin-enriched pasta in healthy young volunteers. *Curr Pharm Des.* 2010;16:825–831.
72. McCartney E. EFSA provokes paradigm shift in EU probiotics marketing. *Nutrafoods.* 2010;9:44–45.
73. Scientific Opinion on the substantiation of a health claim related to oat beta glucan and lowering blood cholesterol and reduced risk of (coronary) heart disease pursuant to Article 14 of Regulation (EC) No 1924/2006.
74. Jones ML, Martoni CJ, Parent M, Prakash S. Cholesterol-lowering efficacy of a microencapsulated bile salt hydrolase-active *Lactobacillus reuteri* NCIMB 30242 yoghurt formulation in hypercholesterolaemic adults. *Br J Nutr.* 2012;107:1505–1513.
75. Jones ML, Martoni CJ, Prakash S. Cholesterol lowering and inhibition of sterol absorption by *Lactobacillus reuteri* NCIMB 30242: a randomized controlled trial. *Eur J Clin Nutr.* 2012;66:1234–1241.
76. Arora T, Loo RL, Anastasovska J, et al. Differential effects of two fermentable carbohydrates on central appetite regulation and body composition. *PLoS One.* 2012;7(8):e43263.
77. Connolly ML, Tzounis X, Tuohy KM, Lovegrove JA. Low glycaemic index wholegrain oat cereal consumption resulted in prebiotic and hypo-cholesterolaemic effects in those 'at risk' of metabolic disease. *Proceedings of The Nutrition Society.* 2011;70(OCE4):E244.
78. Setser CS, Racette WL. Macromolecule replacers in food products. *Crit Rev Food Sci Nutr.* 1992;32:275–297.
79. Redgwell RJ, Fischer M. Dietary fiber as a versatile food component: an industrial perspective. *Mol Nutr Food Res.* 2005;49:521–535.
80. Gibson GR, Roberfroid MB. Dietary modulation of the human colonic microbiota: introducing the concept of prebiotics. *J Nutr.* 1995;125:1401–1412.
81. Ehrmann MA, Korakli M, Vogel RF. Identification of the gene for beta-fructofuranosidase of *Bifidobacterium lactis* DSM10140(T) and characterization of the enzyme expressed in *Escherichia coli. Curr Microbiol.* 2003;46:391–397.
82. Parche S, Amon J, Jankovic I, et al. Sugar transport systems of *Bifidobacterium longum* NCC2705. *J Mol Microbiol Biotechnol.* 2007;12:9–19.
83. Falony G, Calmeyn T, Leroy F, De Vuyst L. Coculture fermentations of *Bifidobacterium* species and *Bacteroides thetaiotaomicron* reveal a mechanistic insight into the prebiotic effect of inulin-type fructans. *Appl Environ Microbiol.* 2009;75:2312–2319.
84. Collins MD, Gibson GR. Probiotics, prebiotics, and synbiotics: approaches for modulating the microbial ecology of the gut. *Am J Clin Nutr.* 1999;69:1052S–1057S.
85. Rumney CJ, Rowland IR. *In vivo* and in vitro models of the human colonic flora. *Crit Rev Food Sci Nutr.* 1992;31:299–331.
86. Gomez E, Tuohy KM, Gibson GR, Klinder A, Costabile A. *In vitro* evaluation of the fermentation properties and potential prebiotic activity of Agave fructans. *J Appl Microbiol.* 2009;108:2114–2121.
87. Salazar N, Ruas-Madiedo P, Kolida S, et al. Exopolysaccharides produced by *Bifidobacterium longum* IPLA E44 and *Bifidobacterium animalis* subsp. *lactis* IPLA R1 modify the composition and metabolic activity of human faecal microbiota in pH-controlled batch cultures. *Int J Food Microbiol.* 2009;135:260–267.
88. Hughes SA, Shewry PR, Li L, Gibson GR, Sanz ML, Rastall RA. *In vitro* fermentation by human fecal microflora of wheat arabinoxylans. *J Agric Food Chem.* 2007;55:4589–4595.
89. Hughes SA, Shewry PR, Gibson GR, McCleary BV, Rastall RA. *In vitro* fermentation of oat and barley derived beta-glucans by human faecal microbiota. *FEMS Microbiol Ecol.* 2008;64:482–493.
90. Tzortzis G, Goulas AK, Baillon ML, Gibson GR, Rastall RA. *In vitro* evaluation of the fermentation properties of galactooligosaccharides synthesised by alpha-galactosidase from *Lactobacillus reuteri. Appl Microbiol Biotechnol.* 2004;64:106–111.
91. Mandalari G, Nueno Palop C, Tuohy K, et al. *In vitro* evaluation of the prebiotic activity of a pectic oligosaccharide-rich extract enzymatically derived from bergamot peel. *Appl Microbiol Biotechnol.* 2007;73:1173–1179.
92. Connolly ML, Lovegrove J, Tuohy KM. Konjac glucomannan hydrolysate beneficially modulates bacterial composition and activity within the faecal microbiota. *J Funct Foods.* 2010;2:219–224.
93. Gibson GR, Beatty ER, Wang X, Cummings JH. Selective stimulation of bifidobacteria in the human colon by oligofructose and inulin. *Gastroenterology.* 1995;108:975–982.
94. Kleessen B, Sykura B, Zunft HJ, Blaut M. Effects of inulin and lactose on fecal microflora, microbial activity and bowel habit in elderly constipated persons. *Am J Clin Nutr.* 1997;65:1397–1402.
95. Brighenti F, Casiraghi MC, Canzi E, Ferrari A. Effect of consumption of a ready-to-eat breakfast cereal containing inulin on the intestinal milieu and blood lipids in healthy male volunteers. *Eur J Clin Nutr.* 1999;53:726–733.
96. Kruse HP, Kleessen B, Blaut M. Effects of inulin on faecal bifidobacteria in human subjects. *Br J Nutr.* 1999;82:375–382.
97. Tuohy KM, Finlay RK, Wynne AG, Gibson GR. A human volunteer study on the prebiotic effects of HP-Inulin – faecal bacteria enumerated using fluorescent *in situ* hybridisation (FISH). *Anaerobe.* 2001;7:113–118.
98. Kolida S, Meyer D, Gibson GR. A double-blind placebo-controlled study to establish the bifidogenic dose of inulin in healthy humans. *Eur J Clin Nutr.* 2007;61:1189–1195.
99. Kleessen B, Schwa S, Boehm A, et al. Jerusalem artichoke and chicory inulin in bakery products affect faecal microbiota of healthy volunteers. *Br J Nutr.* 2007;98:540–549.
100. Yap WKW, Mohamed S, Husni Jamal M, Meyer D, Manap AY. Changes in infants faecal characteristics and microbiota by inulin supplementation. *J Clin Biochem Nutr.* 2008;43:159–166.
101. Ramirez-Farias C, Slezak K, Fuller Z, Duncan A, Holtrop G, Louis P. Effect of inulin on the human gut microbiota: stimulation of *Bifidobacterium adolescentis* and *Faecalibacterium prausnitzii. Br J Nutr.* 2009;101:541–550.
102. Petry N, Egli I, Chassard C, Lacroix C, Hurrell R. Inulin modifies the bifidobacteria population, fecal lactate concentration, and fecal pH but does not influence iron absorption in women with low iron status. *Am J Clin Nutr.* 2012;96:325–331.
103. Costabile A, Kolida S, Klinder A, et al. A double-blind, placebo-controlled, cross-over study to establish the bifidogenic effect of a very-long-chain inulin extracted from globe artichoke (*Cynara scolymus*) in healthy human subjects. *Br J Nutr.* 2010;104:1007–1017.
104. Bouhnik Y, Vahedi K, Achour L, et al. Short-chain fructo-oligosaccharide administration dose dependently increases fecal bifidobacteria in healthy humans. *J Nutr.* 1999;129:113–116.
105. Whelan K, Judd PA, Preedy VR, Simmering R, Jann A, Taylor MA. Fructooligosaccharides and fiber partially prevent the alterations in fecal microbiota and short-chain fatty acid concentrations caused by standard enteral formula in healthy humans. *J Nutr.* 2005;135:1896–1902.

106. Yen CH, Kuo YW, Tseng YH, Lee MC, Chen HL. Beneficial effects of fructo-oligosaccharides supplementation on fecal bifidobacteria and index of peroxidation status in constipated nursing-home residents—a placebo-controlled, diet-controlled trial. *Nutrition*. 2011;27:323–328.
107. Bouhnik B, Raskine L, Simoneau G, Paineau D, Bornet F. The capacity of short-chain fructo-oligosaccharides to stimulate faecal bifidobacteria: a dose–response relationship study in healthy humans. *Nutr J*. 2006;5:8.
108. Tuohy KM, Kolida S, Lustenberger A, Gibson GR. The prebiotic effects of biscuits containing partially hydrolyzed guar gum and fructooligosaccharides – a human volunteer study. *Br J Nutr*. 2001;86:341–348.
109. Moro G, Minoli I, Mosca M, et al. Dosage-related bifidogenic effects of galacto- and fructooligosaccharides in formula-fed term infants. *J Pediatr Gastroenterol Nutr*. 2002;34:291–295.
110. Boehm G, Lidestri M, Casetta P, et al. Supplementation of a bovine milk formula with an oligosaccharide mixture increases counts of faecal bifidobacteria in preterm infants. *Arch Dis Child – Fetal Neonatal Ed*. 2002;86:F178–F181.
111. Knol J, Scholtens P, Kafka C, et al. Colon microflora in infants fed formula with galacto- and fructo-oligosaccharides: more like breast-fed infants. *J Pediatr Gastroenterol Nutr*. 2005;40:36–42.
112. Veereman-Wauters G, Staelens S, Van de Broek H, et al. Physiological and bifidogenic effects of prebiotic supplements in infant formulae. *J Pediatr Gastroenterol Nutr*. 2011;52:763–771.
113. Magne F, Hachelaf W, Suau A, et al. Effects on faecal microbiota of dietary and acidic oligosaccharides in children during partial formula feeding. *J Pediatr Gastroenterol Nutr*. 2008;46:580–588.
114. Ito M, Deguchi Y, Miyamori A, et al. Effects of administration of galactooligosaccharides on the human faecal microflora, stool weight and abdominal sensation. *Microb Ecol Health Dis*. 1990;3:285–292.
115. Ito M, Deguchi Y, Matsumoto K, Kimura M, Onodera N, Yajima T. Influence of galactooligosaccharides on the human fecal microflora. *J Nutr Sci Vitaminol (Tokyo)*. 1993;39:635–640.
116. Bouhnik Y, Flourie B, D'Agay-Abensour L, et al. Administration of transgalacto-oligosaccharides increases fecal bifidobacteria and modifies colonic fermentation metabolism healthy humans. *J Nutr*. 1997;127:444–448.
117. Alles MS, Hartemink R, Meyboom S, et al. Effects of transgalactooligosaccharides on the composition of the human intestinal microflora and on putative risk markers for colon cancer. *Am J Clin Nutr*. 1999;69:980–991.
118. Depeint F, Tzortzis G, Vulevic J, I'Anson K, Gibson GR. Prebiotic evaluation of a novel galactooligosaccharide mixture produced by the enzymatic activity of *Bifidobacterium bifidum* NCIMB 41171, in healthy humans: a randomized, double-blind, crossover, placebo-controlled intervention study. *Am J Clin Nutr*. 2008;87:785–791.
119. Vulevic J, Drakoularakou A, Yaqoob P, Tzortzis G, Gibson GR. Modulation of the fecal microflora profile and immune function by a novel trans-galactooligosaccharide mixture (B-GOS) in healthy elderly volunteers. *Am J Clin Nutr*. 2008;88:1438–1446.
120. Vulevic J, Juric A, Tzortzis G, Gibson. GR. A mixture of trans-galactooligosaccharides reduces markers of metabolic syndrome and modulates the fecal microbiota and immune function of overweight adults. *J Nutr*. 2013;143:324–331.
121. Silk DB, Davis A, Vulevic J, Tzortzis G, Gibson GR. Clinical trial: the effects of a trans-galactooligosaccharide prebiotic on faecal microbiota and symptoms in irritable bowel syndrome. *Aliment Pharmacol Ther*. 2009;29:508–518.
122. Terada A, Hara H, Kataoka M, Mitsuoka T. Effect of lactulose on the composition and metabolic activity of the human faecal flora. *Microb Ecol Health Dis*. 1992;5:43–50.
123. Nagendra R, Viswanatha S, Kumar SA, Murthy BK, Rao SV. Effect of feeding milk formula containing lactulose to infants on faecal bifidobacterial flora. *Nutr Res*. 1995;15:15–24.
124. Ballongue J, Schumann C, Quignon P. Effects of lactulose and lactitol on colonic microflora and enzymatic activity. *Scand J Gastroenterol*. 1997;32(suppl 222):41–44.
125. Tuohy K, Ziemer CJ, Klinder A, Knöbel Y, Pool-Zobel BL, Gibson GR. A human volunteer study to determine the prebiotic effects of lactulose powder on human colonic microbiota. *Microb Ecol Health Dis*. 2002;14:165–173.
126. Bouhnik Y, Attar A, Joly FA, Riottot M, Dyard F, Flourie B. Lactulose ingestion increases faecal bifidobacterial counts: a randomised double-blind study in healthy humans. *Eur J Clin Nutr*. 2004;58:462–466.
127. Kohmoto T, Fukui F, Takaku H, Machida Y, Arai M, Mitsuoka T. Effect of isomalto-oligosaccharides on human faecal flora. *Bifidobacteria Microflora*. 1988;7:61–69.
128. Kaneko T, Kohmoto T, Kikuchi H, Shiota M, Iino H, Mitsuoka T. Effects of isomaltooligosaccharides with different degree of polymerization on human faecal bifidobacteria. *Biosci Biotechnol Biochem*. 1994;58:2288–2290.
129. Gu Q, Yang Y, Jiang G, Chang G. Study on the regulative effect of isomaltooligosaccharides on human intestinal flora. *Wei Sheng Yan Jiu*. 2003;32:54–55.
130. Benno Y, Endo K, Shiragami N, Sayama K, Mitsuoka T. Effects of raffinose intake on human faecal microflora. *Bifidobacteria Microflora*. 1987;6:59–63.
131. Hayakawa K, Mizutani J, Wada K, Masai T, Yoshihara I, Mitsuoka T. Effects of soybean oligosaccharides on human faecal flora. *Microb Ecol Health Dis*. 1990;3:293–303.
132. Dinoto A, Marques TM, Sakamoto K, et al. Population dynamics of Bifidobacterium species in human feces during raffinose administration monitored by fluorescence in situ hybridization-flow cytometry. *Appl Environ Microbiol*. 2006;72:7739–7747.
133. Bouhnik Y, Raskine L, Simoneau G, et al. The capacity of nondigestible carbohydrates to stimulate fecal bifidobacteria in healthy humans: a double-blind, randomized, placebo-controlled, parallel-group, dose–response relation study. *Am J Clin Nutr*. 2004;80:1658–1664.
134. Beards E, Tuohy K, Gibson G. A human volunteer study to assess the impact of confectionery sweeteners on the gut microbiota composition. *Br J Nutr*. 2010;104:701–708.
135. Martínez I, Kim J, Duffy PR, Schlegel VL, Walter J. Resistant starches types 2 and 4 have differential effects on the composition of the fecal microbiota in human subjects. *PLoS One*. 2010;5(11):e15046.
136. Lecerf JM, Dépeint F, Clerc E, et al. Xylo-oligosaccharide (XOS) in combination with inulin modulates both the intestinal environment and immune status in healthy subjects, while XOS alone only shows prebiotic properties. *Br J Nutr*. 2012;108:1847–1858.
137. Cherbut C, Michel C, Raison V, Kravtchenko T, Severine M. Acacia gum is a bifidogenic dietary fiber with high digestive tolerance in healthy humans. *Microb Ecol Health Dis*. 2003;15:43–50.

138. Queipo-Ortuño MI, Boto-Ordóñez M, Murri M, et al. Influence of red wine polyphenols and ethanol on the gut microbiota ecology and biochemical biomarkers. *Am J Clin Nutr*. 2012;95:1323–1334.
139. Clemente-Postigo M, Queipo-Ortuño MI, Boto-Ordoñez M, et al. Effect of acute and chronic red wine consumption on lipopolysaccharide concentrations. *Am J Clin Nutr*. 2013;97:1053–1061.
140. Tzounis X, Rodriguez-Mateos A, Vulevic J, Gibson GR, Kwik-Uribe C, Spencer JP. Prebiotic evaluation of cocoa-derived flavanols in healthy humans by using a randomized, controlled, double-blind, crossover intervention study. *Am J Clin Nutr*. 2011;93:62–72.
141. Elli M, Cattivelli D, Soldi S, Bonatti M, Morelli L. Evaluation of prebiotic potential of refined psyllium (Plantago ovata) fiber in healthy women. *J Clin Gastroenterol*. 2008;42(suppl.3):S174–S176.
142. Costabile A, Klinder A, Fava F, et al. Whole-grain wheat breakfast cereal has a prebiotic effect on the human gut microbiota: a double-blind, placebo-controlled, crossover study. *Br J Nutr*. 2008;99:110–120.
143. Walton GE, Lu C, Trogh I, Arnaut F, Gibson GR. A randomised, double-blind, placebo controlled cross-over study to determine the gastrointestinal effects of consumption of arabinoxylan-oligosaccharides enriched bread in healthy volunteers. *Nutr J*. 2012;11:36.
144. Maki KC, Gibson GR, Dickmann RS, et al. Digestive and physiologic effects of a wheat bran extract, arabino-xylan-oligosaccharide, in breakfast cereal. *Nutrition*. 2012;28:1115–1121.
145. François IE, Lescroart O, Veraverbeke WS, et al. Effects of a wheat bran extract containing arabinoxylan oligosaccharides on gastrointestinal health parameters in healthy adult human volunteers: a double-blind, randomised, placebo-controlled, cross-over trial. *Br J Nutr*. 2012;108:2229–2242.
146. Calame W, Weseler AR, Viebke C, Flynn C, Siemensma AD. Gum arabic establishes prebiotic functionality in healthy human volunteers in a dose-dependent manner. *Br J Nutr*. 2008;100:1269–1275.
147. Cloetens L, Broekaert WF, Delaedt Y, et al. Tolerance of arabinoxylan-oligosaccharides and their prebiotic activity in healthy subjects: a randomised, placebo-controlled cross-over study. *Br J Nutr*. 2010;103:703–713.
148. Robles Alonso V, Guarner F. Linking the gut microbiota to human health. *Br J Nutr*. 2013;109(suppl 2):S21–S26.
149. Davis LM, Martínez I, Walter J, Goin C, Hutkins RW. Barcoded pyrosequencing reveals that consumption of galactooligosaccharides results in a highly specific bifidogenic response in humans. *PLoS One*. 2011;6(9):e25200.
150. Scott KP, Martin JC, Duncan SH, Flint HJ. Prebiotic stimulation of human colonic butyrate-producing bacteria and bifidobacteria, *in vitro*. *FEMS Microbiol Ecol*. 2014;87:30–40.
151. Crittenden RG, Playne MJ. Production, properties and applications of food-grade oligosaccharides. *Trends Food Sci Technol*. 1996;7:353–361.
152. Tuohy KM, Rouzaud GCM, Brück WM, Gibson GR. Modulation of the human gut microbiota towards improved health using prebiotics — assessment of efficacy. *Curr Pharm Des*. 2005;11:75–90.
153. van Loo J, Coussement P, De Leenheer L, Hoebregs H, Smits G. On the presence of inulin and oligofructose as natural ingredients in the Western diet. *Crit Rev Food Sci Nutr*. 1995;35:525–552.
154. Molis C, Flourie B, Ouarne F, et al. Digestion, excretion, and energy value of fructooligosaccharides in healthy humans. *Am J Clin Nutr*. 1996;64:324–328.
155. Elegärd L, Andersson H, Bosaeus I. Inulin and oligofructose do not influence the absorption of cholesterol, or the excretion of cholesterol, Ca, Mg, Zn, Fe or bile acids but increases energy excretion in ileostomy subjects. *Eur J Clin Nutr*. 1997;51:1–5.
156. Rycroft CE, Jones MR, Gibson GR, Rastall RA. A comparative *in vitro* evaluation of the fermentation properties of prebiotic oligosaccharides. *J Appl Microbiol*. 2001;91:878–887.
157. Gibson GR, Wang X. Regulatory effects of bifidobacteria on the growth of other colonic bacteria. *J Appl Bacteriol*. 1994;77:412–420.
158. Sghir A, Chow JM, Mackie RI. Continuous culture selection of bifidobacteria and lactobacilli from human faecal samples using fructooligosaccharide as selective substrate. *J Appl Microbiol*. 1998;85:769–777.
159. Le Blay G, Michel C, Blottière HM, Cherbut C. Prolonged intake of fructo-oligosaccharides induces a short-term elevation of lactic acid-producing bacteria and a persistent increase in cecal butyrate in rats. *J Nutr*. 1999;129:2231–2235.
160. Djouzi Z, Andrieux C. Compared effects of three oligosaccharides on metabolism of intestinal microflora in rats inoculated with a human faecal flora. *Br J Nutr*. 1997;78:313–324.
161. Kleesen B, Hartmann L, Blaut M. Oligofructose and long-chain inulin: influence on the gut microbial ecology of rats associated with a human faecal flora. *Br J Nutr*. 2001;86:291–300.
162. Roberfroid MB, van Loo JAE, Gibson GR. The Bifidogenic nature of chicory inulin and its hydrolysis products. *J Nutr*. 1998;128:11–19.
163. Rossi M, Corradini C, Amaretti A, et al. Fermentation of fructooligosaccharides and inulin by bifidobacteria: a comparative study of pure and fecal cultures. *Appl Environ Microbiol*. 2005;71:6150–6158.
164. Stewart ML, Timm DA, Slavin JL. Fructooligosaccharides exhibit more rapid fermentation than long-chain inulin in an *in vitro* fermentation system. *Nutr Res*. 2008;28:329–334.
165. Kedia G, Vázquez JA, Charalampopoulos D, Pandiella SS. *In vitro* fermentation of oat bran obtained by debranning with a mixed culture of human fecal bacteria. *Curr Microbiol*. 2009;58:338–342.
166. Bourriaud C, Robins RJ, Martin L, et al. Lactate is mainly fermented to butyrate by human intestinal microfloras but inter-individual variation is evident. *J Appl Microbiol*. 2005;99:201–212.
167. Belenguer A, Duncan SH, Calder AG, et al. Two routes of metabolic cross-feeding between *Bifidobacterium adolescentis* and butyrate-producing anaerobes from the human gut. *Appl Environ Microbiol*. 2006;72:3593–3599.
168. Kiefer J, Beyer-Sehlmeyer G, Pool-Zobel BL. Mixtures of SCFA, composed according to physiologically available concentrations in the gut lumen, modulate histone acetylation in human HT29 colon cancer cells. *Br J Nutr*. 2006;96:803–810.
169. Scharlau D, Borowicki A, Habermann N, et al. Mechanisms of primary cancer prevention by butyrate and other products formed during gut flora-mediated fermentation of dietary fibre. *Mutat Res*. 2009;682:39–53.
170. Singh N, Thangaraju M, Prasad PD, et al. Blockade of dendritic cell development by bacterial fermentation products butyrate and propionate through a transporter (Slc5a8)-dependent inhibition of histone deacetylases. *J Biol Chem*. 2010;285:27601–27608.
171. Vinolo MA, Hatanaka E, Lambertucci RH, Newsholme P, Curi R. Effects of short chain fatty acids on effector mechanisms of neutrophils. *Cell Biochem Funct*. 2009;27:48–55.

172. Peng L, Li ZR, Green RS, Holzman IR, Lin J. Butyrate enhances the intestinal barrier by facilitating tight junction assembly via activation of AMP-activated protein kinase in Caco-2 cell monolayers. *J Nutr*. 2009;139:1619–1625.
173. Wright RS, Anderson JW, Bridges SR. Propionate inhibits hepatocyte lipid synthesis. *Proc Soc Exp Biol Med. Soc Exp Biol Med (New York, N.Y.)*. 1990;195:26–29.
174. Ostman E, Granfeldt Y, Persson L, Björck I. Vinegar supplementation lowers glucose and insulin responses and increases satiety after a bread meal in healthy subjects. *Eur J Clin Nutr*. 2005;59:983–988.
175. Freeland KR, Wolever TM. Acute effects of intravenous and rectal acetate on glucagon-like peptide-1, peptide YY, ghrelin, adiponectin and tumour necrosis factor-alpha. *Br J Nutr*. 2010;103:460–466.
176. Al-Lahham SH, Peppelenbosch MP, Roelofsen H, Vonk RJ, Venema K. Biological effects of propionic acid in humans; metabolism, potential applications and underlying mechanisms. *Biochim Biophys Acta*. 2010;1801:1175–1183.
177. Hong YH, Nishimura Y, Hishikawa D, et al. Acetate and propionate short chain fatty acids stimulate adipogenesis via GPCR43. *Endocrinology*. 2005;146:5092–5099.
178. Al-Lahham SH, Roelofsen H, Priebe M, et al. Regulation of adipokine production in human adipose tissue by propionic acid. *Eur J Clin Investig*. 2010;40:401–407.
179. Conterno L, Fava F, Viola R, Tuohy KM. Obesity and the gut microbiota: does up-regulating colonic fermentation protect against obesity and metabolic disease? *Genes Nutr*. 2011;6:241–260.
180. Delzenne NM, Williams CM. Prebiotics and lipid metabolism. *Curr Opin Lipidol*. 2002;13:61–67.
181. Cani PD, Delzenne NM. The role of the gut microbiota in energy metabolism and metabolic disease. *Curr Pharm Des*. 2009;15:1546–1558.
182. Pool-Zobel BL. Inulin-type fructans and reduction in colon cancer risk: review of experimental and human data. *Br J Nutr*. 2005;93 (suppl.1):S73–S90.
183. Boutron-Ruault MC, Marteau P, Lavergne-Slove A, et al. Eripolyp Study Group Effects of a 3-mo consumption of short-chain fructo-oligosaccharides on parameters of colorectal carcinogenesis in patients with or without small or large colorectal adenomas. *Nutr Cancer*. 2005;53:160–168.
184. Griffin IJ, Davila PM, Abrams SA. Non-digestible oligosaccharides and calcium absorption in girls with adequate calcium intakes. *Br J Nutr*. 2002;87(suppl 2):S187–S191.
185. Abrams SA, Griffin IJ, Hawthorne KM, et al. A combination of prebiotic short- and long-chain inulin-type fructans enhances calcium absorption and bone mineralization in young adolescents. *Am J Clin Nutr*. 2005;82:471–476.
186. Tuohy KM, Fava F, Viola R. 'The way to a man's heart is through his gut microbiota' — dietary pro- and prebiotics for the management of cardiovascular risk. *Proc Nutr Soc*. 2014;73(2):172–185.
187. Ley RE, Hamady M, Lozupone C, et al. Evolution of mammals and their gut microbes. *Science*. 2008;320(5883):1647–1651.
188. Demigné C, Morand C, Levrat MA, Besson C, Moundras C, Rémésy C. Effect of propionate on fatty acid and cholesterol synthesis and on acetate metabolism in isolated rat hepatocytes. *Br J Nutr*. 1995;74:209–219.
189. Kleesen B, Hartmann L, Blaut M. Fructans in the diet cause alterations of intestinal mucosal architecture, released mucins and mucosa-associated bifidobacteria in gnotobiotic rats. *Br J Nutr*. 2003;89:597–606.

CHAPTER

6

Bioactivation of High-Molecular-Weight Polyphenols by the Gut Microbiome

Pedro Mena[*,†,‡], Luca Calani*[*,‡], Renato Bruni*[*,‡] and Daniele Del Rio*[*,‡]*

***The φ2 Laboratory of Phytochemicals in Physiology, Department of Food Sciences, University of Parma, Parma, Italy †Phytochemistry Laboratory, Department of Food Science and Technology, CEBAS-CSIC, Murcia, Spain ‡LS9 Bioactives and Health, Interlaboratory Group, Department of Food Sciences, University of Parma, Parma, Italy**

INTRODUCTION

Tannins are phenolic secondary metabolites of plants with molar masses ranging from 300 Da to 3000 Da, and even up to 30,000 Da. Low-molecular-weight (MW) tannins are water-soluble compounds, whereas high-MW plant tannins can also be found in association with proteins or cell wall polysaccharides.[1] They comprise a heterogeneous group of (poly)phenolic substances that are traditionally characterized by conferring astringent taste to different plant organs.[2,3] Based on their structural characteristics, tannins can be classified into four major groups: gallotannins (GTs), ellagitannins (ETs), proanthocyanidins (PAs) or condensed tannins (CTs), and complex tannins.

GTs are galloyl unit derivatives bounded to diverse polyol-, catechin-, or triterpenoid cores, yielding gallic acid (GA) upon hydrolysis. ETs release a hexahydroxydiphenoyl (HHDP) residue after hydrolysis, which undergoes lactonization to form ellagic acid (EA). They contain a glycosidic core but do not contain a glycosidically linked catechin moiety. Both GTs and ETs are termed as "hydrolyzable tannins" (HTs). If catechin, epicatechin or other flavan-3-ols are attached glycosidically to GTs or ETs, these tannins are named "complex tannins," yielding flavan-3-ol and GA or EA moieties after hydrolysis. CTs or PAs are polymeric structures formed by the linkage of catechin monomers and are hardly hydrolyzed; they also lack glycoside residues. Specific tannins in marine brown algae have been also described, the so-called "phlorotannins," which are characterized by the presence of oligomers or polymers of phloroglucinol (1,3,5-trihydroxybenzene) moieties.[1,4]

Plant tannins have manifold biological activities in plant defence. They also have toxic or anti-nutritional effects on herbivores, both mammals and insects, which can reduce nutrient digestibility and protein availability.[5,6] On the other hand, a mounting plethora of different biological features are known to underpin the remarkable role of plant tannins in human health. In particular, the preventive effect of the consumption of tannin-rich foodstuffs on the onset of chronic diseases as well as their microbiological properties has been highlighted.[2,3,7] However, recent evidence points out the fundamental role of the gut microbiota on the metabolism and biological properties of plant tannins. Regardless of the fact that new findings are achieved daily, the information on the metabolic fate and bioavailability of plant tannins in humans is still scarce.

PROANTHOCYANIDINS

Structures and Nomenclature

PAs are complex polyphenolic polymers of plant origin encompassing a large span of polymerization degrees (DP). They are described by the presence of three-ring flavan-3-ol monomer units linked by C–C and in some

Diet-Microbe Interactions in the Gut.
DOI: http://dx.doi.org/10.1016/B978-0-12-407825-3.00006-X

cases *C*–*O* bonds at different positions, and are included in the class of CTs. PAs are characterized by the association of several monomeric units: two to five units comprise oligomers, while molecules composed of more than five units are called polymers. PAs possess a wide range of MWs, with oligomers starting from 578 Da and polymers typically possessing MW averages of 1000–6000 Da, albeit some can attain MW as high as 20,000 Da. The flavan-3-ol units have a typical C15 (C6–C3–C6) flavonoid skeleton, whose three rings are distinguished by the letters A, B and C. In particular, (+)-catechin and (−)-epicatechin are the basic units of this group, but PAs differ in the position and configuration of their monomeric linkages. According to the structure of their monomer, PAs can be classified as procyanidins, prodelphinidins and propelargonidins. More precisely, the degree of hydroxylation on the B-ring is a key feature for their discrimination; while prodelphinidins possess a 3′,4′,5′-trihydroxy substitution on the B-ring, procyanidins have hydroxyl groups on the 3′ and 4′ positions, and propelargonidins have a single OH group on the 4′ position only. The A ring can be described as a resorcinol- or as a phloroglucinol-type ring, but the latter is by far the most common. Monomers are usually linked by C-4→C-8 bonds, but C-4→C-6 linkages can also be found, while the different C-2–C-3 stereochemistry (*cis* or *trans*) of monomers introduces a further element of structural diversity. Usually, 2-3-*trans* stereochemistry is predominant between the chain extenders and oligomers terminate with (+) − catechin subunits. A peculiar classification is used to describe different classes of oligomers according to the position of *C*–*C* or *C*–*O* intermolecular bonds: dimers with *C*–*O* linkages between C2 of the upper unit and the oxygen at C7 position of the lower unit are defined as A-type PAs, while dimers possessing C4–C8 or C4–C6 linkages are classified into B-type and corresponding trimers are classified as C-type PAs. The accurate chemical investigation of PAs was undertaken only in the last few decades due to the lack of suitable methods for isolation, purification and structure elucidation. Therefore, a wide number of complex, heterogeneous structures have been unravelled only recently, making previous description methods insufficient to describe the structural complexity found in nature. Thus, to overcome the limitations of this alphanumeric system and avoid confusing and erroneous assignments in the literature, a new nomenclature was recently introduced. The new classification mimics the same strategy utilized for polysaccharides, as the elementary oligomeric units are designated with the name of the corresponding flavan-3-ol monomers. The linkage between consecutive units, their position and direction are provided in parentheses with an arrow (e.g., 4→) and the configuration at C4 is described as α or β. In type-A PAs, both characteristic linkages are also described within brackets. Therefore, the procyanidin dimer B1 becomes epicatechin-(4β→8)-catechin and dimer A2 is defined as epicatechin-(2β→7, 4β→8)-epicatechin. The entire PA class is extremely populated and a number of structures, varying in terms of DP, monomeric units, linkages and relative abundance, have been described in most vascular and non-vascular plants.

Distribution in the Plant Kingdom: From Ecological Role to Behavior during Gastrointestinal Transit

PAs are the most widespread plant polyphenols after lignin and are considered ubiquitous in the Plant Kingdom. Their presence is more recurrent than that of HTs: they are in fact found in many Pteridophytes, in Gymnosperms and in most woody Angiosperms. In particular, Gymnosperms and Monocotyledons are known to synthesize only CTs but not HTs, which are instead an exclusive product of Dicotyledon secondary metabolism. From a quantitative standpoint, their presence in selected plant tissues can be relevant, as PA content can comprise between 15 and 25% of dry weight in leaves of some tree species, and between 20 and 40% of tannin in bark and between 1 and 35% in roots. As discussed later, they are frequently found in most edible plants, albeit in amounts that are usually extremely low due to palatability constraints. In fact, PAs impart astringency and bitterness to many plant tissue and organs, as their primary ecological role is to act as a defense mechanism against herbivores by reducing the palatability of key plant organs. PAs have several further functions: in planta, they inhibit the development of pathogens, provide protection against UV radiation and oxidative stress, and once leaked in the soil they contribute to shape the environment, inhibiting competitor germination and influencing the growth of specific decay microorganisms. These duties are warranted also by a high degree of resistance to the degradation induced by light, acidic conditions and by most microorganisms. Some of these properties may have a significant connection to the fate of PAs once they are ingested as foods and once they enter the gut. It is in fact widely accepted that ingested PAs, as a result of their resistance to the acidic conditions of the stomach and due to their low absorption in the small intestine, can reach the colon almost intact, where they undergo a plethora of rearrangements, depolymerization and biotranformations through the action of the intestinal microbiota (Figure 6.1). It is speculated that a PA colonic catabolite pool may actually be responsible for physiological,

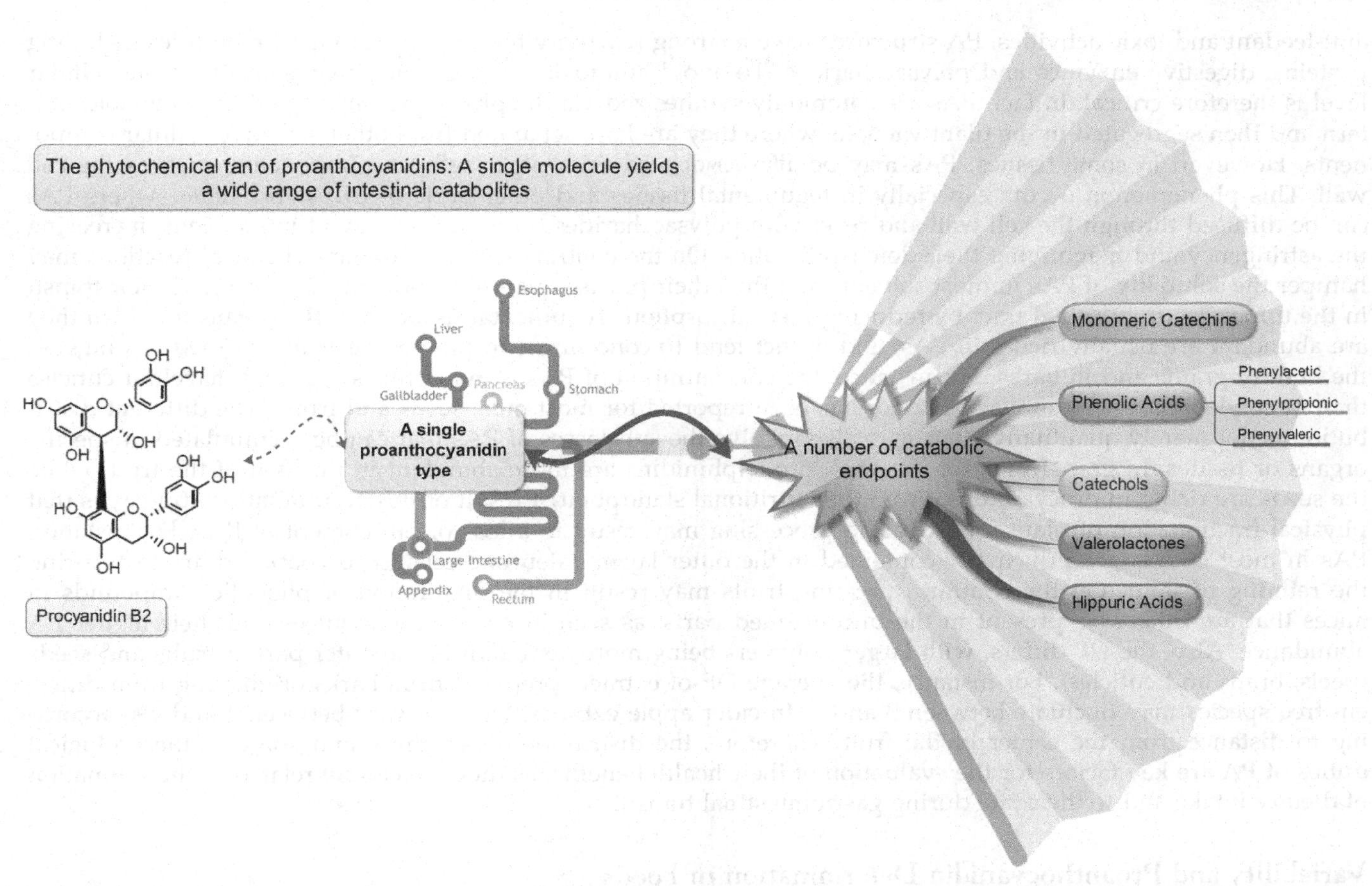

FIGURE 6.1 The phytochemical fan of PAs. A single molecule yields a wide range of intestinal catabolites with potential health benefits.

systemic relapses that can lead to health benefits in humans. This process, at least its catabolic part, is likely to be a close consequence of some of the above-mentioned plant-related factors, hereby included in the possibility to represent a selective carbon source for some bacterial strain not only at colonic level, but also in the environment. An ecological property that may have conspicuous relapses on PA behavior in the digestive tract, and in the gut in particular, is their role in plant–microbe interactions in the soil. In fact, besides their deterrent activity against herbivores, these compounds may influence resource competition, alter nutrient dynamics, and modulate microbial ecology by easing the growth of some friendly bacteria, thus acting as a selective environmental pressure. Indeed, despite being more resistant to microbial degradation than HTs, and notwithstanding their antimicrobial activity on a number of microbial strains, PAs can be actively biodegraded by various decay microorganisms, including bacteria that can use them as a source of carbon. For instance, under particular anaerobic conditions, soil *Bacillus*, *Staphylococcus*, *Esherichia* and *Klebsiella* spp., among others, are capable of using CTs as a source of carbon and degrade them by breaking links between monomers and altering the flavan-3-ol skeleton. Such characteristics represent the ecological and evolutionary basis for the fate of most PAs in the gut of many animals, mammals and humans. Not only are PAs degraded by gut bacteria producing different catabolites, but as a result, they also selectively modulate the growth of those bacteria that can take advantage of them as a source of carbon. Like PA-rich plant litter, peels and brans favor the growth of friendly bacteria in soil and on the surface of fruits and seeds, enjoying their competitive presence against spoilage bacteria, meaning that an increased dietary intake of PAs can positively modulate colonic microflora. For instance, in animals treated with a diet enriched with PAs, *Lactobacillus*, *Bifidobacterium* and *Bacteroides* spp. predominate over *Clostridium* and *Propionibacterium* spp., which are instead the main constituents in the colonic microbiota of animals fed with low-PA diets. In summary, symmetry can be envisaged between the ecological role of PA and their fate and behavior in the digestive tract; the existence of a mutual relationship between PA and intestinal microflora should not come as a surprise when it comes to nutrition and bioavailability.

Regarding storage and accumulation in plant individuals, PA distribution is uneven and follows a precise strategy, which is a consequence of the above-mentioned ecological roles. It is well-known that, to exert their anti-feedant and toxic activities, PA structures have a strong reactivity for a variety of target molecules including proteins, digestive enzymes and polysaccharides. To avoid autotoxic relapses, their management at the cellular level is therefore critical. In fact, PAs are commonly synthesized via the phenylpropanoid and the flavonoid pattern and then segregated in the plant vacuole, where they are kept separated from other sensitive cellular components. However, in some tissues, PAs may be also associated with other cellular components, such as the cell wall. This phenomenon occurs especially in tegumental tissues and outer parts of fruits and stems, where PAs can be diffused through the cell wall and react with polysaccharides through a variety of interactions, increasing the astringency and magnifying their defensive duties. On the contrary, these secondary chemical reactions may hamper the solubility of PAs in most solvents and thus their precise quantification, but may also ease their transit in the upper gastrointestinal tract by reducing their absorption. Tegumental tissues and the organs in which they are abundant are usually richer in PAs, and in fact tend to concentrate in the peel of fruits, in seed coatings, in the bran of grains and in barks. For instance, the concentration of PAs is many times higher in hazelnut cuticles than in cotyledons, and a similar behavior has been reported for most nuts, seeds and fruits. The different distribution is not merely quantitative, but may also involve the subclasses of PAs that can be accumulated in specific organs or tissues. In grapefruits, for instance, prodelphinidins are more abundant in the skin of the fruit, while the seeds are richer in procyanidins. From the nutritional standpoint, a result of this accumulation strategy is that physical fractionation of plant foods during processing may result in a loss or enrichment of PAs. For instance, PAs in most cereals are principally contained in the outer layers (aleurone cells, seed coat) and are lost during the refining of flour. On the contrary, pressing fruits may result in the dissolution of phenolic compounds in juices that are otherwise present in the unconsumed parts, as seen in cloudy apple juices and their higher PA abundance. Also, the DP differs, with larger polymers being more abundant in the outer part of fruits and seeds (peels, brans and cuticles). For instance, the average DP of extracts prepared from barks originating from different tree species may fluctuate between 3 and 8. In cider apple extracts, DP may vary between 4 and 190 according to distance from the center of the fruit. Therefore, the distribution, variability and some of the ecological duties of PA are key factors for the evaluation of their health benefits, as they are closely related to the estimation of dietary intake and to their fate during gastrointestinal transit.

Variability and Proanthocyanidin Determination in Foods

Given their widespread distribution in many edible plants, there has been growing interest in the determination of the presence of PAs in foods and in estimating their actual intake in different dietary habits. The data collected, albeit starting from a lower amount of evidence if compared with other flavonoids, have been correlated with different health benefits with mixed results, due to at least two main flaws: the constraints in their punctual determination and the limited consideration paid to their fate during colonic transit. The first hindrance stems from the combination of the great variability of structures and DP of PAs, the capability to interact irreversibly with proteins and cell walls, the wide range of solubility of oligomers and polymers and the lack of proper, comprehensive standardized methods for PA extraction and analysis. As a consequence, it is extremely difficult to estimate their precise distribution in plant tissues and organs and ultimately evaluate their average daily intake. The simultaneous determination of PAs with a high DP is particularly troublesome and most of the data available for plants and processed foods are thus focused on oligomeric PA, with few exceptions. It must be noted that most literature on food CTs refers to oligomeric procyanidins with good water solubility and MW ranging from 500 to 3000 Da, while a consistent number of non-extractable PA with MW up to 30,000 Da are usually present in whole plant foods, but are not adequately evaluated. In processed foods more than in raw plant materials, the scenario is further complicated by the fact that these substances may be tightly adsorbed on other food ingredients, perturbing analytical procedures. The overall result is that PA content in food plant materials has been underestimated for instance due to the presence of PA tightly bound to cell wall material that is not released by normal extraction methods and due to the fact that most analytical methods do not account for larger polymers. As a result, these evaluations have been only partial and defective and many correlations between PA intake and health benefits in humans and animal models have suffered from consistent biases.

Dietary Sources, Intake and Health Benefits

The average consumption of polyphenols in the diet is 1 g/d and PAs account for a large fraction of the total flavonoid intake in Western populations. PAs are considered responsible for a number of beneficial health effects and are deemed to prevent the onset of various chronic (cardiovascular, cancer, neurodegenerative) diseases, but a reliable correlation requires a robust determination of their actual dietary intake and bioavailability. Indeed, these CTs have been detected in half of 99 plant foods, therefore confirming their dietary relevance. The main dietary sources of PAs are fruits, beverages, such as tea, coffee, wine, legumes and fruit juices, chocolate and, to a lesser extent, vegetables, whole cereals and beer. As described in the "USDA Database for the Proanthocyanidin Content of Selected Foods,"[8] some specific food sources can provide relevant PA contents (cinnamon, 8000 mg/100 g, sorghum bran 4600 mg/100 g), but their limited intake in diets renders them poor or negligible nutritional contributors. On the contrary, berries like chokeberries (660 mg/100 g), cranberries (420 mg/100 g), blackcurrants (140 mg/100 g), blueberries (330 mg/100 g) and grapes (80 mg/100 g) can represent consistent sources in some westernized diets as they can be consumed in sufficient amounts. Common fruits like apples (120–250 mg/100 g according to the cultivar), nectarines (30 mg/100 g), apricots (10 mg/100 g), peaches (70 mg/100 g), strawberries (50 mg/100 g), plums (240 mg/100 g) and even dates (10 mg/100 g) can be considered as relevant contributors too, provided that they are consumed raw, without peeling. Most nuts, in particular if consumed raw, without removing the skin that coats the cotyledons, offer high PA contents, with hazelnuts and pecans being the highest (510 mg/100 g), followed by almonds (180 mg/100 g) and walnuts (70 mg/100 g). Regarding cereals, PAs are not present in corn, rice or wheat, while barley (100 mg/100 g, mostly oligomers) and buckwheat (45 mg/100 g) are potentially good sources. Vegetables are usually scarce sources of PAs and, among legumes, only beans are rich in these polyphenols: red beans can harbor up to 500 mg/100 g and pinto beans can reach 770 mg/100 g. Dark chocolate is likely the most well-known source of PAs, given the favorable combination of consumption potential and content (up to 1700 mg/100 g). PAs are present in adequate amounts only in red wines as a consequence of different winemaking techniques, with an average of 10 mg/100 g for trimers and 30 mg/100 g for dimers, but with maximum values reported at 500 mg/100 g. Due to the presence of PAs in both hops and barley, some beers may contain moderate amounts of PA, at a level of approximately 20 mg/100 g. In raw and processed foods, procyanidins and prodelphinidins are more frequently found, while propelargonidins are relatively rare.

While the amount of data available for the content of PAs in plant foods is adequate, albeit far from complete, evidence concerning PA dietary intake is somewhat lacking as it has been evaluated only in few westernized populations. The mean intake for the general population in the United States is estimated at 53 mg/day/person for polymers and oligomers with DP above 2, while for highly polymerized PA the dietary intake for the Spanish population has been recently estimated at 450 mg/day/person. Instead, in the Finnish population one can expect a daily intake of 139 mg in females and 115 in males. In a detailed overview of a whole Spanish diet, the average intake of procyanidins for legumes (130 mg/g), fruits (308 mg/g) and nuts (11 mg/g) was estimated.

Auger et al.[9] provided an estimation of catechins and procyanidins contents in foods and beverages included in the Mediterranean diet, reporting that fruits and vegetables such as plums, apples, strawberries, other berries, lentils, chocolate and beans may provide a good amount of catechins and procyanidins, with maximum values observed for plums, some apple varieties and broad beans. On average, chocolate and apples contained the largest procyanidin content per serving (164.7 and 147.1 mg, respectively) compared with red wine and cranberry juice (22.0 and 31.9 mg, respectively). However, the procyanidin content varied greatly between apple samples (12.3–252.4 mg/serving) as different cultivars and varieties have distinct PA accumulation patterns. It is evident that only a partial picture is available in terms of PA dietary intake and an evaluation of their actual consumption by different populations is needed, possibly including the wider range possible in terms of DP and populations with different dietary habits. For instance, no evaluations of the dietary intake of PAs have been performed for Asiatic and African populations and, with the sole exception of the Mediterranean diet, no traditional or specific diet has been screened in this regard.

These considerations are strengthened by the peculiar fate of PAs in the gastrointestinal tract. In fact, only polyphenols released from the food matrix by the action of digestive enzymes (small intestine) and bacterial microbiota (large intestine) are bio-accessible in the gut and therefore potentially bioavailable. As mentioned above, the presence of abundant PA bound to cell wall polysaccharides could determine the amount of bioaccessible food polyphenols, which may differ quantitatively and qualitatively from polyphenols included in food databases. Furthermore, the biotransformation of PAs by means of the colonic microflora could impart dramatic changes to the range of molecules that are absorbed at the colonic level and distributed in systemic circulation (Figure 6.1).

Fate of Proanthocyanidins through the Digestive Tract

Epidemiological evidence shows that the consumption of polyphenol-rich foods reduces the risk of CVD and associated conditions, and human intervention studies have supported this association. A human diet containing significant amounts of PAs has been associated with reduced disease risks, but the mechanism of action has not

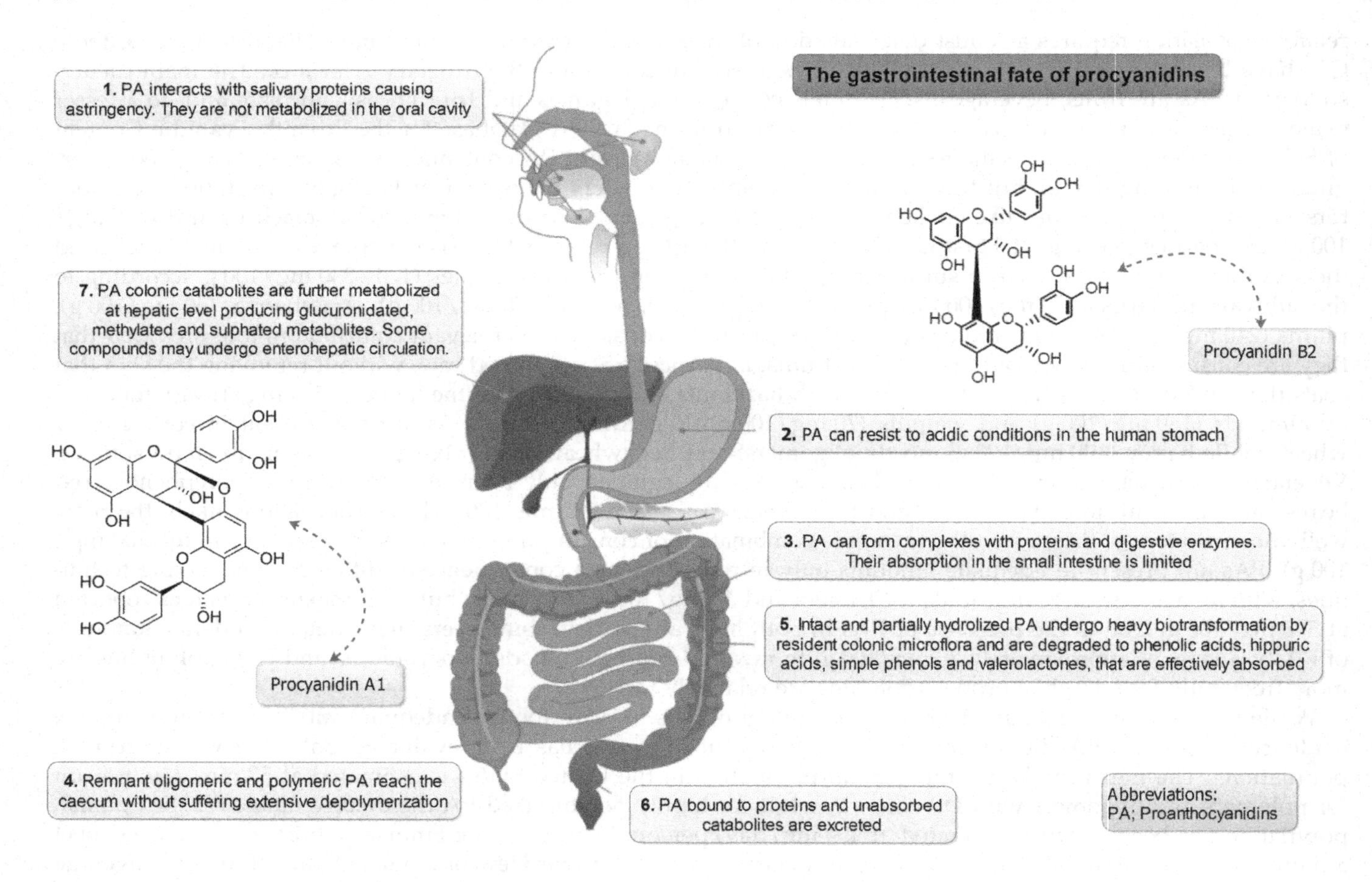

FIGURE 6.2 The metabolic fate of PAs.

been adequately elucidated to date. As mentioned previously, present knowledge suggests that it is very unlikely that these substances, in particular CTs with high DP, are absorbed intact and reach the target tissues; most PAs are not absorbed in the upper gastrointestinal tract and enter the colon where they undergo wide rearrangements and bio-transformations through the action of colonic microbes (Figure 6.2). However, despite various reports that have emerged in the last decade, very little is known about the catabolic and metabolic fate of tannins and their relapses on bioavailability. In particular, various catabolic pathways have been described, but the knowledge regarding the variability of the human microbiome and the effects of bacterial successions in the colon remains scarce. Unlike other flavonoid subclasses, PAs are almost unabsorbed in the upper gastrointestinal tract, especially those with a DP above 3. Thus, under physiological circumstances, more than 90% of the ingested PAs can reach the colon, where their fate is closely related to the presence of resident microflora. Although the absorption of certain other phenols, polyphenols and tannins (PPT) such as procyanidins, chlorogenic acids and anthocyanins has been described in the literature, the levels of the parent compounds in blood after a high dose, or a large amount of PPT-rich food, are very low compared with other flavonoids. In contrast, intervention studies show that procyanidin-rich, chlorogenic acid-rich or anthocyanin-rich foods affect certain biomarkers, but because the concentration of parent compounds to be expected in blood after consumption is too low to affect these biomarkers, other bioactive substances such as metabolites may be responsible and must be identified.

For instance the stability of PAs was followed in humans by regularly sampling the gastric juice with a gastric probe after ingestion of PA-rich chocolate, confirming that they are not degraded in the acidic conditions of the stomach. The extent of biliary excretion of polyphenols has so far not been assessed in humans. Indeed, a very low permeability was observed for PAs when investigating the transport of oligomers across the human intestinal epithelial Caco-2 cells. Moreover, the increase of DP leads to a further decrease of the transport ratios. For instance, the cranberry A-type dimer shows a transport ratio equal to 0.6%, while for trimer and tetramer permeability, ratios of 0.4 and 0.2%, respectively, were observed. Therefore, these oligomers travel further down the

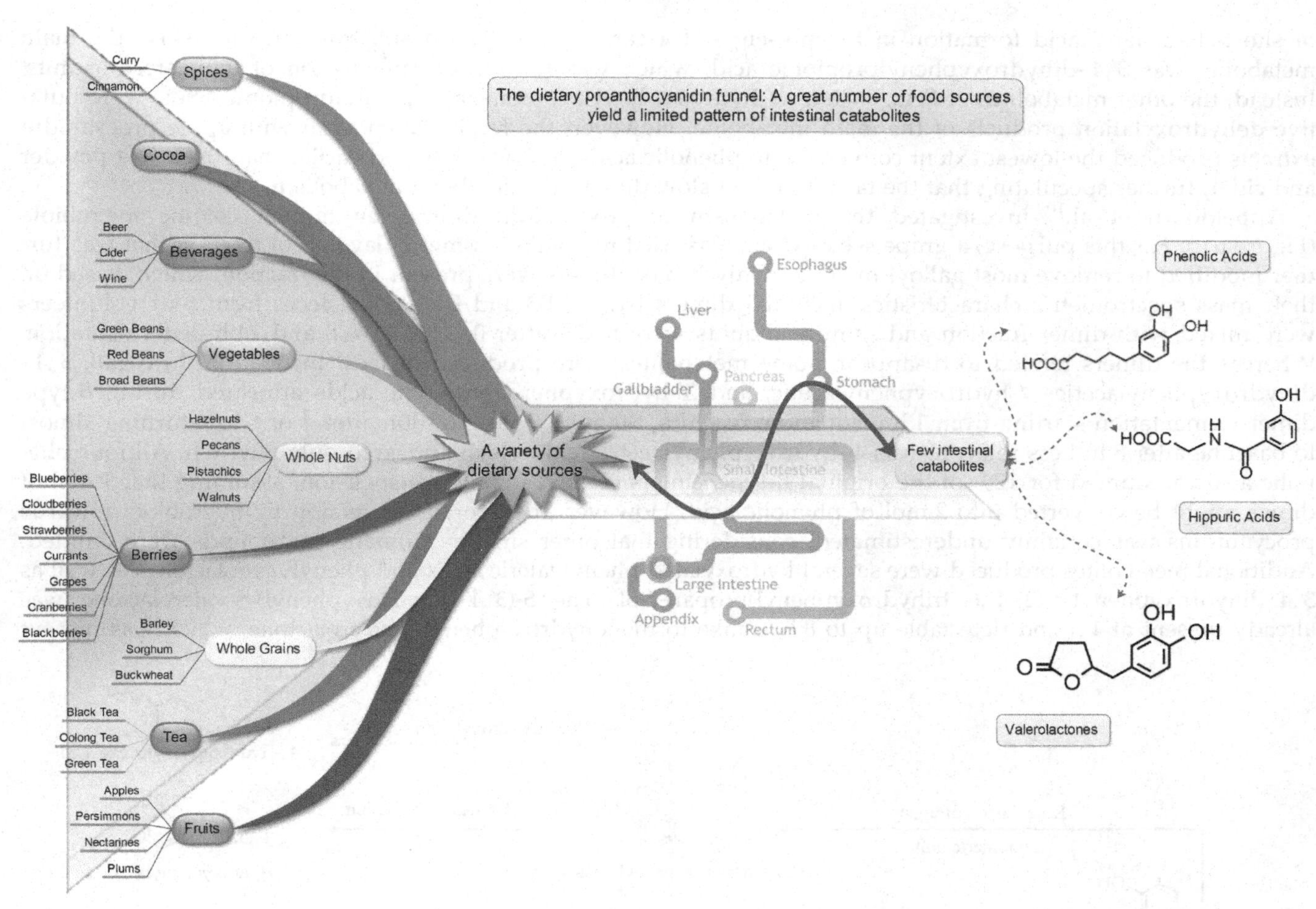

FIGURE 6.3 The dietary PA funnel. Different kinds of PA sources yield a limited pattern of metabolites.

gastrointestinal tract, as confirmed in ileostomists, where 90% of the apple-derived oligomeric procyanidins reach the colon under physiological circumstances. This confirms their availability for colonic biotransformation by resident microflora, which are able to generate several phenolic acids and other smaller aromatic compounds more readily absorbed *in situ*. Therefore, the various health benefits ascribed to PAs, usually unabsorbed as such, might be related not to the direct actions of tannin themselves but to the actions of their colon-derived metabolites. The extreme complexity of this scenario, in which multiple PAs and polymers with a wide range of DP are ingested together in mixed polymers and are then biotransformed, has determined the need for a reductionist approach (Figure 6.3). Most preliminary studies in this regard have therefore been performed with selected microbial strains or with fecal starters and selected molecules.

In Vitro Biotransformation

Several experimental studies performed with human fecal microbes have elucidated several catabolic pathways towards monomeric flavan-3-ols, which are usually shared by CTs. In an *in vitro* fermentation with ^{14}C-labeled polymeric PAs, their degradation by human microbiota was evaluated for up to 48 h incubation. These polymers had an average DP equal to 6 and some dimers or trimers were observed in the ^{14}C-labeled fraction. The colonic microbial degradation decreased the PA level, in contrast to heat-inactivated microflora or without microflora incubation. The fermentation substrates were thus degraded into low-MW aromatic compounds including 3′-hydroxyphenylacetic acid, 3′-hydroxyphenylpropionic acid, 3′-hydroxyphenylvaleric acid, phenylacetic acid, 4′-hydroxyphenylacetic acid and phenylpropionic acid. These products accounted for 9–22% of the initial PA radioactivity, wherein the 3′-hydroxyphenylpropionic acid was the most abundant metabolite throughout 48 h. However, increasing the DP of PAs might inhibit their extension metabolism by colon microflora. This phenomenon appeared when inoculating apple-derived procyanidins with human feces, leading to the suppression of short-chain fatty acid formation in the presence of a carbohydrate-based substrate. In this work, the main metabolite was 3′,4′-dihydroxyphenylpropionic acid, which was formed by ring fission of (epi)catechin units. Instead, the other metabolites of PAs, such as 3′-hydroxyphenylpropionic and phenylpropionic acids, were putative dehydroxylation products of the main metabolite. However, the fecal fermentation with apple procyanidin extracts produced the lowest extent conversion to phenolic acids, and was stopped earlier than the apple powder and cider, further speculating that the tannins might slow down the microbiota metabolism.

Appeldoorn et al.[10] investigated the metabolism of procyanidin dimers by human colonic microbiota (Figure 6.4). For this purpose, a grape seed extract was used to obtain a dimeric flavan-3-ol fraction that was further modified to remove most galloyl moieties. Only B-type dimers were present in the fraction, which, based on their mass spectrometric characteristics, included dimers B1, B2, B3 and B4. Pooled feces from four volunteers were mixed with dimer fraction and sample aliquots were taken after 0, 1, 2, 4, 6, 8 and 24 h of fermentation. Whereas the dimers tended to disappear, some metabolites were produced in fecal incubation. In detail, 3′,4′-dihydroxyphenylacetic, 3′-hydroxyphenylacetic, and 3′-hydroxyphenylpropionic acids appeared during B-type dimer fermentation starting from 1 h incubation, reaching the peak concentration after 4 or 6 h, returning almost to baseline after 8 h. Less relevant was 4′-hydroxyphenylacetic acid, which appeared only after 6 h. All four phenolic acids accounted for 12% of the original B-type dimers content in batch suspension, assuming that 1 mol of dimer might be converted into 2 mol of phenolic acid. However, the overall conversion by microbiota towards procyanidins was certainly underestimated, considering that other smaller aromatic compounds were omitted. Additional metabolites produced were several hydroxylated phenylvaleric acids and phenylvalerolactones as well as 3′,4′-dihydroxyphenyl-3-(2″,4″,6″-trihydroxyphenyl)propan-2-ol. The 5-(3′,4′-dihydroxyphenyl)-γ-valerolactone was already present at 1 h and detectable up to 8 h, unlike to monohydroxyphenyl-γ-valerolactone, which was present

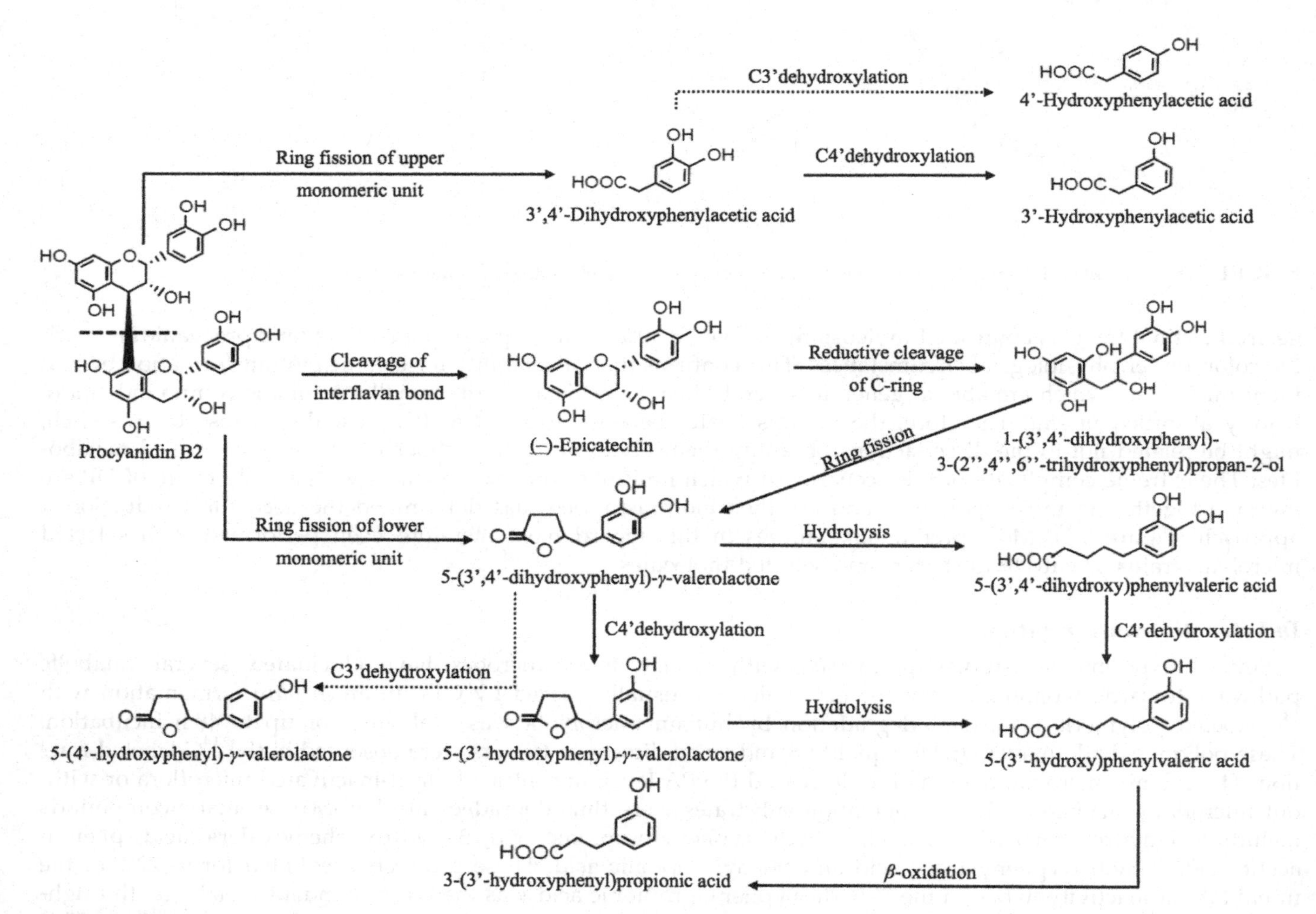

FIGURE 6.4 Tentative catabolism route for human microbial degradation of B-type procyanidin dimmers. The favorite routes are indicated with solid arrows, and minor routes with dotted arrows. *Adapted from Refs 10 and 14.*

only at 6 h, and was the same as 5-(3′,4′-dihydroxyphenyl)valeric acid. Instead, its monohydroxylated phenylvaleric acid appeared also at 8 h. With regard to monohydroxylated phenylvalerolactone and phenylvaleric acid, it was not possible to distinguish the right OH position on the C ring. However, it is plausible that these metabolites retain the 3′-OH position (*meta*), since human microbiota have a preference to dehydroxylate the 4′-OH position (*para*). This phenomenon might be postulated considering the trace amount of 4′-hydroxyphenylpropionic acid compared to its isomer 3′-hydroxyphenylpropionic. Moreover, the favorite dehydroxylation in the *para*-position by human colonic microbiota has been revealed previously for several flavonoids.[11–13] Furthermore, Appeldoorn et al.[10] proposed possible degradation routes for procyanidin B-type dimers. The detection of 3′,4′-dihydroxyphenyl-3-(2″,4″,6″-trihydroxyphenyl)propan-2-ol within 2 h suggests that dimers are cleaved into their monomeric units. However, no additional monomers besides those formed during extraction were detected. Therefore, a slow conversion of dimers into flavan-3-ols followed by a rapid conversion into the 3′,4′-(dihydroxyphenyl)propanol derivative, phenylvalerolactones, and phenolic acids seems to explain the reason for the lack of flavan-3-ol intermediates. However, the degradation pathway which involves the breaking of the interflavan bond seemed to disagree with the recovery of 3′,4′dihydroxyphenylacetic acid among the main metabolites. During the formation of phenylvalerolactones and phenolic acids, cleavage reactions at the level of the C- and A-ring are involved. Another degradation step seems to closely link the phenylvalerolactones with phenylvaleric acids, because of the slower formation of the latter metabolites.

One year later, Stoupi et al.[14] confirmed that microbial biotransformation favored the removal of the 4′-OH rather than the 3′-OH on the C ring, in a study which aimed to compare the catabolism of (−)-epicatechin with that of procyanidin dimer B2. Focusing the attention on dimer compounds, the microbial colonic fermentation led to formation of epicatechin in the incubated samples, derived from cleavage of the C4–C8 interflavan bond. The monomer appeared in anaerobic incubation starting from 3 h and up to 12 h, with maximum recovery at 6 h equal to 6%. The C-ring cleavage formed 1-(3′,4′-dihydroxyphenyl)-3-(2″,4″,6″-trihydroxy phenyl)propan-2-ol and 1-(hydroxyphenyl)-3-(2″,4″,6″-trihydroxyphenyl)propan-2-ol. As for epicatechin, this 3′,4′-dihydroxylated metabolite was recovered at the same time points, unlike its monohydroxylated counterpart, which appeared after 6 h and up to 24 h as a result of a further dehydroxylation step. Both metabolites reached the maximum recovery at 12 h, which was about 13 and 8% for dihydroxylated and monohydroxylated propanol, respectively. The ring fission product 5-(3′,4′-dihydroxyphenyl)-γ-valerolactone was produced after 6 h and up to 24 h, even if the maximum recovery equal to 27% was obtained after 12 h incubation. Despite this metabolite being derived from the cleavage of (epi)catechin units, some differences emerged among (−)-epicatechin and procyanidin dimer B2. Indeed, the monomer was metabolized to a lesser extent and slower than dimers into this valerolactone, considering its appearance between 9 and 48 h after monomer incubation with a maximum recovery of 21% at 24 h. Less relevant after dimer B2 incubation, the 5-(3′-hydroxy phenyl)-γ-valerolactone was recovered only at 24 h with a recovery of 3%. As for the monohydroxylated valerolactone, five other metabolites appeared after 24 h. In detail, phenylacetic acid, 3-(3′,4′-dihydroxy phenyl)propionic acid, 5-(4-hydroxy)-(3′,4′-dihydroxy)phenyl valeric acid, 5-(3′,4′-dihydroxy phenyl)valeric acid and 5-(3′-hydroxy phenyl)valeric acid. Among these late catabolites, 5-(3′-hydroxy phenyl)valeric acid and phenylacetic showed the major peak recovery at 48 h, about 9 and 29%, respectively. However, 3-(3′-hydroxy phenyl)propionic acid was the dominant catabolite among the eleven catabolites derived from the biotransformation of dimer B2, especially during the period from 24–48 h. Propionic acid started to occur at 6 h, but the maximum yield (53%) was obtained at 24 h, with molar recovery that remained high at the last time point with a recovery closed to 49%. This deepened study carried out by Stoupi et al.[14] allowed to them to hypothesize the catabolic pathways involved in the flavan-3-ol biotransformation by human fecal microbiota, including the reaction undergone by dimer B2 (Figure 6.4). As reported by Appeldoorn et al.,[10] the breakdown of the interflavan bond represents a minor metabolic route. However, the findings obtained after procyanidin B2 fermentation indicate that a feature of flavan-3-ol catabolism hinted at the conversion into phenylvalerolactones (C6–C5) and then into phenylvaleric acids (C6–C5). Then, the progressive shortening of the aliphatic chain by α- and β-oxidations would lead to the formation of the C6–C3, C6–C2 and C6–C1 skeletons and repeated dehydroxylations, mainly involving C4′ and the aliphatic side chain. This shortening chain process could be further investigated in an *in vitro* study aimed at evaluating the metabolites produced after microbial colonic fermentation of different polyphenol rich-foods. In particular, the appearance of 3′,4′-dihydroxyphenylacetic acid (C6–C2) after the fermentation of dark chocolate that is rich in oligomeric procyanidins might be linked to the formation of 3,4-dihydroxybenzoic acid (also known as protocatechuic acid) (C6–C1) and its dehydroxylation product hydroxybenzoic acid (at position 3 or 4).

Notwithstanding the fact that the fecal microbiota usually produces procyanidin-derived catabolites with low MW, Stoupi et al.[15] revealed the partial characterization of some procyanidin B2 catabolites with a higher molecular mass than the monomer (290 Da). In detail, 24 high-MW catabolites were detected up to 9 h, with an overall count equal to 20% of the dimer B2. Seven catabolites yielded a 16-Da molecular mass larger than procyanidin B2, suggesting therefore the presence of an additional OH group on the procyanidin aromatic rings. Two catabolites have shown a 6-Da larger MW than dimer B2, which, by means of LC-MS^3 analyses, suggesting the presence of at least one intact reduced form of the epicatechin unit. Five catabolites were 4 Da larger than procyanidin B2, whereas three isomers had mass spectrometry behavior that was consistent with the presence of both reduced flavan-3-ol units; in addition, two compounds kept the epicatechin structure intact, arguing that both reductions occurred in the same monomeric unit. Four catabolites had a molecular mass that were 2 Da larger than dimer B2, as a result of the flavan-3-ol reduction in one of the two epicatechin units. However, these results did not fit with those obtained in a previous work on the metabolism of procyanidin dimers by human fecal microbiota, where no high-MW catabolites or catabolites with an intact A-ring were detected. Therefore, the differences highlighted in this study might be due to different microbial patterns in the human colon which does not show the same capability for splitting the interflavan bond or for reducing the monomeric units.

In a 2012 study, the microbial metabolism of a grape seed flavan-3-ols by human fecal microbiota was evaluated using a batch incubation of three different donors. Non-galloylated and galloylated dimers and trimers represented almost 50% of the total quantified flavan-3-ols, which was slightly less than the level of monomers. Among procyanidins, the dimers B1, B2, B3 and B4, the 3- and 3′-gallate forms of dimer B2, and the two trimers C1 and T2 were used to assess their catabolism by colonic microbiota up to 48h incubation with different sampling times. Several phenolic acids were recovered after fecal fermentation, such as hydroxyphenylpropionic, hydroxyphenylacetic, hydroxybenzoic and hydroxymandelic acids. Moreover, peculiar metabolites exclusively derived from flavan-3-ol catabolism such as diphenylpropan-2-ol, phenyl-γ-valerolactones and phenylvaleric acid derivatives were evaluated. Besides the different metabolites generated, huge differences emerged amongst the fecal donors for: (i) the rate of procyanidin biotransformation; (ii) capability for the generation of specific metabolites; and (iii) amount of metabolites yielded in the fecal suspension. In one of the three donors, a slower degradation rate towards non-galloylated dimers B1, B2 and B3 and trimers C1 and T2 was observed. Also, another volunteer exhibited an astonishing increase of dimeric and trimeric flavan-3-ols after 5 h incubation. Moreover, a relevant temporary increase was observed for procyanidin B4 in all three volunteers during the 5th or 10th hour, depending on the volunteer. Almost all procyanidin dimers and trimers disappeared completely within 24 h of fermentation. Thus, the microbial metabolism led the formation of several ring fission metabolites that allowed their metabolic pathways to be postulated, based on time appearance. The metabolite 5-(3′,4′-dihydroxyphenyl)-γ-valerolactone was derived from heterocyclic C-ring cleavage followed by a further A-ring fission. This metabolite appeared within 5 h, reaching the peak concentration at 10 h and returning to baseline levels at 24 h in two donors, while in one donor it was quantified up to 48 h. The metabolic link between phenylvalerolactones and phenylvaleric acids was considered obvious because of the kinetics shared by them. This was shown for three donors for 5-(3′,4′-dihydroxyphenyl)-γ-valerolactone and 4-hydroxy-5-(3′,4′-dihydroxyphenyl)-valeric acid. At the same time, the ability of the donors to remove the 4′ OH from the C-ring affected the formation of 5-(3′-hydroxyphenyl)-γ-valerolactone and 4-hydroxy-5-(3′-hydroxyphenyl)-valeric acid, as these two compounds had similar generation profiles, as also shown for their dihydroxylated counterparts. For example, the 24 h-faster producer of 5-(3′-hydroxyphenyl)-γ-valerolactone, was also faster for 4-hydroxy-5-(3′-hydroxyphenyl)-valeric acid, while the slower producer showed the peak recovery for these two metabolites at 48 h. In contrast, 4-hydroxy-5-(phenyl)-valeric acid, which showed a slow accumulation in all donors, and led to an increase with fermentation lengthening, peaked at the last time point. Moreover, the formation of the simple γ-valerolactone metabolite, with a similar profile as dihydroxyphenyl-γ-valerolactone, was demonstrated for the first time, therefore indicating the removal of the aromatic moiety. Besides C6–C5 metabolites as phenylvalerolactones and phenylvaleric acids, the *in vitro* fermentation of the grape seed extract led to the production of 3-(3′,4′-dihydroxyphenyl)-propionic acid (also known as dihydrocaffeic acid), starting from 5 h and reaching a peak recovery at 10 h before decreasing after 24 h. However, the dihydrocaffeic acid was produced faster and in a minor amount compared to 3-(3′-hydroxyphenyl)-phenylpropionic acid, which is the metabolite formed by the subsequent dehydroxylation of dihydrocaffeic acid. Other typical metabolites of flavan-3-ol skeleton had a C6–C2 skeleton as phenylacetic acid derivatives. In particular, the 3′,4′-dihydroxyphenylacetic and 3′-hydroxyphenylacetic acids had a marked increase in one of the three donors. These two metabolites might be produced by ring fission of the upper flavan-3-ol unit of procyanidin B-type dimers or by the α-oxidation of the two phenylpropionic acid derivatives cited above. Instead, the 4′-hydroxyphenylacetic and free phenylacetic acids were not relevant, despite their increasing levels remaining for up to 48 h in some cases. Indeed, these two compounds may be derived from colonic microbial metabolism of substances that are already present in the fermentation media, such as phenylalanine and tyrosine.

A similar study involved grape seed extract, and also investigated the microbial biotransformation of monomeric fraction (58% monomers and 42% procyanidins) and rich-oligomeric fraction (7% monomers and 93% procyanidins) using gas chromatography–mass spectrometry analysis. A general trend was observed for the higher formation of some phenolic acids in monomer fermentation with respect to oligomers, indicating a lower availability of oligomeric procyanidins compared to microbial colonic enzymes in breakdown reactions. Focusing on the oligomer fermentation, some small compounds are typical of flavan-3-ol structure, wherein the pathway generation may also be postulated. Of particular relevance are 3′,4′-dihydroxyphenylacetic and 3′-hydroxyphenylacetic acids. The former has not been detected in control samples but was found at a level of 2.1 mg/L after 10 h-incubation with the oligomeric fraction, while the latter increased by 35 times compared to the control groups at the last collection point. GA was already evident after 30 min incubation with a concentration close to 0.5 mg/L, reaching the maximum concentration equal to 1.9 mg/L at 10 h, which was the second and last time point. Trihydroxybenzoic acid was released after the removal of the galloyl group from the (epi)catechin gallates that exist in the grape seed procyanidins. Then, the further decarboxylation of GA led to the formation of pyrogallol, with its appearance only evident at 10 h at a concentration of about 150-times lower than the parent phenolic acid. At the same time, the presence of protocatechuic acid might be related to dehydroxylation of GA. A low concentration was observed for 3′4′-dihydroxyphenylpropionic acid at 10 h fermentation, while the 3′-monoydroxylated counterpart increased by 17 times (2.7 mg/L) compared to control samples at the last time point. However, this study did not control the production of phenylvalerolactones and phenylvaleric acid, unlike a subsequent work carried out by Cueva et al.[16] using UPLC tandem mass detection. In detail, this study aimed to evaluate the small catabolites produced after grape seed fraction fermentation, trying to correlate changes in precursor flavan-3-ols and/or microbial metabolites whit changes in human microbial group populations. The oligomeric fraction was characterized to its procyanidin and monomer content equal to 78 and 21%, respectively. The generation of phenolic catabolites besides the fecal bacteria population was assessed at baseline incubation and 5, 10, 24, 30 and 48 h. Both monomeric and oligomeric fractions led to a decrease in the *Clostridium histolyticum* population compared to control samples, starting from 5 h for monomers and 10 h for oligomers, which was then extended up to the last time point. Moreover, the decrease was also confirmed by comparison with the positive control (prebiotic supplement). An increase in *Lactobacillus/Enterococcus* up to 10 h was observed with both substrates, although this was not significant for oligomers. With regard to the formation of flavan-3-ol catabolites, the peak concentration of 5-(3′,4′-dihydroxyphenyl)-γ-valerolactone and 4-hydroxy-5-(3′,4′-dihydroxyphenyl)-valeric acid was observed 10 h after oligomer fermentation, which was earlier than for monomeric fraction where the highest concentration was recovered. Their monohydroxylated counterparts, 5-(3′-hydroxyphenyl)-γ-valerolactone and 4-hydroxy-5-(3′-hydroxyphenyl)-valeric acid, shared the kinetic generation. After oligomer incubation, these compounds reached the peak formation after 10 h and then decreased before 48 h, unlike the opposite extreme, which occurred with monomers. Instead, 4-hydroxy-5-(phenyl)-valeric acid showed a progressive accumulation from 10 to 48 h for both fractions, while the free form of γ-valerolactone did not appear in incubated oligomer samples. Among C6–C3 metabolites, non-significant differences were seen for dihydroxyphenylpropionic acid and its 3′- and 4′-monoydroxyl counterparts, while for non-hydroxylated phenyl propionic acid, a marked increase was only observed during 24–48 h in the oligomer-derived sample. However, focusing on C6–C2 molecules, the dihydroxyphenylacetic acid showed the largest increase in the presence of grape seed oligomers, which agreed with the results postulated by Appeldoorn et al.,[10] who argued the involvement of the C-ring cleavage of the upper flavan-3-ol unit. Moreover, the same trend was seen for 3′-hydroxyphenylacetic acid, which is the more relevant dehydroxylation product of the dihydroxylated C6–C2 metabolite. In contrast, 4′-hydroxyphenylacetic acid and phenylacetic acid had the same generation profile fermentation for both fractions. Although the interaction of grape seed flavan-3-ols, comprising oligomers, with human colonic microbiota led to the production of interesting changes in the microbial population, Cueva et al.[16] did not observe a correlation with ring fission catabolites produced during fermentation.

The interaction between the food source of polyphenols and the gut microbiota is a critical determinant of polyphenol bioavailability and the related potential health benefits, as it has been demonstrated that the different polyphenolic phytocomplexes obtained from different food sources (e.g., green tea and red wine) may be catabolized with a different pattern by the same intestinal microbiota in gastrointestinal models. Furthermore, their catabolism and the accumulation of different microbial catabolites appear to be dependent on colon location. For instance, during red wine/grape extract feeding, GA and 4-hydroxyphenylpropionic acid remained elevated throughout the colon, whereas they were consumed in the distal colon, where 3-phenylpropionic acid was strongly produced, during black tea extract consumption.[17]

In Vivo Biotransformation

As previously reported, PAs may account for a major fraction of the total flavonoids ingested by Western populations. As cited in the previous section, the interaction of these CTs with human colonic microflora allowed a plethora of catabolites, which may be useful as *in vivo* biomarkers of PA-rich products, to be pinpointed. The cocoa-derived products, such as cocoa powder or dark chocolate, are very interesting because of the dominant content of oligomeric and polymeric procyanidins in respect to monomeric flavan-3-ols. In a study aimed at estimating the urinary change of several microbial phenolic acids in 11 humans who ingested 80 g chocolate containing 439 mg procyanidins and 147 mg monomers, chocolate intake increased the urinary excretion of several phenolic acids: 3′-hydroxyphenylpropionic acid, 3′,4′-dihydroxyphenylacetic, 3′-hydroxyphenylacetic, ferulic, vanillic, and 3-hydroxybenzoic acids. The excretion of these small aromatic compounds was quantified by HPLC-MS/MS, and revealed an increase in urine during 6–48 h, except for vanillic acid, which showed a rapid clearance reaching the peak excretion and complete excretion after 3 and 6 h, respectively. However, the rapid excretion of vanillic acid seemed to be related to the content of vanillin in the chocolate rather than the end-product of microbial conversion of cocoa procyanidins. Also, ferulic acid, which is not usually reported as a flavan-3-ol catabolite, may probably have been derived from the metabolism of clovamide, an amide derivative of caffeic acid in cocoa.

In a subsequent study, 21 healthy humans ingested 40 g of water-soluble cocoa powder. Its flavan-3-ol fraction includes 23% monomers, 13% dimers, and 64% oligomers, with DP comprising between 3 and 8. In humans, all of the screened metabolites increased in urine within 24 h of cocoa consumption, with the exception of phenylacetic acid. However, significantly different excretion, occurring with respect to the control diet, was observed for caffeic, ferulic, 3′-hydroxyphenylacetic, vanillic, 3-hydroxybenzoic, 4-hydroxyhippuric, and hippuric acids. Moreover, as a result of the application of mass spectrometry, two other flavan-3-ol metabolites as 5-(3′, 4′-dihydroxyphenyl)-γ-valerolactone, and its 3′-methoxy counterpart were also identified in human urine. As mentioned above, the dihydroxylated valerolactone is a typical ring fission catabolite of procyanidin, while 5-(3′-methoxy-4′-hydroxyphenyl)-γ-valerolactone, which was not identified as an *in vitro* catabolite, is derived from further interaction with human catechol-*O*-methyl-transferase (COMT). The same research group then evaluated the milk effect on microbial-derived phenolic acids, excreted after a single dose (40 g) of cocoa powder consumed with water or milk. Among the 15 screened phenolic acids, nine showed a urinary increase in both groups. However, the consumption of cocoa with milk reduced the excretion of almost all catabolites, including 3′,4′-dihydroxyphenylacetic, 3,4-dihydroxybenzoic, 4-hydroxybenzoic, 4-hydroxyhippuric, and hippuric, caffeic and ferulic acids. However, the latter two phenolics have little relevance because of their unrelated formation with procyanidin pathway degradation, probably involving modifications of hydroxycinnamic acid amides that are present in cocoa products. Only vanillic and phenylacetic acids showed an upward trend in urine following consumption of the test meal plus milk. A decrease in the urinary excretion of phenolic acid was also observed for other polyphenolic compounds, wherein the subjects ingested orange juice flavanones with yoghurt. These findings confirmed that the food matrix modifies the polyphenols-microbiota interaction at the colonic level.

Forty-two volunteers were recruited in another study, and consumed two daily doses of 20 g with 250 mL of skimmed milk each for a period of 4 weeks. A significant increase in the urinary excretion of different conjugates of 5-(3′,4′-dihydroxyphenyl)-γ-valerolactone was observed in non-hydrolyzed 24-h urine samples after the consumption of cocoa with respect to sole milk consumption. This ring fission product, once absorbed at the colonic level, is extensively conjugated by human phase II enzymes so that it is not recovered in the free form. The main conjugated form was formed after interaction with sulfotransferases to produce sulfate conjugates. The urinary increase in the cocoa-group of dihydroxyphenylvalerolactone-sulfates exceeded 100% and 400%, depending on the isomer. However, also glucuronide conjugates were also formed by the interaction of free dihydroxyphenylvalerolactone with UDP-glucuronosyltransferases. The increase of two glucuronides was close to 190%. Lower excretion levels occurred for methylated forms of dihydroxyphenylvalerolactone-sulfate and dihydroxyphenylvalerolactone-glucuronide, even if they were increased to about 200% the level of the control group. Instead, only glucuronide conjugates of dihydroxyphenylvalerolactones increased in plasma samples. Urinary and plasma samples were then hydrolyzed with β-glucuronidase/sulfatase to obtain the free and methylated forms of microbial catabolites. Cocoa supplementation significantly increased the urinary excretion of colonic microbial-derived phenolic catabolites, including vanillic, 3′,4′-dihydroxyphenylacetic and 3′-hydroxyphenylacetic acids, and dihydroxyphenylvalerolactone. However, the largest excretion of vanillic acid was probably due to cocoa-derived vanillin, as postulated by Rios et al.[18] Focusing on plasma-hydrolyzed samples, only 3′-hydroxyphenylacetic acid and dihydroxyphenylvalerolactone showed a significant increase, to 100 and 336%, respectively. In a subsequent study on cocoa intervention in humans with a high cardiovascular risk, volunteers showing an increase of urinary 3′-hydroxyphenylacetic and vanillic acids also presented a significant improvement in plasma high-density lipoprotein (HDL)-cholesterol concentration compared to their non-excretor counterparts. At the same time, the subjects with a decrease in plasma-oxidized low-density lipoprotein (LDL) concentration had a significant increase in the urinary excretion of 3′-hydroxyphenylacetic and vanillic acids. These findings may focus attention on some microbial-derived catabolites that appeared *in vivo* after cocoa consumption, leading to a relationship between these catabolites and cardiovascular benefits attributed to cocoa flavan-3-ols to be postulated.

A pilot study investigated the plasma and urinary catabolites produced in humans after ingestion of almond skin polyphenol extract. The flavan-3-ol fraction accounted for half of the flavonoid content. PAs accounted for more than 50% of total flavan-3-ols, including procyanidin and propelargonidin dimers (B-type and A-type), besides trimer C1. Focusing on flavan-3-ol catabolites produced by human gut microbiota, dihydroxyphenyl-γ-valerolactone showed an astonishing increase in concentration in urine. The glucuronide derivatives of this ring fission product showed 163- and 218-fold increases after almond skin intake, while the sulfate derivative showed a 100-fold increase compared to baseline urines. Instead, glucuronide and sulfate forms of 5-(methoxy-hydroxyphenyl)-γ-valerolactone rose, reaching a 50-fold increase for one isomer, at best. In hydrolyzed urinary samples, besides the free form of 5-(3′,4′-dihydroxyphenyl)-γ-valerolactone, 5-(hydroxyphenyl)-γ-valerolactone was also detected. The appearance of this catabolite may derive from the dehydroxylation of the 3′,4′-dihydroxylated counterpart, along with ring fission of (epi)afzelechin units of almond skin-derived propelargonidins. The key role of human gut microbiota on almond skin flavan-3-ols was further confirmed by a non-targeted metabolomic study on new urinary flavan-3-ols. The findings unravel several conjugates (mainly sulfates and glucuronides) of hydroxyphenyl-valeric, hydroxyphenyl-propionic, and hydroxyphenyl-acetic acids, which may be associated with almond flavan-3-ol intake. The urinary excretion of glucuronide derivatives of the mono- and dihydroxylated forms of both phenylpropionic and phenylacetic acids mainly correlated with urinary excretion up to 6 h in comparison to the control group, whereas sulfate derivatives significantly contributed to urinary excretion at 10 and 24 h. The excretion profile of nine different phenyl-γ-valerolactone conjugates was related to significant changes in the urine profile from 2–24 h following almond skin consumption. A strong correlation was observed from 6–10 h for most conjugates of both monohydroxyl- and dihydroxyl-phenyl-γ-valerolactones, unlike trihydroxyphenyl-γ-valerolactone glucuronide, which significantly changed during the 10- to 24-h period. This trihydroxy ring fission product was not reported in the previous pilot study, but may be formed via (epi)gallocatechin ring fission, since prodelphinidins up to 6 DP containing only one (epi)gallocatechin unit occur in almond skin. Then, the formed trihydroxyphenyl-γ-valerolactone undergoes glucuronidation by interaction with human UDP-glucuronosyltransferases. As cited above in the "*in vitro* biotransformation" section, the phenyl-γ-valerolactones and 4-hydroxy-5-(phenyl)-valeric acids shared the same pathway generation. Therefore, 4-hydroxy-5-(phenyl)-valeric acid conjugates (methyl, sulfates, glucuronides) presented an excretion pattern similar to phenyl-γ-valerolactones, contributing to urinary changes from 2 to 24 h. However, the excretion profile was different based on the non-, mono- and dihydroxyl groups occurring at the level of phenyl ring of aromatic valeric acids. The changes of dihydroxylated derivatives showed the highest correlation at 6–10 h after ingestion, while, for monohydroxylated counterparts, significant changes were observed up to 24 h. Instead, the sulfate form of 4-hydroxy-5-(phenyl)-valeric acid changed at 10–24 h. These findings clearly suggest that dehydroxylation steps gradually occur as metabolism progresses in the large intestine.

HYDROLYZABLE TANNINS (GALLOTANNINS AND ELLAGITANNINS)

HTs are, together with CTs, the main group of plant tannins and are structurally perhaps the most complex group of tannins.[6] HTs are polyesters of a sugar moiety and phenolic acids. These compounds are easily hydrolyzed by diluted acids, bases, hot water, and enzymatic activity (tannases); because of this, they are termed "hydrolyzable tannins." They are divided into two subclasses according to their structural characteristics: GTs and ETs. If the phenolic acid is GA (3,4,5-trihydroxyl benzoic acid), the compounds are called GTs. On the other hand, ETs are characterized by the presence of at least one HHDP group, which spontaneously rearranges into EA when hydrolyzed. Most ETs are mixed esters with both HHDP and GA (Figure 6.5).[19,20]

Depending on the origin of HTs, their chemistry varies widely, with the molar mass ranging from 300 Da to 3000 Da, and even up to 20,000 Da. It is worth mentioning that some plants produce either ETs or GTs, whereas in other species both of them, and even CTs, have been described.[4] On the other hand, HTs are traditionally thought

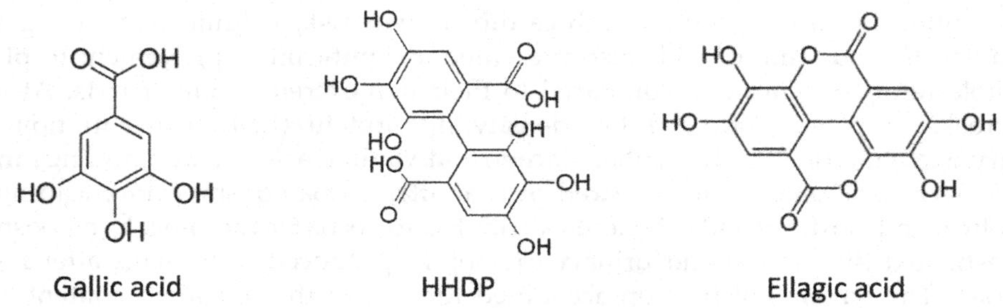

Gallic acid HHDP Ellagic acid

FIGURE 6.5 Structural units of HTs.

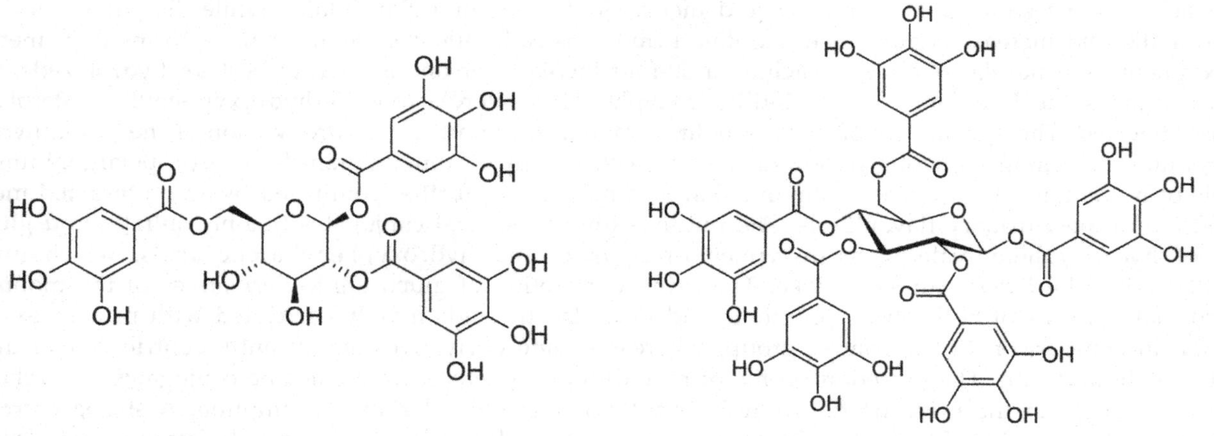

FIGURE 6.6 Some representative GTs. 1,2,6-trigalloyl glucose (left) and 1,2,3,4,6-pentagalloyl glucose (right).

to be an important factor in plant defense against herbivorous insects and have manifold ecological functions.[6] In addition, they may have toxic or antinutritional effects for ruminants as they inhibit gastrointestinal bacteria and reduce ruminant performance, mainly by reducing intake and nutrient digestibility.[5] Tannins commonly comprise from 0.5% to 10% dry weight of tree leaves, but levels reach the 20% of dry weight in some species.[6] Nonetheless, the main interest of HTs in food and nutrition research is related to the several biological activities that their gut-derived metabolites may exert, playing an essential role in a plethora of health-promoting features.

Chemistry of Hydrolyzable Tannins (Gallotannins and Ellagitannins)

GTs are the simplest HTs from a structural point of view. GTs are polygalloyl esters of glucose, that is, they consist of a central polyol, most often glucose, which is surrounded by several GA units. 1,2,3,4,6-penta-*O*-galloyl-β-D-glucose (PGG) is a prototypical GT and the central compound in the biosynthetic pathway of HTs (Figure 6.6).[21] Further GA units can be attached through depside bonds to form more complex structures with high MW that can contain up to 10, and sometimes even more, galloyl residues.[22] However, the depside bonds between galloyl units are considerably less stable than the core ester linkages of PGG. Tannic acid, a generic name for commercially available mixtures of GTs, consists of distinct numbers of galloyl units linked to a PGG core structure. Tannic acid is present at sufficient levels to allow direct isolation from different plants such as *Paeonia suffruticosa*, *Paeonia lactiflora*, *Schinus terebinthifolius*, and the gallnuts of *Rhus chinensis*.[21]

Unlike GTs, ETs are structurally much more complex HTs. ETs are characterized by one or more HHDP units esterified to a sugar core, usually glucose. HHDP groups are constituted by oxidative *C*−*C* bond formation between neighboring galloyl residues of the precursor PGG. This complex class of polyphenols can be categorized according to their structural characteristics into four major groups: monomeric ETs, *C*-glycosidic ETs with an open-chain glycoside core, oligomers, and complex tannins with flavan-3-ols.[19,20] In addition, as noted earlier, HHDP moiety spontaneously undergoes lactonization to yield EA upon hydrolysis. EA is not easily hydrolyzed because of the further *C*−*C* coupling between galloyl units, in such a way that ETs are not so HTs as thought historically.[23] On the other hand, EA can also be found in its free form or as EA derivatives through glycosylation.[24]

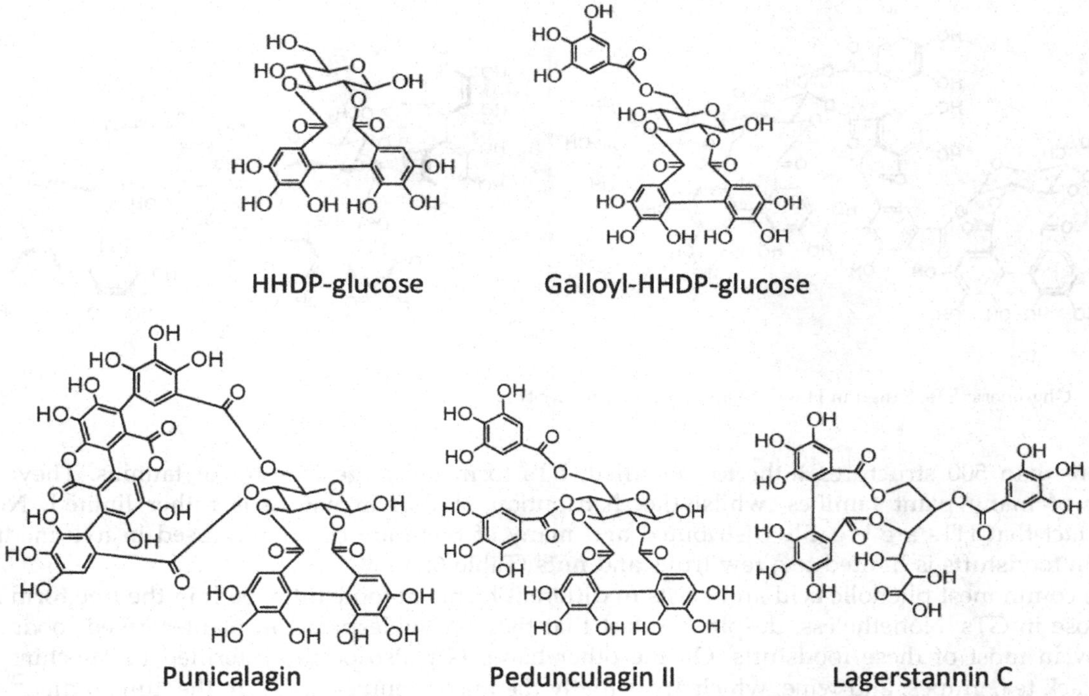

HHDP-glucose Galloyl-HHDP-glucose

Punicalagin Pedunculagin II Lagerstannin C

FIGURE 6.7 Monomeric ETs.

Besides HHDP units, simple ETs can include galloyl, valoneoyl, dehydrohexahydroxydiphenoyl (DHHDP), gallagyl, macaranoyl, tergalloyl and chebuloyl groups. This fact, along with the structural variation achievable among monomers, results in a broad array of diverse structures. The structural variation of simple ETs arises mainly from the linkage between the glucose residue and the chiral HHDP group(s) (i.e., *S*- or *R*-configuration depending on linkage at *O*-2/*O*-3 and *O*-4/*O*-6 of the glucose or at *O*-3/*O*-6, respectively), as well as from the number and position of acyl moieties on the molecule. Casuarictin, pedunculagin, chebulagic acid, corilagin, geraniin, punicalagin, punicalin, tergallagin, and tellimagrandin, among others, are examples of monomeric ETs (Figure 6.7).[19,20]

The sugar moiety of simple ETs can be replaced by other non-aromatic polyhydroxy compounds such as quinic acid and gluconic acid. The *C*-glycosidic ETs with an open-chain glycoside constitute an important subclass of ETs with the structural particularity of having highly characteristic *C*–*C* linkages between the polyol core and the acyl moieties. They are categorized into two types: the castalagin-type contains a flavogalloyl group linked to the gluconic acid such as in vescalagin, castalin and grandinin, whereas the casuarinin-type possesses a HHDP group and is exemplified by casuariin, stachyurin and punicacorteins.[20]

Monomeric ETs can also form various phenolic *C*–*C* and *C*–*O* linkages by subsequent intermolecular oxidative processes to produce additional structural conformations leading to a large number of oligomeric ETs. A myriad of diverse oligomeric structures, up to pentamers, have been isolated.[19,22] Agrimoniin, the first ET dimer discovered, and oenothein B are examples of oligomers linked with a dehydrodigalloyl group and macrocyclic oligomers, respectively (Figure 6.8).[19]

Complex tannins, also called flavano-ETs or flavono-ETs depending on the flavonoid unit, are hybrid tannins composed of a *C*-glycosidic ET and a glycosidically linked flavan-3-ol (like catechin or epicatechin), anthocyanin or procyanidin.[25]

Occurrence and Dietary Sources

In recent years, several works have thoroughly reviewed the dietary sources of HTs.[1,3,26–28] Therefore, only some important facts concerning the occurrence of HTs in foodstuffs are underlined in the present section. In addition, quantitative data are avoided to simplify the evidence showcased, since it is widely conceded that the content of HTs in foodstuffs, as in the case of other phytochemicals, depends on various factors such as genomic differences, environmental conditions, fruit quality, processing, packaging, and storage conditions.

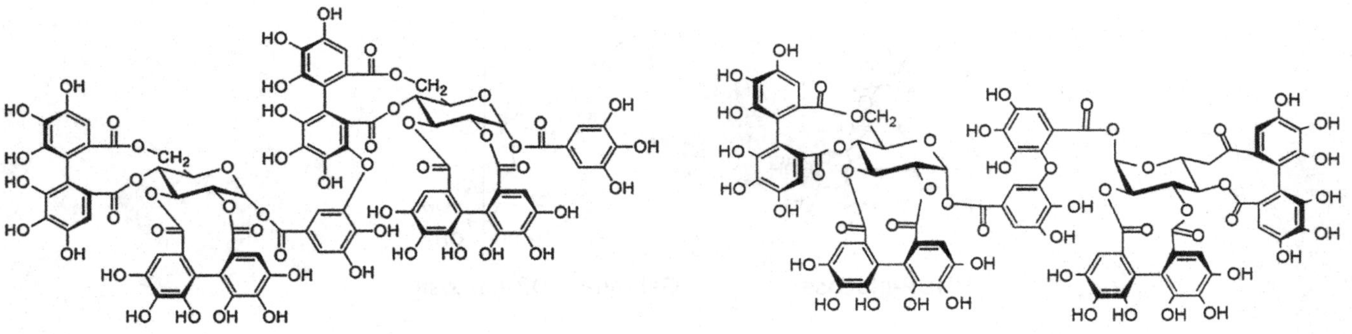

FIGURE 6.8 Oligomeric ETs. Sanguiin H-6 (left) and agrimoniin (right).

With more than 500 structures hitherto identified, ETs form the largest group of tannins. They are typical constituents of many plant families, whilst the distribution of GTs in nature is rather limited. Nonetheless, despite the fact that HTs are broadly distributed and many of them are commonly used in folk medicine, their occurrence in foodstuffs is limited to a few fruits and nuts (Table 6.1).

GA is the commonest phenolic acid and exists in different forms in foodstuffs, both in the free form and esterified to glucose in GTs. Nonetheless, despite the fact that they are widespread in plant-derived foods, quantities are very low in most of these foodstuffs. On the other hand, GA also occurs esterified to catechins or PAs in green tea, black tea, grapes, and wine, which are actually the major sources of GA in the human diet.[29]

GTs occur within clearly defined taxonomic limits in dicotyledons and are found to exist only in some fruits such as grapes, strawberries, raspberries, pecans, in some legumes, cereals and leaf vegetables.[1] Tea leaves also contain two GTs (monogalloyl-glucose and trigalloyl-glucose) as well as an ET (galloyl-HHDP-glucose).[30] GTs and especially tannic acid are added as stabilizing agents in brewing owing to their ability to precipitate proteins. As a result of this procedure, residual GTs can remain in beer.[3]

The occurrence of ETs has been reported, among others, in berries of the *Rubus*, *Ribes* and *Vaccinium* genera, as well as in their derivatives such as juices, jams and jellies. The *Prunus* genus together with pomegranate, persimmon, and mango are also recognized as important sources of ETs. Cloves, oak-aged wines and spirits, and nuts, in particular walnuts, also contain ETs (Figure 6.9; Table 6.1). Many of these plant-derived foodstuffs are also rich sources of PAs, as was mentioned previously.

ETs are known to be abundant in berry fruits. Among them, *Rubus* berries (raspberries, blackberries and cloudberries) and strawberries are those with higher content in these HTs.[27] Sanguiin H-6 is the main ET in raspberry and strawberry.[31] In addition, free EA, as well as different glycosidic derivatives, can be found in their jams.[24] On the other hand, berries are the major contributors to the ET intake in westernized countries. In this sense, strawberries account for the majority (68%) of the total intake of EA derivatives by American adults.[32]

Pomegranate juice is also considered a great source of HTs, in particular ETs, as these compounds are the main class of identified (poly)phenolics.[33,34] A broad array of ET structures has been found in pomegranate juice, with punicalagins and punicalins being the predominant ones.[35] ETs are extensively found in pomegranate husk and membranes and are usually extracted into juice during processing. Thus, the extraction process determines the amounts available in juice, displaying important differences between whole-fruit juices and aril-made ones.[36] However, the migration of these phenolics can confer a bitter and astringent taste to the pomegranate juice if presented to some degree.[37] On the other hand, pomegranate wine also contains ETs, even though its content in punicalagins is drastically depleted because of winemaking.[38]

Nuts are also noticeable sources of ETs. Walnuts are known to be an important source of ETs, with pedunculagin being the main compound.[39] Pecans and chestnuts also account for high levels of EA derivatives.[28]

Muscadine grapes and their wines are recognized as important source of ETs; because of its high content in muscadine wines, EA may induce cloudiness or form precipitates with proteins.[40] On the contrary, grapes do not contain ETs. EA forms part of the phytochemical composition of oak-aged wines as ETs are leached from the oak barrels during wine aging. Vescalagin and castalagin are some of the most abundant ETs extracted from oak wood and can subsequently be transformed into new compounds. Wine content in EA derivatives depends on the barrel used and on the duration of aging. In this sense, the leaching of ETs from oak barrels also occurs

TABLE 6.1 Different Dietary Sources of HTs

Latin Name	Food Group and Foodstuff	ETs and/or EA Derivatives[a]	GTs and/or GA[b,*]
FRUITS			
Ananas comosu	Pineapple	X	
Averrhoa carambola	Starfruit		X
Citrus × aurantifolia	Lime	X	
Citrus × paradisi	Grapefruit		X
Dimocarpus longan	Longan	X	
Diospyros kaki	Persimmon	X	X
Fragaria × ananassa	Strawberry	X	X
Hippophae rhamnoides	Sea buckthorn	X	
Mangifera indica	Mango	X	X
Prunus armeniaca	Apricot	X	
Prunus avium	Sweet cherry	X	
Prunus domestica	Plum	X	
Prunus persica	Peach	X	
Psidium guajava	Guava	X	
Punica granatum	Pomegranate	X	X
Ribes grossularia	Gooseberry	X	
Ribes nigrum	Blackcurrant	X	
Ribes rubrum	Redcurrant	X	
Rubus arcticus	Artic raspberry	X	
Rubus chamaemorus	Cloudberry	X	
Rubus idaeus	Raspberry	X	X
Rubus spp.	Blackberry	X	
Vaccinium oxycoccos	Cranberry	X	
Vaccinium spp.	Blueberry	X	
Vitis rotundifolia	Muscadine grapes	X	
Vitis vinifera	Grapes		X
NUTS			
Anacardium occidentale	Cashew nut	X	
Carya illinoinensis	Pecan	X	X
Castanea sativa	Chesnut	X	X
Corylus avellana	Hazelnut	X	
Juglans regia	Walnut	X	
Pistacia vera	Pistachio	X	
Prunus dulcis	Almonds	X	X
LEGUMES			
Cicer arietinum	Chickpea		X
Vigna sinensis	Cowpea		X

(Continued)

TABLE 6.1 (Continued)

Latin Name	Food Group and Foodstuff	ETs and/or EA Derivatives[a]	GTs and/or GA[b,*]
CEREALS			
Sorghum bicolor	Sorghum		X
VEGETABLES			
Cichorium endivia	Chicory		X
Rheum rhabarbarum	Rhubarb		X
SPICES			
Eugenia caryophillata	Clove	X	
Pimenta dioica	Pepper	X	
BEVERAGES			
Camelia sinensis	Tea	X	X
	Oak-aged wines	X	X
	Whiskey	X	
	Cognac	X	
	Beer		X

[a]*ETs, ellagitannins; EA, ellagic acid.*
[b]*GTs, gallotannins; GA, gallic acid.*
**Other compounds releasing gallic acid after hydrolysis (i.e., flavan-3-ol gallates) are not included.*

throughout the aging period of whiskeys and cognacs.[41] However, the level of HTs in these spirits is very low and should be regarded as anecdotal rather than as a dietary source.

Finally, EA has also been found in several types of honey, although at very low concentrations.[42]

Some reports have estimated the daily intake of HTs by using food composition databases and consumption surveys. Data point to GA as one of the most important contributors to the intake of (poly)phenolic compounds in the French population.[43] Among hundreds of phenolic molecules, GA was third by relevance of consumption indistinct of its form (free or bound to other molecules), and the mean daily intake was 35 mg/day; however, the dietary intake of EA derivatives in this same French cohort was lower, around 3.3 mg/day. Similar data have been reported in American (5.8 mg/day) and Bavarian (5.2 mg/day) populations.[32,44] However, Larrosa et al.[28] disagreed on these data, taking into account that the ET content of several foodstuffs (raspberry, strawberry, pomegranate and walnut) can be quite high. Actually, a strawberry serving can provide 70 mg of ETs whilst a glass of pomegranate juice as much as 1 g. Hence, if these products are introduced into the diet and regularly consumed, the daily intake of EA derivatives may increase substantially. In this case, data on the dietary burden of ETs would match the mean daily intake of 1250 mg/day reported by Saura-Calixto et al.[45] Nevertheless, it is difficult to evaluate the intake of HTs, especially GTs, due to the limited information available and, thus, efforts in this topic should be performed.

Metabolism of Hydrolyzable Tannins in Humans

Currently, understanding the biometabolism of HTs, especially ETs, can contribute to the definition of their bioavailability and potential biological properties on humans. Although there is little information on the absorption, distribution, metabolism and excretion (ADME) of HTs in humans, trials with rodents and pigs, *in vitro* studies and fecal batch fermentations have contributed to understanding tannins metabolism. This fact becomes even more relevant for GTs since, to the best of our knowledge, there is a lack of studies focused on the biometabolism of GTs or GT-rich foodstuffs.

Owing to their basic composition, constricted to galloyl and glucose moieties, the biometabolism of GTs is restricted to the hydrolysis of the polymeric structure to yield GA, which can be further metabolized by both the human enzymatic pool and gut microbiota (Figure 6.10). PGG, a representative GT, was administered to mice by oral gavage and plasma levels were in the sub-micromolar range.[21] The high MW of this compound and its

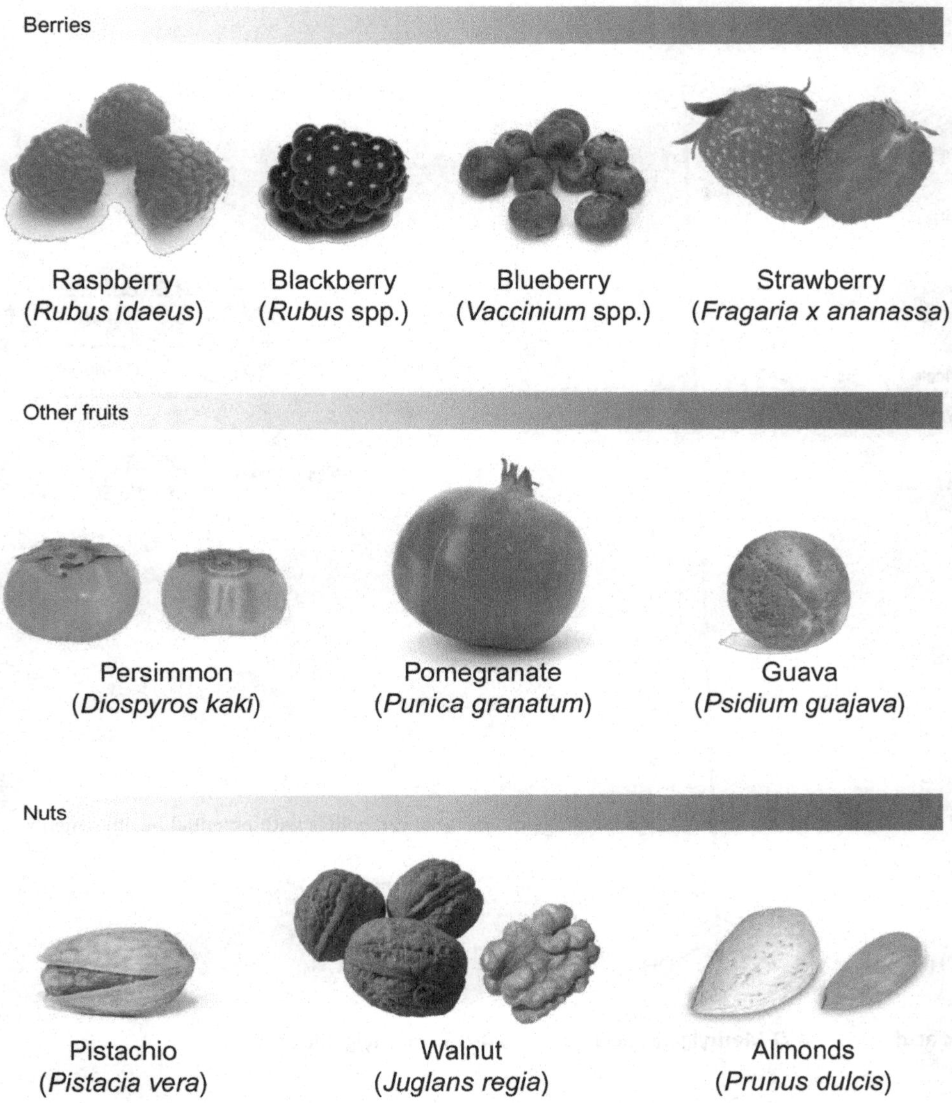

Berries

Raspberry (*Rubus idaeus*)
Blackberry (*Rubus* spp.)
Blueberry (*Vaccinium* spp.)
Strawberry (*Fragaria x ananassa*)

Other fruits

Persimmon (*Diospyros kaki*)
Pomegranate (*Punica granatum*)
Guava (*Psidium guajava*)

Nuts

Pistachio (*Pistacia vera*)
Walnut (*Juglans regia*)
Almonds (*Prunus dulcis*)

FIGURE 6.9 Representative plant-derived food rich in ETs.

ability to bind proteins or other molecules such as phospholipids may slow down its absorption. Moreover, PGG is extensively degraded to tri- and tetra-galloyl glucose when transported across intestinal cells like Caco-2.[46] This degradation occurs in both apical and basolateral directions of the transcellular transport and may be attributed to the presence of human esterases in the intestines. On the other hand, gut microbiota may also cleave certain tannins containing galloyl residues through esterase activity. The GA and gallate moieties released not only from GTs but also from some CTs and ETs can also be metabolized by intestinal microbiota.[47,48] Pyrogallol and pyrocatechol are among the products derived from the metabolism of GA (Figure 6.11).[47] Pyrogallol is formed by decarboxylation of GA whilst pyrocatechol is the dehydroxylated product of the first metabolite, pyrogallol. Furthermore, a point worth mentioning related to the production of gut microbiota metabolites from GA is the high inter-individual variability seen in different batch cultures.[49,50] This fact accounts for different rates of degradation and biotransformation of GTs by colonic microorganisms, which may condition their bioavailability and bioefficacy within the human body.

Gut microbiota catabolites derived from GTs are absorbed into the circulatory system and excreted in urine. GA per se is extremely well absorbed in healthy humans with a reported urinary excretion of 37% of intake.[51,52] GA can be methylated to form urine metabolites including 3-*O*-methylgallic acid, 4-*O*-methylgallic acid, and even 3,4-di-*O*-methylgallic acid (Figure 6.11).[53] Pyrogallol has also been identified as a catabolite of GA derivatives in humans both in its free form and as 2-*O*-sulfate-pyrogallol.[53,54] The biometabolism of flavan-3-ol gallates present in black tea yields 4-*O*-methylgallic acid as a main biomarker of black tea consumption.[47,53] Regarding this, the appearance

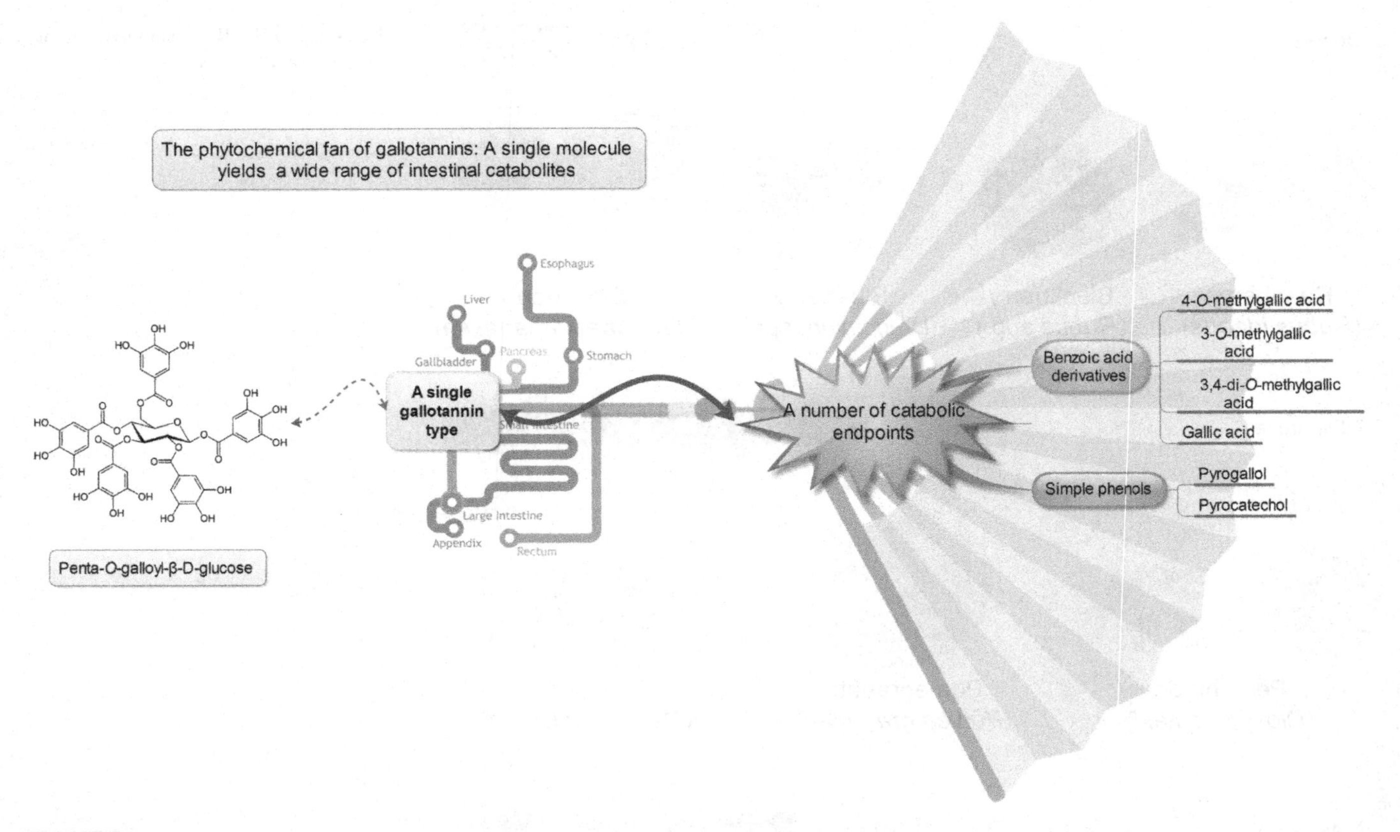

FIGURE 6.10 The phytochemical fan of GTs. A single molecule yields a wide range of intestinal catabolites with potential health benefits.

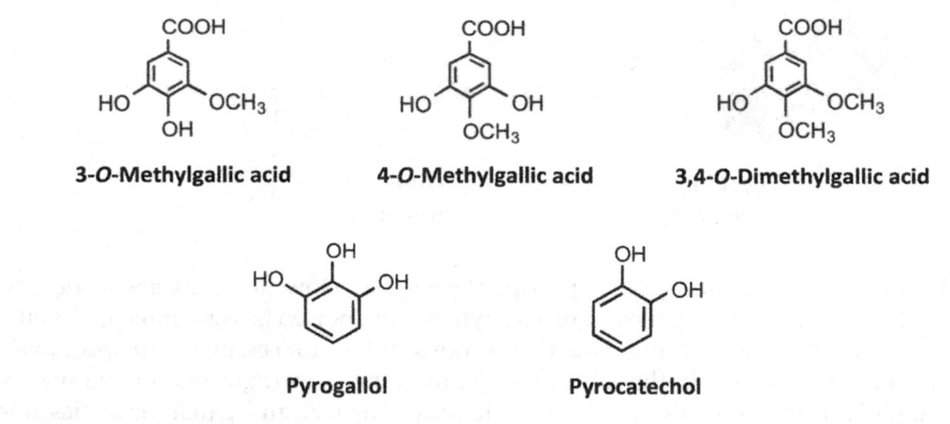

FIGURE 6.11 Metabolites derived from GTs biotransformations.

of 4-*O*-methylgallic acid as a major metabolite derived from GTs could be expected. However, further studies on GT consumption including a significant number of human subjects are required to confirm this hypothesis.

In contrast to GT metabolism, the bioavailability and biometabolism of ETs result in a more complex process (Figure 6.12). The different kinds of units that comprise the ET molecule, as well as the strong *C*–*C* coupling of the HHDP moiety, determine the structural degradation of these HTs. The metabolic fate of ETs after consumption can be divided into two different steps: (1) hydrolysis of the polymeric structure; and (2) metabolism of the structural units, mainly EA and GA.

ETs are generally not absorbed as such. Although punicalagin has been detected in the plasma and urine of rats fed with 6% of their diet as pomegranate husk ETs,[55] the absorption of intact ETs in human biological fluids has not yet been described. ETs are hydrolyzed to release free EA in the small intestine owing to the alkaline physiological pH.[56,57] As a matter of fact, the recovery of EA in the ileal fluid was at a level of 241% for human

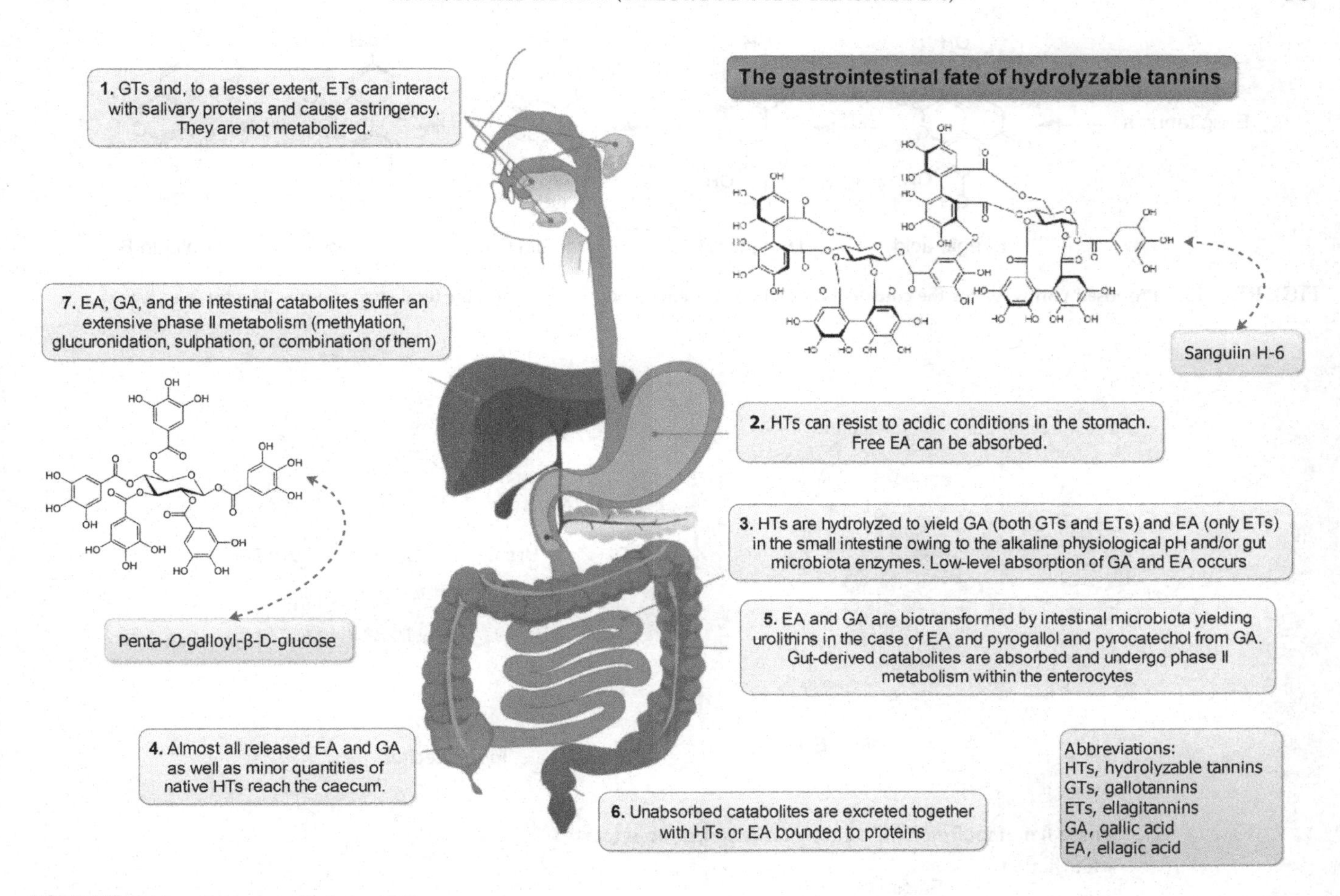

FIGURE 6.12 The metabolic fate of HTs.

ileostomy subjects following raspberry consumption, while the recovery of native ETs was scarce (23% for sanguiin H-6).[58] Gut microflora action might also cleave ester links from the ET core yielding EA (and GA in the case of those ETs containing this phenolic acid), as seen for the hydrolysis of GTs. Before dealing with the biometabolism of EA, it should be mentioned that GA, as part of some ETs, likely follows the same metabolic fate once hydrolyzed as GA from GTs. However, to our knowledge, evidence has not been yet provided, even though fecal incubations of ET geraniin, composed of all common acyl units such as a galloyl, hexahydroxydiphenoyl and dehydrohexahydroxydiphenoyl groups have been performed.[59]

EA has been detected as a circulating metabolite in human plasma at a maximum concentration 1 h after acute ingestion of pomegranate juice.[60] Absorption of free EA contained in food products takes place in the stomach.[57] A small part of the EA released from ETs is transported across the enterocyte apical membrane. Within the enterocyte, EA can be methyl or dimethyl conjugated by the action of catechol ortho-methyl transferase. Then, methylated forms of EA can be subjected to glucuronidation, as in the case of dimethyl-EA glucuronide.[61] In addition, other minor metabolites, isomers of dimethyl-EA sulfates, are produced. Considering these facts, a complex pool of EA metabolites, free and conjugated, is present in systemic circulation. Nonetheless, the poor absorption of free EA (less than 1%) after 7 days' consumption of black raspberry has been reported.[62] Similarly, González-Barrio et al.[58] pointed out that the sum of EA and a glucuronide form excreted in urine was lower than 1% following the consumption of raspberries. Actually, although the stomach and small intestine are sites of absorption and metabolism of EA, the major part of the EA released from ETs is transformed by gut microbiota into urolithins.

Unidentified microorganisms within the intestinal lumen transform EA into a series of metabolites called urolithins, which share a 6*H*-dibenzo[*b*,*d*]pyran-6-one structural nucleus with at least one hydroxyl group. Cleavage and decarboxylation of the lactone ring and different specific dehydroxylations of the EA nucleus account for the whole range of urolithins (Figure 6.13). As a result, urolithins exhibit a different phenolic hydroxylation pattern. Concerning this, the Iberian pig served as an animal model to elucidate urolithin production.[57] Transformation of

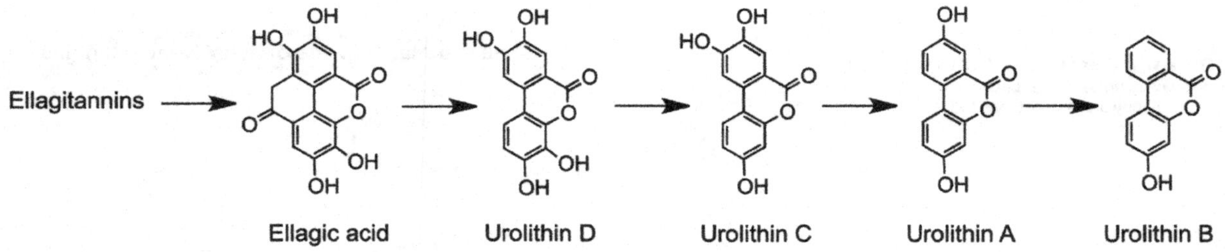

Ellagitannins
Ellagic acid
Urolithin D
Urolithin C
Urolithin A
Urolithin B

FIGURE 6.13 Proposed pathways for the conversion of ETs to EA and urolithins in anaerobic fecal suspensions. *Adapted from Ref. 65.*

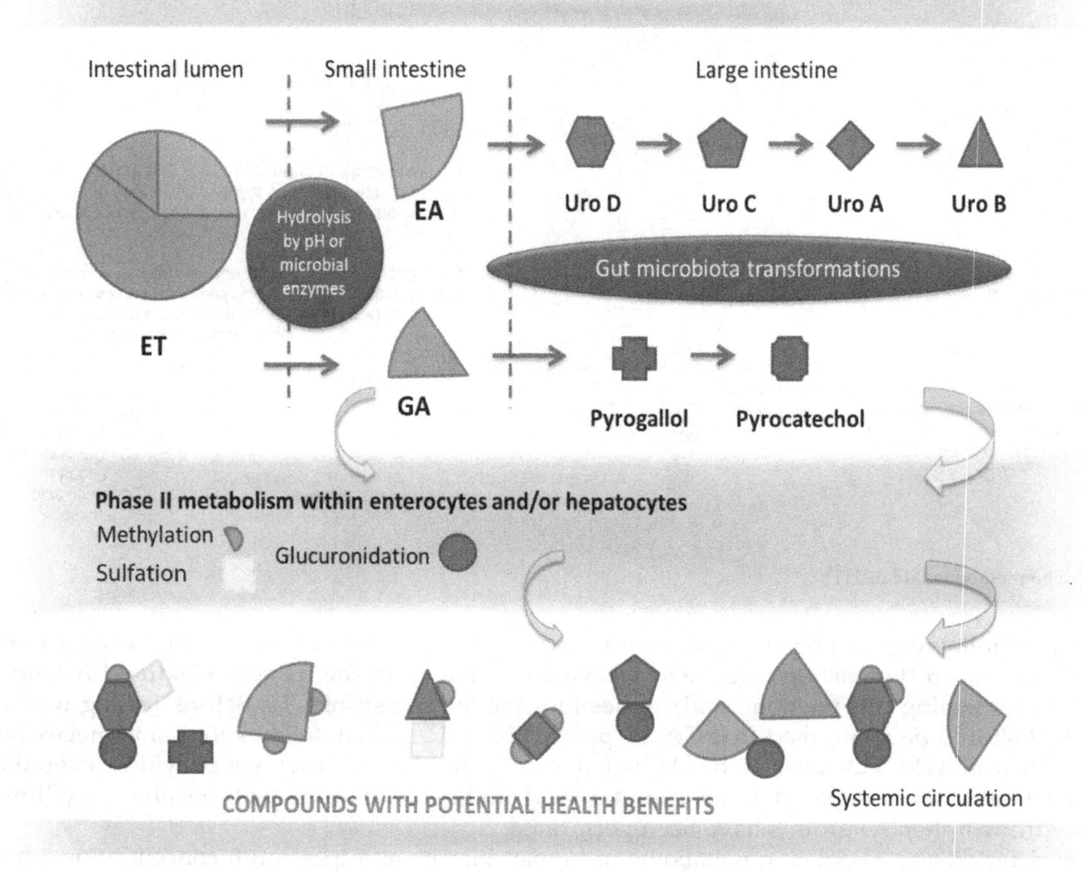

FIGURE 6.14 Schematic representation of the biometabolism of ETs. ET, ellagitannin; EA, ellagic acid; GA, gallic acid; Uro D, urolithin D; Uro C, urolithin C; Uro A, urolithin A; Uro B, urolithin B.

EA starts in the small intestine, specifically in the jejunum, and the first metabolite produced, urolithin D (tetrahydroxydibenzopyranone), retains four phenolic hydroxyls. The metabolism continues along the gastrointestinal tract with the sequential removal of hydroxyl groups, leading to the production of urolithin C (trihydroxydibenzopyranone), urolithin A (dihydroxydibenzopyranone), and finally urolithin B (monohydroxydibenzopyranone) in the distal parts of the colon. Urolithins were absent in biological fluids of ileostomy subjects but present in urine excreted by healthy volunteers, indicating the colon as the site of formation of these ET-derived catabolites in humans.[58] In addition, *in vitro* fecal incubations have served to clarify the microbial origin of these metabolites as well as to demonstrate the production of only urolithin aglycones by microbiota-mediated breakdown of EA.[63–65]

Once absorbed, urolithin aglycones undergo an extensive phase II metabolism (methylation, glucuronidation, sulfation, or a combination of these) either in the wall of the large intestine and/or within the hepatocytes to render more soluble metabolites that may further suffer an intensive enterohepatic recirculation. Finally, these gut microbiota metabolites enter the systemic circulation and are excreted in urine (Figure 6.14, Figure 6.15).[57,65]

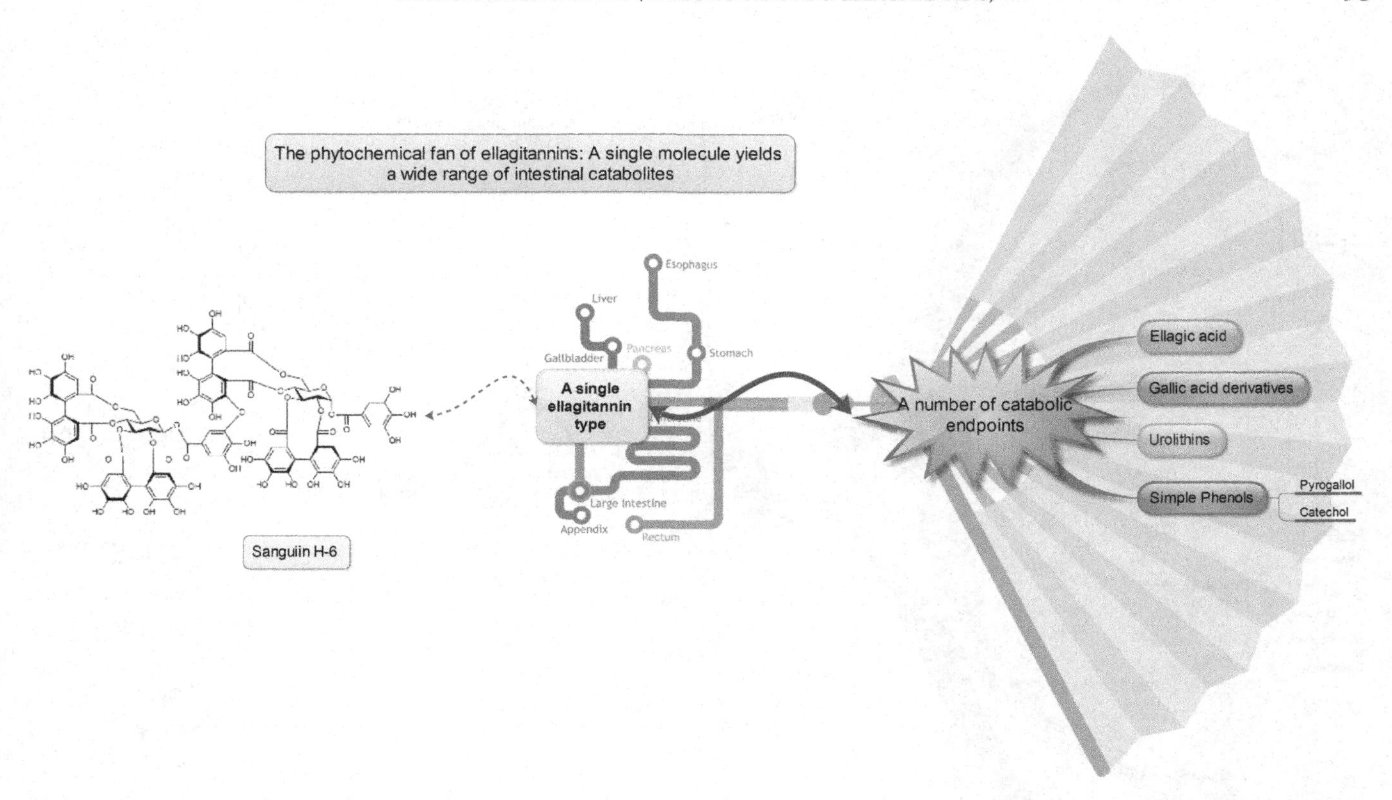

FIGURE 6.15 The phytochemical fan of ETs. A single molecule yields a wide range of intestinal catabolites with potential health benefits.

The metabolites urolithin B, urolithin B-glucuronide, urolithin A, urolithin A-glucuronide, two of its isomers, two urolithin A-sulfoglucuronide isomers, urolithin A-diglucuronide, urolithin C, urolithin C-glucuronide, urolithin C-methyl ether-glucuronide, urolithin C-methyl ether-sulfoglucuronide, urolithin D, urolithin D-glucuronide, and urolithin D-methyl ether-glucuronide have been detected in human biological fluids so far, with urolithin A-glucuronide being the predominant metabolite excreted in urine after ET-rich food administration.[66–69]

Pharmacokinetics of urolithins also supports the involvement of the intestinal microbiota in the production of these catabolites. Urolithins appear in human plasma within a few hours of the consumption of foods containing ETs and reach maximum concentrations at 24–48 h. They remain in biological fluids for up to 92 h and, thus, these circulating metabolites may exert several biological activities that could be responsible for the health-promoting features associated with the consumption of ET-rich foods.[28,58,66,69,70] Concentrations of urolithins achieved in plasma are in the nanomolar (nM) range, although large inter-individual variability has been reported.[28,39,71] In fact, a large person-to-person variation in the timing, quantity, and kind of urolithins excreted in urine has been described.[58] Differences in the colonic microbiota composition seem to be responsible for this large inter-individual variability described in intervention trials with ET-rich foods and may explain the weak statistical significance reported.[39,58,65,66,68] The specific bacteria involved in the metabolic fate of ETs are still unknown; however, patients with metabolic syndrome, which is a pathology that is associated with an altered gut bacterial diversity, showed the same excretion pattern of urolithins as healthy subjects after 12-weeks of walnut consumption.[68] Unfortunately, the fecal bacterial composition of these subjects was not addressed in this study. On the contrary, it is well-known that pomegranate extracts rich in ETs, but not solely punicalagins, can modify the human gut bacteria by enhancing the growth of *Bifidobacterium* spp. and *Lactobacillus* spp., which is a bacterium that is involved in various health benefits.[63]

Processing of foods containing ETs does not affect the production and excretion of urolithin conjugates, since similar levels of these gut microbiota-derived metabolites in human subjects are provided after consumption of the same types and amounts of ETs.[66,72]

Finally, the presence and quantity of urolithins in target organs for the potential biological activity of these ET-derived metabolites, such as the prostate, are independent of the source of ETs (pomegranate juice or walnuts) for similar intakes of these HTs.[67] This fact, together with the nature of ET biometabolism, suggests that diverse

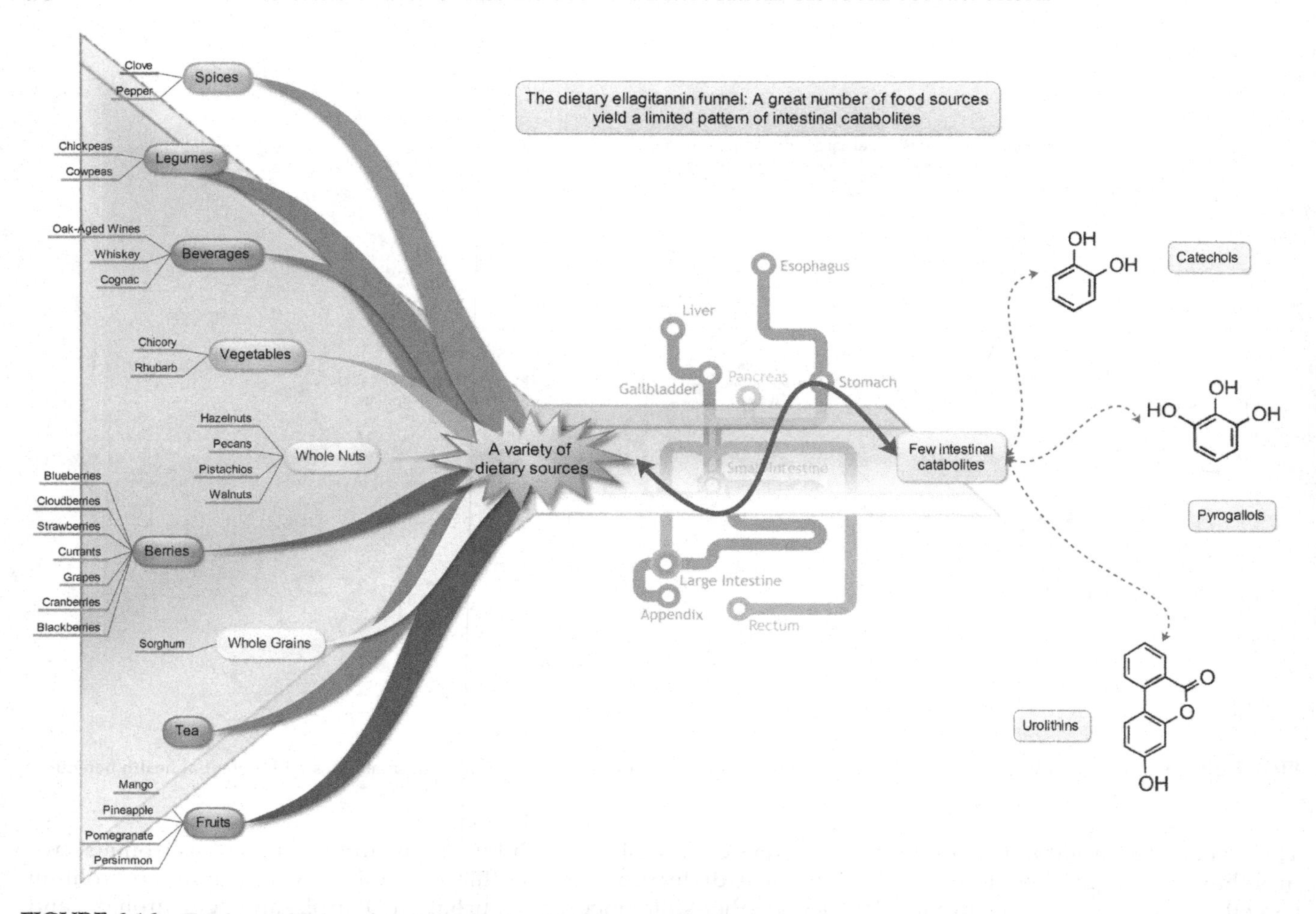

FIGURE 6.16 The dietary ET funnel. Different kinds of ET sources yield a limited pattern of metabolites.

ET-rich foodstuffs render the same limited pattern of intestinal catabolites (Figure 6.16), although with notable differences among subjects.

Protective Effects of Hydrolyzable Tannins Intake in Human Subjects

Although much *in vitro* evidence has been published with regard to the health-promoting features of HTs, it is far from consistent with the beneficial effects observed in humans. In most of the cases, *in vitro* studies have tested the wrong molecules (i.e., those not metabolized *in vivo* or not occurring in foods) at concentrations far from those that are achievable within tissues. On the other hand, animal models tested may offer unsuitable information that is non-translatable to human conditions, further taking into account the diverse metabolite profile derived from ET intake for each animal species.[73] The unique effect of HTs on health management should be tackled through supplementation in human subjects of extracts rich in these compounds. Unfortunately, this approach has been scarcely addressed so far.[70,74] The main achievements in humans have been accomplished by testing the biological properties of food containing HTs. This valuable strategy shows the concomitant effect of other phytochemicals also present in the food matrix to be a major hindrance, which may offer confusing results. In any case, interventions performed with food products as source of HTs have highlighted the promising potential of these phytochemicals on different diseases.

An unregulated balance of reactive oxygen species (ROS) causes oxidation of lipids, proteins and nucleic acids. Damage to these biomolecules leads to a dysregulation of cellular mechanisms and plays an essential role in the onset of several pathologies.[75] Consumption of fruits rich in ETs such as strawberries, walnuts, and pomegranate juice has shown to increase plasma antioxidant capacity and decrease oxidative stress in healthy subjects.[76–79] These findings could be attributed to a great extent to ETs, as the ingestion of a standardized extract of pomegranate juice has displayed similar results.[70] In addition, an improved antioxidant function was noted in an elderly population that consumed pomegranate juice daily for a month.[80] Oxidative injury linked to proteins and lipids was reduced at the same time that plasma glutathione peroxidase and catalase antioxidant enzymes were significantly increased. These antioxidant effects, together with the positive modification of other specific antioxidant biomarkers, were also appreciated after the intake for 3 years of pomegranate juice by patients with carotid artery stenosis.[81] On the contrary, supplementation of this juice in patients with stable chronic obstructive pulmonary disease, where oxidative stress plays a major role in the disease progression, added no benefit to the patient status.[82]

The implications of dietary patterns in the risk reduction of certain kinds of cancers have been well-defined and, hence, chemoprevention through nutritional agents has been recognized as a plausible approach directed to prevent or delay cancer development. In this sense, berry fruits, which are rich in ETs, may have beneficial effects against several types of human cancers.[83] Nonetheless, there are limited human intervention trials of the chemopreventive effects of these particular phytochemicals. A phase II clinical study in men with recurrent prostate cancer and rising prostate-specific antigen (PSA) levels was conducted.[84] Patients were treated with 8 ounces (~240 mL) of pomegranate juice daily until disease progression and results showed a significant prolongation of PSA doubling time, from 15 months at baseline to 54 months post-treatment. This result matched the *in vitro* evidence and animal studies about the chemopreventive properties of pomegranate juice and its constituents on prostate adenocarcinoma. In addition, despite the presence of high levels of anthocyanins in pomegranate juice, the occurrence of urolithins in the human prostate gland upon the consumption of pomegranate juice and walnuts[67] points at ETs as the responsible compounds behind this fact. Several clinical trials are being performed in patients at high risk of cancers of epithelial origin by supplementing a black raspberry extract, which is rich in anthocyanins and EA derivatives.[85,86] Interim findings from 10 patients with Barrett's oesophagus pointed out that the daily consumption of 32 or 45 g (female and male, respectively) of black raspberries promotes a reduction in the urinary excretion of two oxidative damage biomarkers (8-epi-prostaglandin F2α and, to a lesser extent, 8-hydroxy-2′-deoxyguanosine), even though it does not result in a reduction in segment length of Barrett's lesions.[85] Consumption of 20 g of black raspberry extract for an average of 4 weeks by 20 subjects with colorectal cancer and/or polyps also demonstrated a significant drop in proliferation and angiogenesis biomarkers as well as an enhancement of apoptosis in the damaged tissue.[86]

Evidence of the protective effects of ET-containing foodstuffs on cardiovascular diseases has been extensively reviewed for walnuts, pomegranate and strawberries.[87–89] However, the direct link between the ET composition of these foods and its effects in human subjects has not been fully demonstrated so far owing to the likely role of other constituents that could also be involved in the cardioprotective effects of these ET-containing foodstuffs. As a matter of fact, anthocyanins have also displayed a major role in the cardioprotective properties of strawberries and pomegranate,[88,89] whilst phytosterols, tocopherols, and lipids may account for the effects of walnuts on prevention and attenuation of some cardiovascular diseases.[87] All the same, evidence suggests that diets rich in these compounds may decrease the risk of cardiovascular disease development by decreasing lipid peroxidation, the uptake of oxidized-LDL by macrophages, intima media thickness, atherosclerotic lesion areas, angiotensin-converting enzyme activity, systolic blood pressure, and enhancing biological actions of nitric oxide.[27,28,87,90,91]

Finally, the ET fraction of pomegranate may improve the recovery of isometric strength 2–3 days after a damaging eccentric exercise, producing delayed-onset muscle soreness. Nonetheless, serum markers of inflammation and muscle damage did not provide insight regarding possible mechanisms behind this fact.[74]

With respect to GTs, less evidence of their beneficial effects on human subjects has been provided. Despite the fact that tannic acid, PGG, and some plants containing high amounts of GA and GTs have shown several beneficial effects by using *in vitro* and *in vivo* models,[92] insights related to human intervention trials are very scarce.

An Ayurvedic powdered preparation, "Triphala," made by combining *Terminalia chebula*, *Terminalia belarica* and *Emblica officinalis*, which are three plants that are rich in HTs, had good laxative property, helped in the management of hyperacidity and also improved appetite.[93] In addition, another Ayurvedic plant formulation consisting of 5% of GA exhibited hepatoprotective activity in patients with evidence of liver disease after 8 weeks of treatment.[94] Nonetheless, direct relation to its content in HTs should be regarded carefully as the assessment of its composition was not fully accomplished.

On the whole, there is enough evidence that HTs, overall ETs, exert positive effects in the human body and may play a major role in the prevention and management of several diseases. However, more research is needed to fully understand whether ETs and/or likely their *in vivo* metabolites are the responsible compounds linked to the health-promoting features of foodstuffs that are rich in ETs. In this sense, clinical studies with purified extracts may shed light on this question. In addition, human trials should include enough subjects to avoid the high inter-individual variability that exists with regard to the occurrence of ET-derived metabolites, urolithins, which may condition the biological response. For that, monitoring human and gut metabolites in biological fluids is mandatory to associate ET intake with its impact on health status.

CONCLUSIONS

The health benefits associated with the dietary intake of plant tannins are closely related to the transformations of these high-MW compounds by the human gut microbiota. Whereas ETs and GTs can be hydrolyzed and slightly absorbed in the small intestine, the structural units of these HTs and the bulk of PAs reach almost intact the human colon. These molecules undergo a broad array of rearrangements and transformations and actively modulate the composition of the colonic microflora. At the same time, the generated colonic metabolites may act locally or be absorbed into the portal bloodstream and undergo further phase II metabolism, before entering the systemic circulation to exert their effects throughout the human body. Based on these preliminary observations, more focused research should be addressed both at better understanding the modulating action of these plant compounds toward the human microbiota and at identifying which microbial metabolites are able to influence human health at systemic level. This future research should also always take into account the inter-individual variability linked to the presence of specific microbial enterotypes in humans.

References

1. Serrano J, Puupponen-Pimiä R, Dauer A, Aura AM, Saura-Calixto F. Tannins: current knowledge of food sources, intake, bioavailability and biological effects. *Mol Nutr Food Res*. 2009;53:310–329.
2. Santos-Buelga C, Scalbert A. Proanthocyanidins and tannin-like compounds – nature, occurrence, dietary intake and effects on nutrition and health. *J Sci Food Agric*. 2000;80:1094–1117.
3. Clifford MN, Scalbert A. Ellagitannins – nature, occurrence and dietary burden. *J Sci Food Agric*. 2000;80:1118–1125.
4. Mingshu L, Kai Y, Qiang H, Dongying J. Biodegradation of gallotannins and ellagitannins. *J Basic Microbiol*. 2006;46:68–84.
5. Mueller-Harvey I. Unravelling the conundrum of tannins in animal nutrition and health. *J Sci Food Agric*. 2006;86:2010–2037.
6. Salminen JP, Karonen M. Chemical ecology of tannins and other phenolics: we need a change in approach. *Funct Ecol*. 2011;25:325–338.
7. Lila MA. From beans to berries and beyond: taeamwork between plant chemicals for protection of optimal human health. *Ann N Y Acad Sci*. 2007;1114:372–380.
8. U.S. Department of Agriculture. Agricultural Research Service. USDA Database for the Proanthocyanidin Content of Selected Foods. Nutrient Data Laboratory. <http://www.ars.usda.gov/SP2UserFiles/Place/12354500/Data/PA/PA.pdf>; 2004 Accessed 10.04.13.
9. Auger C, Al-Awwadi N, Bornet A, et al. Catechins and procyanidins in mediterranean diets. *Food Res Int*. 2004;37:233–245.
10. Appeldoorn MM, Vincken JP, Aura AM, Hollman PCH, Gruppen H. Procyanidin dimers are metabolized by human microbiota with 2-(3,4-dihydroxyphenyl)acetic acid and 5-(3,4-dihydroxyphenyl)-γ-valerolactone as the major metabolites. *J Agric Food Chem*. 2009;57:1084–1092.
11. Aura AM, Mattila I, Seppänen-Laakso T, Miettinen J, Oksman-Caldentey KM, Orešič M. Microbial metabolism of catechin stereoisomers by human faecal microbiota: comparison of targeted analysis and a non-targeted metabolomics method. *Phytochem Lett*. 2008;1:18–22.
12. Aura AM, O'Leary KA, Williamson G, et al. Quercetin derivatives are deconjugated and converted to hydroxyphenylacetic acids but not methylated by human fecal flora *in vitro*. *J Agric Food Chem*. 2002;50:1725–1730.
13. Jaganath IB, Mullen W, Edwards CA, Crozier A. The relative contribution of the small and large intestine to the absorption and metabolism of rutin in man. *Free Radical Res*. 2006;40:1035–1046.
14. Stoupi S, Williamson G, Drynan JW, Barron D, Clifford MN. A comparison of the *in vitro* biotransformation of (−)-epicatechin and procyanidin B2 by human faecal microbiota. *Mol Nutr Food Res*. 2010;54:747–759.
15. Stoupi S, Williamson G, Drynan JW, Barron D, Clifford MN. Procyanidin B2 catabolism by human fecal microflora: partial characterization of 'dimeric' intermediates. *Arch Biochem Biophys*. 2010;501:73–78.
16. Cueva C, Sánchez-Patán F, Monagas M, et al. *In vitro* fermentation of grape seed flavan-3-ol fractions by human faecal microbiota: changes in microbial groups and phenolic metabolites. *FEMS Microbiol Ecol*. 2013;83:792–805.
17. Van Dorsten FA, Peters S, Gross G, et al. Gut microbial metabolism of polyphenols from black tea and red wine/grape juice is source-specific and colon-region dependent. *J Agric Food Chem*. 2012;60:11331–11342.
18. Rios LY, Gonthier MP, Rémésy C, et al. Chocolate intake increases urinary excretion of polyphenol-derived phenolic acids in healthy human subjects. *Am J Clin Nutr*. 2003;77:912–918.
19. Okuda T, Ito H. Tannins of constant structure in medicinal and food plants—hydrolyzable tannins and polyphenols related to tannins. *Molecules*. 2011;16:2191–2217.
20. Yoshida T, Amakura Y, Yoshimura M. Structural features and biological properties of ellagitannins in some plant families of the order myrtales. *Int J Mol Sci*. 2010;11:79–106.
21. Li L, Shaik AA, Zhang J, et al. Preparation of penta-O-galloyl-β-d-glucose from tannic acid and plasma pharmacokinetic analyses by liquid–liquid extraction and reverse-phase HPLC. *J Pharm Biomed Anal*. 2011;54:545–550.
22. Niemetz R, Gross GG. Enzymology of gallotannin and ellagitannin biosynthesis. *Phytochemistry*. 2005;66:2001–2011.
23. Khanbabaee K, Van Ree T. Tannins: classification and definition. *Nat Prod Rep*. 2001;18:641–649.

24. Koponen JM, Happonen AM, Mattila PH, Törrönen AR. Contents of anthocyanins and ellagitannins in selected foods consumed in Finland. *J Agric Food Chem*. 2007;55:1612–1619.
25. Jourdes M, Pouységu L, Quideau S, Mattivi F, Truchado P, Tomás-Barberán F. *Hydrolyzable Tannins. Handbook of Analysis of Active Compounds in Functional Foods*. CRC Press; 2012:435–460.
26. Bakkalbasi E, Mentes O, Artik N. Food ellagitannins—occurrence, effects of processing and storage. *Crit Rev Food Sci Nutr*. 2009;49:283–298.
27. Landete JM. Ellagitannins, ellagic acid and their derived metabolites: a review about source, metabolism, functions and health. *Food Res Int*. 2011;44:1150–1160.
28. Larrosa M, García-Conesa MT, Espín JC, Tomás-Barberán FA. Ellagitannins, ellagic acid and vascular health. *Mol Aspects Med*. 2010;31:513–539.
29. Crozier A, Jaganath IB, Clifford MN. Dietary phenolics: chemistry, bioavailability and effects on health. *Nat Prod Rep*. 2009;26:1001–1043.
30. Nonaka GI, Sakai R, Nishioka I. Hydrolysable tannins and proanthocyanidins from green tea. *Phytochemistry*. 1984;23:1753–1755.
31. Mullen W, Yokota T, Lean MEJ, Crozier A. Analysis of ellagitannins and conjugates of ellagic acid and quercetin in raspberry fruits by LC-MSn. *Phytochemistry*. 2003;64:617–624.
32. Murphy MM, Barraj LM, Herman D, Bi X, Cheatham R, Randolph RK. Phytonutrient intake by adults in the United States in relation to fruit and vegetable consumption. *J Acad Nutr Diet*. 2012;112:222–229.
33. Borges G, Mullen W, Crozier A. Comparison of the polyphenolic composition and antioxidant activity of European commercial fruit juices. *Food Funct*. 2010;1:73–83.
34. Mena P, Calani L, Dall'Asta C, et al. Rapid and comprehensive evaluation of (Poly)phenolic compounds in pomegranate (*Punica granatum* L.) Juice by UHPLC-MSn. *Molecules*. 2012;17:14821–14840.
35. Mena P, Martí N, Saura D, Valero M, García-Viguera C. Combinatory effect of thermal treatment and blending on the quality of pomegranate juices. *Food Bioprocess Technol*. 2013;6(11):3186–3199.
36. Gil MI, Tomás-Barberán FA, Hess-Pierce B, Holcroft DM, Kader AA. Antioxidant activity of pomegranate juice and its relationship with phenolic composition and processing. *J Agric Food Chem*. 2000;48:4581–4589.
37. Tzulker R, Glazer I, Bar-Ilan I, Holland D, Aviram M, Amir R. Antioxidant activity, polyphenol content, and related compounds in different fruit juices and homogenates prepared from 29 different pomegranate accessions. *J Agric Food Chem*. 2007;55:9559–9570.
38. Mena P, Gironés-Vilaplana A, Martí N, García-Viguera C. Pomegranate varietal wines: phytochemical composition and quality parameters. *Food Chem*. 2012;133:108–115.
39. Cerdá B, Tomás-Barberán FA, Espín JC. Metabolism of antioxidant and chemopreventive ellagitannins from strawberries, raspberries, walnuts, and oak-aged wine in humans: Identification of biomarkers and individual variability. *J Agric Food Chem*. 2005;53:227–235.
40. Lee JH, Talcott ST. Ellagic acid and ellagitannins affect on sedimentation in muscadine juice and wine. *J Agric Food Chem*. 2002;50:3971–3976.
41. Glabasnia A, Hofmann T. Sensory-directed identification of taste-active ellagitannins in American (*Quercus alba* L.) and European oak wood (*Quercus robur* L.) and quantitative analysis in bourbon whiskey and oak-matured red wines. *J Agric Food Chem*. 2006;54:3380–3390.
42. Ferreres F, Andrade P, Gil MI, Tomás-Barberán FA. Floral nectar phenolics as biochemical markers for the botanical origin of heather honey. *Eur Food Res Technol*. 1996;202:40–44.
43. Pérez-Jiménez J, Fezeu L, Touvier M, et al. Dietary intake of 337 polyphenols in French adults. *Am J Clin Nutr*. 2011;93:1220–1228.
44. Radtke J, Linseisen J, Wolfram G. Phenolic acid intake of adults in a Bavarian subgroup of the national food consumption survey. *Z Ernahrungswiss*. 1998;37:190–197.
45. Saura-Calixto F, Serrano J, Goñi I. Intake and bioaccessibility of total polyphenols in a whole diet. *Food Chem*. 2007;101:492–501.
46. Cai K, Hagerman AE, Minto RE, Bennick A. Decreased polyphenol transport across cultured intestinal cells by a salivary proline-rich protein. *Biochem Pharmacol*. 2006;71:1570–1580.
47. Roowi S, Stalmach A, Mullen W, Lean MEJ, Edwards CA, Crozier A. Green tea flavan-3-ols: colonic degradation and urinary excretion of catabolites by humans. *J Agric Food Chem*. 2010;58:1296–1304.
48. van't Slot G, Humpf HU. Degradation and metabolism of catechin, epigallocatechin-3-gallate (EGCG), and related compounds by the intestinal microbiota in the pig cecum model. *J Agric Food Chem*. 2009;57:8041–8048.
49. Hidalgo M, Oruna-Concha MJ, Kolida S, et al. Metabolism of anthocyanins by human gut microflora and their influence on gut bacterial growth. *J Agric Food Chem*. 2012;60:3882–3890.
50. Chen H, Hayek S, Rivera Guzman J, et al. The microbiota is essential for the generation of black tea theaflavins-derived metabolites. *PLoS One*. 2012;7:art. no. e51001.
51. Shahrzad S, Aoyagi K, Winter A, Koyama A, Bitsch I. Pharmacokinetics of gallic acid and its relative bioavailability from tea in healthy humans. *J Nutr*. 2001;131:1207–1210.
52. Shahrzad S, Bitsch I. Determination of gallic acid and its metabolites in human plasma and urine by high-performance liquid chromatography. *J Chromatogr B Biomed Appl*. 1998;705:87–95.
53. Hodgson JM, Morton LW, Puddey IB, Beilin LJ, Croft KD. Gallic acid metabolites are markers of black tea intake in humans. *J Agric Food Chem*. 2000;48:2276–2280.
54. Daykin CA, Van Duynhoven JPM, Groenewegen A, Dachtler M, Van Amelsvoort JMM, Mulder TPJ. Nuclear magnetic resonance spectroscopic based studies of the metabolism of black tea polyphenols in humans. *J Agric Food Chem*. 2005;53:1428–1434.
55. Cerdá B, Llorach R, Cerón JJ, Espín JC, Tomás-Barberán FA. Evaluation of the bioavailability and metabolism in the rat of punicalagin, an antioxidant polyphenol from pomegranate juice. *Eur J Nutr*. 2003;42:18–28.
56. Gil-Izquierdo A, Zafrilla P, Tomás-Barberán FA. An *in vitro* method to simulate phenolic compound release from the food matrix in the gastrointestinal tract. *Eur Food Res Technol*. 2002;214:155–159.
57. Espín JC, González-Barrio R, Cerdá B, López-Bote C, Rey AI, Tomás-Barberán FA. Iberian pig as a model to clarify obscure points in the bioavailability and metabolism of ellagitannins in humans. *J Agric Food Chem*. 2007;55:10476–10485.
58. González-Barrio R, Borges G, Mullen W, Crozier A. Bioavailability of anthocyanins and ellagitannins following consumption of raspberries by healthy humans and subjects with an ileostomy. *J Agric Food Chem*. 2010;58:3933–3939.

59. Ito H, Iguchi A, Hatano T. Identification of urinary and intestinal bacterial metabolites of ellagitannin geraniin in rats. *J Agric Food Chem.* 2008;56:393–400.
60. Seeram NP, Lee R, Heber D. Bioavailability of ellagic acid in human plasma after consumption of ellagitannins from pomegranate (*Punica granatum* L.) juice. *Clin Chim Acta.* 2004;348:63–68.
61. Larrosa M, Tomás-Barberán FA, Espín JC. The dietary hydrolysable tannin punicalagin releases ellagic acid that induces apoptosis in human colon adenocarcinoma Caco-2 cells by using the mitochondrial pathway. *J Nutr Biochem.* 2006;17:611–625.
62. Stoner GD, Sardo C, Apseloff G, et al. Pharmacokinetics of anthocyanins and ellagic acid in healthy volunteers fed freeze-dried black raspberries daily for 7 days. *JClP.* 2005;45:1153–1164.
63. Bialonska D, Ramnani P, Kasimsetty SG, Muntha KR, Gibson GR, Ferreira D. The influence of pomegranate by-product and punicalagins on selected groups of human intestinal microbiota. *Int J Food Microbiol.* 2010;140:175–182.
64. Cerdá B, Periago P, Espín JC, Tomás-Barberán FA. Identification of urolithin a as a metabolite produced by human colon microflora from ellagic acid and related compounds. *J Agric Food Chem.* 2005;53:5571–5576.
65. González-Barrio R, Edwards CA, Crozier A. Colonic catabolism of ellagitannins, ellagic acid, and raspberry anthocyanins: *In vivo* and *in vitro* studies. *Drug Metab Dispos.* 2011;39:1680–1688.
66. Truchado P, Larrosa M, García-Conesa MT, et al. Strawberry processing does not affect the production and urinary excretion of urolithins, ellagic acid metabolites, in humans. *J Agric Food Chem.* 2011;60:5749–5754.
67. González-Sarrías A, Giménez-Bastida JA, Garcíaa-Conesa MT, et al. Occurrence of urolithins, gut microbiota ellagic acid metabolites and proliferation markers expression response in the human prostate gland upon consumption of walnuts and pomegranate juice. *Mol Nutr Food Res.* 2010;54:311–322.
68. Tulipani S, Urpi-Sarda M, García-Villalba R, et al. Urolithins are the main urinary microbial-derived phenolic metabolites discriminating a moderate consumption of nuts in free-living subjects with diagnosed metabolic syndrome. *J Agric Food Chem.* 2012;60:8930–8940.
69. Seeram NP, Henning SM, Zhang Y, Suchard M, Li Z, Heber D. Pomegranate juice ellagitannin metabolites are present in human plasma and some persist in urine for up to 48 hours. *J Nutr.* 2006;136:2481–2485.
70. Mertens-Talcott SU, Jilma-Stohlawetz P, Rios J, Hingorani L, Derendorf H. Absorption, metabolism, and antioxidant effects of pomegranate (*Punica granatum* L.) polyphenols after ingestion of a standardized extract in healthy human volunteers. *J Agric Food Chem.* 2006;54:8956–8961.
71. Cerdá B, Espín JC, Parra S, Martínez P, Tomás-Barberán FA. The potent *in vitro* antioxidant ellagitannins from pomegranate juice are metabolised into bioavailable but poor antioxidant hydroxy-6H-dibenzopyran-6-one derivatives by the colonic microflora of healthy humans. *Eur J Nutr.* 2004;43:205–220.
72. Seeram NP, Zhang Y, McKeever R, et al. Pomegranate juice and extracts provide similar levels of plasma and urinary ellagitannin metabolites in human subjects. *J Med Food.* 2008;11:390–394.
73. González-Barrio R, Truchado P, Ito H, Espín JC, Tomás-Barberán FA. UV and MS identification of urolithins and nasutins, the bioavailable metabolites of ellagitannins and ellagic acid in different mammals. *J Agric Food Chem.* 2011;59:1152–1162.
74. Trombold JR, Barnes JN, Critchley L, Coyle EF. Ellagitannin consumption improves strength recovery 2–3 d after eccentric exercise. *Med Sci Sports Exercise.* 2010;42:493–498.
75. Valko M, Leibfritz D, Moncol J, Cronin MTD, Mazur M, Telser J. Free radicals and antioxidants in normal physiological functions and human disease. *Int J Biochem Cell Biol.* 2007;39:44–84.
76. Hajimahmoodi M, Oveisi MR, Sadeghi N, Jannat B, Nateghi M. Antioxidant capacity of plasma after pomegranate intake in human volunteers. *Acta Med Iran.* 2009;47:125–132.
77. Henning SM, Seeram NP, Zhang Y, et al. Strawberry consumption is associated with increased antioxidant capacity in serum. *J Med Food.* 2010;13:116–122.
78. McKay DL, Chen CYO, Yeum KJ, Matthan NR, Lichtenstein AH, Blumberg JB. Chronic and acute effects of walnuts on antioxidant capacity and nutritional status in humans: a randomized, cross-over pilot study. *Nutr J.* 2010;9:art. no. 21.
79. Tulipani S, Alvarez-Suarez JM, Busco F, et al. Strawberry consumption improves plasma antioxidant status and erythrocyte resistance to oxidative haemolysis in humans. *Food Chem.* 2011;128:180–186.
80. Guo C, Wei J, Yang J, Xu J, Pang W, Jiang Y. Pomegranate juice is potentially better than apple juice in improving antioxidant function in elderly subjects. *Nutr Res.* 2008;28:72–77.
81. Aviram M, Rosenblat M, Gaitini D, et al. Pomegranate juice consumption for 3 years by patients with carotid artery stenosis reduces common carotid intima-media thickness, blood pressure and LDL oxidation. *Clin Nutr.* 2004;23:423–433.
82. Cerdá B, Soto C, Albaladejo MD, et al. Pomegranate juice supplementation in chronic obstructive pulmonary disease: a 5-week randomized, double-blind, placebo-controlled trial. *Eur J Clin Nutr.* 2006;60:245–253.
83. Seeram NP. Berry fruits for cancer prevention: current status and future prospects. *J Agric Food Chem.* 2008;56:630–635.
84. Pantuck AJ, Leppert JT, Zomorodian N, et al. Phase II study of pomegranate juice for men with rising prostate-specific antigen following surgery or radiation for prostate cancer. *Clin Cancer Res.* 2006;12:4018–4026.
85. Kresty LA, Frankel WL, Hammond CD, et al. Transitioning from preclinical to clinical chemopreventive assessments of lyophilized black raspberries: interim results show berries modulate markers of oxidative stress in Barrett's esophagus patients. *Nutr Cancer.* 2006;54:148–156.
86. Wang LS, Arnold M, Huang YW, et al. Modulation of genetic and epigenetic biomarkers of colorectal cancer in humans by black raspberries: a phase I pilot study. *Clin Cancer Res.* 2011;17:598–610.
87. Banel DK, Hu FB. Effects of walnut consumption on blood lipids and other cardiovascular risk factors: a meta-analysis and systematic review. *Am J Clin Nutr.* 2009;90:56–63.
88. Giampieri F, Tulipani S, Alvarez-Suarez JM, Quiles JL, Mezzetti B, Battino M. The strawberry: composition, nutritional quality, and impact on human health. *Nutrition.* 2012;28:9–19.
89. Mena P, Gironés-Vilaplana A, Moreno DA, García-Viguera C. Pomegranate fruit for health promotion: myths and realities. *Funct Plant Sci Biotechnol.* 2011;5:33–42.

90. Aviram M, Dornfeld L, Rosenblat M, et al. Pomegranate juice consumption reduces oxidative stress, atherogenic modifications to LDL, and platelet aggregation: Studies in humans and in atherosclerotic apolipoprotein E-deficient mice. *Am J Clin Nutr*. 2000;71:1062–1076.
91. Rosenblat M, Volkova N, Attias J, Mahamid R, Aviram M. Consumption of polyphenolic-rich beverages (mostly pomegranate and black currant juices) by healthy subjects for a short term increased serum antioxidant status, and the serum's ability to attenuate macrophage cholesterol accumulation. *Food Funct*. 2010;1:99–109.
92. Djakpo O, Yao W. Rhus chinensis and galla Chinensis – folklore to modern evidence: review. *Phytother Res*. 2010;24:1739–1747.
93. Mukherjee PK, Rai S, Bhattacharyya S, et al. Clinical study of 'Triphala' – a well known phytomedicine from India. *Iran J Pharmacol Ther*. 2006;5:51–54.
94. Shah VN, Doshi DB, Shah MB, Bhatt PA. Estimation of biomarkers berberine and gallic acid in polyherbal formulation punarnavashtak kwath and its clinical study for hepatoprotective potential. *Int J Green Pharm*. 2010;4:296–301.

CHAPTER

7

Gut Microbial Metabolism of Plant Lignans: Influence on Human Health

Seth C. Yoder, Samuel M. Lancaster†,‡, Meredith A.J. Hullar* and Johanna W. Lampe**

*Public Health Sciences Division, Fred Hutchinson Cancer Research Center, Seattle, WA, USA †Department of Genome Sciences, University of Washington, Seattle, WA, USA ‡Molecular and Cellular Biology, University of Washington, Seattle, WA, USA

INTRODUCTION

Lignans are organic compounds synthesized by plants and are characterized by the union of two cinnamic acid residues by β-linkage on the C8 of each propyl side-chain.[1–3] They are found in over 70 plant families and are found throughout the plant tissue, in roots, rhizomes, stems, leaves, seeds and fruits.[4,5] Large amounts of lignans and other polyphenols are also found in the knots of soft wood trees, mainly in the form of hydroxymatairesinol.[6,7] Lignans are structurally similar to lignins, three-dimensional polymers that intercalate with cellulose, hemicellulose, and pectin to form the rigid cell wall of plant cells.[1,8] Moreover, both lignans and lignins are created by the same initial phenylpropanoid pathway and are synthesized from monolignols (derived from either phenylalanine or tyrosine), but ultimately enter distinct biochemical pathways.[2,9] Indeed, while lignins are ubiquitous in the plant kingdom, not all plants produce lignans.

Interestingly, despite having been identified over 120 years ago,[7] the botanical properties of lignans still remain unclear. Lignans are thought to play a role in defending the plant against pathogens and pests because they possess some antifungal, antimicrobial, antiviral, and even insecticidal properties.[4,10,11] For example, the lignan haedoxan A isolated from *Phryma leptostachya* has insecticidal properties comparable to that of pyrethrins. Additionally, matairesinol and its related metabolites have been shown to reduce fungal growth of *Fomes annosus* in *Picea abies*.[5,12]

Given the prevalence of lignans in the plant kingdom, they are not surprisingly found in many common foods. The lignans typically present in foods are matairesinol (MAT), pinoresinol (PINO), medioresinol (MED), lariciresinol (LARI), sesamin (SES), syringaresinol (SYR), secoisolariciresinol (SECO), and the glycosylated form of SECO, secoisolariciresinol diglucoside (SDG). Flaxseed is the richest known source of plant lignans with approximately 300,000 μg/100 g fresh weight, the majority of which is in the form of SDG.[13] Other seeds with high lignan concentration include sesame seeds (104,446 μg/100 g), cloudberry seeds (43,876 μg/100 g), hemp seeds (32,473 μg/100 g), and blackberry seeds (23,310 μg/100 g).[14] Cereal grains, including rye, wheat, barley and oats also contain moderate to high amounts of lignans.[13–16] Brassica vegetables, such as kale, cabbage and Brussels sprouts, as well as other vegetables and fruits, such as asparagus, broccoli, garlic, apricots, prunes, and dates, contain moderate amounts of lignans.[13–17] Lignans are even found in small amounts in common beverages, including black tea, green tea, coffee, soy milk, fruit juices, beer and wine.[13,18,19]

Diet-Microbe Interactions in the Gut.
DOI: http://dx.doi.org/10.1016/B978-0-12-407825-3.00007-1

CONVERSION OF PLANT LIGNANS TO ENTEROLIGNANS BY GUT BACTERIA

Despite their antimicrobial, properties, plant lignans can be metabolized and converted to enterolignans (entero- from Greek *enteron* meaning "intestine") by bacteria residing in the intestine.[20] Also referred to as mammalian lignans, enterolignans were discovered independently by two research teams at almost the same time. The two major enterolignans produced by mammalian gut bacteria are enterolactone (ENL) and enterodiol (END). In 1979, Setchell and Adlercreutz first identified what turned out to be ENL, one of the major enterolignans excreted in the urine of both pregnant and non-pregnant women.[21] A year later, Stitch et al. published an article in *Nature* detailing similar findings; they also had isolated what later was identified as ENL from urine in both pregnant and normally ovulating women.[22] Axelson et al. went on to demonstrate in humans that plant lignans are dietary precursors of enterolignans.[23]

Not long after their discovery, it was confirmed that ENL and END were produced by intestinal bacteria. Axelson and Setchell first conducted a study in 1981 comparing germ-free rats and conventional rats showing that the presence of bacteria was necessary for enterolignan formation.[24] Later that same year, Setchell et al. demonstrated in humans that urinary enterolignan excretion falls immediately and significantly following antibiotic treatment.[25] Borriello et al. confirmed the metabolism of plant lignans to ENL and END in vitro using human stool samples.[20]

Following ingestion of SDG or similarly glycated lignans, such as PINO diglucoside or sesaminol triglucoside (STG), the sugar moieties are hydrolyzed in the large intestine by *O*-linked deglycosylation, forming SECO and the other aglycones.[26,27] Studies simulating digestion in the stomach and small intestine show that SDG remains intact in the equivalent of these parts of the gut, suggesting that acid hydrolysis and human intestinal enzymes, such as lactase phlorizin hydrolase (LPH; EC 3.2.1.62), do not play a major role in this initial hydrolysis. [26,28] It is not until the artificial ascending colon, which is inoculated with gut microbes, that SECO is first detected.[28]

In converting SDG to ENL four reactions must take place (Figure 7.1). First, SDG must be converted to SECO by *O*-linked deglycosylation, and secondly SECO is converted to the intermediate dihydroxyenterodiol (DHEND) by *O*-linked demethylation. From this point, DHEND may be converted to END by dehydroxylation, and ultimately to ENL by dehydrogenation of END. Alternatively, DHEND may undergo dehydrogenation to construct a lactone ring, thus forming a second intermediate dihydroxyenterolactone (DHENL), which can then be dehydroxylated to form ENL.[26,27,29–31]

Several investigators have evaluated the capacity of various specific gut bacteria to carry out the reactions necessary for conversion of glycosylated lignans to enterolignans (Table 7.1). Clavel *et al.* determined that *Bacteroides fragilis, Bacteroides ovatus, Clostridium cocleatum, Clostridium saccharogumia, Clostridium ramosum*, and *Bacteroides distasonis* are capable of this *O*-deglycosylation, with the first four able to completely deglycosylate SDG within 20 hours of the experiment.[32–34] Roncaglia *et al.* found ten *Bifidobacterium* strains capable of hydrolyzing SDG.[35] Once deglycosylated, SECO can then be demethylated to produce its intermediate DHEND. Clavel *et al.* demonstrated the ability of *Butyribacterium methylotrophicum, Eubacterium callanderi, Eubacterium limosum*, and *Blautia producta* to catalyze this reaction.[32,34] However, Clavel also noted that the presence of the SECO demethylating species *E. callanderi* and *B. methylotrophicum* in the human intestinal tract has not been reported. (*Blautia* is a newly classified genus, and the *B. producta* taxon replaces *Peptostreptococcus productus* and *Ruminococcus productus* to which some older articles refer.)[40] Jin and Hattori also recently identified *Clostridiaceae bacterium* END-2 (originally referred to simply as "strain END-2") with the ability to demethylate SECO.[36] Following demethylation, *Eggerthella lenta* and *Clostridium scindens* can act on the intermediate by removing a hydroxyl group from each aromatic ring to form END.[34] *Lactonifactor longoviformis* and the aforementioned strain END-2 perform the final step in ENL production by fashioning the lactone ring.[33,36] Alternatively, *L. longoviformis* may create the lactone ring subsequent to SECO demethylation, in which case END does not form but rather a second intermediate, DHENL, which may then continue on to form ENL.[26]

Studies have shown that SES is also capable of being converted to END and ENL in humans, rats, and in in vitro incubations, although the SES biotransformation pathway is still undertermined.[41–44] Due to SES's methylenedioxy functional groups, it is thought that SES follows a conversion pathway (Figure 7.2) different than that of the other major lignans (Figure 7.1). In addition, the organisms responsible for converting SES are not as extensively researched as those involved in SDG conversion. Zhu et al. recently showed that when STG, isolated from sesame cake and the most abundant lignan glycoside in sesame seed, was fermented with human intestinal bacteria, the phylogenetic groups *Bifidobacteria* and *Lactobacillus-Enterococcus* were significantly greater in the STG samples compared to the controls.[45]

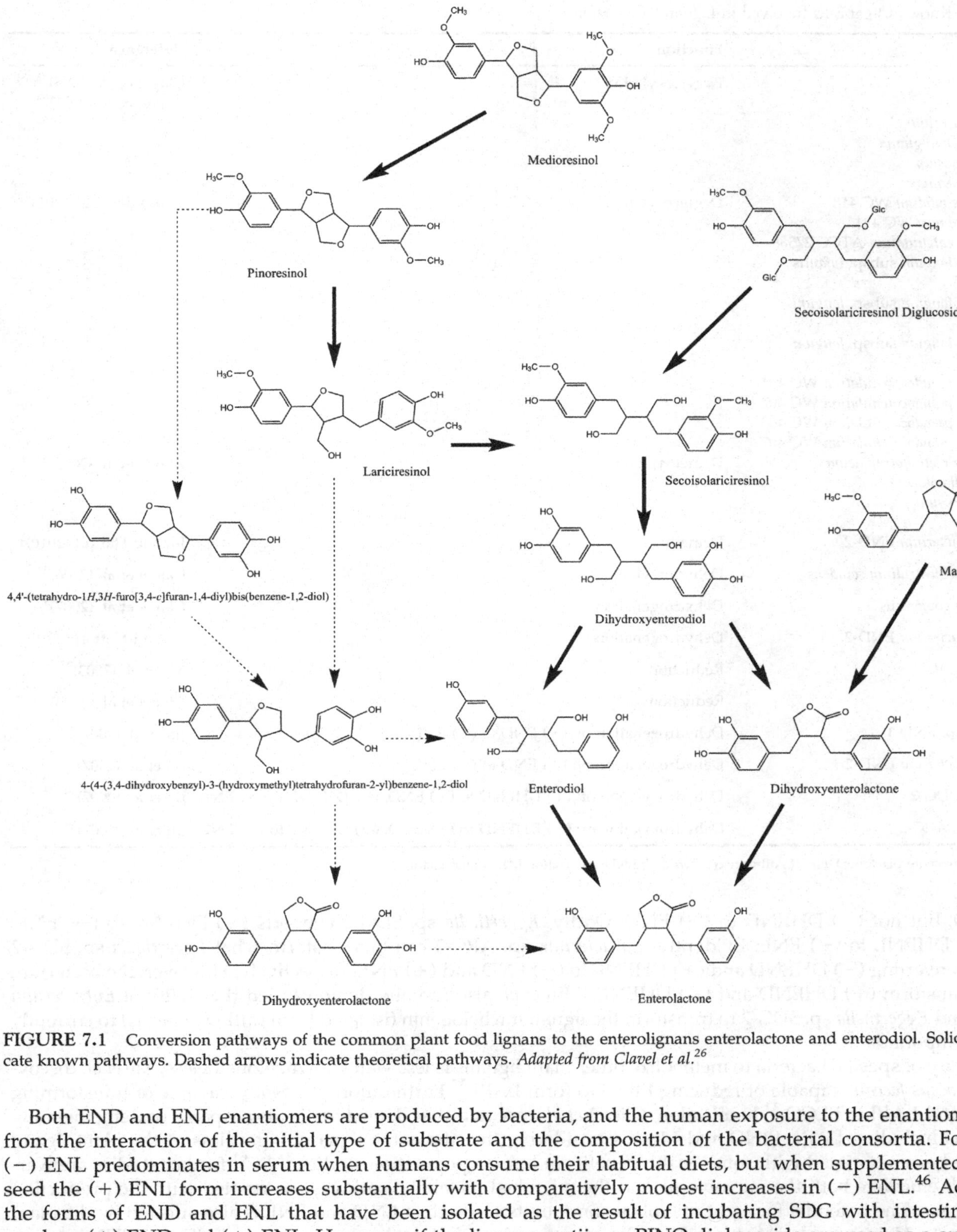

FIGURE 7.1 Conversion pathways of the common plant food lignans to the enterolignans enterolactone and enterodiol. Solid arrows indicate known pathways. Dashed arrows indicate theoretical pathways. *Adapted from Clavel et al.*[26]

Both END and ENL enantiomers are produced by bacteria, and the human exposure to the enantiomers results from the interaction of the initial type of substrate and the composition of the bacterial consortia. For example, (−) ENL predominates in serum when humans consume their habitual diets, but when supplemented with flaxseed the (+) ENL form increases substantially with comparatively modest increases in (−) ENL.[46] Additionally, the forms of END and ENL that have been isolated as the result of incubating SDG with intestinal bacteria produce (+) END and (+) ENL. However, if the lignans arctiin or PINO diglucoside are used as a substrate (−) END and (−) ENL are produced.[47] Subsequently, some bacteria have shown enantiomeric selectivity when converting lignans. *Ruminococcus* sp. END-1 enantioselectively converts (−) END to (−) ENL, while the strain END-2 converts (+) END to (+) ENL.[38] *Eggerthella* sp. SDG-2 can convert (+) dihydroxyenterodiol (DHEND)

TABLE 7.1 Known Organisms Involved in Lignan Conversion

Organism(s)	Function	Reference
Bacteroides fragilis *Bacteroides ovatus* *Clostridium cocleatum* *Clostridium saccharogumia* *Clostridium ramosum* *Bacteroides distasonis*	Deglycosylation	Clavel et al. (2006)[32, 34] Clavel et al. (2007)[33]
(*Bifidobacterium bifidum* WC 418 *Bifidobacterium breve* WC 421 *Bifidobacterium catenulatum* ATCC 27539 *Bifidobacterium longum* subsp. *infantis* ATCC 15697 *Bifidobacterium longum* subsp. *longum* WC 436 *Bifidobacterium longum* subsp. *longum* WC 439 *Bifidobacterium pseudocatenulatum* WC 401 *Bifidobacterium pseudocatenulatum* WC 402 *Bifidobacterium pseudocatenulatum* WC 403 *Bifidobacterium pseudocatenulatum* WC 407	Deglycosylation	Roncaglia et al. (2011)[35]
Butyribacterium methylotrophicum *Eubacterium callanderi* *Eubacterium limosum* *Blautia product*	Demethylation	Clavel et al. (2006)[32,34]
Clostridiaceae bacterium END-2	Demethylation	Jin and Hattori (2010)[36]
Eggerthella lentaClostridium scindens	Dehydroxylation	Clavel et al. (2006)[34]
Lactonifactor longoviformis	Dehydrogenation	Clavel et al. (2007)[33]
Clostridiaceae bacterium END-2	Dehydrogenation	Jin and Hattori (2010)[36]
Enterococcus faecalis	Reduction	Xie et al. (2003)[37]
Eggerthella lenta	Reduction	Clavel et al. (2006)[32]
Ruminococcus sp. END-1	Dehydrogenation of: (−) END to (−) ENL	Jin et al. (2007)[38]
Clostridiaceae bacterium END-2	Dehydrogenation of: (+) END to (+) ENL	Jin et al. (2007)[38]
Eggerthella sp. SDG-2	Dehydroxylation of: (+) DHEND to (+) END; (−) DHENL to (−) ENL	Jin et al. (2007)[39]
Eubacterium sp. ARC-2	Dehydroxylation of: (−) DHEND to (−) END; (+) DHENL to (+) ENL	Jin et al. (2007)[39]

DHEND, dihydroxyenterolactone; DHENL, dihydroxyenterodiol; END, enterodiol; ENL, enterolactone.

to (+) END, but not (−) DHEND to (−) END. Oddly, *Eggerthella* sp. SDG-2 converts (−) DHENL to (−) ENL, but not (+) DHENL to (+) ENL. Evidently, *Eubacterium* sp. ARC-2 can biotransform what *Eggerthella* sp. SDG-2 can not by converting (−) DHEND and (+) DHENL to (−) END and (+) ENL, respectively. However, the bacterium could not transform (+) DHEND and (−) DHENL.[39] Jin *et al.* also recently demonstrated the ability of *Eubacterium* sp. ARC-2 and *Eggerthella* sp. SDG-2 to transform the lignan trachelogenin (isolated from safflower seeds) to currently unnamed compounds.[48]

The capacity of specific bacteria to metabolize other plant lignans is less well known. Nonetheless, Xie *et al.* discovered *Enterococcus faecalis* capable of reducing PINO to form LARI.[37] Furthermore, *E. lenta* is capable of transforming both PINO and LARI to SECO.[32] In addition, Clavel *et al.* reported that *B. producta* is also capable of demethylating other lignans including LARI, PINO and MAT. Considering that the enterolignans, regardless of the plant lignan source, have been associated with some health benefits in human population studies,[31,49–61] further studies are warranted which elucidate both the microbiome and the microbial metabolic pathways associated with the production of enterolignans. The potential for a gut microbial community to product END and ENL is also affected by the plant lignan substrate. Heinonen et al.[62] showed a range of conversion of plant-derived lignans to END and ENL, ranging from no conversion of isolariciresinol to 100% conversion of LARI. Similarly, production of ENL from arctigenin glucoside and MAT ranged from 5 to 62%, respectively. These results further support the importance of the interaction of the plant lignans and the gut microbial community in dictating enterolignan exposure.

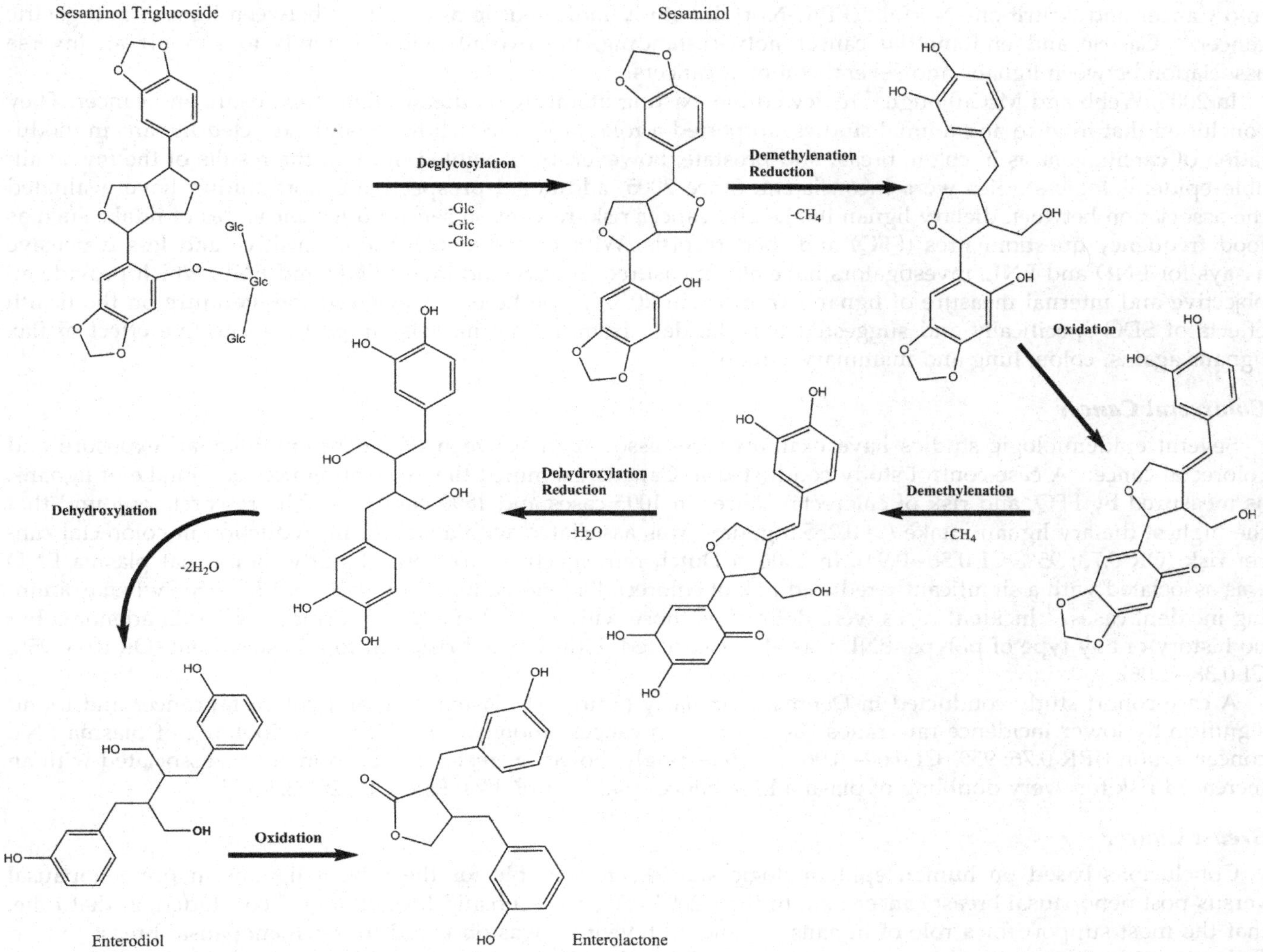

FIGURE 7.2 Overview of potential conversion pathway of the glycated sesame seed lignan, sesaminol triglucoside, to the enterolignans. *Adapted from Jan et al.*[43]

ASSOCIATIONS BETWEEN LIGNAN EXPOSURE AND HUMAN HEALTH

Plant lignans, and particularly their metabolites, the enterolignans, have been shown to have a range of biological activities.[27,31,58,63,64] The structural similarity of END and ENL to 17β-estradiol, a common sex hormone, allows the enterolignans to bind to estrogen receptor alpha (ERα) and exert weak estrogenic or anti-estrogenic effects, which is why they were first classified as "phytoestrogens[63,65–67] However, both in vitro and in vivo studies have identified a variety of other mechanisms by which enterolignans may affect the risk of several chronic diseases. These mechanisms include anti-proliferative, anti-inflammatory, and apoptotic effects.[10,51,58,59,66,68] Several review articles have been published examining the available evidence on lignans and their effects on human health.[31,49–61] Here we summarize the major conclusions of these reviews and provide further updates on the human studies.

Cancer

The three types of cancer most extensively explored in relation to lignan exposure are breast, colorectal and prostate cancer, to which lignans are linked with cancer prevention. Research concerning other cancers, however, is sparse. The handful of human studies on endometrial cancer and lignans tends toward a null association,[49,69–71] and a nested case-control study conducted within the European Prospective Investigation into Cancer and Nutrition–Norfolk (EPIC-Norfolk) study indicated no associations between lignans and gastric cancer.[72] Gastric and endometrial cancer notwithstanding, the overall evidence tends to support an inverse association between lignans and several common cancers.

In 2005, Webb and McCullough[59] reviewed the existing literature on dietary lignan exposure and cancer. They concluded that in vitro and animal studies supported a role for lignan-rich foods and extracted lignans in modulation of carcinogenesis in colon, breast and prostate; however, they pointed out that the results of the few available epidemiologic studies were inconsistent. Since 2005, additional prospective cohort studies have evaluated the association between dietary lignan intake and cancer risk, relying on self-reported measures of intake such as food frequency questionnaires (FFQ) and food records. With the advent of more sensitive and less expensive assays for END and ENL, investigators have also measured urinary and blood END and ENL, which provide an objective and internal measure of lignan exposure. In 2010, Adolphe et al. reviewed the literature on the health effects of SDG specifically and suggested that the data from animal models support a protective effect of flax lignans against colon, lung and mammary tumors.[66]

Colorectal Cancer

Several epidemiologic studies have examined the association between lignan or enterolignan exposure and colorectal cancer. A case-control study conducted in Canada examined the association between intake of lignans, as measured by FFQ, and risk of colorectal cancer in 1095 cases and 1890 controls.[73] The researchers found that the highest dietary lignan intake (>0.255 mg/day) was associated with a significant reduction in colorectal cancer risk (OR 0.73; 95% CI 0.56–0.94). In 2006, a Dutch retrospective case-control study found that plasma END was associated with a significantly reduced risk of colorectal adenoma (OR 0.53; 95% CI 0.32–0.88) when examining incident cases.[74] Incident cases were defined as those with a histologically confirmed colorectal adenoma but no history of any type of polyps. ENL was also associated with reduced risk, but to a lesser extent (OR 0.63; 95% CI 0.38–1.06).

A case-cohort study conducted in Denmark similarly examined plasma ENL and colorectal cancer and found significantly lower incidence rate ratios (IRR) for colon cancer among women for each doubling of plasma ENL concentration (IRR 0.76; 95% CI 0.60–0.96).[75] Interestingly, however, rectal cancer in men was associated with an increased risk for every doubling of plasma ENL concentration (IRR 1.74; 95% CI 1.25–2.44).[75]

Breast Cancer

Conclusions based on human epidemiologic studies are variable for the role of lignans in premenopausal versus postmenopausal breast cancer risk. In their 2005 review, Webb and McCullough[59] concluded, at that time, that the most support for a role of lignans in cancer prevention was observed for premenopausal breast cancer. Following this, the results of two meta-analyses published within a year of each other provide a more systematic evaluation of available studies. A 2009 meta-analysis of 11 studies (four cohort studies and seven case-control studies) assessing lignan exposure and breast cancer risk determined that, though borderline, there was no association between plant lignan intake and breast cancer risk (combined OR 0.93; 95% CI 0.83–1.03) when the relevant observational studies were pooled.[57] However, when the studies were divided by menopausal status, they found a significant reduction in breast cancer risk among the postmenopausal women (combined OR 0.85; 95% CI 0.78–0.93; $p<0.001$). This effect was not observed with pre-menopausal women (combined OR 0.97; 95% CI 0.82–1.15, $p=0.73$). Where data were available, the investigators also systematically examined ENL concentrations in the blood and estimated enterolignan exposure using values produced from food by in vitro fermentation models. Although there was a significant inverse correlation between enterolignan exposure and breast cancer risk (combined OR 0.73; 95% CI 0.57–0.92), the association was no longer significant (combined OR 0.82; 95% CI 0.59–1.14) when blood ENL levels were considered. The authors noted that there was marked heterogeneity between the studies they reviewed making broad conclusions difficult.

Another meta-analysis which included 21 studies (11 prospective cohort studies and 10 case-control studies) concluded that plant lignan intake was not associated with overall breast cancer risk, but if the results are separated by menopausal status, postmenopausal women with a high plant lignan intake have a significantly reduced risk of breast cancer (pooled RE 0.86; 95% CI 0.78–0.94).[50] A similar association was found when four studies were examined assessing dietary enterolignans (pooled RE 0.84; 95% CI 0.71–0.97). In total, these meta-analyses suggest that menopausal status is an important factor to consider in associations of dietary lignans, enterolignans and risk of breast cancer.

More recent studies continue to support the hypothesis that enterolignans have a protective effect on breast cancer, but that this is dependent on menopausal status. A follow-up study of 2653 postmenopausal breast cancer patients diagnosed between 2001 and 2005 concluded that both high estimated ENL (HR 0.60; 95% CI 0.40–0.89; p trend = 0.02) and high estimated END were associated with higher overall survival (HR 0.63; 95% CI 0.42–0.95; p trend = 0.02).[76] Moreover, a retrospective cohort study of 300 breast cancer patients concluded that higher serum ENL levels were associated with both decreased all-cause mortality and decreased breast-cancer mortality.[77] At 5 years, patients with serum ENL levels ≥ 10 nmol/L demonstrated lower all-cause and breast cancer-specific mortality (HR 0.49; 95% CI 0.27–0.91; $p = 0.025$ and HR 0.42; 95% CI 0.22–0.81; $p = 0.009$, respectively). At 10 years and beyond, however, this association became non-significant (all-cause death: HR 0.65; 95% CI 0.40–1.03; $p = 0.07$ and breast cancer-specific (HR 0.67; 95% CI 0.39–1.14; $p = 0.1$). After stratifying the results by menopausal status, the association between serum ENL and both all-cause mortality and breast-cancer mortality decreased and remained statistically significant at 10 years for postmenopausal women (all-cause death: HR 0.48; 95% CI 0.28–0.82; $p = 0.007$ and breast cancer-specific: HR 0.52; 95% CI 0.29–0.94; $p = 0.03$, respectively).[77] After adjusting for menopausal status, these studies show a protective effect of ENL suggesting that different mechanisms may be important to consider in pre- and postmenopausal women.

A case-control study conducted in women in Ontario, Canada, examined the relationship between breast cancer and phytoestrogen intake using a FFQ.[78] The study included 2438 women with breast cancer and 3370 controls and results were stratified by tumor type, menopausal status, and phytoestrogen intake as an adult versus adolescent. Among women with estrogen-receptor-positive and progesterone-receptor-positive (ER+/PR+) tumors, both adult intake of lignans (OR 0.83; 95% CI 0.68–1.00) and adolescent intake of lignans (OR 0.86; 95% CI 0.73–1.00) were associated with reduced risk of breast cancer. After separating the results by menopausal status, the associations in premenopausal women become either null or not significant; however, the association with adolescent lignan intake among postmenopausal women in the ER+/PR+ subgroup remained (OR 0.78; 95% CI 0.64–0.95).[78]

McCann et al. reported on a case-control study analyzing not only lignan intake but also the various types of plant lignans and their relationship to breast tumors. They found that the women in the highest tertile of lignan intake had a 40–50% lower odds of breast cancer compared to those in the lowest tertile of lignan intake.[79] This result was regardless of menopausal status; however, premenopausal women had the strongest associations with PINO and LARI while postmenopausal women had the strongest associations with MAT.

Animal models have been used to examine the underlying mechanism of action between lignans, enterolignans and breast cancer. In animal models, SDG has been shown to promote cell differentiation in mammary glands, delay the progression of *N*-methyl-*N*-nitrosourea-induced mammary tumorigenesis, and produce beneficial mammary gland structural changes during gestation and lactation.[66] More recently, SES was shown to decrease cell proliferation and increased apoptosis of MCF-7 tumors in mice.[80] In addition, a study in germ-free rats showed that, although END and ENL production did not influence the occurrence of breast cancer, the number of tumors, size of the tumors, and tumor cell proliferation decreased, and tumor cell apoptosis increased in rats that were conventionalized with a lignan-converting bacterial consortia. These studies suggest that lignans and bacterially produced enterolignans influence anticancer effects.[81]

Prostate Cancer

The evidence regarding the role that lignans may play in prostate cancer is varied. On the one hand, reviews by McCann et al. and Saarinen et al. cite several in vitro and animal studies that demonstrate the ability of lignans to act as chemopreventive agents in prostate cells.[60,61] Indeed, recent evidence confirms that ENL, even at concentrations achievable in vivo following the intake of lignan precursors, inhibits the proliferation of early-stage prostate cancer cells.[82] However, data available from human studies are inconsistent; some evidence suggests a benefit, while other studies found null associations between lignans and prostate cancer. Since those reviews were published, two case-control studies have been published that actually show higher lignan intake is associated with an increased risk of prostate cancer.[83,84]

A nested case-control study from EPIC-Norfolk examined lignans and prostate cancer.[83,85] Participants completed 7-day food records, which were used to estimate daily lignan intake. In addition to estimating intake of the plant lignans MAT and SECO, the investigators also estimated intake of preformed enterolignans from dairy and other animal products. This was the first study to consider animal sources of enterolignans in the overall estimate of lignan exposure. Given that mammals, particularly ruminants, are capable of producing large amounts of enterolignans in their rumen, some animal products, particularly dairy foods, contain enterolignans.[86] The study included 204 prostate cancer cases and 812 controls. Estimated mean (±SD) intake of preformed enterolignans were 20 μg/day (±9) among cases and 18 μg/day (±9) among controls in this EPIC-Norfolk sample.[83] In an age-adjusted model, total lignan intake was not associated with prostate cancer (OR 1.05; 95% CI

0.81–1.36; $p = 0.72$). However, total enterolignan intake (which included the non-lignan phytoestrogen equol) was positively associated with prostate cancer (OR 1.41; 95% CI 1.12–1.76; $p = 0.003$) as were equol (OR 1.43; 95% CI 1.14–1.80; $p = 0.002$) and ENL (OR 1.39; 95%CI 1.12–1.71, $p = 0.003$) individually. However, after additional adjustment for covariates, such as age, height, weight, physical activity, social class, family history of prostate cancer, and daily energy intake, the associations became non-significant, suggesting that the lignan effect was confounded with other variables.

Another case-control study, conducted in Jamaica, consisted of 175 newly diagnosed prostate cancer cases and 194 controls.[84] The researchers evaluated the relationship of urinary phytoestrogens with total cancer and tumor grade. ENL was measured from spot urine samples. Higher concentrations of ENL were positively associated with both total prostate cancer (OR 1.85; 95% CI 1.01–3.44; $p = 0.027$) and high-grade disease (OR 2.46; 95% CI 1.11–5.46; $p = 0.023$).

The small sample size and limitations inherent in a case-control design advocates caution in espousing the results without follow-up studies, but differential findings based on tumor grade, speak to the importance of considering the heterogeneity of cancers and the potential for different effects depending on tumor grade. To date, only a few studies have been statistically powered to be able to consider the association between lignans and cancers, stratified by tumor type or stage.

Cardiovascular Disease

A review of lignans and cardiovascular disease risk published in 2010 by Peterson et al. summarized the findings of both randomized controlled trials (RCTs) and observational studies.[55] While some of the RCTs showed no effects, many showed favorable effects to blood pressure, C-reactive protein, and lipid profiles. The authors also discussed both observational studies examining lignan intake as well as observational studies examining serum ENL; however, similar to the studies of cancer outcomes, the results of the studies of cardiovascular disease are mixed, and it is difficult to draw definitive conclusions. Five of the 11 observational studies showed decreased cardiovascular risk with either increasing dietary intakes of lignans or increased concentrations of serum ENL, while five studies were described by the authors as having borderline significance, and one was null.[55] The authors noted that a limitation to the capacity to conduct a systematic review of the interventions was the differences in experimental protocols that were used between the studies.

Since that review was published, a 2012 cross-sectional study found that urinary enterolignans were inversely associated with serum triglyceride levels and positively associated with HDL ("good") cholesterol.[87] The authors used three models: an unadjusted model, a model adjusted for age, race/ethnicity, education, income, urinary creatinine (log-transformed), and a third model further adjusted for smoking and alcohol categories, menopause status, use of hormone replacement therapy, BMI, physical activity, and dietary intake of saturated fat, cholesterol and fiber. The results were statistically significant in all three models, with higher urinary excretion of enterolignans (7.84 μmol/L) corresponding to lower serum triglygerides (−0.18 mmol/L) compared to the low levels of enterolignan excretion (0.54 μmol/L) in the third model ($p = 0.003$). High excretion corresponded to 0.06 mmol/L greater serum HDL in the third model ($p = 0.009$).[87]

Other Health Effects

Given the various actions of enterolignans in vivo, there is the potential for these compounds to influence other aspects of health. For example, Pietrofesa et al. tested the effect of isocaloric diets containing three levels of lignans from flaxseed in mice treated with X-rays.[88] The lignan component was designed to mimic the amount of lignans ingested from a 0%, 10% and 20% whole-grain flaxseed diet. The lignan diets significantly mitigated radiation-related animal death (controls demonstrated 36.7% survival 4 months after the treatment compared to 60–73.3% survival in mice fed 10–20% lignans). Additionally, lung fibrosis, radiation-induced lung injury (measured via bronchoalveolar lavage), and inflammation was decreased in the mice fed flaxseed lignans.[88] This would suggest that dietary lignans may help mitigate adverse effects in individuals exposed to radiation.

INTERINDIVIDUAL DIFFERENCES IN LIGNAN METABOLISM

Studying the relationship between lignan exposure and disease risk in human populations is challenging and complex. A variety of foods are sources of lignans and these are not eaten in isolation. Traditionally, cuisines that include dietary sources of lignans also include other foods that may be considered healthy or less healthy. Trying to tease out the association between intake of high-lignan foods in general, which are often also high-fiber foods, or intake of specific lignans, and disease risk is difficult.

In relation to the study of breast cancer, but relevant to other observational studies described in the previous section, Sonestedt and Wirfält described several factors that may contribute to the ambiguous results that have so far been obtained in epidemiologic studies focusing on ENL.[89] They note that several factors play a role in enterolignan exposure, including the composition of the gut microbial community, dietary intake of lignan-containing foods, antibiotic use, smoking status and constipation. The authors also point out the difficulty in accurately and reliably determining enterolignan exposure and the complicated role that genetic factors play in both cancer development and the actions of ENL.

Even under controlled conditions, substantial interindividual variation has been reported regarding enterolignan production. Kuijsten et al.[90] showed in a pharmacokinetic study that a variety of plasma profiles arise with a set dose of SDG, with some individuals producing high amounts of both ENL and END, or higher amounts of one enterolignan as compared to the other (e.g., high ENL, low END or vice versa), or very low amounts of both. All 12 participants ingested the same dose of SDG per kg body weight; however, in five subjects the area under the curve (AUC) of ENL was more than twice that of END; in five others, the AUC of ENL was only one- to two-times the AUC of END; in two participants, the AUC of END exceeded that of ENL; and in one participant, ENL concentrations barely increased. Mean ($\pm$SD) AUC for END and ENL were 966 ± 639 and 1762 ± 1117 nmol/L.h, respectively, the large SD reflecting the wide variation in individual results. Kuijsten et al. noted that the cumulative excretion of END and ENL accounted for about 40% of ingested SDG; however, individual percentages were also variable. Much of this variation is proposed to be due to differences in gut microbial community across individuals.[90] The different patterns of END and ENL appearance in circulation and the range of cumulative excretion values suggest differential capacity of the gut microbes to manage the various steps in SDG metabolism (Figure 7.3), with some unable to effectively hydrolyze the SDG to SECO and others being ineffective at converting END to ENL.

The variation in enterolignan production by gut microbial communities from different individuals is also clearly demonstrated using in vitro incubations. Incubating fresh fecal samples with a flaxseed extract (630 μmol/L) for 72 h, Possemiers et al.[91] found that the production of END and ENL differed strongly among 100 individuals. END was produced in varying amounts in 63% of the samples, whereas ENL was only produced in 39% of the samples. Further, END and ENL production were positively correlated and were associated with higher β-glucuronidase activity, supporting the importance of the capacity to carry out the initial steps in SDG conversion.

Another potential source of variation in urine or blood measures of ENL and END, may stem from interindividual differences in biotransformation of the enterolignans[92] although no controlled evaluation of the effects of genetic variants in biotransformation enzymes on lignan availability has been conducted. In humans, plant lignans and their metabolites are efficiently conjugated with glucuronic acid or, to a lesser extent, sulfate. Conjugation takes place in the gut epithelium and liver by UDP-glucuronosyltransferases (UGT) and sulfotransferases, and the conjugates are excreted in urine and bile. Those that are re-excreted in bile undergo enterohepatic recycling.[93] In urine, ENL and END are excreted primarily as monoglucuronides (95 and 85%, respectively), with small percentages being excreted as monosulfates (2–10%) and free aglycones (0.3–1%). [94] Colon cell lines have been shown to rapidly glucuronidate enterolignans, suggesting that the majority of conjugation of lignans likely occurs in the colon.[95] Further, hepatic microsomal oxidation is also much slower compared to glucuronidation, suggesting that oxidation products of END and ENL are minor metabolites of enterolignans.[96] Effects of enterolignans on intestinal transporters have not been evaluated, but SES has been shown to increase the mRNA expression of several transporters, thereby possibly affecting absorption of lignans and other phytochemicals.[97]

In addition to the variation in microbial and host metabolism, several physiologic, demographic and lifestyle characteristics of individuals appear to influence the production of enterolignans in vivo, including constituents of diet and other non-dietary factors.

Diet

Diet plays a large role in enterolignan production, both due to the differences in food content of the precursors, but also due to effects of food matrices and other dietary factors on availability and conversion of plant lignans to enterolignans. An early study by Adlercreutz et al. evaluated urinary enterolignan excretion in women consuming macrobiotic, omnivorous and lactovegetarian diets and reported that followers of macrobiotic diets, as compared to the lactovegetarians, had substantially greater END and ENL excretion.[98] The lactovegetarians in turn, as compared to the omnivores, had greater excretion of enterolignans.

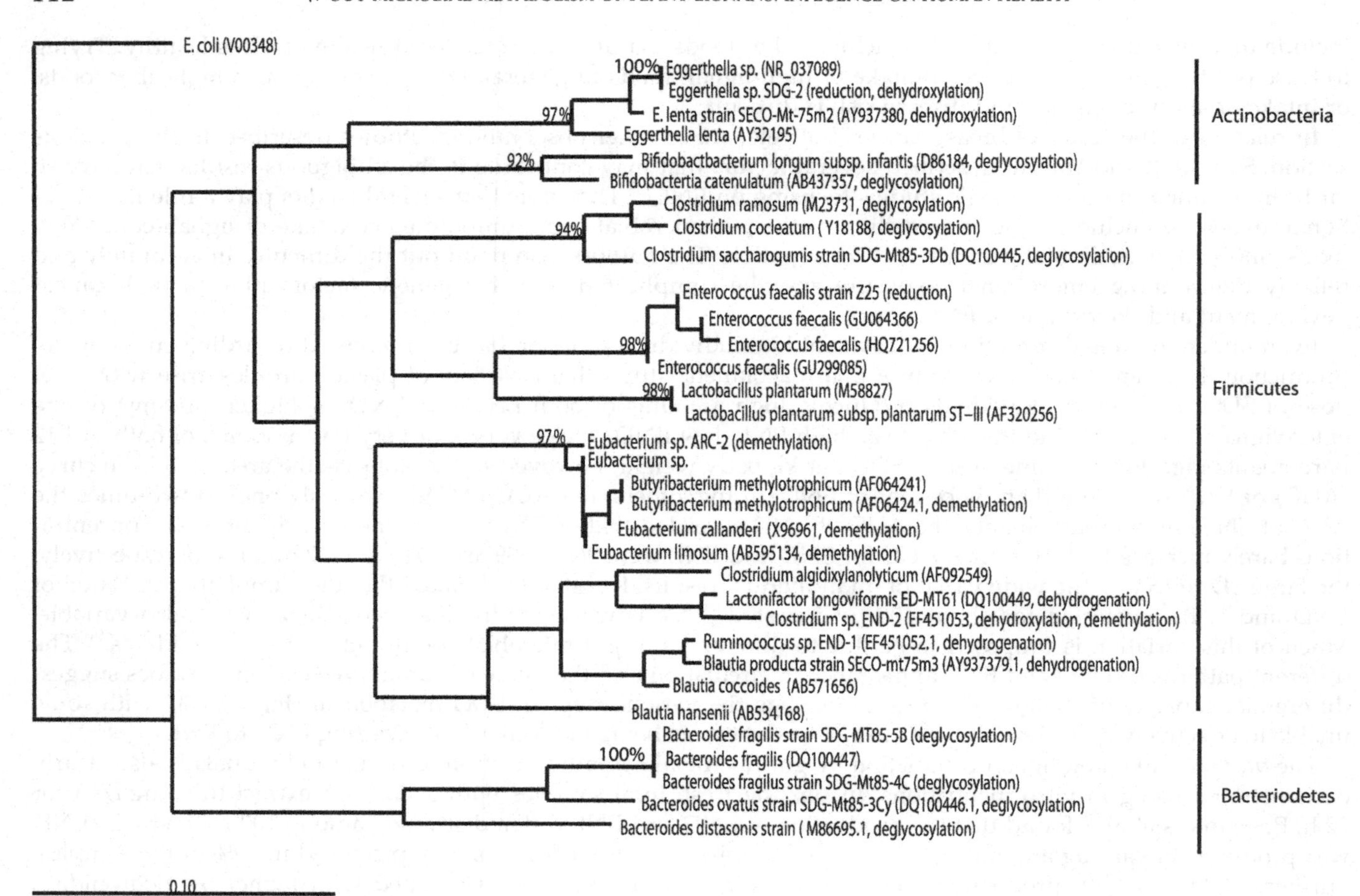

FIGURE 7.3 Phylogenetic tree of lignan-converting bacteria showing that the organisms involved in enterolignan production are distributed across Phyla. Phylogenetic distances were calculated using Neighbor Joining with bootstrap values in percentage (using 1000 replicates) in ARB (www.arb-silve.de) on aligned 16 S rRNA genes from pure cultures of bacteria associated with enterolignan production. The metabolic reactions involved in enterolignan production catalyzed by the bacteria and the accession number for the 16S rRNA gene sequence is in parentheses (see Table 7.1 for more information). Ten percent change in sequence indicated by bar.

Since that early work, many dietary intervention studies have been conducted, further solidifying the link between diet and enterolignan production. Fruits, vegetables, and berries have been demonstrated to increase ENL production.[99–101] Rye, a rich source of plant lignans, has been shown to increase ENL production in three dietary intervention studies.[102–104] However, a similar intervention study involving rye products showed no such increase in ENL production.[105] The reason for this discrepancy is unclear; however, the authors note that their participants were generally younger and may have consumed more dietary fiber in their habitual diets than participants in the other studies. This may have contributed to faster lignan transit through the gut, and therefore incomplete biotransformation of the plant lignans.[105] Other dietary trials involving lignan-rich flaxseed and sesame seed revealed significant ENL production.[41,42,106–109]

Moreover, it appears that the food matrix and food processing may affect the bioavailability of lignans. In a randomized, crossover study, 12 participants were fed whole, crushed and ground flaxseed daily (0.3 g/kg body weight). The results indicated that, compared to ground flaxseed, the mean relative bioavailability of enterolignans from whole flaxseed was 28% and crushed flaxseed was 43%.[110] Another dietary study found no effect of baking on the urinary or plasma concentrations of enterolignans when participants were fed ground flaxseed mixed in applesauce or ground flaxseed baked in bread and muffins, suggesting that these compounds are heat stable.[109]

Although it has yet to be studied systematically in humans, the presence of dietary fiber may affect the production of enterolignans. In vitro fecal suspensions incubated with 1 g each of extractable and non-extractable fractions of rye produced starkly different results.[111] The extractable fraction contained mainly soluble fiber and

46 mg/100 g of total plant lignans. The non-extractable fraction contained mainly insoluble fiber and 13 mg/100 g of total plant lignans. Protein and starch amounts were very similar among the fractions. After 48 h of incubation, suspensions containing the non-extractable fraction produced significantly more ENL. The extractable fractions rapidly produced short-chain fatty acids (SCFAs) which quickly decreased the pH of the samples below 5.0. The non-extractable fractions, on the other hand, produced SCFAs at a much slower rate, thus allowing pH to stay in a more neutral range favoring ENL production. The extractable fraction also contained more ferulic giving the samples containing it a lower initial pH.[111] This suggests that the type of fiber consumed with the lignan may affect its bioavailability.

Sex Differences in Enterolignan Production

The evidence regarding sex differences in lignan metabolism is mixed, but there appears to be a tendency for women to produce more enterolignans than men. Several cross-sectional studies examining the associations between diet and enterolignan production show no sex differences.[87,112–115] However, a cross-sectional study of 2380 Finnish men and women showed that women had higher concentrations of serum ENL than did men.[116] Indeed, plasma ENL and END both appeared earlier and also achieved a higher maximum concentration in women compared to men when following a single SDG dose (1.31 μmol/kg body weight).[90] Jacobs et al. conducted a feeding study involving a whole-grain diet and found that women had a higher mean baseline serum ENL concentration, although the rise in serum ENL after the whole-grain diet was similar in men and women.[117] Another randomized cross-over trial that involved feeding men and women flaxseed and measuring urinary lignan excretion showed no differences between men and women.[118] Similar trials involving a lignan dose have showed no difference in enterolignan production between men and women.[102,110] In contrast, in one controlled feeding study, in which vegetables were fed, men excreted more ENL than women during the experimental diets.[101] The tendency here for women to produce more enterolignans than their male counterparts may be explained by Clavel et al. who noted that women tended to harbor more ENL-producing and END-producing bacteria.[119] Further, women have been shown to have longer gastrointestinal transit times than men[120]; this longer residence time of fiber-containing foods and plant lignans in the gut may further contribute to greater enterolignan production.

Other Factors Associated with Enterolignan Production

Physiologic and sociodemographic factors have also been shown to be associated with enterolignan levels. Because of the importance of the gut microbial community to enterolignan production, the use of oral antimicrobials is inversely associated with serum ENL concentrations. In a cross-sectional study, Kilkkinen et al. showed, in a sample of 2753 Finnish men and women, that ENL concentrations in antimicrobial users, as compared to non-users of antimicrobials, was significantly lower, even if the time elapsed since antimicrobial treatment was 12–16 months prior to blood sampling.[121] The number of antimicrobial treatments also correlated inversely with ENL concentration.

Gut residence time appears to be another factor that influences ENL production, although most studies rely on non-quantitative measures to assess this factor. For example, Kilkkinen et al., also in a Finnish population, observed positive associations in men with serum ENL concentration and constipation, perhaps demonstrating an association between longer residence time and more ENL production.[116] The same study also found that, in women, serum ENL concentration was positively and independently associated with age and constipation, while ENL was inversely associated with smoking. Moreover, female subjects of normal weight had a significantly higher ENL concentration than their underweight or obese counterparts. Other studies in European populations have supported Kilkkinen's findings, reporting inverse associations between body mass index (BMI), smoking, and frequency of bowel movements and plasma ENL concentration.[115,122]

Recently, the association between sociodemographic and other lifestyle variables and urine ENL levels was evaluated in a large sample of men and women ≥20 years in the US. In a subset (n = 3000) of the 2003–2006 National Health and Nutrition Examination Survey (NHANES), which collects cross-sectional data on the health and nutritional status of the U.S. population, Rybak et al. found that age, income and physical activity were positively correlated with urinary ENL.[123] Further, similar to the studies in European populations, smoking and BMI were inversely correlated with urinary ENL. Despite these associations, the selected sociodemographic and lifestyle variables only explained a limited amount of the total variability ($R^2 \leq 4\%$), suggesting that, compared to other factors, their effects on enterolignan levels are modest.

CONCLUSIONS

Human exposure to END and ENL, bioactive metabolites resulting from the gut microbial metabolism of a variety of plant lignans, is highly variable. From the research available to date, we know that it is dependent on several environmental and physiologic factors including: the types of plant lignans consumed; the foods and food matrices of which the lignans are a part; intake of other foods or medications that may alter gut microbial activity (e.g., dietary fiber, anti-microbial drugs); gut microbial community composition and activity; and gut transit time and factors that affect it. Given this range of factors, characterizing lignan exposure is challenging. Relying on database measures of plant lignan intake does not capture the contribution of the gut microbial community to END and ENL exposure. In contrast, measuring urinary or circulating END and ENL provides an internal measure of exposure to the bioactive compounds, but as Sonestedt and Wirfält point out, also provides a biomarker of dietary lignan intake, a biomarker of a healthy lifestyle and possibly a marker of gut microbial capacity.[89] It is the complexity of these measures as biomarkers that may contribute to the uncertainties related to the association between lignans and chronic disease risk in observational studies. With the advent of new techniques that allow for rapid and efficient characterization of gut microbial community, adding information on the gut microbial community structure to the statistical models may allow us to better characterize the modifying effect of microbes on lignan–disease risk associations.

Acknowledgements

This work was supported in part by US National Cancer Institute grants U01 CA161809 and U01 CA162077, and Fred Hutchinson Cancer Research Center.

References

1. Ayres D, Loike J. *Lignans: Chemical, Biological and Clinical Properties*. Cambridge: Cambridge University Press; 1990.
2. Umezawa T. Diversity in lignan biosynthesis. *Phytochem Rev*. 2003;2:371–390.
3. Hearon WM, MacGregor WS. The naturally occurring lignans. *Chemical Rev*. 1955;55:957–1068.
4. Pan J-Y, Chen S-L, Yang M-H, Wu J, Sinkkonen J, Zou K. An update on lignans: natural products and synthesis. *Nat Prod Rep*. 2009;26:1251–1292.
5. Gang DR, Dinkova-Kostova AT, Davin LB, Lewis NG. Phylogenetic links in plant defense systems: lignans, isoflavonoids, and their reductases. In: Hedin PA, Hollingworth RM, Masler EP, Miyamoto J, eds. *Phytochemicals for Pest Control*. Washington, DC: American Chemical Society; 1997:59–89.
6. Mäki-Arvela P, Holmbom B, Salmi T, Murzin DY. Recent progress in synthesis of fine and specialty chemicals from wood and other biomass by heterogeneous catalytic processes. *Catal Rev*. 2007;49:197–340.
7. Holmbom B, Eckerman C, Eklund P, et al. Knots in trees – a new rich source of lignans. *Phytochem Rev*. 2003;2:331–340.
8. Cesarino I, Araújo P, Domingues Júnior AP, Mazzafera P. An overview of lignin metabolism and its effect on biomass recalcitrance. *Braz J Bot*. 2012;35:303–311.
9. Davin LB, Jourdes M, Patten AM, Kim K-W, Vassão DG, Lewis NG. Dissection of lignin macromolecular configuration and assembly: comparison to related biochemical processes in allyl/propenyl phenol and lignan biosynthesis. *Nat Prod Rep*. 2008;25:1015–1090.
10. Cunha WR, Luis M, Sola RC, et al. Lignans: chemical and biological properties. In: Rao V, ed. *Phytochemicals – A Global Perspective of Their Role in Nutrition and Health*. Rijeka, Croatia: InTech; 2012:213–234.
11. Harmatha J, Dinan L. Biological activities of lignans and stilbenoids associated with plant–insect chemical interactions. *Phytochem Rev*. 2003;2:321–330.
12. Lewis NG, Kato MJ, Lopes N, Davin LB. Lignans: diversity, biosynthesis, and function. In: Seidl PR, Gottlieb OR, Kaplan MAC, eds. *Chemistry of the Amazon*. Washington, DC: American Chemical Society; 1995:135–167.
13. Milder IEJ, Arts ICW, Van de Putte B, Venema DP, Hollman PCH. Lignan contents of Dutch plant foods: a database including lariciresinol, pinoresinol, secoisolariciresinol and matairesinol. *Br J Nutr*. 2005;93:393–402.
14. Smeds AI, Eklund PC, Willför SM. Content, composition, and stereochemical characterisation of lignans in berries and seeds. *Food Chem*. 2012;134:1991–1998.
15. Smeds AI, Eklund PC, Sjöholm RE, et al. Quantification of a broad spectrum of lignans in cereals, oilseeds, and nuts. *J Agric Food Chem*. 2007;55:1337–1346.
16. Peñalvo JL, Haajanen KM, Botting N, Adlercreutz H. Quantification of lignans in food using isotope dilution gas chromatography/mass spectrometry. *J Agric Food Chem*. 2005;53:9342–9347.
17. Blitz CL, Murphy SP, Au DLM. Adding lignan values to a food composition database. *J Food Compos Anal*. 2007;20:99–105.
18. Mazur WM, Wähälä K, Rasku S, Salakka A, Hase T, Adlercreutz H. Lignan and isoflavonoid concentrations in tea and coffee. *Br J Nutr*. 1998;79:37–45.
19. Nurmi T, Heinonen S, Mazur W, Deyama T, Nishibe S, Adlercreutz H. Lignans in selected wines. *Food Chem*. 2003;83:303–309.
20. Borriello SP, Setchell KD, Axelson M, Lawson AM. Production and metabolism of lignans by the human faecal flora. *J Appl Bacteriol*. 1985;58:37–43.

21. Setchell KDR, Adlercreutz H. The excretion of two new phenolic compounds (Compound 180/442 and Compound 180/410) during the human menstrual cycle and in pregnancy. *J Steroid Biochem.* 1979;11:xv–xvi.
22. Stitch SR, Toumba JK, Groen MB, et al. Excretion, isolation and structure of a new phenolic constituent of female urine. *Nature.* 1980;287:738–740.
23. Axelson M, Sjövall J, Gustafsson BE, Setchell KDR. Origin of lignans in mammals and identification of a precursor from plants. *Nature.* 1982;298:659–660.
24. Axelson M, Setchell KD. The excretion of lignans in rats – evidence for an intestinal bacterial source for this new group of compounds. *FEBS Lett.* 1981;123:337–342.
25. Setchell KD, Lawson AM, Borriello SP, et al. Lignan formation in man—microbial involvement and possible roles in relation to cancer. *Lancet.* 1981;2:4–7.
26. Clavel T, Doré J, Blaut M. Bioavailability of lignans in human subjects. *Nutr Res Rev.* 2006;19:187–196.
27. Lampe JW, Atkinson C, Hullar MA. Assessing exposure to lignans and their metabolites in humans. *J AOAC Int.* 2006;89:1174–1181.
28. Eeckhaut E, Struijs K, Possemiers S, Vincken J-P, Keukeleire DD, Verstraete W. Metabolism of the lignan macromolecule into enterolignans in the gastrointestinal lumen as determined in the simulator of the human intestinal microbial ecosystem. *J Agric Food Chem.* 2008;56:4806–4812.
29. Wang LQ, Meselhy MR, Li Y, Qin GW, Hattori M. Human intestinal bacteria capable of transforming secoisolariciresinol diglucoside to mammalian lignans, enterodiol and enterolactone. *Chem Pharm Bull.* 2000;48:1606–1610.
30. Wang L. Mammalian phytoestrogens: enterodiol and enterolactone. *J Chromatogr, B: Anal Technol Biomed Life Sci.* 2002;777:289–309.
31. Landete JM. Plant and mammalian lignans: a review of source, intake, metabolism, intestinal bacteria and health. *Food Res Int.* 2012;46:410–424.
32. Clavel T, Borrmann D, Braune A, Doré J, Blaut M. Occurrence and activity of human intestinal bacteria involved in the conversion of dietary lignans. *Anaerobe.* 2006;12:140–147.
33. Clavel T, Lippman R, Gavini F, Doré J, Blaut M. *Clostridium saccharogumia* sp. nov. and *Lactonifactor longoviformis* gen. nov., sp. nov., two novel human faecal bacteria involved in the conversion of the dietary phytoestrogen secoisolariciresinol diglucoside. *Syst Appl Microbiol.* 2007;30:16–26.
34. Clavel T, Henderson G, Engst W, Doré J, Blaut M. Phylogeny of human intestinal bacteria that activate the dietary lignan secoisolariciresinol diglucoside. *FEMS Microbiol Ecol.* 2006;55:471–478.
35. Roncaglia L, Amaretti A, Raimondi S, Leonardi A, Rossi M. Role of bifidobacteria in the activation of the lignan secoisolariciresinol diglucoside. *Appl Microbiol Biotechnol.* 2011;92:159–168.
36. Jin J-S, Hattori M. Human intestinal bacterium, strain END-2 is responsible for demethylation as well as lactonization during plant lignan metabolism. *Biol Pharm Bull.* 2010;33:1443–1447.
37. Xie L-H, Akao T, Hamasaki K, Deyama T, Hattori M. Biotransformation of pinoresinol diglucoside to mammalian lignans by human intestinal microflora, and isolation of Enterococcus faecalis strain PDG-1 responsible for the transformation of (+)-pinoresinol to (+)-lariciresinol. *Chem Pharm Bull.* 2003;51:508–515.
38. Jin J-S, Kakiuchi N, Hattori M. Enantioselective oxidation of enterodiol to enterolactone by human intestinal bacteria. *Biol Pharm Bull.* 2007;30:2204–2206.
39. Jin J-S, Zhao Y-F, Nakamura N, et al. Enantioselective dehydroxylation of enterodiol and enterolactone precursors by human intestinal bacteria. *Biol Pharm Bull.* 2007;30:2113–2119.
40. Liu C, Finegold SM, Song Y, Lawson PA. Reclassification of *Clostridium coccoides, Ruminococcus hansenii, Ruminococcus hydrogenotrophicus, Ruminococcus luti, Ruminococcus productus* and *Ruminococcus schinkii* as *Blautia coccoides* gen. nov., comb. nov., *Blautia hansenii* comb. nov., *Blautia hydroge. Int J Syst Evol Microbiol.* 2008;58:1896–1902.
41. Coulman KD, Liu Z, Hum WQ, Michaelides J, Thompson LU. Whole sesame seed is as rich a source of mammalian lignan precursors as whole flaxseed. *Nutr Cancer.* 2005;52:156–165.
42. Peñalvo JL, Heinonen S-M, Aura A-M, Adlercreutz H. Dietary sesamin is converted to enterolactone in humans. *J Nutr.* 2005;135:1056–1062.
43. Jan K-C, Hwang LS, Ho C-T. Biotransformation of sesaminol triglucoside to mammalian lignans by intestinal microbiota. *J Agric Food Chem.* 2009;57:6101–6106.
44. Liu Z, Saarinen NM, Thompson LU. Sesamin is one of the major precursors of mammalian lignans in sesame seed (Sesamum indicum) as observed in vitro and in rats. *J Nutr.* 2006;136:906–912.
45. Zhu X, Zhang X, Sun Y, et al. Purification and fermentation in vitro of sesaminol triglucoside from sesame cake by human intestinal microbiota. *J Agric Food Chem.* 2013;61:1868–1877.
46. Saarinen NM, Smeds AI, Peñalvo JL, Nurmi T, Adlercreutz H, Mäkelä S. Flaxseed ingestion alters ratio of enterolactone enantiomers in human serum. *J Nutr Metab.* 2010. Available from: http://dx.doi.org/10.1155/2010/403076.
47. Jin J-S, Hattori M. Further studies on a human intestinal bacterium *Ruminococcus* sp. END-1 for transformation of plant lignans to mammalian lignans. *J Agric Food Chem.* 2009;57:7537–7542.
48. Jin J-S, Tobo T, Chung M-H, Ma C, Hattori M. Transformation of trachelogenin, an aglycone of tracheloside from safflower seeds, to phytoestrogenic (−)-enterolactone by human intestinal bacteria. *Food Chem.* 2012;134:74–80.
49. Adlercreutz H. Lignans and human health. *Crit Rev Clin Lab Sci.* 2007;44:483–525.
50. Buck K, Zaineddin AK, Vrieling A, Linseisen J, Chang-Claude J. Meta-analyses of lignans and enterolignans in relation to breast cancer risk. *Am J Clin Nutr.* 2010;92:141–153.
51. Cardoso Carraro JC, Dantas MI de S, Espeschit ACR, Martino HSD, Ribeiro SMR. Flaxseed and human health: reviewing benefits and adverse effects. *Food Rev Int.* 2012;28:203–230.
52. Cassidy A, Hanley B, Lamuela-Raventos RM. Isoflavones, lignans and stilbenes – origins, metabolism and potential importance to human health. *J Sci Food Agric.* 2000;80:1044–1062.
53. Cornwell T, Cohick W, Raskin I. Dietary phytoestrogens and health. *Phytochemistry.* 2004;65:995–1016.
54. Gikas PD, Mokbel K. Phytoestrogens and the risk of breast cancer: a review of the literature. *Int J Fertil Women's Med.* 2005;50:250–258.

55. Peterson J, Dwyer J, Adlercreutz H, Scalbert A, Jacques P, McCullough ML. Dietary lignans: physiology and potential for cardiovascular disease risk reduction. *Nutr Rev*. 2010;68:571–603.
56. Bedell S, Nachtigall M, Naftolin F. The pros and cons of plant estrogens for menopause. *J Steroid Biochem Mol Biol*. 2014;139:225–236.
57. Velentzis LS, Cantwell MM, Cardwell C, Keshtgar MR, Leathem AJ, Woodside JV. Lignans and breast cancer risk in pre- and post-menopausal women: meta-analyses of observational studies. *Br J Cancer*. 2009;100:1492–1498.
58. Touré A, Xueming X. Flaxseed lignans: source, biosynthesis, metabolism, antioxidant activity, bio-active components, and health benefits. *Compr Rev Food Sci Food Saf*. 2010;9:261–269.
59. Webb AL, McCullough ML. Dietary lignans: potential role in cancer prevention. *Nutr Cancer*. 2005;51:117–131.
60. McCann MJ, Gill CIR, McGlynn H, Rowland IR. Role of mammalian lignans in the prevention and treatment of prostate cancer. *Nutr Cancer*. 2005;52:1–14.
61. Saarinen NM, Tuominen J, Pylkkänen L, Santti R. Assessment of information to substantiate a health claim on the prevention of prostate cancer by lignans. *Nutrients*. 2010;2:99–115.
62. Heinonen S, Nurmi T, Liukkonen K, et al. In vitro metabolism of plant lignans: new precursors of mammalian lignans enterolactone and enterodiol. *J Agric Food Chem*. 2001;49:3178–3186.
63. Aehle E, Müller U, Eklund PC, Willför SM, Sippl W, Dräger B. Lignans as food constituents with estrogen and antiestrogen activity. *Phytochemistry*. 2011;72:2396–2405.
64. Kurzer MS, Xu X. Dietary phytoestrogens. *Annu Rev Nutr*. 1997;17:353–381.
65. Mueller SO, Simon S, Chae K, Metzler M, Korach KS. Phytoestrogens and their human metabolites show distinct agonistic and antagonistic properties on estrogen receptor alpha (ERalpha) and ERbeta in human cells. *Toxicol Sci*. 2004;80:14–25.
66. Adolphe JL, Whiting SJ, Juurlink BHJ, Thorpe LU, Alcorn J. Health effects with consumption of the flax lignan secoisolariciresinol diglucoside. *Br J Nutr*. 2010;103:929–938.
67. Penttinen P, Jaehrling J, Damdimopoulos AE, et al. Diet-derived polyphenol metabolite enterolactone is a tissue-specific estrogen receptor activator. *Endocrinology*. 2007;148:4875–4886.
68. Saleem M, Kim HJ, Ali MS, Lee YS. An update on bioactive plant lignans. *Nat Prod Rep*. 2005;22:696–716.
69. Lof M, Weiderpass E. Epidemiologic evidence suggests that dietary phytoestrogen intake is associated with reduced risk of breast, endometrial, and prostate cancers. *Nutr Res*. 2006;26:609–619.
70. Zeleniuch-Jacquotte A, Lundin E, Micheli A, et al. Circulating enterolactone and risk of endometrial cancer. *Int J Cancer*. 2006;119:2376–2381.
71. Bandera EV, Williams MG, Sima C, et al. Phytoestrogen consumption and endometrial cancer risk: a population-based case-control study in New Jersey. *Cancer Causes Control*. 2009;20:1117–1127.
72. Zamora-Ros R, Agudo A, Luján-Barroso L, et al. Dietary flavonoid and lignan intake and gastric adenocarcinoma risk in the European Prospective Investigation into Cancer and Nutrition (EPIC) study. *Am J Clin Nutr*. 2012;96:1398–1408.
73. Cotterchio M, Boucher BA, Manno M, Gallinger S, Okey A, Harper P. Dietary phytoestrogen intake is associated with reduced colorectal cancer risk. *J Nutr*. 2006;136:3046–3053.
74. Kuijsten A, Arts ICW, Hollman PCH, van't Veer P, Kampman E. Plasma enterolignans are associated with lower colorectal adenoma risk. *Cancer Epidemiol Biomarkers Prev*. 2006;15:1132–1136.
75. Johnsen NF, Olsen A, Thomsen BLR, et al. Plasma enterolactone and risk of colon and rectal cancer in a case-cohort study of Danish men and women. *Cancer Causes & Control*. 2010;21:153–162.
76. Buck K, Zaineddin AK, Vrieling A, et al. Estimated enterolignans, lignan-rich foods, and fibre in relation to survival after postmenopausal breast cancer. *Br J Cancer*. 2011;105:1151–1157.
77. Guglielmini P, Rubagotti A, Boccardo F. Serum enterolactone levels and mortality outcome in women with early breast cancer: a retrospective cohort study. *Breast Cancer Res Treat*. 2012;132:661–668.
78. Anderson LN, Cotterchio M, Boucher BA, Kreiger N. Phytoestrogen intake from foods, during adolescence and adulthood, and risk of breast cancer by estrogen and progesterone receptor tumor subgroup among Ontario women. *Int J Cancer*. 2013;132:1683–1692.
79. McCann SE, Hootman KC, Weaver AM, et al. Dietary intakes of total and specific lignans are associated with clinical breast tumor characteristics. *J Nutr*. 2012;142:91–98.
80. Truan JS, Chen J-M, Thompson LU. Comparative effects of sesame seed lignan and flaxseed lignan in reducing the growth of human breast tumors (MCF-7) at high levels of circulating estrogen in athymic mice. *Nutr Cancer*. 2012;64:65–71.
81. Mabrok HB, Klopfleisch R, Ghanem KZ, Clavel T, Blaut M, Loh G. Lignan transformation by gut bacteria lowers tumor burden in a gnotobiotic rat model of breast cancer. *Carcinogenesis*. 2012;33:203–208.
82. McCann MJ, Rowland IR, Roy NC. Anti-proliferative effects of physiological concentrations of enterolactone in models of prostate tumourigenesis. *Mol Nutr Food Res*. 2013;57:212–224.
83. Ward HA, Kuhnle GGC. Phytoestrogen consumption and association with breast, prostate and colorectal cancer in EPIC Norfolk. *Arch Biochem Biophys*. 2010;501:170–175.
84. Jackson MD, McFarlane-Anderson ND, Simon GA, Bennett FI, Walker SP. Urinary phytoestrogens and risk of prostate cancer in Jamaican men. *Cancer Causes Control*. 2010;21:2249–2257.
85. Ward HA, Kuhnle GGC, Mulligan AA, Lentjes MAH, Luben RN, Khaw K. Breast, colorectal, and prostate cancer risk in the European Prospective Investigation into Cancer and Nutrition—Norfolk in relation to phytoestrogen intake derived from an improved database. *Am J Clin Nutr*. 2010;91:440–448.
86. Kuhnle GGC, Dell'Aquila C, Aspinall SM, Runswick SA, Mulligan AA, Bingham SA. Phytoestrogen content of foods of animal origin: dairy products, eggs, meat, fish, and seafood. *J Agric Food Chem*. 2008;56:10099–10104.
87. Peñalvo JL, López-Romero P. Urinary enterolignan concentrations are positively associated with serum HDL cholesterol and negatively associated with serum triglycerides in U.S. adults. *J Nutr*. 2012;142:751–756.
88. Pietrofesa R, Turowski J, Tyagi S, et al. Radiation mitigating properties of the lignan component in flaxseed. *BMC Cancer*. 2013;13:179.
89. Sonestedt E, Wirfält E. Enterolactone and breast cancer: methodological issues may contribute to conflicting results in observational studies. *Nutr Res*. 2010;30:667–677.

90. Kuijsten A, Arts ICW, Vree TB. Hollman PCH. Pharmacokinetics of enterolignans in healthy men and women consuming a single dose of secoisolariciresinol diglucoside. *J Nutr*. 2005;135:795–801.
91. Possemiers S, Bolca S, Eeckhaut E, Depypere H, Verstraete W. Metabolism of isoflavones, lignans and prenylflavonoids by intestinal bacteria: producer phenotyping and relation with intestinal community. *FEMS Microbiol Ecol*. 2007;61:372–383.
92. Lampe JW, Chang J-L. Interindividual differences in phytochemical metabolism and disposition. *Semin Cancer Biol*. 2007;17:347–353.
93. Setchell K, Adlercreutz H. Mammalian lignans and phytoestrogens. Recent studies on their formation, metabolism, and biological role in health and disease. In: Rowland IR, ed. *Role of the Gut Flora in Toxicity and Cancer*. London: Academic Press; 1988:316–345.
94. Adlercreutz H, Van der Wildt J, Kinzel J. Lignan and isoflavonoid conjugates in human urine. *J Steroid Biochem Mol Biol*. 1995;52:97–103.
95. Jansen GHE, Arts ICW, Nielen MWF, Müller M, Hollman PCH, Keijer J. Uptake and metabolism of enterolactone and enterodiol by human colon epithelial cells. *Arch Biochem Biophys*. 2005;435:74–82.
96. Jacobs E, Kulling SE, Metzler M. Novel metabolites of the mammalian lignans enterolactone and enterodiol in human urine. *J Steroid Biochem Mol Biol*. 1999;68:211–218.
97. Okura T, Ibe M, Umegaki K, Shinozuka K, Yamada S. Effects of dietary ingredients on function and expression of P-Glycoprotein in human intestinal epithelial cells. *Biol Pharm Bull*. 2010;33:255–259.
98. Adlercreutz H, Fotsis T, Bannwart C, et al. Determination of urinary lignans and phytoestrogen metabolites, potential antiestrogens and anticarcinogens, in urine of women on various habitual diets. *J Steroid Biochem*. 1986;25:791–797.
99. Stumpf K, Pietinen P, Puska P, Adlercreutz H. Changes in serum enterolactone, genistein, and daidzein in a dietary intervention study in Finland. *Cancer Epidemiol Biomarkers Prev*. 2000;9:1369–1372.
100. Mazur WM, Uehara M, Wähälä K, Adlercreutz H. Phyto-oestrogen content of berries, and plasma concentrations and urinary excretion of enterolactone after a single strawberry-meal in human subjects. *Br J Nutr*. 2000;83:381–387.
101. Kirkman LM, Lampe JW, Campbell DR, Martini MC, Slavin JL. Urinary lignan and isoflavonoid excretion in men and women consuming vegetable and soy diets. *Nutr Cancer*. 1995;24:1–12.
102. Juntunen KS, Mazur WM, Liukkonen KH, et al. Consumption of wholemeal rye bread increases serum concentrations and urinary excretion of enterolactone compared with consumption of white wheat bread in healthy Finnish men and women. *Br J Nutr*. 2000;84:839–846.
103. Vanharanta M, Mursu J, Nurmi T, et al. Phloem fortification in rye bread elevates serum enterolactone level. *Eur J Clin Nutr*. 2002;56:952–957.
104. McIntosh GH, Noakes M, Royle PJ, Foster PR. Whole-grain rye and wheat foods and markers of bowel health in overweight middle-aged men. *Am J Clin Nutr*. 2003;77:967–974.
105. Kristensen MB, Hels O, Tetens I. No changes in serum enterolactone levels after eight weeks' intake of rye-bran products in healthy young men. *Scandinavian Journal of Food & Nutrition*. 2005;49:62–67.
106. Knust U, Spiegelhalder B, Strowitzki T, Owen RW. Contribution of linseed intake to urine and serum enterolignan levels in German females: a randomised controlled intervention trial. *Food Chem Toxicol*. 2006;44:1057–1064.
107. Lampe JW, Martini MC, Kurzer MS, Adlercreutz H, Slavin JL. Urinary lignan and isoflavonoid excretion in premenopausal women consuming flaxseed powder. *Am J Clin Nutr*. 1994;60:122–128.
108. Kurzer MS, Lampe JW, Martini MC, Adlercreutz H. Fecal lignan and isoflavonoid excretion in premenopausal women consuming flaxseed powder. *Cancer Epidemiol Biomarkers Prev*. 1995;4:353–358.
109. Nesbitt PD, Lam Y, Thompson LU. Human metabolism of mammalian lignan precursors in raw and processed flaxseed. *Am J Clin Nutr*. 1999;69:549–555.
110. Kuijsten A, Arts ICW, van't Veer P, Hollman PCH. The relative bioavailability of enterolignans in humans is enhanced by milling and crushing of flaxseed. *J Nutr*. 2005;135:2812–2816.
111. Aura A-M, Myllymäki O, Bailey M, Penalvo JL, Adlercreutz H, Poutanen K. Interrelationships between carbohydrate type, phenolic acids and initial pH on in vitro conversion of enterolactone from rye lignans. In: Salovaara H, Gates F, Tenkanen M, eds. *Dietary Fibre: Components and Functions*. Wageningen, The Netherlands: Wageningen Academic Publishers; 2007:235–245.
112. Adlercreutz H, Honjo H, Higashi A, et al. Urinary excretion of lignans and isoflavonoid phytoestrogens in Japanese men and women consuming a traditional Japanese diet. *Am J Clin Nutr*. 1991;54:1093–1100.
113. Horner NK, Kristal AR, Prunty J, Skor HE, Potter JD, Lampe JW. Dietary determinants of plasma enterolactone. *Cancer Epidemiol Biomarkers Prev*. 2002;11:121–126.
114. Lampe JW, Gustafson DR, Hutchins AM, et al. Urinary isoflavonoid and lignan excretion on a Western diet: relation to soy, vegetable, and fruit intake. *Cancer Epidemiol Biomarkers Prev*. 1999;8:699–707.
115. Milder IEJ, Kuijsten A, Arts ICW, et al. Relation between plasma enterodiol and enterolactone and dietary intake of lignans in a Dutch endoscopy-based population. *J Nutr*. 2007;137:1266–1271.
116. Kilkkinen A, Stumpf K, Pietinen P, Valsta LM, Tapanainen H, Adlercreutz H. Determinants of serum enterolactone concentration. *Am J Clin Nutr*. 2001;73:1094–1100.
117. Jacobs DR, Pereira MA, Stumpf K, Pins JJ, Adlercreutz H. Whole grain food intake elevates serum enterolactone. *Br J Nutr*. 2002;88:111–116.
118. Cunnane SC, Hamadeh MJ, Liede AC, Thompson LU, Wolever TM, Jenkins DJ. Nutritional attributes of traditional flaxseed in healthy young adults. *Am J Clin Nutr*. 1995;61:62–68.
119. Clavel T, Henderson G, Alpert C-A, et al. Intestinal bacterial communities that produce active estrogen-like compounds enterodiol and enterolactone in humans. *Appl Environ Microbiol*. 2005;71:6077–6085.
120. Lampe JW, Fredstrom SB, Slavin JL, Potter JD. Sex differences in colonic function: a randomised trial. *Gut*. 1993;34:531–536.
121. Kilkkinen A, Pietinen P, Klaukka T, Virtamo J, Korhonen P, Adlercreutz H. Use of oral antimicrobials decreases serum enterolactone concentration. *Am J Epidemiol*. 2002;155:472–477.
122. Johnsen NF, Hausner H, Olsen A, et al. Intake of whole grains and vegetables determines the plasma enterolactone concentration of Danish women. *J Nutr*. 2004;134:2691–2697.
123. Rybak ME, Sternberg MR, Pfeiffer CM. Sociodemographic and lifestyle variables are compound- and class-specific correlates of urine phytoestrogen concentrations in the U.S. population. *J Nutr*. 2013;1–9.

CHAPTER

8

Gut Microbiome Modulates Dietary Xenobiotic Toxicity: The Case of DON and Its Derivatives

Martina Cirlini, Renato Bruni and Chiara Dall'Asta

Department of Food Science, University of Parma, Parma, Italy

INTRODUCTION

As recently demonstrated,[1,2] the gut microbiota play a key role in the metabolism of xenobiotics, including environmental pollutants, dietary supplements, food additives and toxic natural substances. These compounds are metabolized and eliminated through the urine, bile and faeces. Although polar substance can be eliminated without structural changes, most commonly these compounds undergo metabolic transformations such as enzymatic functionalization and/or conjugation reactions to facilitate their elimination. Thus, a deeper investigation of the catabolic fate of xenobiotics involving both metabolic pathways and microbiota degradation may lead to a better understanding of the *in vivo* toxicological relevance of these compounds.

Among the most common transformations affecting xenobiotics, dehydroxylation, decarboxylation, dealkylation, dehalogenation and deamination reactions have been reported as gut microflora-mediated processes.[3] In addition, gut microflora stability itself can be affected by xenobiotics.[4,5] As an example, natural compounds found in dark chocolate,[5] pomegranate by-products[6] or probiotics[7] have also been shown to modulate the gut microflora environment.

Among xenobiotics, those naturally present in food are of particular concern as their occurrence can not be easily avoided. Furthermore, these compounds are often stable under technological conditions, and thus they accumulate along the food supply chain. Mycotoxins, one of the most important classes of toxic natural substances in food and feed, are secondary metabolites produced by different fungal species (e.g., *Aspergillus*, *Penicillium* and *Fusarium* genera) in crops, among them cereals and nuts.[8]

The presence of mycotoxins in food and feed may affect human and animal health on account of many different adverse health effects such as induction of cancer and mutagenicity, as well as estrogenic, gastrointestinal and kidney disorders.[8] National and international organizations are constantly evaluating the risk that such mycotoxins pose to humans, resulting in statutory or guideline maximum permissible limits for the most widespread and toxic ones.[9]

Among mycotoxins, those produced by *Fusarium* fungi are particularly relevant for grains. In particular, deoxynivalenol (DON) is considered the most common contaminant in cereal worldwide.[10,11]

At a molecular level, the toxic activity of DON is exerted through the disruption of the normal cell functioning by inhibiting protein synthesis and by activating critical cellular kinases involved in signal transduction related to proliferation, differentiation and apoptosis.[12] In general, monogastric animals are more affected than polygastric ones and effects on humans could be compared to those reported for pigs. With respect to chronic effects, growth, immune function and reproduction are all adversely affected by DON in animals.[12]

If ingested at a concentration of 0.15 mg/kg bw/day, DON is able to cause vomiting in swine, while it has been estimated that levels up to 1–2 mg/kg in the diet can cause partial feed refusal in livestock, mainly pigs.[12] Unfortunately, an accurate acute dose–response evaluation in humans has not been reported so far. However, it has been possible to ascribe the induction of acute gastroenteritis to vomiting in both mammals and humans to

Diet-Microbe Interactions in the Gut.
DOI: http://dx.doi.org/10.1016/B978-0-12-407825-3.00008-3

the dysregulation of the immune and/or neuroendocrine functions. However, chronic consumption of lower doses may not trigger such a physiological defensive response and without determining consistent damages in various organs. In addition, chronic effects on growth, immune function and reproduction are also suggested, based on animal studies.[13] The toxic effects exerted by DON in humans are summarized in Figure 8.1.

Besides DON, its fungal metabolites – i.e., 3- and 15-acetyl DON – can be concomitantly present in cereals in significant amounts (10–20%). In addition, the co-occurrence of masked forms such as DON-3-glucoside (D3G) has recently been demonstrated, which are enzymatically formed in planta as a detoxification strategy.[14] Data regarding D3G occurrence in cereals show that this conjugate is commonly present at levels up to 20% of DON and tends to accumulate in bran, thus being predominant in fiber-enriched products.

Very recently, other conjugates have been detected in grains and cereal-products, such as DON-oligoglycosides, DON-glutathione, DON-S-cysteine, DON-S-cysteinyl-glycine and DON-sulfonate.[15]

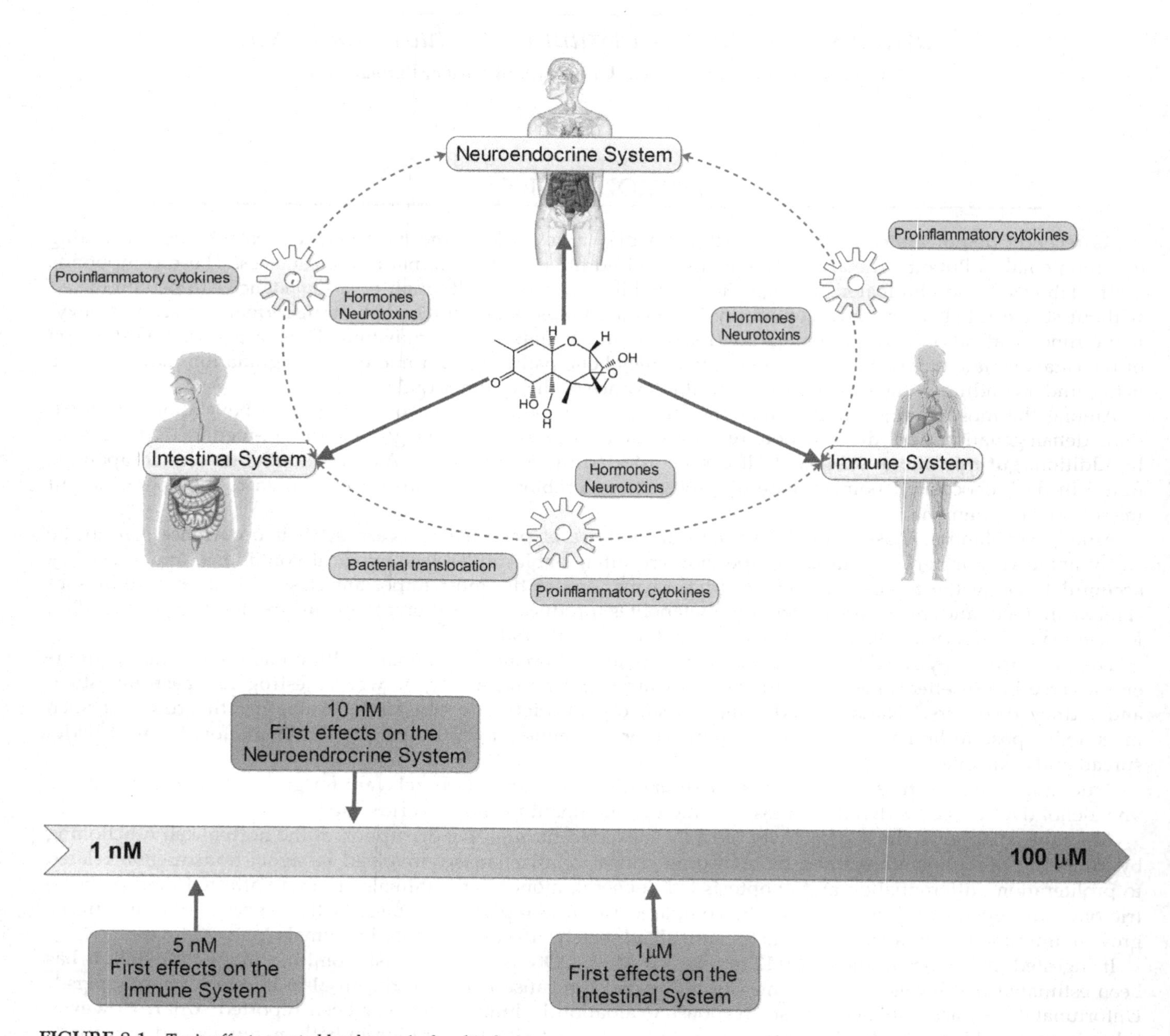

FIGURE 8.1 Toxic effects exerted by deoxynivalenol in humans.

The stability of these masked derivatives upon digestion is a crucial point in risk assessment, because a possible release of the parent compound in the gastrointestinal tract may lead to an underestimation of the actual human exposure.[16,17]

In particular, the intestinal tract of animals and humans contains vast amounts of bacteria forming the commensal microbiota that lives in symbiosis with the host.[18] At present, the gut microbiome could be considered as an additional organ system, playing important roles in the functionality of the intestinal and immune systems. In particular, gut bacteria can hydrolyze glycosides, glucuronides, sulfates, amides, esters and lactones, and can also perform reactions such as reduction, decarboxylation and demethylation.[19] As recently demonstrated, these complex modifications may act on nutrients, thus generating low-molecular-weight metabolites that can be efficiently absorbed in situ. These compounds may undergo further phase-II metabolism, before entering the systemic blood circulation and finally being excreted in urine.[20]

A number of recent publications addressed the gut microbiome-driven chemical modification of nutrients in humans, while similar processes involving xenobiotics have still rarely been studied. For this reason, this chapter is focused on the role played by the gut microbiome on the catabolic fate of DON derivatives.

GASTRIC STABILITY OF DON DERIVATIVES

The potential hydrolysis of masked mycotoxins to parent forms was tested by *in vitro* models under different conditions, mimicking those of human/animal digestion. All the studies confirmed the stability of D3G and other conjugates under gastric acidic conditions.[16,21] Similarly, their stability to gastric enzymatic activity was also reported:[21] D3G was subjected to enzymatic hydrolysis applying several glycosylhydrolases dissolved in artificial gut juice such as amylase, β-glucosidase (from almond and from human cytosol) and different β-glucuronidases, but results showed that D3G remained unchanged, suggesting that this molecule remained stable after passage through the small intestine.

In addition, Dall'Erta et al.[16] described the stability of D3G and two zearalenone conjugates upon digestion by application of an *in vitro* assay that simulated human digestion. In particular, the salivary step (5 min), the gastric step (120 min) and the duodenal step (120 min) were mimicked, performing various sequential enzymatic treatments.

The stability of masked forms under digestion demonstrated that the parent compound, which is highly absorbed in the small intestine, is not released at this stage. Since the higher polarity of these conjugates strongly affects their ability to enter the cell and thus to be absorbed, it is possible to infer the lower toxicity of the masked forms compared to DON. However, the bacterial transformation of these metabolites should be carefully considered since DON could also be reabsorbed in the large intestine or could exert its toxic activities in loco.

BACTERIAL TRANSFORMATION AND INTESTINAL ABSORPTION OF DON AND ITS DERIVATIVES

The bacterial transformation and intestinal adsorption of DON in humans and animals has been recently reviewed.[22,23] The gastrointestinal fate of DON is shown in Figure 8.2.

Having entered the body through the ingestion of contaminated food, DON and its derivatives may pass through the gut wall depending on the bacterial metabolism. Monogastric species such as humans and pigs are characterized by a high bacterial content (microorganisms: 10^9-10^{12}) located in their colon, while polygastric animals such as ruminants present a high bacterial content both before and after the small intestine. This localization strongly affects the intestinal absorption and metabolism of ingested DON and its metabolites, thus being responsible for significant catabolic degradation and greatly impacting the bioavailability of these molecules.[24–26]

In monogastric animals, large amounts of ingested DON can cross the intestinal epithelium and reach the blood compartment. In pigs, a fast and efficient absorption of the toxin through the proximal small intestine[24–26] has been shown, probably involving the jejunum tract.[27] Similarities between the human and pig intestines suggest that humans could also efficiently absorb the ingested DON with the same mechanism. Although few data are available regarding the mechanism of intestinal absorption of DON, *in vitro* studies suggested that it may take place through passive transcellular and/or paracellular diffusion.[28] Accordingly, more polar DON conjugates such as D3G are supposed to be less adsorbed in the small intestine compared to their parent form.

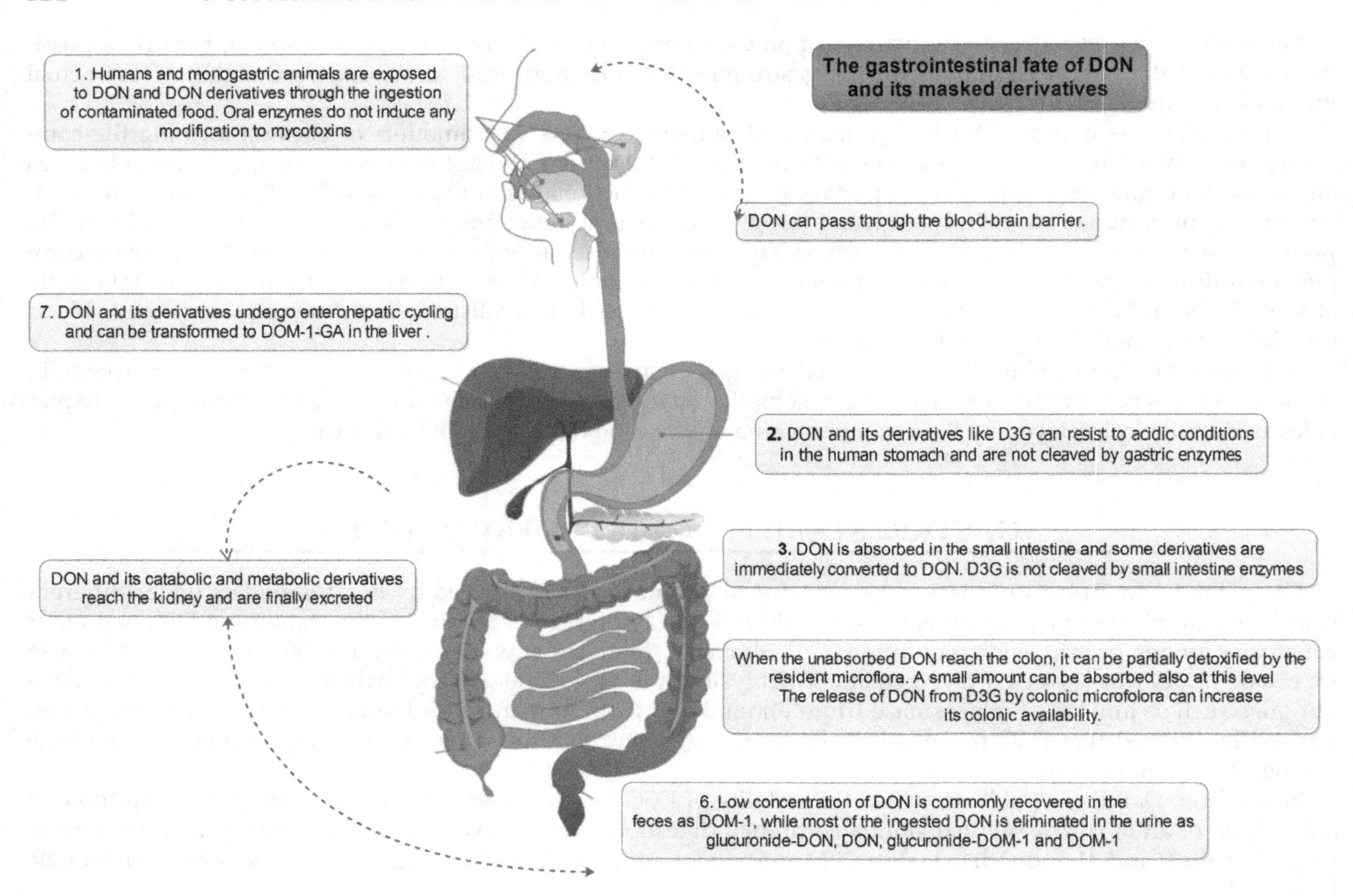

FIGURE 8.2 Description of the gastrointestinal fate of DON in humans.

This thesis was supported by results reported by De Nijs et al.,[29] who tested *in vitro* the transformation and possible absorption of D3G in the small intestine. In particular, D3G was submitted to a treatment with human Caco-2 cells in a transwell system. The results showed that Caco-2 cells were not able to convert D3G into DON after 24 h of treatment, while it seemed that when DON was submitted to the same process, a part of it was absorbed by human cells (about 23% of the initial added amount). On the basis of these experiments, it was possible to conclude that D3G is not hydrolyzed or transformed during its passage through the upper tract of the human intestine, while DON can be partially absorbed by human cells in this tract.[29]

Concerning polygastric animals, DON seems to be efficiently detoxified by rumen microflora through a de-epoxidation reaction under anaerobic conditions,[30,31] this reaction leading to the formation of lower immunosuppressive catabolites compared to DON.[32] A similar degradation step may take place in the colon of monogastric species, although to a lower extent. In particular, a low concentration of DON is commonly recovered in the feces of monogastric animals as DOM-1, while most of the ingested DON is eliminated in the urine as DON, glucuronide-DON, glucuronide-DOM-1 and DOM-1.[33,34]

It is well known that different species show different de-epoxidation efficiencies, but this capability in humans is quite low.[35] In particular, from the data reported in the literature, it could be argued that the pre-exposure of the microbiota to DON is a key factor for inducing the appearance of the bacterial detoxification activity, either through the induction of the expression of particular enzymes and/or the selection of particular detoxifying bacterial species.[35] According to very recent studies,[36] experiments conducted with human feces from five volunteers showed that only one spontaneously possessed bacteria able to transform DON in DOM-1.

Starting from these results, further studies might outline whether different dietary habits or the higher abundance of some microbiota-constituent species may affect DOM-1 formation in humans.

The possible degradation of DON derivatives in the colon tract of monogastric animals was also explored. In the first model study, D3G was submitted to *in vitro* incubation with intestinal bacteria cultures, mainly

Lactobacillus, *Enterococcus*, *Enterobacter* and *Bifidobacterium* types. The authors reported a partial hydrolysis of the masked form, thus outlining that DON conjugates could be cleaved to the parent form after bacterial hydrolysis in the human colon tract.[21]

Very recently, D3G degradation by the human gut microbiome was described by Dall'Erta et al.[16] using a fecal fermentation assay based on five healthy volunteers. D3G was fully recovered after 30 min, but it was completely deglucosylated and degraded – mainly to DON – after 24 h. Traces of de-epoxylated DON (DOM-1) were found in the samples after 24 h of treatment. A similar experiment was reported by Gratz et al.[36] The results obtained in these two independent studies were in agreement, demonstrating that the release of DON from D3G can increase its colonic availability, although DON absorption is supposed to be higher in the duodenum and in the small intestine.

DON AND DON-CONJUGATES IMPACT ON THE HUMAN GUT

When the impact of DON on the intestinal function is taken into consideration, recent results about the ability of the human gut microbiome to release DON from its conjugates are of great concern. Although the intestine is the less-sensitive target regarding DON bioactivity, it is also exposed to higher doses of DON compared to other organs.

In vitro and *in vivo* experiments have shown that DON inhibits the intestinal absorption of nutrients including glucose and amino acids, by human[37] and animal intestinal epithelial cells (IEC),[38] and it also affects the permeability of the intestinal epithelium through modulation of the tight junction complexes.[39] It must be noted that IEC are the first target of DON in cases of natural exposure through ingestion of contaminated food.

Regarding the well-described toxic activity of DON against the immune system, alterations of the immune cells could affect the intestinal and brain functions, as recently described.[13] In particular, the intestinal and systemic production of cytokines could participate in the growth retardation, feed refusal and emesis caused by DON on account of the effects on the neuroendocrine system. The consequent inflammation could lead to an increased permeability of the intestinal and blood barriers, thus affecting the xenobiotic absorption.

In consideration of all these points, the cleavage of DON derivatives in the colon tract may increase the toxic effects already attributed to DON, thus representing an additional source of exposure in humans.

As recently stated by several studies,[22,40,41] the safety factor calculated for DON according to the actual provisional maximum tolerable daily intake (PMTDI) seems to be too low, in consideration of the doses affecting cell functions. Further contributions to this underestimation may be found in the DON synergistic effect with other xenobiotics and in the higher exposure in particular populations such as vegans, children or patients suffering from immune-system-related pathologies, neurological disorders or inflammatory bowel disease.[40,42] In this context, a release of DON from its conjugates occurring in the human colon may result in an increased overall effect that can not be explained on the basis of the DON amount detectable with routine/official methods.

Besides the detoxifying role played by gut bacteria towards mycotoxins and other xenobiotics, the maintenance of a correct balance in the gastrointestinal microflora is crucial for proper nutrient uptake and for an efficient protection against pathogens, according to the so-called "intestinal homeostasis." Thus an impaired balance of the intestinal microbiome, such as in a dysbiosis condition, could adversely affect human health (Figure 8.3).

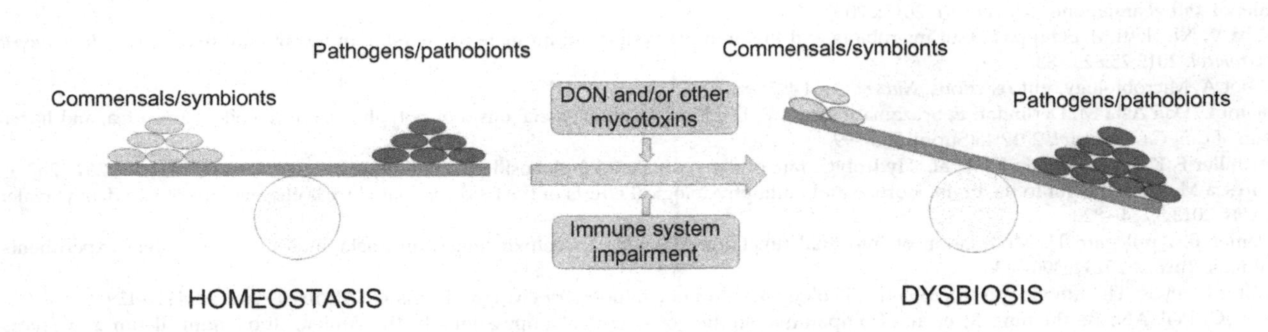

FIGURE 8.3 The impaired balance of the intestinal microbiome adversely affect human health.

Unfortunately, the knowledge about the effects of mycotoxin exposure on the human microbiome is still limited and reported studies mainly refer to the role played by intestinal microflora in mycotoxin detoxification in animals.[43] Nonetheless, mycotoxins may actually affect the gut microflora, as some of them exhibit antimicrobial activities in animals.[44] In addition, it has been proven that chronic exposure to low doses of DON may induce a shift towards intestinal aerobic bacteria in pigs.[45] Since the number and composition of intestinal microflora are significantly modified in inflammatory bowel diseases in humans, with an increase in the number of aerobic bacteria and a parallel decrease in the number of anaerobic bacteria, mycotoxin exposure might represent a potential risk factor for chronic intestinal inflammatory diseases.[40] Since data supporting or contrasting these hypothesis are still poor, further studies should be addressed to better understand the role played by mycotoxins in the imbalance of intestinal microflora in humans.

According to such observations, a re-evaluation of the actual PMTDI for DON by food agencies should be advised, especially in consideration of the increasing number of DON metabolites reported in the literature. On the other hand, human gut microbiome modulation through the diet should be considered as a possible approach to clarify inter-individual differences affecting the overall exposure, as well as a possible strategy for diminishing DON toxicity throughout a bacterial-driven detoxification.

References

1. Clayton TA, Baker D, Lindon JC, Everett JR, Nicholson JK. Pharmacometabonomic identification of a significant host–microbiome metabolic interaction affecting human drug metabolism. *Proc Nat Acad Sci USA*. 2009;106:14728–14733.
2. Sousa T, Paterson R, Moore V, Carlsson A, Abrahamsson B, Basit AW. The gastrointestinal microbiota as a site for the biotransformation of drugs. *Int J Pharm*. 2008;363:1–25.
3. Wilson ID, Nicholson JK. The role of gut microbiota in drug response. *Curr Pharm Des*. 2009;15:1519–1523.
4. Lindenbaum J, Rund DG, Butler Jr VP, Tse-Eng D, Saha JR. Inactivation of digoxin by the gut flora: reversal by antibiotic therapy. *N Engl J Med*. 1981;305:789–794.
5. Sekirov I, Tam NM, Jogova M, et al. Antibiotic-induced perturbations of the intestinal microbiota alter host susceptibility to enteric infection. *Infect Immun*. 2008;76:4726–4736.
6. Bialonska D, Ramnani P, Kasimsetty SG, Muntha KR, Gibson GR, Ferreira D. The influence of pomegranate by-product and punicalagins on selected groups of human intestinal microbiota. *Int J Food Microbiol*. 2010;140:175–182.
7. Martin FP, Wang Y, Sprenger N, et al. Probiotic modulation of symbiotic gut microbial–host metabolic interactions in a humanized microbiome mouse model. *Mol Syst Biol*. 2008;4:157–172.
8. Desjardins A, Maragos C, Norred W, et al. Mycotoxins: Risks in Plant, Animal, and Human System. Ames, IA, USA: Council for Agricultural Science and Technology; 2003.
9. Van Egmond HP, Schothorst RC, Jonker MA. Regulations relating to mycotoxins in food: perspectives in a global and European context. *Anal Bioanal Chem*. 2007;389:147–157.
10. Wegulo SN. Factors influencing deoxynivalenol accumulation in small grain cereals. *Toxins*. 2012;4:1157–1180.
11. Miller JD. Mycotoxins in small grains and maize: old problems, new challenges. *Food Addit Contam Part A, Chem Anal Control Expo Risk Assess*. 2008;25:219–230.
12. Pestka JJ. Deoxynivalenol: mechanisms of action, human exposure, and toxicological relevance. *Arch Toxicol*. 2010;84:663–679.
13. Bonnet MS, Roux J, Mounien L, Dallaporta M, Troadec JD. Advances in deoxynivalenol toxicity mechanisms: the brain as a target. *Toxins*. 2012;4:1120–1138.
14. Berthiller F, Crews C, Dall'Asta C, et al. Masked mycotoxins: a review. *Mol Nutr Food Res*. 2012;57:165–186.
15. Kluger B, Bueschl C, Lemmens M, et al. Stable isotopic labelling-assisted untargeted metabolic profiling reveals novel conjugates of the mycotoxin deoxynivalenol in wheat. *Anal Bioanal Chem*. 2013;405:5031–5036.
16. Dall'Erta A, Cirlini M, Dall'Asta M, Del Rio D, Galaverna G, Dall'Asta C. Masked mycotoxins are efficiently hydrolysed by the humancolonic microbiota, releasing their toxic aglycones. *Chem Res Toxicol*. 2013;26:305–312.
17. Warth B, Sulyok M, Berthiller F, Schuhmacher R, Krska R. New insights into the human metabolism of the *Fusarium* mycotoxins deoxynivalenol and zearalenone. *Toxicol Lett*. 2013;220:88–94.
18. Iebba V, Nicoletti M, Schippa S. Gut microbiota and the immune system: an intimate partnership in health and disease. *Int J Immunopathol Pharmacol*. 2012;25:823–833.
19. Abbot A. Microbiology: gut reactions. *Nature*. 2004;427:284–286.
20. Calani L, Dall'Asta M, Derlindati E, Scazzina F, Bruni R, Del Rio D. Colonic metabolism of polyphenols from coffee, green tea, and hazelnut skins. *J Clin Gastroenterol*. 2012;46(Suppl):S95–99.
21. Berthiller F, Krska R, Domig KJ, et al. Hydrolytic fate of deoxynivalenol-3-glucoside during digestion. *Toxicol Lett*. 2011;206:264–267.
22. Maresca M. From the gut to the brain: journey and pathophysiological effects of the food-associated trichothecene mycotoxin deoxynivalenol. *Toxins*. 2013;5:784–820.
23. Grenier B, Applegate TJ. Modulation of intestinal functions following mycotoxin ingestion: meta-analysis of published experiments in animals. *Toxins*. 2013;5:396–430.
24. Walter J, Ley R. The human gut microbiome: Ecology and recent evolutionary changes. *Annu Rev Microbiol*. 2011;65:411–429.
25. Frey JC, Pell AN, Berthiaume R, et al. Comparative studies of microbial populations in the rumen, duodenum, ileum and faeces of lactating dairy cows. *J Appl Microbiol*. 2010;108:1982–1993.

26. Smith HW. Observations on the flora of the alimentary tract of animals and factors affecting its composition. *Journal of Pathology and Bacteriology*. 1965;89:95–122.
27. Avantaggiato G, Havenaar R, Visconti A. Evaluation of the intestinal absorption of deoxynivalenol and nivalenol by an *in vitro* gastrointestinal model, and the binding efficacy of activated carbon and other adsorbent materials. *Food Chem Toxicol*. 2004;42:817–824.
28. Awad WA, Aschenbach JR, Setyabudi FM, Razzazi-Fazeli E, Böhm J, Zentek J. *In vitro* effects of deoxynivalenol on small intestinal D-glucose uptake and absorption of deoxynivalenol across the isolated jejunal epithelium of laying hens. *Poult Sci*. 2007;86:15–20.
29. De Nijs M, Van den Top HJ, Portier L, et al. Digestibility and absorption of deoxynivalenol-3-ß-glucoside in in vitro models. *World Mycotoxin J*. 2013;5:319–324.
30. King RR, McQueen RE, Levesque D, Greenhalgh R. Transformation of deoxynivalenol (vomitoxin) by rumen microorganisms. *J Agric Food Chem*. 1984;32:1181–1183.
31. Kollarczik B, Gareis M, Hanelt M. *In vitro* transformation of the *Fusarium* mycotoxins deoxynivalenol and zearalenone by the normal gut microflora of pigs. *Nat Toxins*. 1994;2:105–110.
32. Volkl A, Vogler B, Schollenberger M, Karlvosky P. Microbial detoxification of mycotoxin deoxynivalenol. *J Basic Microbiol*. 2004;44:147–156.
33. Eriksen GS, Pettersson H, Johnsen K, Lindberg JE. Transformation of trichotecenes in ileal digesta and faeces from pigs. *Arch Tierernahr*. 2002;56:263–274.
34. Eriksen GS, Pettersson H, Lindberg JE. Absorption, metabolism and excretion of 3-acetyl DON in pigs. *Arch Tierernahr*. 2003;57:335–345.
35. Karlovsky P. Biological detoxification of the mycotoxin deoxynivalenol and its use in genetically engineered crops and feed additives. *Appl Microbiol Biotechnol*. 2011;91:491–504.
36. Gratz SW, Duncan G, Richardson AJ. Human fecal microbiota metabolize deoxynivalenol and deoxynivalenol-3-glucoside and may be responsible for urinary de-epoxy deoxynivalenol. *Appl Environ Microbiol*. 2013;79:1821–1825.
37. Maresca M, Mahfoud R, Garmy N, Fantini J. The mycotoxin deoxynivalenol affects nutrient absorption in human intestinal epithelial cells. *J Nutr*. 2002;132:2723–2731.
38. Awad WA, Razzazi-Fazeli E, Böhm J, Zentek J. Effects of B-trichothecenes on luminal glucose transport across the isolated jejunal epithelium of broiler chickens. *J Anim Physiol Anim Nutr (Berl)*. 2008;92:225–230.
39. Pinton P, Nougayrède JP, del Rio JC, et al. The food contaminant deoxynivalenol, decreases intestinal barrier permeability and reduces claudin expression. *Toxicol Appl Pharmacol*. 2009;237:41–48.
40. Maresca M, Fantini J. Some food-associated mycotoxins as potential risk factors in humans predisposed to chronic intestinal inflammatory diseases. *Toxicon*. 2010;56:282–294.
41. Muri SD, van der Voet H, Boon PE, van Klaveren JD, Brüschweiler BJ. Comparison of human health risks resulting from exposure to fungicides and mycotoxins via food. *Food Chem Toxicol*. 2009;47:2963–2974.
42. Wild CP, Gong YY. Mycotoxins and human disease: a largely ignored global health issue. *Carcinogenesis*. 2010;31:71–82.
43. Schatzmayr G, Zehner F, Täubel M, et al. Microbiologicals for deactivating mycotoxins. *Mol Nutr Food Res*. 2006;50:543–551.
44. Dänicke S, Hegewald AK, Kahlert S, et al. Studies on the toxicity of deoxynivalenol (DON), sodium metabisulfite, DON-sulfonate (DONS) and de-epoxy-DON for porcine peripheral blood mononuclear cells and the Intestinal Porcine Epithelial Cell lines IPEC-1 and IPEC-J2, and on effects of DON and DONS on piglets. *Food Chem Toxicol*. 2010;48:2154–2162.
45. Wache YJ, Valat C, Postollec G, et al. Impact of deoxynivalenol on the intestinal microflora of pigs. *Int J Mol Sci*. **10**: 1–17.
46. Creppy EE. Update of survey, regulation and toxic effects of mycotoxins in Europe. *Toxicol Lett*. 2002;127:19–28.

CHAPTER

9

Gut Microbiota–Immune System Crosstalk: Implications for Metabolic Disease

Francesca Fava

Department of Food Quality and Nutrition, Research and Innovation Centre, Fondazione Edmund Mach, Trento, Italy

GUT MICROBIAL RECOGNITION BY THE IMMUNE SYSTEM

The human gut microbiota is a highly diverse and evolutionary plastic bacterial community, with up to 1000 different species, each one with a unique profile of immunogenic molecules, which may evolve at a much faster rate than mammalian cells.[1,2] This complex microbial ecosystem constitutes a constant stimulus and a highly allergenic challenge for the host immune system. To cope with this diversity the host immune system recognizes only certain highly conserved allergens shared between related groups of microorganisms. Immune sensing of bacteria is mainly mediated by epithelial cells (ECs) at the surface of the mucosa and underlying dendritic cells (DCs) which extend dendrites across the epithelial layer to sample luminal bacteria. Microorganisms and their immunological components (usually nucleic acids, cell wall fragments or flagella) are recognized by the host innate immune system through microbial antigen receptors, the pattern recognition receptors (PRRs), detecting invariant microbial motifs (pathogen-associated molecular patterns, PAMPs or microbe-associated molecular patterns, MAMPs).[3] MAMPs include conserved microbial cell surface molecules like lipopolysaccharide (LPS), teichoic acid, peptidoglycan, mannan extracellular polysaccharides, flagellin, as well as bacterial (CpG motif) DNA and double-stranded RNA. These microbial components serve as immunological barcodes or recognition features with which the immune system differentiates different microorganisms. Any given microorganism will present with different profiles of these MAMPs which combine to form, in effect, a microbial fingerprint which can be read by the immune system. Moreover, microorganisms at the intestinal mucosa appear to possess many different channels of communication with the host immune system and activate different PRRs, indicating synergistic action or a degree of functional redundancy between PRRs. PRRs include of the extracellular toll-like receptors (TLRs), the intracellular nucleotide-binding oligomerization domain (Nod)-like receptor family, and the CD14 receptor for bacterial lipopolysaccharide. TLRs are the best characterized signaling PRRs which lead to the activation of the transcription factor NF-κβ and recognize cell wall components of Gram-positive and Gram-negative bacteria including lipopolysaccharide (TLR-4), teichoic acid and peptidoglycan (TLR-2), and bacterial flagellin, a common feature of intestinal pathogens (TLR-5) and unmethylated CpG DNA (TLR-9).[3] Table 9.1 lists the common PRRs and their microbial ligands. In contrast to the TLRs, Nod proteins occur cytoplasmically rather than on the cell surface, allowing recognition of microbial components translocated into the cell by carrier molecules or upon cellular invasion by pathogens. Three important Nod-like receptors, Nod1, Nod2 and ice protease-activating factor (IPAF), are encoded by caspase recruitment domain (CARD) 4, CARD15 and CARD12, respectively. Nod1 recognizes gamma-D-glutamyl-meso-diamino-pimelic acid, a degradation product of peptidoglycan; Nod2 binds muramyl dipeptide (MDP) another peptidoglycan constituent, and IPAF binds flagellin. The CARD domain of Nod1 and Nod2 interacts with receptor interacting protein-2 (RIP2/RICK) protein kinase which mediates inhibitor of kappa B (IκB) and NF-κB-dependent expression of proinflammatory cytokines.[6] Nod2 also

Diet-Microbe Interactions in the Gut.
DOI: http://dx.doi.org/10.1016/B978-0-12-407825-3.00009-5

TABLE 9.1 The Common TLR and NODs Thought to be Involved in Microbe : Immune System Crosstalk

Receptor	Ligands
TLR 1	Bacterial triacyl lipopeptides
TLR 2	Gram positive bacterial cell wall component lipoteichoic acid.
	Host cell heat shock protein HSP70.
	Fungal zymosan (Beta-glucan).
	Multiple bacterial glycolipids, lipopeptides and lipoproteins.
TLR 3	Viral double-stranded RNA, poly I:C
TLR 4	Gram negative cell wall component lipopolysaccharide (LPS).
	Bacterial and host heat shock proteins.
	Host cellular components: fibrinogen, heparan sulfate, hyaluronic acid.
	Nickle and opioid drugs
TLR 5	Bacterial flagellin
TLR 6	Mycoplasma (bacteria without cell walls), multiple diacyl lipopeptides
TLR 7	Viral single-stranded RNA
TLR 8	Viral single-stranded RNA
TLR 9	Bacterial and viral unmethylated CpG DNA
TLR 10	Unknown
TLR 11	Toxoplasma gondii Profilin
TLR 12	Toxoplasma gondii Profilin
TLR 13[4,5]	Bacterial ribosomal RNA and viral sequence "CGGAAAGACC"
NOD1	Gamma-D-glutamyl-meso-diamino-pimelic acid from bacterial peptidoglycan
NOD2	Muramyl dipeptide from bacterial peptidoglycan
IPAF	Bacterial flaggellin

mediates NF-κB through genes associated with Retinoid-IFN-induced Mortality (Grim)-19 and TGF-β.[7,8] IPAF mediates IL-1β production in a caspase-1 dependant manner.[9]

Certain microbial ligands can be sensed by more than one PRR. Typical examples are bacterial flagellin and peptidoglycan, which can be sensed by TLR5 and TLR2, respectively, on the cell surface, and by IPAF and Nod intracellularly. Nods and TLRs have been shown to synergize the release of cytokines from myeloid cells, resulting in amplification of the resultant immune response.[10,11] Individual species of both Gram-positive and Gram-negative bacteria will, therefore, possess more than one TLR and Nod ligand and induce immunological signaling through more than one host recognition system.[12]

Stimulation of PRRs immediately triggers activation of signal cascades for antimicrobial immune responses and downstream transcription of inflammatory modulators. This consequently alerts both innate and acquired immunity, via recruitment/stimulation of immune effector cells through signaling molecules and antigen presentation by DC to naïve B cells and T cells in the MLN and subsequent sIgA production, respectively. PRR detection of MAMP in the gastrointestinal tract, has also the function of maintaining the vital mechanism of immune tolerance and intestinal homeostasis.[13]

Immune Effectors of Intestinal Microbiota–Host Crosstalk

First contact between the intestinal microbiota and the host occurs via mucosal ECs (or modified epithelial cells called M cells in the Peyer's Patches), and underlying or intercalating DCs.[1,3,6–8] Mucosal ECs express polymeric immunoglobulin receptor (PigR) and secretary component, MHC class I and II molecules, other adhesion molecules and a variety of cytokines and chemokines.[14] These molecules play important roles in signaling microbial encounters to underlying immune cells, and the induction of appropriate immune responses. Polarized ECs act as sensors of microbial encounters that in health, initiate appropriate defensive responses. DCs, present in both organized lymphoid tissues and along the length of the lamina propria, are involved in potentiating immune responses, enhanced antigen capture via induction of TLR and development of active immunity.[15] They are potent antigen presenting cells, critical for the induction of primary or initial immune responses at the mucosa and generation of T-cell-dependent B-cell responses. DCs express co-stimulatory molecules (CD80, CD86), TLR, C-type lectins such as dendritic cell-specific intercellular adhesion molecule 3-grabbing nonintegrin (DC-SIGN) and mannose receptor, which recognize carbohydrate receptors on microorganisms and other accessory ligands, necessary for induction of immune response or induction of tolerance, on their cell surfaces. These molecules mediate DC antigen presenting function. DCs themselves play an important role in differentiation of naïve T cells in the mesenteric lymph nodes (MLNs) into T regulatory cells which subsequently express homing proteins for the gastrointestinal mucosa where they mediate immune homeostasis.[16] They also direct helper T-cell responses toward Th1 and Th2, or regulatory patterns. Th1 immune responses are dependant on DC production of IL-12 and are characterized by production of interferon (IFN)-γ and IL-12. Th2 responses involve IL-4, IL-5, IL-6 and IL-13 and induce humoral immunity. As mentioned above, DCs can induce T cell differentiation into gut homing Tregs which assume regulatory functions, prevent excessive expansion of inflammatory T cells and thereby induce oral tolerance. DCs therefore play an important role in sequestering effector immune cells to the mucosal surface to respond to microbial challenge and also are involved in oral tolerance. One mechanism by which DCs induce T reg cells, is the release of IL-10 or TGF-β resulting in Tr1 and Th3 cells, regulatory T cells that act through secretion of IL-10 and TGF-β. Other mechanisms involve production of immunosuppressive INF-α or the induction of indoleamine 2,3-dioxygenase (IDO), which is an immunoregulatory enzyme with key functions in the interaction between DCs and Treg cells. DC immaturity or partial maturation at the time of antigen presentation has also been implicated in tolerance induction, particularly as a result of their interaction with apoptotic cells and probiotics.[12,15,16]

INTESTINAL BARRIER, GUT PERMEABILITY AND METABOLIC INFLAMMATION

The gut mucosal surface is a principal site of entry of pathogens into the human organism but in health the cells of the intestinal wall, both human and microbial, combine to provide an efficient barrier to invading microorganisms. The intestinal epithelial barrier, made up of a continuous layer of epithelial cells, tightly connected one to another and in contact with the lamina propria underlying immune cells (i.e., organized lymphoid tissues, such as the Peyer's patches), mucus-producing goblet cells and overlying mucus layer, and the resident microbiota all cooperate to form an essential system for protecting the host from pathogen invasion, colonization and overgrowth. The thick layer of mucin overlying the intestinal wall has long been considered to limit exposure of the intestinal surfaces to the luminal microorganisms in a non-specific manner. However, recent studies have indicated a more functional and active role of mucin in immune signaling and optimal function of intestinal immune cells, aiding DC sampling of luminal antigens through a porus small intestinal mucus layer and dampening of DC inflammatory responses to bacterial antigens upon co-stimulation with hyperglycosylated mucin MUC2. This carbohydrate-dependent attenuation of DC proinflammatory profiles and conditioning of DC and EC tolerogenic respones to luminal antigens appears to play an important role in the development of oral tolerance.[17] Similarly, mucin production by goblet cells has been sown to be under the regulation of NOD-like receptor family pyrin domain containing 6 (NLRPF6) inflammasome which also plays a role in shaping the composition of the gut microbiota.[18,19] The intestinal resident microbiota constitute an important obstacle to invading pathogens, through several mechanisms, including competition for metabolic growth substrates, production of antimicrobial compounds, occupancy of intestinal niches, and potentiation of host immune responses (i.e., production of IgA and IgM).[20] A well-controlled and regulated bacterial contact is the basis for maintenance of healthy gut immune functions, while excessive bacterial exposure would lead to stimulation of proinflammatory immune response. Similarly, the gut microbiota, particularly through butyrate production following dietary fiber fermentation, contributes to the integrity of the gut wall, providing about 50% of the mucosal energy requirement ensuring both optimal mucosal cell proliferation and also contributing to epigenetic regulation of cellular differentiation and programmed cell death, ensuring efficient cell cycling and sloughing off of aged or damaged epithelial cells from the tips of intestinal villi.[21] Recently, also, butyrate has been shown to induce differentiation of colonic Treg cells and also peripheral Treg cell generation, establishing a role for microbial colonic fermentation in both local (intestine) and global immune homeostasis.[22–24]

Compromised intestinal epithelial barrier function and tight junction integrity have been recently linked to the chronic inflammation associated with obesity. The bacterial LPS component of Gram-negative gut bacteria was found to play a major role in the inflammatory state associated with obese-related metabolic disorders.[25] High levels of LPS (2 to 3-fold increased plasma LPS levels, 10- to 50-times lower than that found in septicemia or infections) were detected in mice, where metabolic syndrome was induced upon high-fat (HF) feeding. Concomitantly with this "metabolic endotoxemia," changes in gut microbial composition were observed in mice after HF diet, in particular a depletion of the beneficial *Bifidobacterium* spp., which led to increased intestinal permeability, increased body weight, insulin sensitivity and liver, adipose tissue and systemic inflammation (i.e., elevated TNF-α, IL-1, IL-6 and plasminogen activator inhibitor – PAI-1).[25] Bifidobacteria are known to help in maintaining a healthy gut and, more specifically, the integrity of the epithelial barrier.[26,27] Potentiation of bifidobacterial populations through supplementation with the prebiotic oligofructose concomitant with the HF feeding induced improved intestinal barrier through increased tight junction protein (increased mRNA of ZO-1 and occluding) and increased proglucagone-derived peptide (GLP)-2 levels.[28] GLP-2 supports gut barrier function by promoting intestinal growth and the insulin-like growth factor (IGF)-1 and beta-catenin pathways. An increase in the enteroendocrine hormone GLP-1 was also found after prebiotic supplementation in HF-fed mice, this leading to stimulation of the central nervous system, increased leptin sensitivity, increased satiety, decreased food intake, decreased fat mass (especially visceral fat), increased insulin sensitivity, decreased postprandial glycemia and diabetes.[29]

The mechanism of induction of low-grade chronic inflammation that accompanies HF-diet-induced obesity was demonstrated to be dependent on TLR-4. Interestingly, in the absence of TLR-4/CD14 receptor for LPS, no low-grade chronic inflammation and insulin resistance were induced by HF feeding in mice.[25,30,31] Recently also TLR-2 and TLR-5 were shown to be involved in the innate immune system activation that is responsible for the inflammation induced after HF diet.[32,33] These results confirm gut microbiota implication in the onset of metabolic disorders associated with obesity. Moreover, in animal studies at least, this high-fat-induced metabolic endotoxemia and subsequent metabolic disease may be reversed using prebiotic dietary fibers and certain bacterial supplements.[34,35]

EFFECTS OF INTESTINAL BACTERIAL SHORT-CHAIN FATTY ACIDS (SCFAs) ON INFLAMMATION AND METABOLISM

SCFAs are end-products of bacterial metabolism, mainly derived from anaerobic fermentation and hydrolysis of undigested dietary polysaccharides and proteins, which exert important physiological effects.[36] The main SCFA products in the colon are acetate, propionate and butyrate in the ratio 2:1:1, can be absorbed by the host and provide for up to 10% of the basal energy requirements.[36,37] While acetate predominantly enters the bloodstream to the liver and peripheral tissues where it is used for respiration and cholesterol synthesis in the liver and brain, propionate and butyrate are used up in large part by colonocytes and the liver as a source of energy.[4] Beside their role in energy supply, SCFA can influence leukocyte recruitment, in particular neutrophils chemiotaxis, through binding to the G protein-coupled receptor 43 (GPR 43) and activation of the mitogen-activated protein kinase (MAPK), protein kinase C (PKC) cascade and activating transcriptional factor-2 (ATF-2).[5,38,39] SCFAs, especially butyrate, were shown to reduce the production of the chemokine macrophage chemoattractant protein-1 (MCP-1), an important factor that binds to monocytes and macrophages. Macrophages are the predominant immune cells involved in the secretion of proinflammatory mediators in states of chronic inflammation, such as insulin resistance and obesity, atherosclerosis, and also rheumatoid arthritis and neurodegenerative diseases. Some authors reported that butyrate administration is able to lower MCP-1 production by the adipose tissue in diet-induced obese animals, therefore lowering immune cells recruitment, inflammation and consequent insulin resistance. Butyrate also lowers MCP-1, IL-6 and INF-gamma production by intestinal epithelial cells, and therefore influences the intestinal inflammatory and immune response. SCFAs, especially butyrate and propionate, also lower pro-inflammatory cytokines such as TNF-α and nitric oxide (NO) in neutrophils. SCFAs also influence phagocytosis and the production of reactive oxygen species by neutrophils and macrophages.[40–42] Similarly, propionate, and probably other SCFAs, have been shown to regulate the production of inflammatory molecules from adipose tissue, an important contributor to the low-grade chronic systemic inflammation characteristic of metabolic syndrome, obesity and old age.[43,44]

Blockade of inflammatory signals by butyrate and propionate was also shown to be carried out through another mechanism, which is independent from GPR and acts on dendritic cells (DCs) development. DCs are found in peripheral and lymphoid tissues, where they act as a guardian against pathogens, with the role of continuous antigen sampling and antigen presentation to naïve T cells for mounting an adaptive immune response towards invading microorganisms. Recent studies have demonstrated that butyrate and also propionate may inhibit histone deacetylase (HDAC) not only in colon cells, but also in DCs, with a mechanism that necessitates involvement of the Na^+-coupled monocarboxylate transporter SLC_5A8. According to current findings butyrate and propionate block DC maturation and, therefore, arrest intestinal immune response and inflammation, through SLC_5A8-dependent DC entry and HDAC inhibition.[45] Gut bacteria were shown to be necessary to induce the expression of the SCFA transporter SLC_5A8 on DC, since germ-free mice appeared to have suppressed SLC_5A8 expression. Inhibition of precursor DC development might be a significant contributor of bacterial butyrate and propionate in reducing obesity-associated inflammation and potentiating bacterial fermentation of dietary fibers might represent an important target in the management of low-grade chronic inflammation characteristic of the metabolic syndrome and indeed other chronic diseases where unresolved inflammation plays a pathobiological role including autoimmune diseases and the aging process.[23,46,47]

DIETARY FAT METABOLISM, BILE ACIDS AND GUT MICROBIOTA

Bile acids are an important class of metabolites derived from cholesterol, synthesized in the liver into their primary forms, the cholic acid (CA) and the chenodeoxicholic acid (CDCA). In response to fat intake, BA are conjugated with taurine or glycine and excreted via the bile ducts into the lumen of the small intestine, where they help fat digestion by acting as detergents. Once in the gut lumen, BA enter the so-called enterohepatic circulation, which consists of the modification of primary BA into secondary BA through deconjugation, dehydrogenation, dehydroxylation, and sulfation by the gut microbiota, and re-absorption via the portal vein to the liver for their recycling.[48,49] Besides their role in dietary fat metabolism and in cholesterol homeostasis thanks to the enterohepatic circulation, BA also carry out a much wider range of actions as signaling molecules. BA bind the intracellular farnesoid X receptor (FXR), that is mainly expressed in the enterohepatic tissues, and to the plasma membrane-bound TGR-5 receptor, which is expressed in a broad variety of cells in different organs and with different degrees of expression. As ligands for these cellular receptors BA influence in a variety of physiological processes including lipid and bile metabolism, glucose homeostasis, thermogenesis, inflammation and bone–mineral homeostasis.[49] Interestingly, all these physiological processes have been studied, with some success especially in animal models, as targets for prebiotic and probiotic dietary functional foods.[50]

FXR activation contributes to maintenance of intestinal integrity and reduces intestinal permeability. FXR expression in the intestine is stimulated by activation of the microbiota-sensor TLR-9[51] through recruitment of the response element interferon regulatory factor (IRF)-7 (i.e., a member of the interferon regulatory family of transcription factors involved in the antiviral innate immune response) on FXR promoter. Microbially derived CpG stimulation of TLR-9 induces increased expression of FXR and its target gene SHP, which inhibits the transcription factor AP-1, a positive regulator of inflammation. Therefore, TLR9/FXR interaction contributes to regulate intestinal inflammation. TLR-9 knock-out mice were shown to be more susceptible to TNBS-induced colitis and the severity of colitis was reduced in the same mice after administering an FXR agonist. On the other hand, FXR knock-out mice were shown to be unresponsive to CpG therapy to reduce the severity of TNBS-induced colitis, thus demonstrating that CpG activation of TLR-9 lowers inflammation only through FXR activation.

TGR-5 is highly expressed in brown adipose tissue, muscle, macrophages, monocytes, intestine, gallbladder and spleen. Binding of BA to the G-protein-coupled cell membrane receptor TGR-5 induces increased energy expenditure in muscle and brown adipose tissue.[49,52,53] TGR-5 activation was shown to be able to induce a significant body weight reduction in mice fed a HF diet, while some studies show TGR5 −/− mice to have increased body weight and fat content.[53] TGR-5 activation also improves glucose metabolism, insulin sensitivity, and also gastrointestinal motility and appetite,[54] since it was observed that *in vitro* stimulation of TGR-5 induces GLP-1 secretion by enteroendocrine cells and human intestinal cells.[55–57] Others have shown that the role of TGR5 in metabolic disease is both dependent on gender and diet, especially in terms of insulin sensitivity and fatty liver disease.[58] TGR-5 also plays a role in lowering the low grade chronic inflammation associated with the metabolic syndrome. In particular BA activation of TGR-5 on monocytes and macrophages reduces their phagocytic activity and the production of proinflammatory cytokines in response to LPS stimulation.[59] Activation of TGR-5 in Kupffer cells, where this receptor is highly expressed, might be responsible for the protective effect of TGR-5 in liver steatosis and non-alcoholic fatty liver disease (NAFLD).[60,61] These data strongly support the potential of using activation of this BA membrane receptor for ameliorating several aspects of the metabolic syndrome.[57] Beside its anti-inflammatory effect expleted on macrophage function (i.e., phagocytic capability and cytokine secretion) TGR-5 was also recently shown to regulate intestinal homeostasis and barrier integrity.[62] TGR-5 −/− mice were shown to display altered distribution and maturation of intestinal mucus cells together with disrupted organization of tight junction proteins (e.g,. especially altered distribution of zonulin-1), which rendered the animals more susceptible to DSS-induced colitis.[62] TGR-5 expression was upregulated in wild-type mice challenged with DSS to induce colitis. Interestingly, single nucleotide polymorphism (SNP) in TGR-5 gene (i.e., an SNP that causes loss of function) was found in association with intestinal inflammatory diseases, such as ulcerative cholitis and sclerosing cholangitis (PSC), both characterized by increased intestinal permeability.[63]

Studies highlighting the multiple roles of the BA-receptors TGR-5 and FXR suggest a complex interplay between metabolic and immune functions, which have important consequences in pathologies characterized by chronic inflammation, such as inflammatory bowel disease (IBD) and the metabolic syndrome and indeed aging.[49,50,53] Diet, especially fermentable fibers or prebiotics, probiotics with bile salt hydrolase activity or bile acid chelating plant polyphenols, will therefore mediate significant physiological effects through the gut microbiota bile acid handling, on both metabolic and inflammatory processes. Morever, as discussed below, the gut microbiota plays a central role in coordinating immunity and metabolism through the intestinal epithelium gene expression pathways.

The gut microbiota deeply influences the pool of BA and also BA profiles of tissues and organs outside the enterohepatic compartments, such as the kidneys, heart, plasma, adipose tissue, muscle, where they have a signaling function relevant to glucose homeostais, lipid metabolism and energy homeostasis.[64,65] The main part of extra-enterohepatic BA are the unconjugated forms of BA, which are microbially derived. Unconjugated BA contribute to increased basal metabolic rate and energy expenditure. Interestingly, it has long been known that germ-free (GF) mice have a lower basal metabolic rate compared to wild type mice.[66] GF mice were also shown to have higher taurine-conjugated and lower glycine-conjugated or unconjugated BA, due to increased taurine availability as a consequence of the absence of gut microbes, responsible for the taurine degradation to sulfate and sulfate catabolites.[64] Taurine-conjugated BA are more efficient at fat emulsification, digestion and absorption. Studies conducted in IBD patients support the role of the gut microbiota in influencing the pool of BA. The condition of intestinal dysbiosis characteristic of IBD has been linked to BA dysmetabolism and to the chronic inflammatory state.[67] BA biotrasformation by the faecal microbiota of IBD patients was impaired, with reduced fecal secondary BA and concomitant increased conjugated and sulfated BA found in IBD patients compared to healthy subjects. Fecal gut microbiota composition showed reduced *Clostridium coccoides*, *Clostridium leptum* and *Faecalibacterium prausnitzii* species within the *Firmicutes* phylum in IBD versus healthy individuals. The primary BA-enriched profile characteristic of IBD was not able to counteract the proinflammatory effect of IL-1β on Caco-2 intestinal epithelial cells *in vitro*, while secondary BA exerted an anti-inflammatory effect, with decreased IL-1β induced IL-8 production.[67]

While on one side the gut microbiota can influence BA pool, the pool of BA produced in response to the diet can in turn shape gut microbial composition.[68] Recent studies have shown that high intake of milk-derived fat was able to preferably induce the formation of taurine-conjugated BA, which increased gut bioavailability of sulfur, a direct product of taurine microbial metabolism and a growth substrate for sulfite-reducing bacteria.[69] The bloom of *Bilophila wadsworthia*, a member of the *Deltaproteobacteria* family, was observed in association with the increased taurocholic acid following high milk-fat intake in colitis-susceptible IL10 −/− mice. Together with increased *B. wadsworthia*, the mice showed increased intestinal inflammation after a high milk-fat diet or after a low-fat diet supplemented with taurocholic acid, while no inflammation nor BA and gut microbial composition changes were seen in the same type of mice after a low-fat diet. The same gut microbial, BA and immune effect was not observed either when other dietary fat different from milk fat was administered to IL10 −/− mice, thus demonstrating that the type of fat is important in shaping our gut microbiota and the host metabolic and immune response.[69]

DIET, TMAO, GUT MICROBIOTA AND ATHEROSCLEROSIS

Metabolism of the dietary lipid phosphatidylcholine by gut bacteria generates trimethylamine (TMA), which is then oxidized into trimethylamine N-oxide (TMAO) by hepatic flavin mono-oxygenase (FMO). FMO expression is induced by the BA nuclear receptor FXR, upon stimulation with BA. TMAO is implicated in the formation of atherosclerotic plaque through induction of upregulation of the expression of two macrophage scavenger receptors, CD36 and SR-A1.[70] Additionally TMAO were recently shown to inhibit reverse cholesterol transport (RCT), the mechanism through which cholesterol is removed from peripheral macrophages and transported to the liver where it is recycled for cholesterol synthesis.[71] Importantly, RCT was demonstrated to be afftected by TMAO formation through inhibition of BA synthesis, which represents one important mechanism regulating cholesterol levels in circulation.

In murine experiments, high choline diet was shown to be associated with atherosclerosis through a microbiota-dependant TMAO formation. Atherosclerotic prone C57BL/6J.ApoE −/− mice on a high choline diet (1% choline supplementation, wt/wt) showed high aortic lesions and expression of CD36 macrophage scavenger receptors compared to mice on a control diet (around 0.1% choline, wt/wt).[70] Suppression of the gut microbiota through administration of broad-spectrum antibiotics inhibited choline-mediated atherosclerosis. Similar studies in humans confirmed the necessity of gut microbiota-mediated TMAO production to induce atherosclerosis and cardiovascular disease (CVD). In particular a large follow-up study on 4007 subjects at high risk of CVD showed that repression of gut microbiota by antibiotic treatment efficiently reduced TMAO formation and decreased the risk of severe cardiovascular events.[72]

Interestingly, recent studies also showed that gut microbial metabolism of dietary L-carnitine, a trimethylamine predominantly found in red meat, also results in the production of TMAO and in promotion of atherosclerosis in mice.[71] Gut microbiota ability to form TMAO from L-carnitine varies among individuals in relation to their dietary habits. In particular it was shown that vegan and vegetarian human subjects have extremely reduced TMAO formation compared to omnivores after intake of L-carnitine through red meat consumption. Experiments were carried out employing d3-(methyl)-carnitine along with red meat ingestion. Long-term vegan and vegetarian subjects were shown to be virtually uncapable of generating TMAO from L-carnitine. The fasting basal levels of TMAO were also lower in vegan and in vegetarians compared to omnivores. Postprandial plasma concentrations of d3-(methyl)-carnitine were found to be higher in vegans and vegetarians, maybe due to decreased microbial metabolism and consequent higher absorption of carnitine. Long-term vegan/vegetarian dietary habits bring about alteration in gut microbiota composition. It was observed that high TMAO producers presented with higher abundance of the genus *Prevotella*, while lower TMAO producers had higher abundance of bacteria belonging to the *Bacteroides* genus. Prevalence of several other bacterial populations, expressed as operational taxonomic units (OTUs), were found to differ significantly between vegans/vegetarians and omnivores, with the omnivores generally presenting with higher *Clostridium*, *Clostridiaceae* and *Peptostreptococcaceae* and lower *Lachnospira* and *Sporobacter*. Gut microbiota-dependent TMAO formation was shown to be inducible. Experiments in GF mice, which were initially unable to produce d3-(methyl)-TMA (and TMAO) from d3-(methyl)-carnitine, eventually acquired this capability after being microbially colonized in conventional cages.[71]

IMMUNE VERSUS METABOLIC FUNCTIONS IN INTESTINAL EPITHELIAL CELLS GENE NETWORKS

The mammalian gut constitutes a multifunctional organ intimately involved both in immunity and metabolism. Moreover, these physiological functions appear to be coordinated by host genetics, diet and the resident microbiota. In health the human gut explicates both immunity and metabolic homeostasis, while in conditions of impaired immunity the intestinal epithelium has been shown to be able to substitute missing immune functions by switching on immune genes and switching off metabolic genes, such as genes involved in lipid absorption (i.e., GATA4).[73] Recent studies in mice mimicking immunodeficiency (i.e., B cells knock-out, BcKO) showed that in the absence of B cell produced-IgA, the commensal bacteria send a stronger signal to the gut epithelium and this causes increased expression of IFN-dependent immunity genes. Concomitantly it was observed that genes linked to lipid metabolism, such as GATA4, a transcription factor that promotes lipid absorption, were downregulated when the immunity genes were upregulated. Interestingly, around 60% of the gene expression changes observed in BcKO mice, were also common to GATA4KO mice.[73] In another study GATA4KO mice were shown to be resistant to diet-induced obesity and insulin resistance, also through increased GLP-1 release.[74] GATA4 is also involved in differentiation of human enterocytes, particularly inducing the expression of tight junction proteins.[75] It was postulated that in conditions of immunodeficiency there is a shift of intestinal functions, which is reflected in lipid malabsorption, decreased body fat deposition, but also increased intestinal permeability (through decreased GATA4-regulated tight junctions expression), which leads to increased plasma LPS. The gut epithelium possesses differential gene networks, one that regulates immunity and one that governs lipid metabolism, the two being inversely expressed.

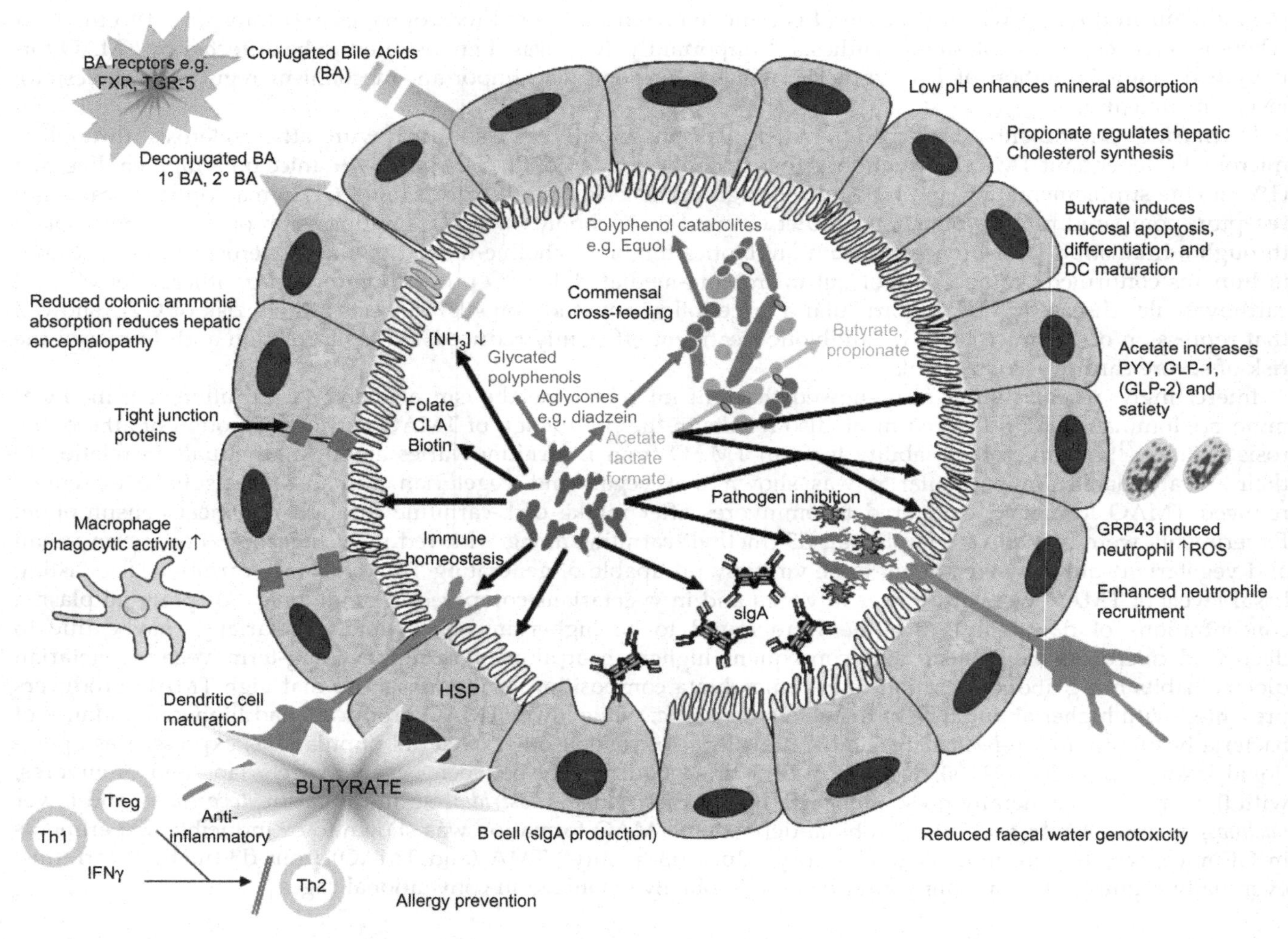

FIGURE 9.1 A schematic representation of diet:microbe interactions and how they shape immune function within the gut. Key metabolic processes within the human gut microbiota, especially carbohydrate fermentation, the enterohepatic circulation of bile acids and biotransformation of plant bioactive polyphenols by the gut microbiota play important roles in regulating both inflammatory and metabolic processes within the intestine, but also in other body tissues, like the liver, adipose tissue and brain, which are intimately involved in regulating whole-body glucose, lipid and energy metabolism, and also the chronic low-grade inflammation characteristic of metabolic diseases like diabetes, CVD, Alzheimer's and metabolic syndrome.

CONCLUSION

Diet plays a critical role in shaping the gut microbiota and also in shaping its ability to regulate host metabolism and immune function as summarized in Figure 9.1. Beginning in early life, diet controls the programmed successional development of the gut microbiota and through it, maturation of gut mucosal architecture and digestive function, immune tolerance and co-metabolic pathways, especially those related to SCFA and BA metabolism. Obesity, and the diseases of obesity, the metabolic syndrome, type 2 diabetes, non-alcoholic fatty liver disease, early dementia and Alzheimer's disease, all share abberant gut microbiota profiles and immune malfunction. The low-grade unresolved systemic inflammation associated with the onset of these diseases appears to originate, at least in part, from the gut, thereafter acting systemically through the liver, adipose tissue and brain to mediate both pathogenesis and accelerated aging. Dietary fat, but probably also high-fat, Western-style diets appear to break down co-regulated metabolic and inflammatory processes within the gut. Conversely, supporting saccharolytic fermentation by the gut microbiota or their interaction with bile acids, in particular through dietary supplementation with prebiotics or bile salt hydrolyzing probiotics, appears to reinforce optimal function of the gut barrier as a first line defense against metabolic and inflammatory disease.

However, key mechanistic studies in models systems supported by real-life data from human dietary interventions are necessary to uncover the physiological processes at play within the gut environment/ecosystem responsible for regulating this interplay between metabolism and inflammation with the ultimate aim of developing dietary strategies to reducing chronic disease risk at the population level.

References

1. Koboziev I, Reinoso Webb C, Furr KL, Grisham MB. Role of the enteric microbiota in intestinal homeostasis and inflammation. *Free Radic Biol Med*. 2014;68C:122–133.
2. Goldsmith JR, Sartor RB. The role of diet on intestinal microbiota metabolism: downstream impacts on host immune function and health, and therapeutic implications. *J Gastroenterol*. 2014;49(5):785–798.
3. Suresh R, Mosser DM. Pattern recognition receptors in innate immunity, host defense, and immunopathology. *Adv Physiol Educ*. 2013;37:284–291.
4. Fava F, Lovegrove JA, Gitau R, Jackson KG, Tuohy KM. The gut microbiota and lipid metabolism: implications for human health and coronary heart disease. *Curr Med Chem*. 2006;13:3005–3021.
5. Cox MA, Jackson J, Stanton M, et al. Short-chain fatty acids act as antiinflammatory mediators by regulating prostaglandin E(2) and cytokines. *World J Gastroenterol*. 2009;15:5549–5557.
6. Kobayashi K, Inohara N, Hernandez LD, et al. RICK/Rip2/CARDIAK mediates signalling for receptors of the innate and adaptive immune systems. *Nature*. 2002;416:194–199.
7. Chen CM, Gong Y, Zhang M, Chen JJ. Reciprocal cross-talk between Nod2 and TAK1 signaling pathways. *J Biol Chem*. 2004;279:25876–25882.
8. Barnich N, Hisamatsu T, Aguirre JE, Xavier R, Reinecker HC, Podolsky DK. GRIM-19 interacts with nucleotide oligomerization domain 2 and serves as downstream effector of anti-bacterial function in intestinal epithelial cells. *J Biol Chem*. 2005;280:19021–19026.
9. Warren SE, Mao DP, Rodriguez AE, Miao EA, Aderem A. Multiple Nod-like receptors activate caspase 1 during Listeria monocytogenes infection. *J Immunol*. 2008;180:7558–7564.
10. Le Bourhis L, Benko S, Girardin SE. Nod1 and Nod2 in innate immunity and human inflammatory disorders. *Biochem Soc Trans*. 2007;35:1479–1484.
11. Kawai T, Akira S. Toll-like receptors and their crosstalk with other innate receptors in infection and immunity. *Immunity*. 2011;34:637–650.
12. Fava F, Danese S. Intestinal microbiota in inflammatory bowel disease: friend of foe? *World J Gastroenterol*. 2011;17:557–566.
13. Cerf-Bensussan N, Gaboriau-Routhiau V. The immune system and the gut microbiota: friends or foes? *Nat Rev Immunol*. 2010;10:735–744.
14. Saenz SA, Taylor BC, Artis D. Welcome to the neighborhood: epithelial cell-derived cytokines license innate and adaptive immune responses at mucosal sites. *Immunol Rev*. 2008;226:172–190.
15. Wells JM, Rossi O, Meijerink M, van Baarlen P. Epithelial crosstalk at the microbiota–mucosal interface. *Proc Natl Acad Sci USA*. 2011;108 (suppl 1):4607–4614.
16. Mann ER, Landy JD, Bernardo D, et al. Intestinal dendritic cells: their role in intestinal inflammation, manipulation by the gut microbiota and differences between mice and men. *Immunol Lett*. 2013;150:30–40.
17. Shan M, Gentile M, Yeiser JR, et al. Mucus enhances gut homeostasis and oral tolerance by delivering immunoregulatory signals. *Science*. 2013;342:447–453.
18. Chen GY. Role of Nlrp6 and Nlrp12 in the maintenance of intestinal homeostasis. *Eur J Immunol*. 2014;44:321–327.
19. Wlodarska M, Thaiss CA, Nowarski R, et al. NLRP6 Inflammasome Orchestrates the Colonic Host–Microbial Interface by *Regulating Goblet Cell Mucus Secretion*. *Cell*. 2014;156:1045–1059.
20. Britton RA, Young VB. Role of the intestinal microbiota in resistance to colonization by Clostridium difficile. *Gastroenterology*. 2014;146 (6):1547–1553.
21. Plöger S, Stumpff F, Penner GB, et al. Microbial butyrate and its role for barrier function in the gastrointestinal tract. *Ann N Y Acad Sci*. 2012;1258:52–59.
22. Smith PM, Howitt MR, Panikov N, et al. The microbial metabolites, short-chain fatty acids, regulate colonic Treg cell homeostasis. *Science*. 2013;341:569–573.
23. Arpaia N, Campbell C, Fan X, et al. Metabolites produced by commensal bacteria promote peripheral regulatory T-cell generation. *Nature*. 2013;504:451–455.
24. Kugelberg E. Mucosal immunology: bacteria get T(Reg) cells into shape. *Nat Rev Immunol*. 2014;14:2–3.
25. Cani PD, Amar J, Iglesias MA, et al. Metabolic endotoxemia initiates obesity and insulin resistance. *Diabetes*. 2007;56:1761–1772.
26. Lindfors K, Blomqvist T, Juuti-Uusitalo K, et al. Live probiotic *Bifidobacterium lactis* bacteria inhibit the toxic effects induced by wheat gliadin in epithelial cell culture. *Clin Exp Immunol*. 2008;152:552–558.
27. Bergmann KR, Liu SX, Tian R, et al. Bifidobacteria stabilize claudins at tight junctions and prevent intestinal barrier dysfunction in mouse necrotizing enterocolitis. *Am J Pathol*. 2013;182:1595–1606.
28. Cani PD, Possemiers S, Van de Wiele T, et al. Changes in gut microbiota control inflammation in obese mice through a mechanism involving GLP-2-driven improvement of gut permeability. *Gut*. 2009;58:1091–1103.
29. Cani PD, Knauf C, Iglesias MA, Drucker DJ, Delzenne NM, Burcelin R. Improvement of glucose tolerance and hepatic insulin sensitivity by oligofructose requires a functional glucagon-like peptide 1 receptor. *Diabetes*. 2006;55:1484–1490.
30. Cani PD, Bibiloni R, Knauf C, et al. Changes in gut microbiota control metabolic endotoxemia-induced inflammation in high-fat diet-induced obesity and diabetes in mice. *Diabetes*. 2008;57:1470–1481.
31. Kim KA, Gu W, Lee IA, Joh EH, Kim DH. High fat diet-induced gut microbiota exacerbates inflammation and obesity in mice via the TLR4 signaling pathway. *PLoS One*. 2012;7:e47713.

32. Vijay-Kumar M, Aitken JD, Carvalho FA, et al. Metabolic syndrome and altered gut microbiota in mice lacking Toll-like receptor 5. *Science*. 2010;328:228–231.
33. Caricilli AM, Picardi PK, de Abreu LL, et al. Gut microbiota is key modulator of insulin resistance in TLR 2 knockout mice. *PLoS Biol*. 2011;9:e1001212.
34. Cani PD, Neyrinck AM, Fava F, et al. Selective increases of bifidobacteria in gut microflora improve high-fat-diet-induced diabetes in mice through a mechanism associated with endotoxaemia. *Diabetologia*. 2007;50:2374–2383.
35. Everard A, Belzer C, Geurts L, et al. Cross-talk between *Akkermansia muciniphila* and intestinal epithelium controls diet-induced obesity. *Proc Natl Acad Sci USA*. 2013;110:9066–9071.
36. Macfarlane GT, Macfarlane S. Fermentation in the human large intestine: its physiologic consequences and the potential contribution of prebiotics. *J Clin Gastroenterol*. 2011;45(suppl):S120–S127.
37. Cummings JH, Pomare EW, Branch WJ, Naylor CP, Macfarlane GT. Short chain fatty acids in human large intestine, portal, hepatic and venous blood. *Gut*. 1987;28:1221–1227.
38. Eftimiadi C, Buzzi E, Tonetti M, et al. Short-chain fatty acids produced by anaerobic bacteria alter the physiological responses of human neutrophils to chemotactic peptide. *J Infect*. 1987;14:43–53.
39. Eftimiadi C, Tonetti M, Cavallero A, Sacco O, Rossi GA. Short-chain fatty acids produced by anaerobic bacteria inhibit phagocytosis by human lung phagocytes. *J Infect Dis*. 1990;161:138–142.
40. Mills SW, Montgomery SH, Morck DW. Evaluation of the effects of short-chain fatty acids and extracellular pH on bovine neutrophil function *in vitro*. *Am J Vet Res*. 2006;67:1901–1907.
41. Vinolo MA, Hatanaka E, Lambertucci RH, Newsholme P, Curi R. Effects of short chain fatty acids on effector mechanisms of neutrophils. *Cell Biochem Funct*. 2009;27:48–55.
42. Vinolo MA, Rodrigues HG, Hatanaka E, Sato FT, Sampaio SC, Curi R. Suppressive effect of short chain fatty acids on production of proinflammatory mediators by neutrophils. *J Nutr Biochem*. 2011;22:849–855.
43. Al-Lahham S, Roelofsen H, Rezaee F, et al. Propionic acid affects immune status and metabolism in adipose tissue from overweight subjects. *Eur J Clin Invest*. 2012;42:357–364.
44. Layden BT, Yalamanchi SK, Wolever TM, Dunaif A, Lowe Jr. WL. Negative association of acetate with visceral adipose tissue and insulin levels. *Diabetes Metab Syndr Obes*. 2012;5:49–55.
45. Singh N, Thangaraju M, Prasad PD, et al. Blockade of dendritic cell development by bacterial fermentation products butyrate and propionate through a transporter (Slc5a8)-dependent inhibition of histone deacetylases. *J Biol Chem*. 2010;285:27601–27608.
46. Wang B, Morinobu A, Horiuchi M, Liu J, Kumagai S. Butyrate inhibits functional differentiation of human monocyte-derived dendritic cells. *Cell Immunol*. 2008;253:54–58.
47. Nascimento CR, Freire-de-Lima CG, da Silva de Oliveira A, Rumjanek FD, Rumjanek VM. The short chain fatty acid sodium butyrate regulates the induction of CD1a in developing dendritic cells. *Immunobiology*. 2011;216:275–284.
48. Chiang JY. Bile acid metabolism and signaling. *Compr Physiol*. 2013;3:1191–1212.
49. Joyce SA, Gahan CG. The gut microbiota and the metabolic health of the host. *Curr Opin Gastroenterol*. 2014;30:120–127.
50. Tuohy KM, Fava F, Viola R. 'The way to a man's heart is through his gut microbiota' – dietary pro- and prebiotics for the management of cardiovascular risk. *Proc Nutr Soc*. 2014;4:1–14.
51. Renga B, Mencarelli A, Cipriani S, et al. The bile acid sensor FXR is required for immune-regulatory activities of TLR-9 in intestinal inflammation. *PLoS One*. 2013;8(1):e54472.
52. Kawamata Y, Fujii R, Hosoya M, et al. A G protein-coupled receptor responsive to bile acids. *J Biol Chem*. 2003;278:9435–9440.
53. Pols TW. TGR5 in inflammation and cardiovascular disease. *Biochem Soc Trans*. 2014;42:244–249.
54. Pols TW, Noriega LG, Nomura M, Auwerx J, Schoonjans K. The bile acid membrane receptor TGR5: a valuable metabolic target. *Dig Dis*. 2011;29:37–44.
55. Parker HE, Wallis K, le Roux CW, Wong KY, Reimann F, Gribble FM. Molecular mechanisms underlying bile acid-stimulated glucagon-like peptide-1 secretion. *Br J Pharmacol*. 2012;165:414–423.
56. Harach T, Pols TW, Nomura M, et al. TGR5 potentiates GLP-1 secretion in response to anionic exchange resins. *Sci Rep*. 2012;2:430.
57. Stepanov V, Stankov K, Mikov M. The bile acid membrane receptor TGR5: a novel pharmacological target in metabolic, inflammatory and neoplastic disorders. *J Recept Signal Transduct Res*. 2013;33:213–223.
58. Vassileva G, Hu W, Hoos L, et al. Gender-dependent effect of Gpbar1 genetic deletion on the metabolic profiles of diet-induced obese mice. *J Endocrinol*. 2010;205:225–232.
59. Haselow K, Bode JG, Wammers M, et al. Bile acids PKA-dependently induce a switch of the IL-10/IL-12 ratio and reduce proinflammatory capability of human macrophages. *J Leukoc Biol*. 2013;94:1253–1264.
60. Keitel V, Donner M, Winandy S, Kubitz R, Häussinger D. Expression and function of the bile acid receptor TGR5 in Kupffer cells. *Biochem Biophys Res Commun*. 2008;372:78–84.
61. McMahan RH, Wang XX, Cheng LL, et al. Bile acid receptor activation modulates hepatic monocyte activity and improves nonalcoholic fatty liver disease. *J Biol Chem*. 2013;288:11761–11770.
62. Cipriani S, Mencarelli A, Chini MG, et al. The bile acid receptor GPBAR-1 (TGR5) modulates integrity of intestinal barrier and immune response to experimental colitis. *PLoS One*. 2011;6(10):e25637.
63. Hov JR, Keitel V, Schrumpf E, Häussinger D, Karlsen TH. TGR5 sequence variation in primary sclerosing cholangitis. *Dig Dis*. 2011;29:78–84.
64. Swann JR, Want EJ, Geier FM, et al. Systemic gut microbial modulation of bile acid metabolism in host tissue compartments. *Proc Natl Acad Sci USA*. 2011;108(suppl 1):4523–4530.
65. Dekaney CM, von Allmen DC, Garrison AP, et al. Bacterial-dependent up-regulation of intestinal bile acid binding protein and transport is FXR-mediated following ileo-cecal resection. *Surgery*. 2008;144:174–181.
66. Sewell DL, Wostmann BS, Gairola C, Aleem MI. Oxidative energy metabolism in germ-free and conventional rat liver mitochondria. *Am J Physiol*. 1975;228(2):526–529.

67. Duboc H, Rajca S, Rainteau D, et al. Connecting dysbiosis, bile-acid dysmetabolism and gut inflammation in inflammatory bowel diseases. *Gut*. 2013;62:531–539.
68. Islam KB, Fukiya S, Hagio M, et al. Bile acid is a host factor that regulates the composition of the cecal microbiota in rats. *Gastroenterology*. 2011;141:1773–1781.
69. Devkota S, Wang Y, Musch MW, et al. Dietary-fat-induced taurocholic acid promotes pathobiont expansion and colitis in Il10 −/− mice. *Nature*. 2012;487:104–108.
70. Wang Z, Klipfell E, Bennett BJ, et al. Gut flora metabolism of phosphatidylcholine promotes cardiovascular disease. *Nature*. 2011;472:57–63.
71. Koeth RA, Wang Z, Levison BS, et al. Intestinal microbiota metabolism of L-carnitine, a nutrient in red meat, promotes atherosclerosis. *Nat Med*. 2013;19:576–585.
72. Tang WH, Wang Z, Levison BS, et al. Intestinal microbial metabolism of phosphatidylcholine and cardiovascular risk. *N Engl J Med*. 2013;368:1575–1584.
73. Shulzhenko N, Morgun A, Hsiao W, et al. Crosstalk between B lymphocytes, microbiota and the intestinal epithelium governs immunity versus metabolism in the gut. *Nat Med*. 2011;17:1585–1593.
74. Patankar JV, Chandak PG, Obrowsky S, et al. Loss of intestinal GATA4 prevents diet-induced obesity and promotes insulin sensitivity in mice. *Am J Physiol Endocrinol Metab*. 2011;300:E478–E488.
75. Guillemot L, Spadaro D, Citi S. The junctional proteins cingulin and paracingulin modulate the expression of tight junction protein genes through GATA-4. *PLoS One*. 2013;8(2):e55873.

CHAPTER

10

The Interplay of Epigenetics and Epidemiology in Autoimmune Diseases: Time for Geoepigenetics

Carlo Selmi,†,‡ and Angela Ceribelli*,†*

*Division of Rheumatology and Clinical Immunology, Humanitas Clinical and Research Center, Milan, Italy
†BIOMETRA Department, University of Milan, Milan, Italy ‡Division of Rheumatology, Allergy, and Clinical Immunology, University of California, Davis, California, USA

THE ETIOLOGY AND PATHOGENESIS OF AUTOIMMUNE DISEASE

Commensal microbiota plays a key role for immune tolerance in our organism. Recent reports show that alterations in microbiota and dietary factors may be involved in the pathogenesis of autoimmune diseases because of a mechanism of molecular mimicry and reaction to self-antigens as if they were considered as external antigens.[1] The etiology of autoimmune diseases is unknown and it is now clear that many factors play a role in disease onset. The influence of diet and microbiota is clearly present in diseases with autoimmune background such as type 1 diabetes, inflammatory bowel diseases, celiac disease, and asthma,[2] but not in other autoimmune diseases such as Systemic Lupus Erythematosus (SLE), Rheumatoid Arthritis (RA) and Systemic Sclerosis (SSc). The study of twins has been a powerful method to understand the influence of genetic and non-genetic factors in SLE, RA and SSc. In fact, the low concordance rate for disease onset in monozygotic (MZ) twins shows that genetic factors are not enough to induce the disease, but also external, non-genetic, environmental factors are needed to develop the clinical phenotype of autoimmune diseases.[3] Autoimmune diseases are characterized by the abnormal response of the immune system against self-antigens, leading to persistent systemic or organ-specific manifestations, in some cases life-threatening. The autoimmune response is based on both adaptive and innate immunity arms, with inflammatory mediators attacking structures such as joints and renal glomeruli, and self-reacting antibodies directed against self-antigens derived from cellular apoptosis or post-translational modifications. Epidemiology studies have confirmed that autoimmune diseases are in general more frequent in female patients, and their incidence and prevalence increase in middle-age, with the exception of lupus and pediatric diseases. The reason for this epidemiological distribution remains unknown, but numerous hypotheses stress the role played by genetic (i.e., sex chromosomes) and/or environmental factors (i.e., gut microbiota, diet, infections, exposure to chemical and physical agents favoring apoptosis) in the onset of autoimmune diseases. Interestingly, recent studies have shown that early colonization of the gut by commensal microorganisms may be the key in the development of a tolerant and well-functioning immune system, protecting from autoimmune disease development both in infancy and later in life, and this process may be mediated through microbiota regulation of sex hormones.[4,5]

Epigenetics is a phenomenon characterized by stable and potentially heritable changes in gene expression that do not involve variations in the DNA sequence.[6] Some of these changes are heritable and can affect gene expression or cellular phenotype, leading to different disease expressions and modifying cell functions for multiple generations. Several epigenetic mechanisms are currently studied, but DNA methylation, histone modifications and

Diet-Microbe Interactions in the Gut.
DOI: http://dx.doi.org/10.1016/B978-0-12-407825-3.00010-1

microRNAs (miRNAs) profiles are universally considered to be able to affect the epigenome of a cell. Importantly, many of these epigenetic mechanisms have been shown to be modulated by diet:microbiota interactions with implications for immune function both locally within the gut and systemically in diverse organs including the liver, adipose tissue and brain. In particular, short-chain fatty acid butyrate and propionate produced by commensal bacteria are able to induce histone deacetylase inhibition and thus regulate the function of macrophages in the intestinal lamina propria to maintain immune tolerance.[7] Another epigenetic mechanism that can be influenced by gut microorganisms is miRNAs expression, and also this process can be involved in the regulation of host genes expression that control immune response.[8] Dietary elements (i.e., folate, betaine, choline) and gut bacteria (i.e., folate-producing bifidobacteria) can influence carbon metabolism and the methylation status detected both in the gut and systemically,[9] and this further supports the ability of diet:microbiota interaction to influence epigenetics and thus immune gene expression.

DNA methylation consists of the addition of a methyl group to the adenine or cytosine of a DNA sequence generally leading to the reduction of gene expression in a specific sequence, and it is a fundamental process not only for normal cellular development, but also for carcinogenesis or X-chromosome inactivation in women.[10]

Histone modifications are another epigenetic phenomenon characterized by several changes such as lysine acetylation and methylation, serine and threonine phosphorylation, lysine ubiquitination and sumoylation. These can occur within the histone amino-terminal tails protruding from the surface of the nucleosome and also on the globular core region.[11] These epigenetic events on the histones can affect chromosome function in two ways: a structural change in the histone itself and its link with the chromosomes, or variation in the protein binding sites of the histone.[11]

MiRNAs are 21-nucleotide long, non-coding, single-stranded RNAs able to influence gene expression in a variety of physiological or pathological conditions and have been studied for their epigenetic potential. They normally act by binding to a target messenger RNA (mRNA) to induce its silencing and recent studies have shown that miRNAs are also able to influence the DNA methyltransferases (DNMTs) enzymes that control *de novo* or existing DNA methylation.[12] For example, this is the case of miR-126, miR-21 and miR-148 that are positive or negative regulators of DNMTs, and thus influence the expression of specific genes in SLE.[12]

A recent workshop was held in order to discuss the role of environmental factors on the human autoimmune response, and the major theories that emerged from this workshop were that: (i) innate immunity contributes to the onset of autoimmune diseases via Toll-Like Receptor pathways and autoimmune adaptive response; (ii) environmental factors are able to alter B cell tolerance and induce the production of pathogenic autoantibodies; (iii) Th17 cells are an immunity checkpoint that can be modulated by xenobiotics, allergens and micronutrients; (iv) Treg cells can be influenced in their number or activation status by environmental agents; (v) post-transcriptional modifications can induce the formation of neoantigens and lead to tolerance breakdown; (vi) epigenetic changes such as DNA methylation can cause CD4 + and B cell changes and thus initiate the autoimmune response.[13]

THE RATIONALE FOR GEOEPIGENETICS

As mentioned before, epigenetic mechanisms play a key role in many different diseases, from cancer[14] to inflammation[15] and autoimmunity.[16] As for autoimmune diseases, epidemiology and clinical data vary greatly in different world regions and cohorts studied, and this may be due to different case finding methods, but also to a possible role of environmental factors.[16] In fact, the epigenome is very vulnerable to environmental stimuli from embryogenesis, when DNA synthesis is enhanced and epigenetic changes can accumulate. These variations can be stable during mitosis and they can also affect the germ line, so that they are inheritable. Other epigenetic changes can occur during adult life and aging, involving somatic cells and possibly explaining the phenotypic difference of diseases that can occur in MZ twins.[17,18] Genome-wide association studies (GWAS) were performed in recent years aimed at finding the genes predisposing to diseases such as systemic autoimmune diseases, but they failed in this objective because of gaps between genetic signature and phenotypic expression. This gap may be filled by further and more specific genetic studies or by the role played by epigenetic factors. From these observations, a great interest has now developed in what we may define as "geoepigenetics," a field that aims at studying the influence of non-genetic factors in specific diseases in different areas of the world. Among the non-genetic factors, the "hygiene hypothesis" has been largely developed in recent years, supporting the notion that lack of early childhood exposure to antigenic factors and to gut flora or probiotics can cause defects in immune tolerance and development of the immune system.[19] Also the change from a traditional low-energy to a fat-rich

TABLE 10.1 Epidemiological and Epigenetic Factors (i.e., Geoepigenetics) of Autoimmune Diseases

	Epidemiology		Epigenetic Factors			
	Prevalence	**Incidence**	**Genetic**	**Chemical**	**Physical**	**Biologic**
SLE	40–53/100,000 in Europe and US, 159/100,000 in Afro-Caribbean descent[23,24]	1.8–7.6 cases/ 100,000 per year	X chromosomes,[25] HLA region on chromosome 6[26]	Drugs (procainamide, hydralazine, isoniazide[24]), xenobiotics,[27,28] pollutants[29,30]	UV light[31] (also causing low levels of 25–OH vitamin D[32,33])	Bacterial DNA infections[34,35]
RA	0.5–1% adults in developed countries[36]	41 cases/100,000/ per year	HLA-DR3 and -DR4,[37,38] PTPN22[39]	Tobacco smoke (>10 packs/year)[39,40]	Not known	Periodontitis pathogens,[41] EBV and HHV6 [42,43]
SSc	19–75/100,000; 469/100,000 in Choctaw Native American Indians[44]	20 cases/million per year[44]	Not known; Fetal–maternal microchimerism?[45]	Xenobiotics;[46] tobacco smoke for clinical worsening [46]	Not known	CMV, HBV, EBV and toxoplasmosis[47,48]

Different epigenetic factors may influence the onset of SLE, RA and SSc, leading to a different epidemiologic expression of the disease.

diet seems to play a role in the development of the immune system in different areas of the world,[20,55] and in Western countries this seems to be responsible for an increased incidence of allergic and autoimmune diseases, while the proposed role of helminth parasites is certainly fascinating but awaits solid confirmation also from therapeutic interventions.[21,22]

The aim of the present chapter is to analyze the role of the main epigenetic factors involved in specific autoimmune diseases, along with the epidemiologic epigenetic risk factors described in previous studies for arbitrarily chosen paradigmatic autoimmune diseases. The epidemiological and epigenetic factors (i.e., geoepigenetics) of autoimmune diseases are shown in Table 10.1.

GEOEPIGENETICS OF SYSTEMIC LUPUS ERYTHEMATOSUS (SLE)

SLE is a systemic autoimmune disease characterized by the presence of autoantibodies that react against self-antigens (i.e., dsDNA, chromatin) leading to heterogeneous clinical manifestations such as malar rash, arthritis, glomerulonephritis.[49,50] The rate of SLE varies considerably between countries, ethnicity, gender[23] with prevalence ranging between 53 per 100,000 in the US and 40 per 100,000 in Northern Europe.[24] SLE occurs more frequently and more severely in patients of non-European descent, with a rate three-times higher (159 per 100,000) among patients of Afro-Caribbean descent.[23] Similar to most autoimmune diseases, SLE affects women more frequently than men (9 to 1), and this could be secondary to the presence of two X chromosomes that carry immunological related genes, which can mutate and contribute to the onset of SLE.[25] The Y chromosome has no identified mutations associated with autoimmune disease, and this may explain why males have such a lower risk of developing SLE. An interesting model to study SLE is represented by male patients with Klinefelter syndrome. This syndrome is defined by a 47, XXY karyotype that is clinically characterized by poor development of secondary sexual features and infertility.[51] We know that SLE is more frequent in female than male populations, but recent studies revealed that the risk of SLE among men with Klinefelter's syndrome is equal to the risk of SLE among women[52] and about 14-fold higher than 46, XY men.[53] One possible explanation is that the risk of SLE is related to a gene dose effect for the X chromosome mediated by the abnormal inactivation of genes on the X chromosome or by a gene without an SLE-associated polymorphism. In this way, a gene that is not affected by X inactivation will have higher levels of expression in those patients with Klinefelter's syndrome and two X chromosomes.[54] Accordingly, it is now clear that many genetic and non-genetic predisposing factors are involved in SLE onset. The observation that SLE is not transmitted among families leads us to think that predisposing genes, mainly located in the HLA region on chromosome 6,[26] increase the probability to develop SLE but are not sufficient, and environmental factors are the "second hit" required to trigger the onset of SLE. This opens up the world of "geoepigenetics," because non-genetic factors play a key role in SLE onset with different weight in the ethnic groups analyzed.

Chemical, biological and physical agents are the proposed environmental factors acting on epigenetic mechanisms in SLE, and their importance changes in different areas of the world where they are more or less prevalent, as shown by epidemiology studies.[23] Possible mechanisms through which these environmental factors may lead to the breakdown of immune tolerance, and thus autoimmunity, include polyclonal B cell activation, direct effect impairing the immune response, effects on innate immunity, direct interaction with regulatory cells, modification of self antigens, and alterations of DNA methylation.[13,55]

Chemical agents are widely considered as risk factors for SLE onset, and previous reports identified drugs, xenobiotics and pollutants among the most dangerous chemicals able to induce SLE.[56–72] Drug-induced lupus is a clinical condition that mimics SLE and that can be solved by interrupting the use of drugs responsible for this abnormal reaction, such as procainamide, isoniazid and hydralazin.[24] This is a clear example of what we call "epigenetic influence mediated by a non-genetic factor" because the use of a drug is the environmental factor able to trigger lupus, with a reversible effect.

Xenobiotics are chemical compounds identified in organisms that do not normally produce or require them. In most cases the term "xenobiotic" is referred to pollutants and to their effect on biological systems, where they are considered as antigenic substances. Examples of xenobiotics are mercuric chloride, platinum salts, and silica (nano)particles, but xenobiotics are also antibiotics and many other substances produced by humans that are incorporated in microorganisms and human body even if they are not part of a normal diet.[27] The role of xenobiotics has recently been studied for its link with antigen degradation and presentation, and the consequent antigen-driven autoimmune response.[28]

Previous reports showed an increase of SLE cases in residential areas situated near industrial sites or in places with severe environmental contamination.[29,30,73] In particular, pollution caused by high concentrations of oil field waste, pristane, mercury and phytane seems to be associated with higher risk of developing SLE for the general population.[64] Pristane has already been used in animal models to demonstrate the induction of autoantibodies and clinical features similar to human SLE.[74]

Another hypothesis trying to explain the onset of SLE is the trigger induced by infections. In particular previous reports focused on the specificity of anti-dsDNA antibodies in SLE patients suggested that bacterial DNA components may be the DNA particle recognized by human antibodies and causing lupus manifestations such as glomerulonephritis.[34,35]

The analysis of levels of 25-OH vitamin D (25-OH vD) provided some solid hints in the pathogenesis of SLE. In fact, lupus patients frequently have low 25-OH vD levels that may be due to their reduced exposure to UV light, a very potent physical risk factor for lupus onset, and also due to lupus renal disease and to the use of drugs that interfere with 25-OH vD metabolism like glucocorticoids.[75] The identification of 25-OH vD receptors in cells of the immune system and the production of 25-OH vD by dendritic cells has opened to the possibility that also 25-OH vD may be a factor influencing the epigenetic mechanisms responsible for SLE, as the lowest 25-OH vD levels are usually detected with lupus flares, low complement levels and increase in anti-dsDNA antibody titer.[32] Alterations in 25-OH vD levels may also depend on the area of exposure and they also have seasonal changes, according to some reports reflecting different waves of SLE onset or worsening of clinical cases.[33]

However, not only chemical factors but also physical agents are considered as risk factors for SLE onset, and this is true in particular for UV light exposure. In fact, it is well known that direct exposure to UV light can trigger SLE, not only for the onset of cutaneous lesions but also for the systemic involvement.[31] The reason is that UV exposure may trigger the redistribution of nuclear antigens to the cellular surface or induce the production of new antigens that are recognized by the altered immune system as foreign particles, leading to lupus manifestations.[31]

As previously mentioned, several agents can induce epigenetic alterations and they are considered responsible for changes in DNA methylation and histone modifications, the two major epigenetic changes in SLE. Indeed, human SLE still represents the best example of how DNA methylation can regulate immune cell functions and lead to SLE disease manifestations.[76] This can be seen clearly and reversibly in drug-induced lupus where procainamide and hydralazine cause DNA demethylation and thus autoreactivity and autoantibody overstimulation.[77] From the analysis of these conditions we can conclude that genetic predisposition is not enough for the onset of SLE, because a "second hit" from environmental factors is necessary to influence epigenetics and induce lupus manifestations.

MiRNAs are extensively studied in SLE because they seem to play a role in the disease pathogenesis and in the regulation of inflammatory process mediated by cytokines like interferon type I (IFN-I). Because the expression of miRNAs changes in the different phases of SLE disease activity, many groups aim at identifying miRNA(s) that could be used as biomarkers.[78] For example this is the case of miR-146a that has been extensively studied in SLE patients thanks to its correlation with IFN-I production, disease susceptibility and activity.[79,80]

GEOEPIGENETICS OF RHEUMATOID ARTHRITIS (RA)

RA is a chronic systemic autoimmune disease that predominantly affects the joints and in some cases also internal organs.[81] The etiopathogenesis of this disease is not clear, but autoantibodies like the Rheumatoid Factor (RF) and anti-cyclic citrullinated peptide (CCP) play a key role in the joint damage.[82] In terms of etiology, RA is nowadays considered as a multi-factorial disease where multiple "hits" are necessary to trigger the damage and clinical manifestations that characterize RA.

Numerous studies have tried to identify the main risk factors leading to RA, and some genetic and epigenetic factors have been identified so far. In particular, genetic variations in human leukocyte antigen (HLA) -DR3 and -DR4 are considered able to influence the clinical expression, prognosis and autoantibody production in RA patients.[37,38] In fact, beside genetic predisposition, environmental factors have also been reported previously as triggers for RA, with different weight in the onset or disease progression of RA. This is confirmed by the fact that first-degree relatives have a prevalence rate for RA of 2–3%, and the concordance rate in MZ twins is only 15–20%, so other non-genetic factors must play a key role in RA.[83,84] Heritability is estimated to account for 60% in RA, and only a small number of genetic components have been identified so far.[85] A recent epidemiologic study conducted in Denmark shows that shared and non-shared environmental triggers and/or epigenetic events may be even more important than genetic variations leading to RA, and authors also hypothesize that different risk factors may be associated with specific RA clinical manifestations;[85] nonetheless, the role of dietary or microbiota differences were not addressed in particular.

One of the most important chemical factors associated with RA is tobacco smoke,[40] which seems to be responsible for severe inflammation and seropositive (that is, manifesting positivity for serum rheumatoid factor or anti-CCP antibody) RA. In fact, heavy cigarette smoking seems to influence the expression of susceptibility genes such as PTPN22[39] as shown in a study performed in Caucasian women affected by RA. The influence of smoke on RA is defined as multiplicative by the authors because smoking more than 10 packs of cigarettes/year and the PTPN22 risk allele give together a three-fold higher risk of developing RA.[39] This association was not confirmed for other genes of interest in RA, such as CTLA-4 and PADI-4. A possible hypothesis is that heavy smoking triggers a biological response through the citrullination of proteins mediated by specific enzymes, and this induces the production of anti-CCP antibodies that are involved in bone damage in RA patients.[86]

As for SLE and production of anti-dsDNA antibodies, microbial exposure is also considered an important risk factor for RA. Recently, many reports suggest the role of pathogens responsible for periodontal disease in RA patients.[41] In fact, they show the existence of a link between periodontitis and RA onset or clinical worsening, even if the biochemical processes and clinical link between these two aspects are not clear.[87] Some studies focused on the analysis of miRNAs in maxilla and spleen cells of animal models treated with a microbial complex composed of *Porphyromonas gingivalis, Tannerella forsythia* and *Treponema denticola*, the three main pathogens involved in periodontal disease.[88] MiR-146a seems to be involved in the regulation of the inflammatory response in periodontal disease, and the loss of tolerance against these microbes plus the continuous antigenic stimulation may be a trigger for RA onset and worsening.[88] Both large epidemiological studies[89–91] and small case-control studies have shown an association between periodontal disease and RA, but a uniform evaluation with specific parameters to allow a common conclusion is still lacking.[92] Similarly, in Gram-negative species which possess lipopolysaccharide (LPS) as a major cell wall component, leakage of this inflammatory compound across mucosal surfaces has been suspected to mediate not only RA but also other diseases linked to low-grade chronic inflammation, such as diabetes, metabolic syndrome and dementia. Unresolved chronic inflammation may be a likely candidate for triggering autoimmune diseases where an intricate connection to mucosal microbiota and diet seems to be present. Data supporting this hypothesis are mainly derived from experimental studies, albeit on clinical samples. This is the case, for example, of the representation and different response to LPS by monocytes in seniors and in patients with RA.[93,94] Further, the association between RA and periodontal disease by Gram-negative bacteria[95] with the growing role of IL22 in both RA and the immune response to Gram-negative infections[96] warrant future investigation.

Beside these periodontal pathogens, epidemiological studies have shown the potential association between RA and two herpes viruses: the Epstein–Barr virus (EBV) and Human Herpes Virus 6 (HHV6).[42] In fact, higher levels of antibodies against these two viruses were detected in RA patients, and also *in vitro* studies demonstrated an abnormal response of RA individuals against EBV in particular.[43] However, despite these observations, it is not clear why the widespread infections by EBV and HHV6 would be able to induce RA only in a small subset of people, as the worldwide prevalence of RA is 0.5–1% of adults in the developed countries.[36]

As for SLE, RA genomics can explain only up to 50% of heritability, at least with the present knowledge of the disease, and the most common risk loci are PTPN22, PADI4, CCR6, IL2RA and TNFAIP3.[97] Studies performed in MZ twins demonstrate a largely incomplete concordance rate for RA onset that further supports the multifactorial etiopathogenesis of RA and the role played by non-genetic environmental factors.[98] As for other systemic autoimmune diseases, RA epigenome could be influenced by environmental factors through histone modification and DNA methylation that influence gene expression. Recent reports demonstrated a global hypomethylation in T cells of RA cases and in RA-fibroblast like-synoviocytes compared to healthy controls.[99,100]

In recent years, miRNAs have been studied for their role in mRNA degradation or inhibition of translation. In particular miR-115 and miR-203 are overexpressed in RA fibroblast-like synoviocytes and their increase correlates with increased production of metalloproteinase-6 and IL-6 but it is inversely correlated with the degree of DNA methylation.[101,102] The expression of several other miRNAs is aberrant in RA patients, as for miR-146a and miR-155, and it changes according to cell type, availability of *in vivo* or *in vitro* models and experimental settings with different influence on the inflammatory setting.[103]

GEOEPIGENETICS OF SYSTEMIC SCLEROSIS

Systemic Sclerosis (SSc) is a chronic systemic autoimmune disease characterized by fibrosis of the skin and internal organs.[104] Epidemiology studies estimate an annual incidence of 20 cases per million for SSc, and a prevalence of 19–75 cases per 100,000 people with a female:male ratio of 3:1 that appears to increase with age. Population studies suggest that SSc develops more frequently in the US than in European countries and Asia.[44] As for SLE, SSc also has a higher, and almost double incidence in African Americans and in a group of Indian Americans called "Choctaw Native" in Oklahoma, with the highest prevalence in the world of 469 cases in 100,000 people.[44] All these observations support the importance of genetic factors in some population groups or areas, but as for the other autoimmune diseases like SLE and RA, environmental factors are also required to trigger the onset of SSc.[105–107] In fact, genetic factors are not sufficient to induce SSc as shown in studies on MZ twins where the concordance rate for the onset of SSc was as low as 30%,[108] suggesting the need for other factors to trigger this disease in predisposed patients. This led to the interest in epigenetic studies, also because epigenetic alterations can be found in the skin of SSc patients, where a global DNA hypomethylation status is a frequent feature.[109] Experiments performed on cultured SSc fibroblasts show that these cells maintain a profibrotic status through their growth and also when transferred out of the SSc disease environment, showing the presence of a profibrotic cellular phenotype influenced by epigenetic factors.[110]

Xenobiotics are proposed to influence the DNA methylation status of SSc fibroblast, and the main suspected elements are silica dust, organic solvents, heavy metals and some drugs. Both animal and *in vitro* models have shown that these xenobiotic agents can alter the structure and function of antigens in the cell nucleus, leading to an abnormal immune response.[27] Other environmental pollutants considered in the last years as potential risk factors for SSc are vinyl chloride, toxic oil, tryptophan, gadolinium, bleomycin, and pentazocine.[46] Smoking was studied in SSc patients after the observation that it has a strong influence on RA worsening, as mentioned before, but no role was demonstrated in SSc, with only a possible role in clinical worsening.[111] A meta-analysis on silica exposure showed an increased relative risk for men to develop SSc, but this was not significant in women.[112]

Beside chemical risk factors, biological agents also may be responsible for SSc onset. In particular, previous reports analyzed the role played by Citomegalovirus (CMV), a single-stranded DNA virus, as the original epitope of the immune response in SSc, but no clear conclusions have been drawn so far.[47] In fact, it is not clear why such a widespread viral infection by CMV is able to induce SSc only in a small subset of the population. Similar studies were also performed on the analysis of Hepatis B virus, EBV and toxoplasmosis in SSc patients, but no clear conclusion has been obtained so far.[48]

Another interesting mechanism that seems to be behind the autoimmune reaction in SSc is called "microchimerism" which consists of the fact that during pregnancy fetal cells circulating in the maternal blood may trigger an immune reaction to what is considered as foreign and antigenic material.[45,47] This field of investigation derives from the observation that SSc can become clinically manifest just after pregnancy, so fetal components may alter the immune response and induce SSc onset.

The role played by epigenetics in SSc is certainly less studied and understood compared to other autoimmune diseases such as RA and SLE, for many reasons such as the rarity of this disease and the difficulty of creating good *in vitro* and *in vivo* models. However, the lowest twin concordance rate among all autoimmune disease was identified for SSc, suggesting again that genetic predisposition is largely insufficient for disease onset and that environmental factors are crucial for the disease manifestations.[108] Our group recently studied DNA methylation in MZ twins concordant and discordant for SSc, showing that the most significant differences were in the methylation status of genes located on the X chromosome.[113] This could mean that X chromosome genes may contribute to SSc susceptibility independently from other genetic variants, and this could also explain the reason for female predisposition to SSc. Previous groups identified other epigenetic phenomena in samples of SSc patients, such as DNA hypomethylation in the CD4+ T cells of SSc patients compared to healthy controls.[109]

Only in recent years has research focused on the role played by miRNAs in the onset and disease expression of SSc. In particular the up- or downregulation of miRNAs such as miR-29, miR-7, miR-150 contributes to the deposition of collagen in SSc skin and lungs, leading to skin and pulmonary fibrosis which characterize SSc.[114–116]

CONCLUSIONS

We herein described our case to propose "geoepigenetics" as a new term and a new branch of investigation focused on the analysis of how epigenetic and epidemiologic factors can interact to trigger an autoimmune disease. In fact, it is now clear that paradigmatic systemic autoimmune diseases like SLE, RA and SSc are the result of a combination of genetic and non-genetic risk factors. Some predisposing genes have been identified so far, but studies conducted on MZ twins show that the concordance rates for autoimmune diseases is quite low, so genetic predisposition is only one of the factors involved in the onset of these diseases. Many environmental factors (chemical, biological, physical) can influence epigenetic mechanisms and thus gene expression, triggering the autoimmune disease. A very recent field of investigation in epigenetics is also represented by miRNAs, that can silence gene expression by influencing their target mRNA(s), and they may represent the near future of geoepigenetics studies.

This chapter has discussed the importance of environmental factors in the pathobiology of three autoimmune diseases through epigenetic mechanisms. SLE, RA and SSc have not been studied largely from the aspect of gut microbiota or dietary involvement but these two factors may be related to them as with other diseases with autoimmune background and more intricately related to the gastrointestinal environment, such as inflammatory bowel diseases or diabetes. In fact, it may well be that aberrant microbiota profiles mediated by lack of appropriate environmental exposures at the right time in life, or by radical shifts in diet, contribute significantly to the environmental factors thought to drive autoimmune disease onset and progression. Understanding how gut microbiota:diet interactions influence autoimmune processes will shed new light, not only on how these diseases arise allowing population-based preventative steps, but also on how gut microbiota is readily modified by diet, and this could open the intriguing possibility of efficacious therapy for these devastating diseases.

References

1. Chervonsky AV, et al. *Cold Spring Harb Perspect Biol.* 2013;5(3):a007294.
2. Manzel A, et al. *Curr Allergy Asthma Rep.* 2014;14(1):404.
3. Ballestar E. Epigenetics lessons from twins: prospects for autoimmune disease. *Clin Rev Allergy Immunol.* 2010;39(1):30–41.
4. Olszak T, et al. Microbial exposure during early life has persistent effects on natural killer T cell function. *Science.* 2012;336(6080):489–493.
5. Markle JG, et al. Sex differences in the gut microbiome drive hormone-dependent regulation of autoimmunity. *Science.* 2013;339 (6123):1084–1088.
6. Bird A. Perceptions of epigenetics. *Nature.* 2007;447(7143):396–398.
7. Chang PV, et al. The microbial metabolite butyrate regulates intestinal macrophage function via histone deacetylase inhibition. *Proc Natl Acad Sci USA.* 2014;111(6):2247–2252
8. Dalmasso G, et al. Microbiota modulate host gene expression via microRNAs. *PLoS One.* 2011;6(4):e19293.
9. Rossi M, Amaretti A, Raimondi S. Folate production by probiotic bacteria. *Nutrients.* 2011;3(1):118–134.
10. Jaenisch R, Bird A. Epigenetic regulation of gene expression: how the genome integrates intrinsic and environmental signals. *Nat Genet.* 2003;33(Suppl):245–254.
11. Cosgrove MS, Boeke JD, Wolberger C. Regulated nucleosome mobility and the histone code. *Nat Struct Mol Biol.* 2004;11(11):1037–1043.
12. Ceribelli A, Yao B, Dominguez-Gutierrez PR, Chan EK. Lupus T cells switched on by DNA hypomethylation via microRNA? *Arthritis Rheum.* 2011;63(5):1177–1181.
13. Selmi C, Leung PS, Sherr DH, et al. Mechanisms of environmental influence on human autoimmunity: a National Institute of Environmental Health Sciences expert panel workshop. *J Autoimmun.* 2012;39(4):272–284.
14. Hatzimichael E, Crook T. Cancer epigenetics: new therapies and new challenges. *J Drug Deliv.* 2013;2013:529312.
15. Oppermann U. Why is epigenetics important in understanding the pathogenesis of inflammatory musculoskeletal diseases? *Arthritis Res Ther.* 2013;15(2):209.
16. De Santis M, Selmi C. The therapeutic potential of epigenetics in autoimmune diseases. *Clin Rev Allergy Immunol.* 2012;42(1):92–101.

17. Faulk C, Dolinoy DC. Timing is everything: the when and how of environmentally induced changes in the epigenome of animals. *Epigenetics*. 2011;6(7):791–797.
18. Skinner MK. Environmental epigenomics and disease susceptibility. *EMBO Rep*. 2011;12(7):620–622.
19. Garn H, Renz H. Epidemiological and immunological evidence for the hygiene hypothesis. *Immunobiology*. 2007;212(6):441–452.
20. Salonen A, de Vos WM. Impact of diet on human intestinal microbiota and health. *Annu Rev Food Sci Technol*. 2014;5:239–262.
21. McSorley HJ, Maizels RM. Helminth infections and host immune regulation. *Clin Microbiol Rev*. 2012;25(4):585–608.
22. Maizels RM, Yazdanbakhsh M. Immune regulation by helminth parasites: cellular and molecular mechanisms. *Nat Rev Immunol*. 2003; 3(9):733–744.
23. Danchenko N, Satia JA, Anthony MS. Epidemiology of systemic lupus erythematosus: a comparison of worldwide disease burden. *Lupus*. 2006;15(5):308–318.
24. Rahman A, Isenberg DA. Systemic lupus erythematosus. *N Engl J Med*. 2008;358(9):929–939.
25. Invernizzi P, Miozzo M, Oertelt-Prigione S, et al. X monosomy in female systemic lupus erythematosus. *Ann N Y Acad Sci*. 2007; (1110):84–91.
26. Morris DL, Taylor KE, Fernando MM, et al. Unraveling multiple MHC gene associations with systemic lupus erythematosus: model choice indicates a role for HLA alleles and non-HLA genes in Europeans. *Am J Hum Genet*. 2012;91(5):778–793.
27. Chen M, von Mikecz A. Xenobiotic-induced recruitment of autoantigens to nuclear proteasomes suggests a role for altered antigen processing in scleroderma. *Ann N Y Acad Sci*. 2005;1051:382–389.
28. von Mikecz A. Xenobiotic-induced autoimmunity and protein aggregation diseases share a common subnuclear pathology. *Autoimmun Rev*. 2005;4(4):214–218.
29. Kardestuncer T, Frumkin H. Systemic lupus erythematosus in relation to environmental pollution: an investigation in an African-American community in North Georgia. *Arch Environ Health*. 1997;52(2):85–90.
30. Balluz L, Philen R, Ortega L, et al. Investigation of systemic lupus erythematosus in Nogales, Arizona. *Am J Epidemiol*. 2001;154 (11):1029–1036.
31. Cooper GS, Wither J, Bernatsky S, et al. Occupational and environmental exposures and risk of systemic lupus erythematosus: silica, sunlight, solvents. *Rheumatology (Oxford)*. 2010;49(11):2172–2180.
32. Abou-Raya A, Abou-Raya S, Helmii M. The effect of vitamin D supplementation on inflammatory and hemostatic markers and disease activity in patients with systemic lupus erythematosus: a randomized placebo-controlled trial. *J Rheumatol*. 2013;40(3):265–272.
33. Birmingham DJ, Hebert LA, Song H, et al. Evidence that abnormally large seasonal declines in vitamin D status may trigger SLE flare in non-African Americans. *Lupus*. 2012;21(8):855–864.
34. Hamilton KJ, Schett G, Reich 3rd CF, Smolen JS, Pisetsky DS. The binding of sera of patients with SLE to bacterial and mammalian DNA. *Clin Immunol*. 2006;118(2–3):209–218.
35. Chowdhry IA, Kowal C, Hardin J, Zhou Z, Diamond B. Autoantibodies that bind glomeruli: cross-reactivity with bacterial antigen. *Arthritis Rheum*. 2005;52(8):2403–2410.
36. Scott DL, Wolfe F, Huizinga TW. Rheumatoid arthritis. *Lancet*. 2010;376(9746):1094–1108.
37. Van Jaarsveld CH, Otten HG, Jacobs JW, Kruize AA, Brus HL, Bijlsma JW. Association of HLA-DR with susceptibility to and clinical expression of rheumatoid arthritis: re-evaluation by means of genomic tissue typing. *Br J Rheumatol*. 1998;37(4):411–416.
38. Zhou Y, Tan L, Que Q, et al. Study of association between hla-dr4 and dr53 and autoantibody detection in rheumatoid arthritis. *J Immunoassay Immunochem*. 2013;34(2):126–133.
39. Costenbader KH, Chang SC, De Vivo I, Plenge R, Karlson EW. Genetic polymorphisms in PTPN22, PADI-4, and CTLA-4 and risk for rheumatoid arthritis in two longitudinal cohort studies: evidence of gene–environment interactions with heavy cigarette smoking. *Arthritis Res Ther*. 2008;10(3):R52.
40. Arnson Y, Shoenfeld Y, Amital H. Effects of tobacco smoke on immunity, inflammation and autoimmunity. *J Autoimmun*. 2010;34(3): J258–J265.
41. Demmer RT, Molitor JA, Jacobs Jr. DR, Michalowicz BS. Periodontal disease, tooth loss and incident rheumatoid arthritis: results from the First National Health and Nutrition Examination Survey and its epidemiological follow-up study. *J Clin Periodontol*. 2011;38(11):998–1006.
42. Alvarez-Lafuente R, Fernandez-Gutierrez B. Potential relationship between herpes viruses and rheumatoid arthritis: analysis with quantitative real time polymerase chain reaction. *Ann Rheum Dis*. 2005;64(9):1357–1359.
43. Balandraud N, Roudier J, Roudier C. Epstein–Barr virus and rheumatoid arthritis. *Autoimmun Rev*. 2004;3(5):362–367.
44. Mayes MD. Scleroderma epidemiology. *Rheum Dis Clin North Am*. 2003;29(2):239–254.
45. Bianchi DW. Fetomaternal cell trafficking: a new cause of disease? *Am J Med Genet*. 2000;91(1):22–28.
46. Barnes J, Mayes MD. Epidemiology of systemic sclerosis: incidence, prevalence, survival, risk factors, malignancy, and environmental triggers. *Curr Opin Rheumatol*. 2012;24(2):165–170.
47. Jimenez SA, Derk CT. Following the molecular pathways toward an understanding of the pathogenesis of systemic sclerosis. *Ann Intern Med*. 2004;140(1):37–50.
48. Arnson Y, Amital H, Guiducci S, et al. The role of infections in the immunopathogensis of systemic sclerosis—evidence from serological studies. *Ann NY Acad Sci*. 2009;1173:627–632.
49. Hochberg MC. Updating the American College of Rheumatology revised criteria for the classification of systemic lupus erythematosus. *Arthritis Rheum*. 1997;40(9):1725.
50. Tan EM, Cohen AS, Fries JF, et al. The 1982 revised criteria for the classification of systemic lupus erythematosus. *Arthritis Rheum*. 1982;25 (11):1271–1277.
51. Lanfranco F, Kamischke A, Zitzmann M, Nieschlag E. Klinefelter's syndrome. *Lancet*. 2004;364(9430):273–283.
52. Dillon S, Aggarwal R, Harding JW, et al. Klinefelter's syndrome (47, XXY) among men with systemic lupus erythematosus. *Acta Paediatr*. 2011;100(6):819–823.
53. Scofield RH, Bruner GR, Namjou B, et al. Klinefelter's syndrome (47, XXY) in male systemic lupus erythematosus patients: support for the notion of a gene–dose effect from the X chromosome. *Arthritis Rheum*. 2008;58(8):2511–2517.

54. Sawalha AH, Harley JB, Scofield RH. Autoimmunity and Klinefelter's syndrome: when men have two X chromosomes. *J Autoimmun.* 2009;33(1):31–34.
55. Selmi C, Tsuneyama K. Nutrition, geoepidemiology, and autoimmunity. *Autoimmun Rev.* 2010;9(5):A267–A270.
56. Richardson B, et al. Effect of an inhibitor of DNA methylation on T cells. I. 5-Azacytidine induces T4 expression on T8+ T cells. *J Immunol.* 1986;137(1):35–39.
57. Lu Q, Kaplan M, Ray D, Zacharek S, Gutsch D, Richardson B. Demethylation of ITGAL (CD11a) regulatory sequences in systemic lupus erythematosus. *Arthritis Rheum.* 2002;46(5):1282–1291.
58. Zhao M, Sun Y, Gao F, et al. Epigenetics and SLE: RFX1 downregulation causes CD11a and CD70 overexpression by altering epigenetic modifications in lupus CD4 + T cells. *J Autoimmun.* 2010;35(1):58–69.
59. Zhao S, Long H, Lu Q. Epigenetic perspectives in systemic lupus erythematosus: pathogenesis, biomarkers, and therapeutic potentials. *Clin Rev Allergy Immunol.* 2010;39(1):3–9.
60. Gorelik G, Fang JY, Wu A, Sawalha AH, Richardson B. Impaired T cell protein kinase C delta activation decreases ERK pathway signaling in idiopathic and hydralazine-induced lupus. *J Immunol.* 2007;179(8):5553–5563.
61. Lu Q. The critical importance of epigenetics in autoimmunity. *J Autoimmun.* 2013;41:1–5.
62. Biggioggero M, Meroni PL. The geoepidemiology of the antiphospholipid antibody syndrome. *Autoimmun Rev.* 2010;9(5):A299–A304.
63. Borchers AT, Naguwa SM, Shoenfeld Y, Gershwin ME. The geoepidemiology of systemic lupus erythematosus. *Autoimmun Rev.* 2010;9(5): A277–A287.
64. Dahlgren J, Takhar H, Anderson-Mahoney P, Kotlerman J, Tarr J, Warshaw R. Cluster of systemic lupus erythematosus (SLE) associated with an oil field waste site: a cross sectional study. *Environ Health.* 2007;6:8.
65. Park SK, et al. Traffic-related particles are associated with elevated homocysteine: the VA normative aging study. *Am J Respir Crit Care Med.* 2008;178(3):283–289.
66. Rai K, Huggins IJ, James SR, Karpf AR, Jones DA, Cairns. BR. DNA demethylation in zebrafish involves the coupling of a deaminase, a glycosylase, and gadd45. *Cell.* 2008;135(7):1201–1212.
67. Cornelius LA, Sepp N, Li LJ, et al. Selective upregulation of intercellular adhesion molecule (ICAM-1) by ultraviolet B in human dermal microvascular endothelial cells. *J Invest Dermatol.* 1994;103(1):23–28.
68. Li Y, Zhao M, Yin H, et al. Overexpression of the growth arrest and DNA damage-induced 45alpha gene contributes to autoimmunity by promoting DNA demethylation in lupus T cells. *Arthritis Rheum.* 2010;62(5):1438–1447.
69. Salvador JM, Hollander MC, Nguyen AT, et al. Mice lacking the p53-effector gene Gadd45a develop a lupus-like syndrome. *Immunity.* 2002;16(4):499–508.
70. Grolleau-Julius A, Ray D, Yung RL. The role of epigenetics in aging and autoimmunity. *Clin Rev Allergy Immunol.* 2010;39(1):42–50.
71. Huck S, Deveaud E, Namane A, Zouali M. Abnormal DNA methylation and deoxycytosine-deoxyguanine content in nucleosomes from lymphocytes undergoing apoptosis. *FASEB J.* 1999;13(11):1415–1422.
72. Huck S, Zouali M. DNA methylation: a potential pathway to abnormal autoreactive lupus B cells. *Clin Immunol Immunopathol.* 1996;80(1):1–8.
73. Fessel WJ. Systemic lupus erythematosus in the community. Incidence, prevalence, outcome, and first symptoms; the high prevalence in black women. *Arch Intern Med.* 1974;134(6):1027–1035.
74. Satoh M, Richards HB, Shaheen VM, et al. Widespread susceptibility among inbred mouse strains to the induction of lupus autoantibodies by pristane. *Clin Exp Immunol.* 2000;121(2):399–405.
75. Cutolo M, Otsa K. Review: vitamin D, immunity and lupus. *Lupus.* 2008;17(1):6–10.
76. Richardson B. Primer: epigenetics of autoimmunity. *Nat Clin Pract Rheumatol.* 2007;3(9):521–527.
77. Ballestar E, Esteller M, Richardson BC. The epigenetic face of systemic lupus erythematosus. *J Immunol.* 2006;176(12):7143–7147.
78. Miao CG, Yang YY, He X, et al. The emerging role of microRNAs in the pathogenesis of systemic lupus erythematosus. *Cell Signal.* 2013.
79. Lofgren SE, Frostegard J, Truedsson L, et al. Genetic association of miRNA-146a with systemic lupus erythematosus in Europeans through decreased expression of the gene. *Genes Immun.* 2012;13(3):268–274.
80. Luo X, Yang W, Ye DQ, et al. A functional variant in microRNA-146a promoter modulates its expression and confers disease risk for systemic lupus erythematosus. *PLoS Genet.* 2011;7(6):e1002128.
81. Klareskog L, Catrina AI, Paget S. Rheumatoid arthritis. *Lancet.* 2009;373(9664):659–672.
82. Firestein GS. Evolving concepts of rheumatoid arthritis. *Nature.* 2003;423(6937):356–361.
83. Silman AJ, MacGregor AJ, Thomson W, et al. Twin concordance rates for rheumatoid arthritis: results from a nationwide study. *Br J Rheumatol.* 1993;32(10):903–907.
84. Bellamy N, Duffy D, Martin N, Mathews J. Rheumatoid arthritis in twins: a study of aetiopathogenesis based on the Australian twin registry. *Ann Rheum Dis.* 1992;51(5):588–593.
85. Svendsen AJ, Kyvik KO, Houen G, et al. On the origin of rheumatoid arthritis: the impact of environment and genes—a population based twin study. *PLoS One.* 2013;8(2):e57304.
86. Klareskog L, Stolt P, Lundberg K, et al. A new model for an etiology of rheumatoid arthritis: smoking may trigger HLA-DR (shared epitope)-restricted immune reactions to autoantigens modified by citrullination. *Arthritis Rheum.* 2006;54(1):38–46.
87. Kaur S, White S, Bartold PM. Periodontal disease and rheumatoid arthritis: a systematic review. *J Dent Res.* 2013;92(5):399–408.
88. Nahid MA, Rivera M, Lucas A, Chan EK, Kesavalu. L. Polymicrobial infection with periodontal pathogens specifically enhances microRNA miR-146a in ApoE −/− mice during experimental periodontal disease. *Infect Immun.* 2011;79(4):1597–1605.
89. Berthelot JM, Le Goff B. Rheumatoid arthritis and periodontal disease. *Joint Bone Spine.* 2010;77(6):537–541.
90. Han JY, Reynolds MA. Effect of anti-rheumatic agents on periodontal parameters and biomarkers of inflammation: a systematic review and meta-analysis. *J Periodontal Implant Sci.* 2012;42(1):3–12.
91. Rutger Persson G. Rheumatoid arthritis and periodontitis — inflammatory and infectious connections. Review of the literature. *J Oral Microbiol.* 2012;4.
92. Bingham 3rd CO, Moni M. Periodontal disease and rheumatoid arthritis: the evidence accumulates for complex pathobiologic interactions. *Curr Opin Rheumatol.* 2013;25(3):345–353.

93. Krasselt M, et al. CD56+ monocytes have a dysregulated cytokine response to lipopolysaccharide and accumulate in rheumatoid arthritis and immunosenescence. *Arthritis Res Ther*. 2013;15(5):R139.
94. Liou LB. Different monocyte reaction patterns in newly diagnosed, untreated rheumatoid arthritis and lupus patients probably confer disparate C-reactive protein levels. *Clin Exp Rheumatol*. 2003;21(4):437–444.
95. Monsarrat P, et al. Effect of periodontal treatment on the clinical parameters of patients with rheumatoid arthritis: study protocol of the randomized, controlled ESPERA trial. *Trials*. 2013;14:253.
96. Pan HF, et al. Emerging role of interleukin-22 in autoimmune diseases. *Cytokine Growth Factor Rev*. 2013;24(1):51–57.
97. Viatte S, Plant D, Raychaudhuri S. Genetics and epigenetics of rheumatoid arthritis. *Nat Rev Rheumatol*. 2013;9(3):141–153.
98. Bogdanos DP, Smyk DS, Rigopoulou EI, et al. Twin studies in autoimmune disease: genetics, gender and environment. *J Autoimmun*. 2012;38(2–3):J156–J169.
99. Richardson B, Scheinbart L, Strahler J, Gross L, Hanash S, Johnson M. Evidence for impaired T cell DNA methylation in systemic lupus erythematosus and rheumatoid arthritis. *Arthritis Rheum*. 1990;33(11):1665–1673.
100. Karouzakis E, Gay RE, Michel BA, Gay S, Neidhart. M. DNA hypomethylation in rheumatoid arthritis synovial fibroblasts. *Arthritis Rheum*. 2009;60(12):3613–3622.
101. Stanczyk J, Pedrioli DM, Brentano F, et al. Altered expression of MicroRNA in synovial fibroblasts and synovial tissue in rheumatoid arthritis. *Arthritis Rheum*. 2008;58(4):1001–1009.
102. Stanczyk J, Ospelt C, Karouzakis E, et al. Altered expression of microRNA-203 in rheumatoid arthritis synovial fibroblasts and its role in fibroblast activation. *Arthritis Rheum*. 2011;63(2):373–381.
103. Ceribelli A, Nahid MA, Satoh M, Chan. EK. MicroRNAs in rheumatoid arthritis. *FEBS Lett*. 2011;585(23):3667–3674.
104. Hachulla E, Launay D. Diagnosis and classification of systemic sclerosis. *Clin Rev Allergy Immunol*. 2011;40(2):78–83.
105. Takagi K, Kawaguchi Y, Kawamoto M, et al. Activation of the activin A-ALK-Smad pathway in systemic sclerosis. *J Autoimmun*. 2011;36 (3–4):181–188.
106. Arora-Singh RK, Assassi S, del Junco DJ, et al. Autoimmune diseases and autoantibodies in the first degree relatives of patients with systemic sclerosis. *J Autoimmun*. 2010;35(1):52–57.
107. Gourh P, Agarwal SK, Martin E, et al. Association of the C8orf13-BLK region with systemic sclerosis in North-American and European populations. *J Autoimmun*. 2010;34(2):155–162.
108 Feghali-Bostwick C, Medsger Jr. TA, Wright TM. Analysis of systemic sclerosis in twins reveals low concordance for disease and high concordance for the presence of antinuclear antibodies. *Arthritis Rheum*. 2003;48(7):1956–1963.
109. Lei W, Luo Y, Lei W, et al. Abnormal DNA methylation in CD4+ T cells from patients with systemic lupus erythematosus, systemic sclerosis, and dermatomyositis. *Scand J Rheumatol*. 2009;38(5):369–374.
110. Maxwell DB, Grotendorst CA, Grotendorst GR, LeRoy EC. Fibroblast heterogeneity in scleroderma: Clq studies. *J Rheumatol*. 1987;14 (4):756–759.
111. Chaudhary P, Chen X, Assassi S, et al. Cigarette smoking is not a risk factor for systemic sclerosis. *Arthritis Rheum*. 2011;63 (10):3098–3102.
112. McCormic ZD, Khuder SS, Aryal BK, Ames AL, Khuder SA. Occupational silica exposure as a risk factor for scleroderma: a meta-analysis. *Int Arch Occup Environ Health*. 2010;83(7):763–769.
113. Selmi C, Feghali-Bostwick CA, Lleo A, et al. X chromosome gene methylation in peripheral lymphocytes from monozygotic twins discordant for scleroderma. *Clin Exp Immunol*. 2012;169(3):253–262.
114. Etoh M, Jinnin M, Makino K, et al. MicroRNA-7 down-regulation mediates excessive collagen expression in localized scleroderma. *Arch Dermatol Res*. 2013;305(1):9–15.
115. Peng WJ, Tao JH, Mei B, et al. MicroRNA-29: a potential therapeutic target for systemic sclerosis. *Expert Opin Ther Targets*. 2012;16 (9):875–879.
116. Honda N, Jinnin M, Kira-Etoh T, et al. miR-150 down-regulation contributes to the constitutive type I collagen overexpression in scleroderma dermal fibroblasts via the induction of integrin beta3. *Am J Pathol*. 2013;182(1):206–216.

CHAPTER

11

Obesity-Associated Gut Microbiota: Characterization and Dietary Modulation

Qing Shen and Vatsala Maitin

School of Family and Consumer Sciences, Nutrition and Foods Program, Texas State University, San Marcos, Texas, USA

THE OBESITY PANDEMIC

Obesity is defined as excessive fat accumulation in the body, often causing adverse effects on health. It is considered a risk factor for several chronic diseases including cardiovascular disease (CVD), hypertension, type 2 diabetes (T2D), non-alcoholic fatty liver disease (NAFLD), and colon cancer.[1] According to the World Health Organization (WHO), obesity is classified as having a Body Mass Index (BMI) over 30 kg/m^2 while a BMI of 25–30 kg/m^2 is considered overweight.[2] The prevalence of obesity has sky-rocketed worldwide in the past two or three decades, with increasing obese populations not only in developed countries but also in developing countries. This has been linked to changes in diet and lifestyle, as urban communities in developing countries gain greater access to "Western-style" energy-dense foods, along with amenities that lead to reduced physical activity. In the United States, more than one-third of adults and almost 17% of youth were categorized as obese in 2009–2010. Adults aged 60 and over are more likely to be obese than younger adults indicating a long-term exacerbation of body fat accumulation due to an unhealthy diet and a sedentary lifestyle.[3] In China, from 1993 to 2009, the prevalence of obesity increased from 2.9 to 11.4% in men and from 5.0 to 10.1% in women,[4] with similar trends reported in India.[5] The cost of obesity is societal as well as personal. In the United States, obesity accounted for 5.5% of total medical expenditures, or about $39 billion in 1986[6] but 21% or about $190 billion in 2005.[7] If this trend continues, researchers have estimated that by 2030, obesity-related medical costs could rise by $48–$66 billion a year in the USA.[8]

GENETIC DETERMINANTS OF OBESITY

The most common cause of obesity is an energy imbalance, with energy intake exceeding energy expenditure. When this happens, adipocytes (fat cells) tend to store the surplus energy in the form of triglycerides (TG) and in the long term, the adipose tissue becomes hypertrophic and hyperplasic leading to excess fat accumulation.

Genetic variations may also predispose an individual to obesity. To date, variations in 11 genes have been identified as monogenic causes of obesity; including genes for leptin, leptin receptor, melanocortin 3 receptor and melanocortin 4 receptor (*MC4R*).[9] The most well-known obesity-related genes are those for leptin (*ob*) and its receptor (*db*). Zhang et al. (1994)[10] at Rockefeller University first discovered mice with a mutation in the ob gene (*ob/ob*) and lack of ob gene product led to severe obesity with increased energy intake (hyperphagia) but decreased energy expenditure (reduced metabolic rate, thermogenesis and physical activity). The gene product was later characterized as a circulating factor and named leptin derived from the Greek word leptos meaning thin.[11] Leptin is primarily secreted by adipocytes, circulates throughout the body and crosses the blood–brain

Diet-Microbe Interactions in the Gut.
DOI: http://dx.doi.org/10.1016/B978-0-12-407825-3.00011-3

barrier to convey a satiety signal and thus reduce food intake. Basically, the greater the number of adipocytes, the higher the circulating leptin concentrations, and the smaller the amount of food that the brain signals should be consumed.[11] In humans, genetic deficiency of leptin is very rare.[12] In most cases of obesity, the leptin signaling pathway seems to be blunted despite a high level of circulating leptin (corresponding to the large number of adipocytes), a state commonly termed "leptin resistance."[13] The discovery of the leptin receptor indicated that one possible mechanism for leptin resistance is receptor dysfunction, such as that caused by a loss of function mutation in the receptor gene.[14] However, a genetic deficiency of leptin receptor is not common in obese humans either. Therefore, several alternative mechanisms have been proposed for "leptin resistance," including self-regulation (i.e., leptin may regulate its own receptor), limited tissue access through the blood–brain barrier, and cellular and circulating molecules that may inhibit the action of leptin.[13] Despite the differences with humans, genetically modified experimental animals with leptin receptor deficiency such as diabetic mice (*db/db*) and Zucker fatty rats (*fa/fa*) are useful models of obesity-associated metabolic diseases.[15]

The exploration of obesity-related genes remains an extensive research area and the latest updates on human obesity gene map reports 127 candidate genes, many of which function cooperatively, indicating a polygenic etiology of obesity development.[16] The most well-studied obesity genes are: *MC4R*, a deficiency of which accounts for the most common form of monogenic obesity with prevalence rates ranging from 0.5 to 5.8%; *PPAR*γ, with a role in polygenic obesity and interaction with nutrients and the gut microbiota; and *FTO*, another player in polygenic obesity and most strongly linked to obesity susceptibility.[9] While an individual's genetic make-up is a contributing factor, both the extent and the rate of increase in obesity worldwide in genetically very distinct populations strongly discount genetics as the primary determinant. Environmental factors are therefore more likely to be the root cause of the modern obesity pandemic, an observation that offers hope as unlike genes, environmental factors may be modulated to reduce disease risk. Aside from diet, considered a key environmental factor underlying obesity, the composition of an individual's gut microbiota has also recently been identified as a possible contributor to the obesogenic environment.

OBESITY ASSOCIATED GUT MICROBIOTA

The human microbiota hosts up to 10^{14} microbial cells, with the colon being the most densely populated area. There are ~1500 microbial species overall with ~160 species present per individual.[17] This co-evolved microbiota plays a mutualistic role in host–microbe interactions by regulating the host immune system, metabolizing nutrients not digested and absorbed in the upper gut (i.e., fermentation of dietary fiber and plant polyphenols), promoting epithelial barrier function while also supporting its own growth and thriving in high numbers.[18]

Characterizing the microbiota composition has long been a research focus toward understanding host–microbe interactions. The methods of investigation have evolved over time, from culture-based techniques to culture-independent approaches such 16 S rRNA gene quantification by fluorescence *in situ* hybridization (FISH) or quantitative PCR (qPCR); and presently have been revolutionized by next-generation deep sequencing/pyrosequencing (NGS). FISH is considered the most quantitative method and enumerates bacterial cells *in situ* without DNA extraction and/or PCR; it is limited however by the choice of probes and is also time-consuming. Compared to FISH, qPCR is a little less accurate and associated with more experimental variability, but it is faster. Next-generation deep sequencing is semi-quantitative but potentially all inclusive in terms of coverage. However, qPCR as well as NGS are prone to bias related to DNA extraction (Gram-negative cells are more easily ruptured than Gram-positive cells) and/or the PCR (GC-rich sequence amplification may be compromised during competitive PCR by universal primers).

These culture-independent molecular microbiology techniques have shown that the gut microbiota of obese individuals differs from that of lean individuals, and that the obese microbiota can be modified through weight loss, diet, drugs or surgery. Although most studies confirm a relationship between body weight and gut microbiota, there is little agreement on the bacterial species or even groups that constitute an "obese" or "lean" microbiota. This lack of consensus means that a causative or protective role in obesity can not yet be confidently assigned to specific microorganisms. In the following sections, we present data from both animal and human studies examining the microbiology of obesity and how it may be modulated using probiotics, prebiotics and synbiotics.

Gut Microbiota and Obesity: Evidence from Mice

Differences in Bacterial Composition at a Phylum Level

The main feature found in obese microbiota from experimental mice is a higher abundance of Firmicutes than Bacteroidetes.[19,20] By using Sanger sequencing 16 S rRNA gene surveys, Ley et al. (2005)[19] reported that genetically obese (*ob*/*ob*) mice (leptin deficient) possessed 50% fewer Bacteroidetes and a concomitantly higher proportion of Firmicutes compared to their lean siblings. This higher ratio of Firmicutes:Bacteroidetes in *ob*/*ob* mice as well as conventionally raised mice fed a Western diet was later confirmed by Turnbaugh *et al.* using both Sanger sequencing and NGS platform.[20,21] These very important seminal papers demonstrating a link between gut microbiota and obesity set the stage for further research into elucidating the compositional differences in gut microbiota between lean and obese states.

Differences in Bacterial Composition at a Genus/Group Level

Genetically obese mouse models or high-fat-diet (HFD) -induced obese mice are commonly used for investigating diet–obesity interactions and the role of microbiota. A metagenomics approach using NGS platform revealed that in mice that were subjected to high-fat-diet-induced obesity (DIO), a single clade of Firmicutes, i.e., Mollicutes, was enriched relative to lean, chow-fed mice.[21] The Mollicutes bloom was reduced by restriction of dietary fats or carbohydrates but could be transferred to lean germ-free (GF) mice through microbial transplantion from mice with DIO.[21] The Mollicutes sequences retrieved from this study were re-classified as *Erysipelotrichaceae*.[22] Another research group using the same mouse model found that the *Erysipelotrichaceae* family contained four sub-groups, each responding differently to diet or host health status rather than uniformly.[23] A study conducted in a genetically obese rat model Zucker *fa*/*fa* (a leptin receptor deficiency that mimics human obesity caused by leptin resistance) showed that total bacteria, *Bifidobacterium* and *Atopobium* were significantly lower in obese *fa*/*fa* rats but *Eubacterium rectale*/*Clostridium coccoides* group and *Lactobacillus*/*Entercoccus* levels were significantly higher, by using FISH for bacterial enumeration.[24]

Cani et al. (2007)[25] reported a decrease of *Bifidobacterium*, *E. rectale*/*C. coccoides* group and *Bacteroides*-like mouse intestinal bacteria in HFD-fed mice by FISH quantification. Recently, using qPCR in the same HFD mice, they found a decrease in *Bacteroides*/*Prevotella*, *E. rectale*/*C. coccoides*, *Lactobacillus* and *Roseburia*, but an increase in *Bifidobacterium*.[26] However, most studies regarding the impact of high-fat diets on the gut microbiota report a lowering of bifidobacteria using FISH or qPCR. Murphy et al. (2010)[27] analyzed the fecal microbiota of HFD-fed, ob/ob and lean control mice over 8 weeks (at 7, 11 and 15 weeks of age) using NGS, and reported a progressive increase of Firmicutes in both HFD and *ob*/*ob* mice along with a decrease in Bacteroidetes in both groups of mice as well as lean controls, with the decrease being significant in the ob/ob group. Interestingly, when the data was scrutinized at the genus level, bifidobacteria increased significantly from 7 weeks to 11 weeks in *ob*/*ob* mice but the HFD group had lower bifidobacteria content at 11 weeks compared to the lean controls. The same research group later reported a decrease in the genus *Bacteroides*, but increase in *Lactobacillus* in the diet-induced obese mice after 20 weeks on the same HFD as earlier.[28] Serino *et al.* (2011)[29] characterized the gut microbiota composition in mice that developed diabetes after a HFD (HFD-D group) or were resistant to diabetes after a HFD (HFD-DR) by using both q-PCR and NGS. A higher Firmicutes to Bacteroidetes ratio was found in HFD-D compared to HFD-DR. However, *Lachnospiraceae*, *Oscillibacter* and *Ruminococcaceae*, all belonging to Firmicutes, were reduced in HFD-D but S24-7 from Bacteroidetes tripled in number, further highlighting the fluctuation within each phylum at a lower taxonomic level such as family or genus. The variability was partly related to the specifics of the diet formulation, animal model used and the method of microbiota composition analysis.

Gut Microbiota and Obesity: Evidence from Human Studies

Difference in Bacterial Composition at Phylum Level

Ley et al. (2006),[30] in a weight-loss study with 12 obese subjects, reported a high Firmicutes to Bacteroidetes ratio in obese volunteers compared to lean controls. After weight loss on either a fat-restricted or carbohydrate-restricted diet, the microbiota of obese subjects started to resemble their lean counterparts, with an increase in Bacteroidetes and a decrease in Firmicutes. This change was only division-wide and no significant difference was detected at lower taxonomic levels and the bacterial diversity remained stable. The authors also indicated that an increase in Bacteroidetes was positively correlated with of the percentage loss of body weight, but only in subjects who had lost at least 6% of their body weight on a fat-restricted diet and 2% on a carbohydrate-restricted diet. Turnbaugh et al. (2009)[31] characterized the gut microbiota composition of 31 monozygotic twins, 23 dizygotic twins and their mothers (n = 46) by both Sanger sequencing and NGS. Full-length 16 S rRNA gene PCR products were used for Sanger sequencing while PCR amplicons from V2 and V6 regions as well as shot-gun DNA fragments were subject to NGS. A low Bacteroidetes to Actinobacteria ratio but no difference in Firmicutes was detected in the obese microbiota when sequences from different methods were analyzed in combination. The bacterial diversity was reduced in the obese microbiota, which was not observed in Ley's study.[30] This may be due to the larger sample size and the deep sequencing ability of NGS. The extent of microbial diversity and gene-richness has been proposed to have implications for metabolic disease risk in overweight or obese individuals.[32]

Difference in Bacterial Composition at Genus/Groups Level

The notion of a distinct obesity-associated microbiota becomes more complicated when viewed at lower taxonomic levels. Duncan et al. (2008),[33] using FISH, demonstrated that the abundance of *Bacteroides* as detected by probe Bac303 (covering many gram-negative human gut bacteria that belong to the Bacteroidetes phylum), did not differ significantly between the obese microbiota and the lean microbiota, and no significant correlation was detected between BMI and the proportions of *Bacteroides*. Moreover, after 8 weeks on reduced-carbohydrate diets, most individuals showed a decreased percentage of *Bacteroides*. A potentially beneficial bacterial group belonging to Firmicutes, i.e., *Roseburia* and *Eubacterium rectale* also decreased significantly. Their previous findings also showed a significant decrement in *Bifidobacterium* after 4 weeks on a carbohydrate-restricted diet indicating a high responsiveness of these bacterial groups to fermentable carbohydrates.[34] Another research group, using qPCR, reported significantly lower *C. leptum*, *Ruminococcus flavefaciens*, *Bifidobacterium* and *Methanobrevibacter* in obese individuals while the ratio of Firmicutes:Bacteroidetes was actually significantly lower in obese subjects compared to lean.[35] Interestingly, Raoult's group reported more *Lactobacillus* in obese individuals especially *Lactobacillus reuteri*, while *Bifidobacterium animalis* and *Methanobrevibacter smithii* were negatively correlated with BMI.[36–38] It should be noted that the analyzed cohort included anorexic patients in which gut hormone levels and diet could be very different compared to healthy subjects, thus impacting their gut microbiota composition. A recent study reported an opportunistic pathogen *Enterobacter cloacae* B29 from a morbidly obese volunteer as the root cause of obesity as determined in GF mice.[39] However, the results showed that only the high fat fed + B29 mice developed obesity but normal chow fed + B29 mice remained lean, indicating diet as the overbearing factor in the development of obesity.

A recent large-scale screening study in 345 T2D Chinese subjects reported that *Bacteroides ceccae*, *C. hathewayi*, *C. ramosum*, *C. symbiosum*, *Eggerthella lenta*, *Escherichia coli*, *Akkermansia muciniphila* and *Desulfovibro* sp. were enriched in T2D patients while the healthy controls harbored mainly butyrate producers from clostridial cluster XIVa and IV including *E. rectale*, *Roseburia intestinalis* and *F. prausnitzii*. The lower abundance of these butyrate-producers and increased opportunistic pathogens in T2D microbiota was the major characteristic of T2D related gut dysbiosis.[40] This is in agreement with what we know about the role of fiber fermentation in obesity and T2D. Fiber has been consistently shown in population-based epidemiological studies to be inversely related to obesity and to be protective against the diseases of obesity such as T2D, cardiovascular disease and certain cancers. As well as having an impact on satiety in the upper gut, fiber is fermented into short-chain fatty acids (SCFA) acetate, butyrate and propionate, which have been shown to mediate many physiological processes linked to controlling body weight, glucose and fat metabolism, gut permeability and inflammation. Therefore, a lean microbiota dominated by fiber-degrading, butyrate-producing bacteria agrees with evidence gathered over the past 50 years from population-based nutritional studies.

In summary, the variability reported in the obesity-associated gut microbiota in the representative studies listed above suggests a need for detailed characterization at a lower and "functional" taxonomic level rather than at the level of the phylum as in earlier studies, to potentially serve as a better target for modulation.

Archaea Methanogens

Besides bacteria, human gut microbiota also contains Archaea, *Methanobrevibacter smithii*, which removes the end-product H_2 from bacterial fermentation, thus facilitating the fermentation rate and colonic energy production in the form of SCFA. It has been reported that *M. smithii* enhanced the degradation of dietary fructans to acetate by *Bacteroides thetaiotaomicron*, thereby promoting energy harvest and weight gain in germ-free (GF) mice.[41] Turnbaugh et al. (2006)[20] also demonstrated that the cecal microbiome of obese (*ob/ob*) mice hosted more gene tags that matched Archaea than the microbiomes of lean (*ob*/ + and + / +) littermates implying a higher population of the archaeon or its gene content in the obese microbiome. However, human studies revealed a different picture. Although Zhang et al. (2009)[42] detected higher number of Archaea 16 S rRNA gene copies in the obese human microbiome compared to normal-weight or post-gastric-bypass individuals using qPCR, Armougom et al. (2009),[36] also via qPCR, showed a high concentration of *M. smithii* in anorexic patients with a very low BMI of 12.73 on average. Although the obese group had a slightly higher concentration of *M. smithii* in this study, this conclusion was revised in another report by the same group to state that there was in fact a decrease in *M. smithii* load in the obese group, when the data was analyzed differently.[43] In this recent study with a larger samples size of 68 obese subjects and 47 controls, they demonstrated a depletion of *M. smithii* in the obese group. Similar results were obtained by Schwiertz et al. (2010)[35] using qPCR with a decrease in *Methanobrevibacter* in overweight (but metabolically healthy) persons.

With respect to evidence presented from mouse studies, most of which employ GF/gnotobiotic mice, one must consider that the functioning of digestive processes in GF animals is likely much different from their conventional counterparts. Every animal in nature with an organized intestine, with the exception of gnotobiotic animals, employs microorganisms to break down complex polysaccharides through fermentation. Studies associating germ-free animals with one or a few strains of microorganisms are very useful at elucidating microbe–microbe interactions within the gut but less useful at modeling mammalian–microbiota interactions, since the germ-free state creates an abnormal intestinal environment both in terms of microbial activities and mucosal development and immune function. Colonization of germ-free adult animals with exogenous microorganisms or even conventionalization with whole microbiota bypasses the natural succession-based development of the gut microbiota, the mucosal architectural development and ontology, immune-education and establishment of co-metabolic processes between the mammalian host and microbiota. We now know that these processes are programmed through a combination of host genetic factors and environmental processes, such as diet, breastfeeding and weaning, and by the sequence of microbial encounters in early life.[44]

Cause or Consequence

Some fundamental differences in obese and lean gut microbiota raise the question of whether the altered gut microbiota is a cause or the consequence of obesity. Although many studies point to a causal role of gut microbiota by demonstrating that GF mice develop obesity if inoculated with obese microbiota while remaining lean if inoculated with lean microbiota,[20,21,45,46] one must keep in mind that the obese microbiota in the donor mice was induced by either HFD or knock-out of certain obesity-related genes. Also, as mentioned above, GF animals mature in terms of their gut physiology, immune-homeostasis and inter-kingdom metabolic pathways in a manner discrete from conventional, microbiota-replete animals. Their response to sudden colonization or their ability to compete for dietary nutrients may, therefore, also be different from that of conventionally reared animals.

Some studies do provide evidence to suggest that unlike diet, which is an independent mediator of obesity, gut microbiota's role in the process is diet-dependent rather than in a stand-alone capacity. Fleissner et al. (2010)[22] demonstrated this fact through an interesting set of observations from GF and conventionalized (CV) mice fed three different isocaloric diets differing in fat and carbohydrate composition. No difference was observed between the two groups of mice on LFD (carbohydrate–protein–fat ratio: 41:42:17). Response to a HFD (carbohydrate–protein–fat ratio: 41:16:43) differed based on the diet's formulation. HFD1 containing coconut oil promoted more weight gain in GF mice while HFD2 formulated with beef tallow and vegetable shortening promoted more weight gain in CV mice. The differences in the saturated and trans-fat content in the two HFDs, along with higher sucrose in HFD2 were presumably behind the contradictory responses. Hildebrandt et al. (2009)[47] demonstrated that a HFD induced similar phyla fluctuations in wild-type mice that became obese on HFD-feeding, as well in resistin-like molecule beta (RELM)-KO mice that remained lean in response to a HFD, reiterating that the compositional changes were in response to diet and not body weight. Even the seminal paper on weight loss-induced phylum-wide shifts in gut microbiota in human subjects suggested a role of diet in shaping the gut microbiota composition, as the weight-loss was achieved by either a fat-restricted diet or carbohydrate-restricted diet.[30] It is likely then, that both the diet and the gut microbiota co-determine an obese or lean status and that any alterations in gut microbiota composition could be both a cause and consequence of obesity (Figure 11.1).

To further clarify the causal role of gut microbiota in obesity, we need to answer two questions: first, does the microbiota influence obesogenic pathways directly or through its components/metabolic products; and second, are there specific host–microbe signaling mechanisms that lead to obesity?

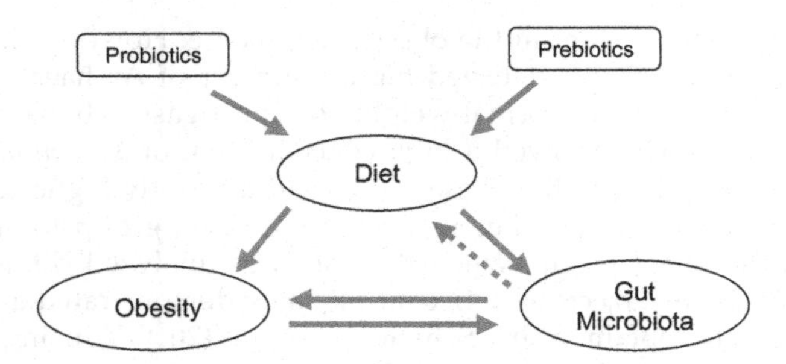

FIGURE 11.1 A schematic summary of the relationship between diet, obesity and gut microbiota.

The first question has been well addressed by Cani et al. (2007)[25] who demonstrated that lipopolysaccharide (LPS) from Gram-negative bacteria can translocate from the gut lumen to the blood stream causing "metabolic endotoxemia" leading to a low-grade systemic inflammation and obesity, a state that may also be triggered in response to HFD feeding (discussed in the section Interactions between Gut Microbes and Obesity: "The Inflammation Theory").

The second question on host–microbe molecular signaling mechanisms has attracted much research interest, and been addressed by using knock out (KO) mice lacking certain signaling receptors such as toll-like receptors 2 (TLR2), CD14/TLR4, TLR9, NOD-like receptors (NOD1/NLRC1), NOD2/NLRC2 as well as inflammasome proteins NLRP3, NLRP6 (discussed in the section Interactions between Gut Microbes and Obesity: "The Inflammation Theory"). However, the gene knockout protocols by themselves may impact gut homeostasis and gut microbiota composition. The deficiency of a certain receptor may also have a physiological side effect, impact gut hormone production, gut–brain axis signaling and some KO mice may become hyperphagic.[48] Thereby the altered microbiota composition could be due to the alterations in gut homeostasis.

One way to rule out "genetically induced," or "HFD-induced" factors when trying to delineate the role of gut microbiota in obesity development is to carry out a fecal transplantation from "naturally occurring" obese mice to healthy, lean, "conventionally reared" recipients. A similar approach was used in a recent human study where fecal microbiota transplantation from lean donors to obese patients with metabolic syndrome improved insulin sensitivity and increased the abundance of butyrate-producing bacteria.[49]

In short, it is hard to conclude whether the gut microbiota is playing an active causative role in the development of obesity, or whether the obesity-associated profile of microbiota is simply in response to an obese state or the underlying diet. However, considering that the gut microbiota has been implicated in the onset of obesity in several animal studies, is modified by obesity and obesity-inducing diets, and appears to mediate some mechanisms of diet-related obesity and associated conditions, there is reasonable evidence to suggest its importance in metabolic health. The "chicken and egg" question therefore seems inconsequential compared to the significance of elucidating the role of the gut microbiota in host nutritional metabolism.

INTERACTIONS BETWEEN GUT MICROBES AND OBESITY: "THE ENERGY EXTRACTION THEORY"

Gut Microbiota and Dietary "Energy-Harvest"

A role for the gut microbiota in enhanced caloric extraction from the diet or "energy-harvest" was proposed based on observations that conventionally raised (CONV-R) C57BL/6 J male mice had 42% more total body fat than their GF counterparts despite the fact that the CONV-R animals actually consumed 29% less food.[45] Similarly, a 14-day period of "conventionalization" of GF mice with CONV-R cecal microbiota also led to a 57% increase in their total body fat and a 7% decrease in lean body mass despite reduced food intake. The effect was even more pronounced in female mice, with an 85% increase in total body fat and a 9% decrease in lean body mass. Conventionalization of GF mice also led to a decrease in insulin sensitivity. Interestingly, the host metabolic rate was higher in conventionalized mice than their GF counterparts indicating that the leanness of GF mice was not owing to higher energy expenditure. An exploration into the processes via which gut microbiota may be influencing dietary energy-harvest and storage revealed two key mechanisms: breakdown of indigestible dietary plant polysaccharides by bacterial glycoside hydrolases; and suppression of the intestinal gene expression of a lipoprotein lipase (LPL) inhibitor called fasting-induced adipose factor (FIAF) or Angiopoietin-like 4 (ANGPTL4). Bacterial polysaccharide hydrolysis generated lipogenic substrates, viz. monosaccharides and SCFA. Gut microbiota also promoted efficient monosaccharide uptake by enhancing the expression of the sodium/glucose transporter-1 (SGLT1)[50] along with doubling the density of capillaries in the small intestine.[51] Arrival of these lipogenic substrates in the liver then promoted hepatic triglyceride synthesis. In line with studies on GF vs. conventionalized mice, a study comparing the functional meta-genomes of *ob/ob* mice and their lean littermates (+ / + and *ob*/ +) found that the obese microbiome was enriched in eight glycoside hydrolase families capable of hydrolyzing dietary polysaccharides.[20] The second mechanism involving microbial suppression of FIAF results in increased fat accumulation via effects on LPL activity. LPL plays a role in fatty acid uptake into adipocytes by hydrolyzing triglycerides from circulating lipoproteins. The fatty acids can then be stored as triglycerides in adipocytes by re-esterification. Suppression of FIAF, an LPL-inhibitor in response to conventionalization of germ-free mice would thus translate to an increase in LPL activity and resulting adipocytic fat-storage.

Backhed et al. (2007)[46] further demonstrated that GF mice fed a high-fat/high-sugar Western diet were protected from obesity or insulin resistance compared to their conventional counterparts implicating the gut microbiota in the development of diet-induced obesity. Here, they proposed another mechanism preventing diet-induced obesity in GF mice, the AMP-activated protein kinase (AMPK) pathway. AMPK is a fuel gauge monitoring cellular energy levels in the liver, the brain and skeletal muscles. Activation of AMPK in response to metabolic stress (exercise, hypoxia, glucose deprivation) promotes catabolic pathways such as fatty acid oxidation and inhibits anabolic pathways such as glycogen synthesis, thereby resulting in an increased intracellular ratio of AMP to ATP.[52] The author further demonstrated that GF mice resistant to diet-induced obesity had 40% higher phospho-AMPK (active) concentration, 43% higher phospho-ACC (acetylCoA carboxylase) concentration and 17% more CPT1 (carnitine palmitoyltransferase-1) concentration in the gastrocnemius muscle as compared to conventional mice. Similar increases in phospho-AMPK were also observed in the livers of GF mice. Thereby they concluded that the resistance of GF mice to diet-induced obesity was at least in part due to increased AMPK activity and AMPK-driven fatty acid oxidation in peripheral tissues.[46] However, the mechanism of the suppressive effect of gut microbiota on AMPK activity remains elusive. Furthermore, butyrate, which is considered a beneficial SCFA produced by gut microbiota fermentation, actually upregulates AMPK activity *in vitro*[53] and *in vivo*.[54]

In contrast to the above study, where absence of gut microbiota protected against Western diet-induced obesity, such protection was not afforded against HFD-induced obesity in another germ-free mouse model.[22] Although conducted in a different mouse model (C3H) rather than the C57BL/6 J mice above, the authors demonstrated that GF mice in this case gained *more* body fat than the conventional mice on a high-fat diet.[22] However, when fed a Western diet exactly as defined by Backhed et al. (2007)[46] the same C3H GF mice resisted diet-induced obesity.[22] The authors thus suggested that the absence of gut microbiota does not protect against diet-induced obesity in every instance, and that more investigation is needed into the specific dietary components affecting the gut microbiota and the host. They also questioned the regulatory role of FIAF in fat storage, as the plasma levels of FIAF in their study were the same between GF and conventional mice, and also remained unaltered among various diet groups. Although the intestinal *Fiaf* mRNA levels were higher in GF than conventional mice, FIAF protein was undetectable in the intestinal mucosa. Based on these results, the authors suggest that FIAF's role as an LPL inhibitor in peripheral tissues may be relatively limited.

With respect to phylum-level compositional changes, Murphy et al. (2010)[27] showed that changes in the proportion of Firmicutes, Bacteroidetes and Actinobacteria in the gut of HFD-fed and obese (*ob/ob*) mice were not associated with alterations in energy-harvesting capacity of the gut microbiota (as assessed by fecal energy content measured using bomb calorimetry, and fecal SCFA amounts). The results suggest that the ratio of Firmicutes to Bacteroidetes is not a definite indicator of the energy extraction capacity of the gut microbiota. As in other studies with GF mice, the model itself may be partly responsible for observed discrepancies as GF-mice are artificial experimental models that do not exist in nature. Conventionalization of GF-mice can thus be expected to produce dramatic effects on physiological processes, as the mice cope with the rapid appearance of microbial metabolites, energy substrates, antigens and signaling molecules.[44] This may be quite disparate from the real-life situation, where our co-evolved microbiota collaborates and co-operates at many levels with mammalian physiology, metabolism and immune system, in a complementary fashion.[18]

The "energy extraction theory" was recently also put to the test in human subjects.[55] Twelve lean and nine obese adult white males were recruited and assigned to either a 2400- or 3400-kcal/day diet for 3 days in a randomized crossover manner, after an initial 3-day period on a weight-maintenance diet. Interestingly, no significant differences in the three dominant bacterial phyla (Firmicutes, Bacteroidetes and Actinobacteria) were found between lean and obese individuals on the initial weight-maintenance diet, in contrast to previous studies. In response to an altered calorie load, a 20% increase in gut Firmicutes and a 20% decrease in Bacteroidetes were associated with ~150 kcal increase in energy absorption by the host. The lean subjects seemed to be more sensitive to an increased caloric intake, exhibiting a significant reduction in stool energy loss and phylum-level changes in microbiota composition on a 2400- vs. 3400-kcal/day diet. Similar changes were not seen with obese subjects.[55] The authors suggest that the extent of divergence in energy intake compared to a weight-maintaining diet may influence the efficiency of energy absorption from the diet.

Role of SCFA and their Receptors in Dietary "Energy-Harvest"

SCFA produced by bacterial fermentation of dietary carbohydrates are the primary form of energy absorbed in the colon, accounting for 5–10% of human energy needs.[56] Absorption of SCFA in the colon is a very efficient process, with less than 5% being excreted in the feces.[57] Acetate, propionate and butyrate, are the three major SCFA in the colon, present in a molar ratio of about 57:22:21;[58] with branched chain fatty acids mainly from protein fermentation present in lower amounts.[59,60] Recent studies have shown that SCFA are absorbed by both inactive and active mechanisms via transporters like monocarboxylic acid transporter 1(MCT1). Gene expression of MCT1 has been shown to be upregulated by butyrate in the intestine. The activity of this transporter protein is induced by elevated SCFA concentrations in the lumen in a rapid nutrient sensing mechanism wherein MCT1 presents itself on the luminal surface of epithelial cells when SCFA concentrations increase.[61,62] Therefore, greater the amount of SCFA produced by the gut microbiota in the colon, the more SCFA are absorbed from the gut, possibly lowering lumenal concentrations. Activity of MCT1 is inhibited by the primary bile salt chenodeoxycholic acid,[63] which may explain why SCFA concentrations in the feces of obese individuals or animals on high-fat diets may be higher than lean individuals or those on high-fiber diets.[64] However, as in the case of microbiota composition, reports relating fecal SCFA, diet and body weight are also conflicting. While higher than average SCFA concentrations were reported in obese compared to lean individuals on a Western diet;[65] a study comparing the effect of diet of rural African and urban European children on gut microbiota reported higher fecal SCFA concentrations in the feces of African children on traditional high-fiber diets compared to Italian children on low-fiber, Western-style diets.[66] SCFA concentrations in feces therefore, are not necessarily indicative of colonic fermentation patterns of the gut microbiota[65,66] except when measured in an isolated environment with no host: microbe co-utilization such as fermentation vessels or using dynamic methods to measure SCFA production and absorption following test meals, e.g., using stable isotope dilution in human postprandial blood.

Colonic SCFA may contribute to host adiposity by serving as energy substrates or as signaling molecules. SCFA receptors in the colon are present on enteroendocrine L cells as G protein-coupled receptors GPR41 (later renamed FFA3) and GPR43 (later renamed FFA2).[67–69] The potency order of the major SCFA for GPR41 activation is propionate > butyrate > acetate, while GPR43 is equally sensitive to each. In terms of specificity, acetate is more selective for GPR43 and propionate is a more effective ligand for GPR41.[70,71] Besides the colon, GPR41 and GPR43 can be found on various cell types including adipocytes.[72]

Studies on GPR41 and GPR43 have yielded interesting yet complex results with respect to modulating host adiposity, with therapeutic potential. Samuel et al. (2008)[73] demonstrated that GPR41 was required for developing microbiota-induced obesity. Conventionally raised *Gpr41 −/−* mice as well as gnotobiotic *Gpr41 −/−* mice cocolonized with *B. thetaiotaomicron and M. smithii* were found to be leaner than the wild-type (+ / +) mice on the same standard chow diet. The leaner phenotype of *B. thetaiotaomicron/M. smithii*-colonized *Gpr41 −/−* mice was related to faster intestinal transit, due to reduced expression of PYY, a hormone inhibitor of transit rate. This was associated with a greater SCFA excretion and reduced host absorption of SCFA, suggesting a reduced capacity for "energy-harvest." GPR43 has also been reported to promote energy storage by increasing adipogenesis[74,75] and inhibiting adipocyte lipolysis.[76] An increase in *Gpr43* gene expression in adipose tissues has been reported in response to HFD-feeding in mice.[74] *Gpr43*-deficient mice were protected from high-fat-diet-induced obesity by an increase in energy expenditure through increased body temperature and by reduced macrophage infiltration in white adipose tissue.[77]

While both GPR41 and GPR43 have been implicated in promoting energy-storage, paradoxically, SCFA generated from oligofructans have been shown to mitigate body fat accumulation and the severity of diabetes in mice[74] (discussed later). Remarkably opposite results to the above studies with respect to GPR43 were recently also reported by Kimura et al.,[78] wherein overexpression rather than deletion of *GPR43* conferred protection against HFD-induced obesity. The study specifically explored the role of GPR 43 in the adipose tissue and found that SCFA mediated activation of adipose-specific GPR43 promoted leanness by suppressing insulin-mediated fat deposition in the adipose, promoting energy expenditure and preferential utilization of lipids for energy needs.[78] GPR41/43 have also been shown to play a role in modulating circulating leptin-levels, which has implications in regulating food consumption. Increased *Gpr41* expression in the adipose tissue in mice was shown to stimulate leptin secretion[70,72] in response to the SCFA propionate. However, some studies failed to detect *Gpr41* in mouse adipose tissue, instead reporting a high expression of *Gpr43*, and implied that SCFA can promote leptin secretion from the adipose not necessarily via GPR41[75,79] but perhaps also via GPR43. Differential regulation of *Gpr41* has also been reported in the nervous system, where it is expressed at a high level in sympathetic ganglia and promotes sympathetic nervous system activity when activated by the SCFA propionate but suppresses sympathetic nervous system activity when activated by ketone bodies (representative of starvation or diabetes).[79] The results point to varying signaling pathways and functions involving GPR41/43 in different cell types such as enteroendocrine cells, adipose cells and neurogenic cells, in line with the divergent physiological roles performed by the GPCR family.

INTERACTIONS BETWEEN GUT MICROBES AND OBESITY: "THE APPETITE CONTROL THEORY"

Overeating is a major driver of obesity and results from an imbalance in the processes that regulate food intake, including "hunger," "appetite" and "satiety," along with environmental cues. Normal human eating behavior is episodic, stopping when we feel full (satiation) and starting again after a period of time when we feel hungry. The period between the two meals is an index of satiety.[80] There are a number of gut hormones that convey a "full" or "hungry" signal to the brain. Cholecystokinin (CCK) secreted by enteroendorcine I cells and glucagon-like peptide 1 (GLP-1) secreted by enteroendorcrine L cells are satiation-related gut hormones and convey a "full" signal. Peptide YY (PYY) secreted by enteroendorcrine L cells and glucose-dependent insulinotropic polypeptide (GIP) secreted by enteroendorcrine K cells also convey a sense of "fullness," while ghrelin secreted by gastric cells conveys the "hungry" signal that determines the initiation of a meal.[80] Therefore, dietary or pharmacological approaches able to increase the levels of GLP-1 and PYY, and decrease those of ghrelin, can be useful means of increasing satiety and reducing energy intake.

Recent studies have highlighted that the gut microbiota and its fermentation products, SCFA, can impact gut hormone levels thereby regulating satiety. The same enteroendorcrine L cells that secrete GLP-1and PYY also house the SCFA receptors GPR41 and GPR43, implying an association between colonic SCFA production and appetite or food intake.[81]

One way to study the correlation between SCFA and gut hormone production is direct infusion of SCFA into animal or human intestines. In rats, intra-colonic infusions of propionate and butyrate alone, as well as a mixture of three SCFA (acetate, propionate and butyrate) increased PYY secretion in a dose-dependent manner. Acetate alone, however, did not have a significant impact.[82] In humans, acute increases of GLP-1 and PYY were found in hyperinsulinemic patients after rectal or intravenous infusion of acetate.[83] Intravenous infusions of acetate stimulated GLP-1 to a greater extent than rectal infusions while the reverse was observed for PYY.[83]

An indirect way to influence the colonic SCFA profiles is via increased digestion of dietary fiber, especially the prebiotics, dietary fibers which selectively support the growth of beneficial bacteria and SCFA production in the gut (discussed in the section Gut Microbiota as a Therapeutic Target of Probiotics, Prebiotics and Synbiotics). Rats fed a high-fiber diet had higher levels of plasma GLP-1 and PYY, lower plasma GIP, a five-fold increase in colonic PYY mRNA levels and an eleven-fold increase in proglucagon mRNA levels compared to rats fed a control diet or a high-protein diet.[84] Another mouse feeding study, with type 2 resistant starch (RS), showed lowered plasma GIP concentrations but increases in GLP-1 and PYY, reducing weight-gain partly due to reduced energy intake.[85] Rats fed galactooligosaccharides (GOS) also showed increased plasma levels of PYY but not GLP-1, increased colonic gene expression of PYY and proglucagon,[86] reduced energy intake and fat pad weight. Inulin-type fructans have been shown to increase plasma levels of GLP-1 and colonic gene expression of proglucagon,[87,88] protecting rats from HFD-induced obesity.[89] A human study showed that a 21-g/day intake of fructooligosaccharides (FOS) reduced ghrelin and increased PYY in overweight adults[90] associated with a reduced caloric intake, supporting weight loss. Another clinical study examining the effects of a high wheat-fiber cereal reported increased GLP-1 in hyperinsulinemic patients after 1 year of intake and increased plasma concentrations of acetate and butyrate,[91] but without alterations in body weight. Favorable effects of inulin were also seen when administered to healthy human subjects as inulin-rich high-fructose corn syrup (HFCS) compared to HFCS without inulin, with increased postprandial serum acetate, propionate, and butyrate, plasma GLP-1 and reduced ghrelin.[92]

To further elucidate microbial impact on gut satiety hormone levels, a few studies have characterized the composition of the associated microbiota. No changes in the cecal microbiota were detected in pectin-supplemented rats using culture-based viable counting techniques, although cecal acetate and total SCFA levels increased. Pectin feeding also increased plasma GLP-2 concentrations, but did not affect the GLP-2R mRNA levels in the small intestine.[93] In a type 2 diabetes model (Goto–Kakizaki rats), feeding type 2 RS increased cecal butyrate producing bacteria as quantified by qPCR, along with cecal SCFA and GLP-1 levels.[94] A role for gut microbiota alterations was demonstrated in modulating glycemia in HFD-induced diabetic mice. HFD notably modified the cecal bacterial composition in these mice, with enrichment reported in *Parabacteroides jonsonii*, *Alistipes putredinis* and *Bacteroides vulgatus*, as determined by DGGE. The cecal microbiota was restored to "normal" upon administration of resveratrol, a natural polyphenol in grapes and red wine with myriad health benefits including anti-diabetic properties. Resveratrol feeding also increased portal and intestinal GLP-1 and colonic mRNA levels of proglucagon (GLP-1 precursor),[95] improving glucose tolerance. The role of fungal chitin-glucan (CG) was explored in modulating glucose and lipid metabolism in HFD-fed obese mice. CG treatment helped restore the bacterial numbers from clostridial cluster XIVa, including *Roseburia*, a major butyrate-producer, which were decreased due to HFD feeding. This was correlated with decreases in weight gain, fasting hyperglycemia, glucose intolerance, hepatic triglyceride accumulation and hypercholesterolemia, independent of GLP-1 levels as portal GLP-1 and gut proglucagon mRNA levels remained unchanged.[26] Another study looking at the role of age-dependent microbiota changes in health found that feeding type 2 RS from high-amylose maize to aged mice increased the numbers of *Bacteroidetes* and *Bifidobacterium*, *Akkermansia* and *Allobaculum* species, detected by deep sequencing and q-PCR. The increments in *Bifidobacterium* and *Akkermansia* species were positively correlated with cecal proglucagon mRNA levels, indicating their potential to favorably impact health by GLP-1 production.[96] While GPR41 is a key receptor mediating SCFA effects on energy-storage, and also implicated in regulating energy intake,[97] a recent study reported that butyrate and propionate are able to protect against HFD-induced obesity and stimulate GLP and GIP via a GPR41-independent mechanism, suggesting that alternate mechanisms may be involved.[98] Overall, the variety of animal models employed, including GF, normal and conventionalized mice highlight the importance of microbial fermentation in fine tuning mammalian energy metabolism. Microbial metabolites may serve as potent mediators of appetite regulation, diet-related obesity and diabetes.

Another factor that regulates food intake and obesity is the interaction between gut microbiota and the innate immune system. A well-studied receptor in this context is Toll like-receptor 5 (TLR5). TLR5 is highly expressed in the mouse gut mucosa, recognizes bacterial flagellin as a pathogen-associated molecular pattern (PAMP) and has been implicated in intestinal inflammation and related changes in gut microbiota. One recent study showed that TLR5-knockout mice (T5KO) became hyperphagic and developed obesity, metabolic syndrome and related hyperlipidemia, hypertension and insulin resistance.[48] The hyperphagic/obese phenotype of T5KO mice was correlated with changes in the gut microbiota composition at the bacterial species level, with 116 bacterial phylotypes from various phyla enriched or reduced in T5KO mice compared to WT mice. Transplantation of the T5KO gut microbiota to GF WT mice induced metabolic syndrome in GF WT mice, pointing to a causative role of T5KO gut microbiota.[48] In summary, the gut microbiota plays an important role in regulating appetite, obesity and diabetes through SCFA signaling or interactions with the innate immune system. Dietary components such as fermentable fiber and prebiotics, and microbial metabolites generated in the gut may thus serve as simple yet potent agents to combat these diseases.

INTERACTIONS BETWEEN GUT MICROBES AND OBESITY: "THE INFLAMMATION THEORY"

Obesity is associated with a state of chronic, low-grade systemic inflammation. The adipocytes themselves, as well as the infiltrated macrophages in the adipose tissue in an obese state, express inflammatory cytokines.[99,100] This low-grade systemic inflammation is caused, at least in part, by translocation of bacterial lipopolysaccharide

(LPS) from the intestinal lumen to the circulation. An elevation of two- to three-times in plasma LPS concentrations, termed "Metabolic endotoxemia"[25] has been implicated in the pathology of several chronic diseases including obesity, insulin resistance, diabetes and atherosclerosis. Intestinal LPS, when crossing the gut wall, are preferentially carried by chylomicrons, lipoproteins particles that transport dietary lipids. A HFD may thus induce or exacerbate metabolic endotoxemia. LPS is detected and internalized by endocytosis involving TLR4, assisted by the pattern recognition receptor CD14. CD14 has been identified as an essential trigger for LPS-mediated metabolic endotoxemia as CD14 − / − mice were resistant to the adverse metabolic states induced by LPS or HFD.[25] Gut permeability may also contribute to LPS translocation, with increased permeability driven by local inflammation at the intestinal wall or alterations in gut microbial ecology responsible for maintaining mucosal architecture. Several strains of *Bifidobacterium* and *Lactobacillus*, and their metabolic end products, including SCFA, have been shown to induce expression of the tight junction proteins responsible for maintaining effective intestinal barrier activity.[101] Intestinal permeability, on the other hand, has been associated with low relative abundance of *Bifidobacterium* in the intestine. Similar to the "metabolic endotoxemia" promoted by bacterial cell fragments, HFD can cause inflammation and metabolic disorders by promoting translocation of whole bacterial cells across the intestinal wall to blood and tissues such as adipose, termed "metabolic bacteremia." The process was mediated via CD14 and Nod1 receptors and the lack of either one protected mice from HFD-induced obesity and diabetes.[102]

A key phenomenon causing elevated adipose tissue inflammation has been proposed to be macrophage infiltration. The pro-inflammatory cytokines produced in the adipose tissue are largely from infiltrated macrophages. Macrophage accumulation was shown to be directly proportional to adiposity in an animal study with less than 10% adipose macrophage content in lean mice and over 50% in obese mice.[103] Macrophages themselves exhibit phenotypic versatility, and can polarize to either pro-inflammatory "M1" or anti-inflammatory "M2" phenotype. An obese state alters the ratio of M1 and M2 macrophages, with M1 macrophages accumulating.[104] Interestingly, a deficiency of TLR4 was shown to attenuate adipose tissue inflammation and promote M2 polarization of adipose tissue macrophages as well as peritoneal macrophages but this did not impact systemic insulin sensitivity.[105] Another study comparing GF mice against those monocolonized with *E. coli* showed an LPS-dependent macrophage accumulation in adipose tissue of the monocolonized mice. Colonization with *E. coli* also enhanced macrophage polarization to a pro-inflammatory M1 phenotype and induced impairment of glucose and insulin tolerance; both of which effects were LPS-independent.[106] Gut microbiota thus utilize a multitude of mechanisms to exert their influence on host metabolism, mediated by whole cells, cell-components or metabolites through interactions with host receptors or influence on gut barrier function/permeability.

The gut barrier is formed by intercellular tight junction (TJ) proteins (Zonula Occludens-1 and Occludin) which act as gatekeepers for substances passing from the gut lumen to the circulation.[107] Zonulin is the only human protein discovered to date that can reversibly regulate intestinal permeability by modulating intercellular tight junction proteins. Zonulin release has shown to be triggered by small intestinal exposure to bacteria and gluten[108] and increase permeability. Bacterial overgrowth in small intestine or a "dysbiosis" in the gut microbiota may thus increase zonulin release and gut permeability, triggering systemic inflammation. On the other hand, prebiotics, by virtue of supporting a favorable pattern of gut microbial composition, are associated with increased TJ mRNA expression and improved TJ protein distribution (apical localization) thus improving the gut barrier. The mechanism was thought to be GLP-2 dependent as it was blunted in the presence of a GLP-2 antagonist.[101] GLP-2 is produced by the enteroendocrine L cells and secreted along with GLP-1 upon nutrient ingestion. GLP-2 promotes intestinal growth and enhances gut barrier function via Insulin-like Growth Factor-1 IGF-1 and beta-catenin pathways.[109–111] A role in promoting TJ function has also been demonstrated for secreted bioactive compounds from certain probiotics, i.e., *Lactobacillus casei*, *Lactobacillus rhamnosus* GG and *Bifidobacterium infantis*, via an increase in Zonula Occludens-1 protein levels.[112,113]

Another pathway that has been studied in the context of gut barrier function is TLR-signaling. TLR2 has been reported to preserve tight junctions and its deficiency was shown to cause tight junctions disruption owing to anti-apoptotic failure of the intestinal epithelium.[114] TLR2 KO mice developed the features of metabolic syndrome correlated with a three-fold increase in Firmicutes, a decrease in bifidobacteria, and an increase in LPS absorption. Wide-spectrum antibiotics were able to relieve the TLR2KO mice of these symptoms, while the same symptoms could be introduced into wild-type mice by microbiota transplantation from TLR2KO mice.[115] In contrast to the above reports, TLR2-deficiency in some studies was found to be beneficial, and protected mice against HFD-induced obesity, due to an elevated metabolic rate and preferential use of lipids as a metabolic substrate.[116,117]

The intracellular NOD-like receptors (NLRs) also play important roles in gut bacteria-mediated inflammation. NOD1 and NOD2 recognize peptidoglycan motifs from bacterial cells. NOD1 is more specialized in responding to Gram-negative bacteria by recognizing meso-diaminopimelic acid while NOD2 recognizes muramyl dipeptide in both Gram-positive and Gram-negative bacteria.[118] A study exploring the role of NOD2 in gut microbiota development and composition reported that NOD2 KO mice had significantly higher numbers of fecal Bacteroidetes, significantly higher ileum Firmicutes and marginally higher ileum Bacteroidetes, compared to wild-type mice. NOD2's impact on gut microbiota composition was also seen in humans. Human subjects carrying the Nod2 frame shift mutation (SNP13) had significantly increased loads of Bacteroidetes and Firmicutes but reduced *Faecalibacterium* prausnitzii.[119] NOD2 may also be important in mediating specific bacterial effects on the host. For example, in a model of experimental colitis, the strain-specific anti-inflammatory capacity of *Lactobacillus salivarius* Ls33 was NOD2-dependent. NOD-2 recognized a specific peptidoglycan-derived muropeptide on the bacteria and generated an immune response leading to local anti-inflammatory IL-10 production.[120]

Compared to the anti-inflammatory role of NOD2, NOD1 acts more like a pro-inflammatory mediator by recognizing Gram-negative bacteria, thereby facilitating their translocation and "metabolic bacteremia" described earlier.[102] The authors investigated the potential role of leptin and a probiotic strain in preventing this state. Intestinal delivery of leptin by an engineered *Lactococcus lactis* strain, or oral intake of the probiotic *Bifidobacterium animalis* subsp. *lactis* 420 alleviated this bacterial translocation and reduced pro-inflammatory cytokine production by intestinal cells, suggesting probiotic-intervention as a promising approach to manage bacteremia-induced metabolic disorders.[102] The involvement of both NOD1 as well as CD14/TLR4 in mediating bacterial translocation indicates a synergistic role of immune receptors in controlling local and systemic inflammation.[102]

Another participant in the inflammation-related pathways linking gut microbiota and metabolic disease is a caspase-activating multiprotein complex termed the "inflammasome," through its roles in pathogen-sensing and the innate immune system.[121] A recent study pointed out a metabolic role for inflammasome proteins in the progression of NAFLD and obesity via gut microbiota associated mechanisms.[122] The NLRP6 and NLRP3 inflammasomes, through activation of the effector protein IL-18, negatively regulated NAFLD progression. NLRP6- and NLRP3-deficient mice exhibited elevated numbers of *Prevotellaceae* and *Porphyromonadaceae* in the intestine and released agonists of TLR4 and TLR9 into systemic circulation, promoting hepatic steatosis and inflammation.[122]

The latest mechanism via which gut bacteria have been shown to exert their effect on inflammation is via impact on regulatory T cells. Butyrate produced by commensal bacterial fermentation, as well as propionate, are reportedly able to promote regulatory T-cell generation in the colon, which are anti-inflammatory in nature.[123,124]

An integrated view of the three mechanisms underlying gut microbiome's influence on obesity and related metabolic states, i.e., energy-extraction, appetite-regulation and inflammation, points to cooperative rather than discrete roles for each pathway. At the heart of each of these regulatory processes is a complex and dynamic three-way interaction between the host, the microbiome and diet. Dietary components are processed in the gut synergistically by mammalian digestive processes and bacterial action. The profile of the metabolic products generated is dependent upon host genetics, host physiology and the composition of the gut microbiome. The microbes in conjunction with the products of nutrient digestion can exert varying effects on host metabolism (desirable or undesirable) by specific interactions with host cellular receptors or in a non-specific manner via entry into general blood circulation (Figure 11.2). An example of a common thread that ties the three processes together is a role for SCFAs, another being enrichment or depletion in specific "beneficial" bacterial groups such as *Bifidobacterium* sp. While this new found role for gut microbiota in metabolic disease makes it an exciting novel therapeutic target; as with any emerging field, there are conflicting data in literature which presents some challenges with regards to what comprises an "ideal" microbiota and how best to achieve it via dietary means. A major scientific endeavor in this context is the NIH Human Microbiome Project. The recently concluded first phase of the project was mainly intended to answer the question of "who's there?" in terms of the identity of the gut microbiome while the second phase is focused on "what are they (the microbes) doing?" with regards to impact on human health. Combined with the mechanistic understanding already generated through studies in a variety of mouse models, some of which have been highlighted above, a clearer picture should emerge in future regarding the potential of tweaking our gut microbial composition as a means to achieve better metabolic health. In the meantime, we summarize in the next section, the results from selected human studies exploring the role of known strategies of gut microbiota modulation, i.e., probiotics, prebiotics and synbiotics, in modulating obesity.

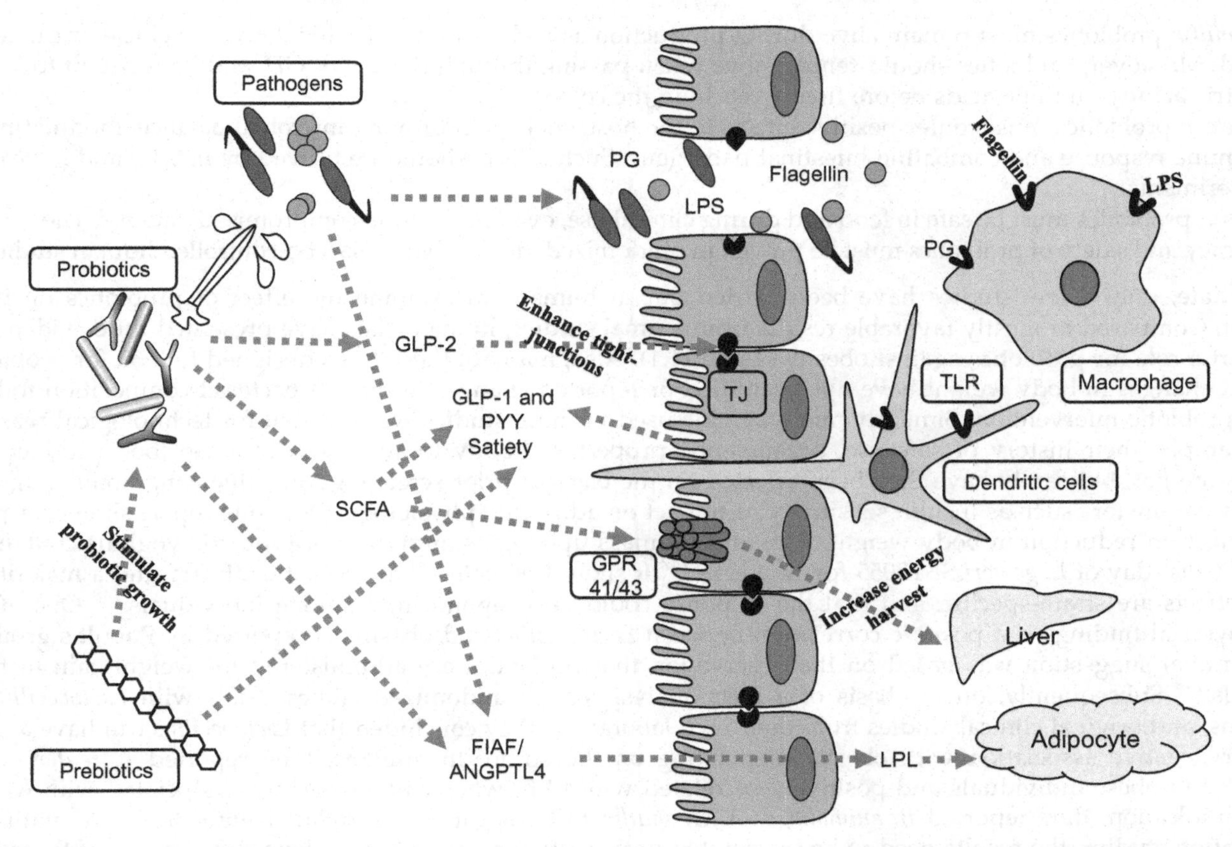

FIGURE 11.2 A schematic summary of microbial activities in gut energy absorption, and satiety regulation and systemic inflammation. FIAF, fasting-induced adipose factor; LPL, lipoprotein lipase; GLP-1, glucagon-like peptide-1; GLP-2, glucagon-like peptide-2; GPR41/43, G protein-coupled receptor 41/43; LPS, lipopolysaccharides; PG, peptidoglycan; TJ, tight junctions; TLR, toll-like protein.

GUT MICROBIOTA AS A THERAPEUTIC TARGET OF PROBIOTICS, PREBIOTICS AND SYNBIOTICS

Obesity has a multifactorial etiology that may involve a genetic predisposition, a poor-quality diet, excess energy intake, insulin resistance, systemic inflammation, or all the above. Once an individual becomes obese, it is extremely difficult to permanently lose the excess weight, as the mechanisms regulating energy balance are reset in favor of weight maintenance. It is estimated that the rate of successful weight loss regimens is only about 15%.[125,126] Additional means to aid the traditional approaches of caloric restriction and physical activity in maintaining a healthy body weight are thus highly desirable. With the identification of gut microbiota as a novel therapeutic target for obesity, existing dietary means of "beneficially" altering gut microbiota composition by probiotics, prebiotics and synbiotics have garnered fresh interest with respect to their potential in modulating obesity.

Probiotics are live microorganisms usually consumed in the diet through fermented dairy foods and beverages. Fuller (1991)[127] first defined probiotics as "live microbial food supplements that beneficially affect the host by improving the intestinal microbial balance." An updated definition was released in 2002 by the FAO/WHO,[128] which is the most current version: "Probiotics are live microorganisms which, when administered in adequate amounts, confer a health benefit to the host." Several criteria must be met when a microorganism is proposed as a probiotic strain for human use:[129]

- *Human origin*: a probiotic strain isolated from the human gastrointestinal tract (GIT) is believed to be safer for human consumption and more effective in its adapted intestinal ecosystem.
- *GRAS (generally regarded as safe) status*: so far, only *Bifidobacterium* sp. and *Lactobacillus* sp. have received GRAS status from the US FDA to be used for human consumption.

- *Viability*: probiotics must remain alive during production as well as in the specific delivery vehicle or carrier food. Moreover, probiotics should remain alive when passing through the upper GIT and be resistant to gastric acidity and bile acids before finally reaching the colon.
- *Efficacy*: probiotics must confer positive effects to the host, such as improving microbial balance, modulating immune response and combating intestinal pathogens. Such effects should be proved by *in vitro* and *in vivo* experiments.
- *Safety*: probiotics must be safe in food and during clinical use, even in immune-compromised patients. The efficacy and safety of probiotics must be proven in randomized, double-blind placebo-controlled human studies.

To date, only a few studies have been carried out in humans to examine the effect of probiotics on body weight. Compared to mostly favorable results from animal studies, human trials have presented little evidence to support a role for probiotics against obesity (Table 11.1). Also, many of these trials designed to look for probiotic-related changes in body weight have not quantified or reported changes in the gut bacterial composition following a probiotic intervention. Similarly, many strains used in these studies were chosen for technological reasons, for example, their history of safe use, organoleptic properties, ability to survive within the food product and within the gut, but rarely have they been selected on the basis of prior screening for influencing body weight or related parameters such as insulin sensitivity or impact on adipocyte physiology. Only one report showed a positive effect on reduction in body weight, BMI and fat mass upon consumption of a probiotic yoghurt containing $\sim 10^{10}$ CFU/day of *L. gasseri* SBT2055 for 12 weeks.[147] It should be noted that probiotic efficacy and a majority of their effects are strain-specific; and that the response could vary significantly among individuals.[148] One of the controversial findings is a positive correlation between *Lactobacillus* and obesity as reported by Raoult's group.[36] Their initial suggestion was based on the observation that probiotics are administered for weight-gain in farm animals.[149] Subsequently, on the basis of a meta-analysis of 17 randomized clinical trials with *Lactobacillus* in humans, and original clinical studies from their own laboratory, they concluded that *Lactobacillus* can have a positive or negative association with obesity, depending on the strain. In humans, they reported *L. reuteri* to be enriched in obese individuals and positively correlated with BMI, whereas *L. gasseri* was associated with weight loss. In addition, they reported *B. animalis*, and *M. smithii* to be negatively correlated with BMI.[38] As with any association studies, the results need to be interpreted with caution as no direct evidence for the role of these bacteria in modulating obesity or leanness was presented.

Another set of interesting data has been obtained from studies investigating the impact of antibiotics on adiposity and diabetes. Cani et al. (2008)[150] and Murphy et al. (2013)[28] demonstrated that antibiotic treatment of HFD-induced obese mice led to a reduction in weight and fat and a reduction in glucose intolerance, related to a reduction in LPS levels and metabolic endotoxemia. Conversely, Cho et al. (2012)[151] showed that antibiotic use early in life predisposed the mice to adiposity accompanied by an increase in Firmicutes to Bacteroidetes ratio and increase in SCFA synthesis. While we currently lack enough information to translate such findings into clinical practice, it is certainly tantalizing to imagine a future role for targeted antibiotic therapy to address obesity and diabetes in diseased individuals. At the same time, because of the adverse impacts from early life, caution should be used against indiscriminate antibiotic use, especially in infants and children.

Prebiotics were first defined as "non-digestible dietary ingredients that beneficially affect the host by selectively stimulating the growth and/or activity of one or a limited number of bacteria in the colon, thus improving host health" by Gibson and Roberfroid in 1995.[152] Several modifications since have led to the current definition, i.e., "a dietary prebiotic is a selectively fermented ingredient that results in specific changes in the composition and/or activity of the GI microbiota thus conferring benefit(s) upon host health."[153] Based on this definition, a dietary ingredient which has the ability to resist digestion in the upper gut and reach the colon has to meet two main criteria to be considered a prebiotic:[129]

- *Fermentability*: both *in vitro* and *in vivo* studies have to be performed to test fermentability by gut microbiota.
- *Selectivity*: a prebiotic substance should selectively promote the growth of one or more beneficial bacteria and/or enhance metabolic activity of such beneficial microbes. *In vitro* and *in vivo* studies must therefore be performed on prebiotic candidates to characterize their effects on modulating microbiota composition and/or activities.

The prebiotics most commonly consumed and studied include fructooligosaccharides (FOS), inulin, galactooligosaccharides (GOS) and lactulose. Finally, synbiotics are defined as "mixtures of pro- and prebiotics, which beneficially affect the host, by improving the survival and implantation of live microbial dietary supplements in the gastrointestinal tract."[129]

TABLE 11.1 Obesity-Related Human Intervention Trials of Probiotics, Prebiotics and Synbiotics in Adult Subjects*

Product or Placebo and Dosage	Study Design	Numbers of Subjects Included for Data Analysis	Duration	Gut Microbiota Profiles	Anthropometric Measures and Metabolic Biomarkers	Reference
PROBIOTICS						
Fermented milk with: Group 1: *S. thermophilus*, 10^7 CFU/mL and *L. acidophilus*, 10^7 CFU/mL ($\sim10^9$ CFU/day) Group 2: *S. thermophilus*, 10^8 CFU/mL and *L. rhamnosus*, 10^8 CFU/mL ($\sim10^{10}$ CFU/day) Group 3: *E. faecium*, 10^7 CFU/mL ($\sim10^9$ CFU/day) and *S. thermophilus* 10^9 CFU/mL ($\sim10^{11}$ CFU/day) Group 4: Chemically fermented milk Group 5: Two placebo pills	Randomized, double-blind, placebo-controlled, *parallel* intervention	Group 1, n = 16 (4 males and 12 females), age: 38.6 ± 8.40, BMI: 30.0 ± 2.80 Group 2, n = 14 (4 males and 10 females), age: 37.9 ± 8.98, BMI: 30.2 ± 2.62 Group 3, n = 16 (4 males and 12 females), age: 37.8 ± 8.00, BMI: 30.1 ± 2.40 Group 4, n = 14 (5 males and 9 females), age: 39.4 ± 7.86, BMI: 30.0 ± 2.37 Group 5, n = 10 (3 males and 7 females), age: 38.3 ± 10.12, BMI: 29.9 ± 3.48	8 weeks	ND	The fermented milk group 3 had reduced LDL-cholesterol and increased fibrinogen levels No effects on body weight and fat mass	130
Probiotic (*L. acidophilus* NCFM, $\sim10^{10}$ CFU/day) vs. SiO_2/lactose (placebo)	Randomized, double-blind, placebo-controlled, *parallel* intervention	Healthy males or male subjects with glucose intolerance and/or diabetes mellitusProbiotic group, n = 21, age: 55 ± 15.2, BMI: 28.1 ± 7.0 Placebo group, n = 24, age: 60 ± 12.9, BMI: 28.7 ± 6.1	4 weeks	In treatment group, *L. acidophilus* NCFM was detectable by DGGE in 2 out of 20 subjects at baseline and 15 out of 20 subjects after intervention; in placebo group, it was detected in 6 out of 24 at baseline and 3 out of 24 after intervention	Insulin sensitivity preserved in probiotic group and decreased in placebo group No effects on plasma inflammatory markers LPS challenge induced systemic inflammation in both groups	131
Probiotic yoghurt (*L. gasseri* SBT2055, $\sim10^{10}$ CFU/day) vs. conventional yoghurt (placebo)	Randomized, multicenter, double-blind, placebo-controlled, *parallel* intervention	Probiotic group, n = 43 (29 males and 14 females), age: 48.3 ± 9.3, BMI: 27.5 ± 1.7 Placebo group, n = 44 (30 males and 14 females), age: 49.2 ± 9.1, BMI: 27.2 ± 1.7	12 weeks	ND	Reduced body weight, BMI, waist and hip circumference, visceral and subcutaneous fat mass; increased serum adiponectin levels in the probiotic group	132
Probiotic yoghurt with *B. lactis* Bb12 and *L. acidophilus* La5 ($\sim10^7$ CFU) Conventional yoghurt (control) No yoghurt (control)	Randomized, triple-blind, controlled, *parallel* intervention	Healthy females, Probiotic group, n = 30, age: 60.7 ± 7.0, BMI: 24.0 ± 2.4 Conventional group, n = 30, age: 58.5 ± 6.8, BMI: 23.0 ± 2.4 No yoghurt group, n = 30, age: 59.3 ± 7.3, BMI: 23.8 ± 3.0	6 weeks	ND	No effects on body weight, BMI, or serum lipid levels	133
Probiotic yoghurt with *B. lactis* Bb12 and *L. acidophilus* La5 (10^6 CFU/g, $\sim10^8$ per day) vs. conventional yoghurt (control)	Randomized, double-blind, controlled, *parallel* intervention	T2D patients, Probiotic group, n = 30 (12 males and 18 females), age: 51.00 ± 7.32, BMI: 29.14 ± 4.30	6 weeks	ND	Reduced fasting glucose and hemoglobin A1c Increased erythrocyte superoxide dismutase and glutathione peroxidase	134

(*Continued*)

TABLE 11.1 (Continued)

Product or Placebo and Dosage	Study Design	Numbers of Subjects Included for Data Analysis	Duration	Gut Microbiota Profiles	Anthropometric Measures and Metabolic Biomarkers	Reference
		Control group, n = 30 (11 males and 19 females), age: 50.87 ± 7.68, BMI: 28.95 ± 3.65			activities and total antioxidant status; decreased serum malondialdehyde concentration No effects on insulin and erythrocyte catalase activity	
Probiotic yoghurt with *L. casei* Shirota (10^8 CFU/mL, ~10^{10} CFU/day) Standard therapy with no placebo/yoghurt Healthy control with no placebo/yoghurt	Randomized, open label, controlled, *parallel* intervention	Metabolic syndrome (MS) patients, probiotic group, n = 13, age: 51.5 ± 11.4, BMI: 35.4 ± 5.3 MS patients, standard therapy group, n = 15, age: 54.5 ± 8.9, BMI: 31.6 ± 3.6 Healthy control group, n = 10, age: 40.6 ± 15.2, BMI: 25.2 ± 2.6	3 months	ND	No effects on any tested parameters	135
Synbiotic shake (*L. acidophilus* and *B. bifidum*: 10^8 CFU/mL, ~10^{10} CFU/day; FOS: 2 g/day) vs. conventional shake (placebo)	Randomized, double-blind, placebo-controlled, *parallel* intervention	Female T2D patients Synbiotic group, n = 9, age: 55.47 ± 2.0, BMI: 27.70 ± 0.78 Placebo group, n = 9, age: 56.89 ± 1.7, BMI: 28.21 ± 0.85	30 days	ND	Increased serum HDL and reduced fasting glycemia in the synbiotic group	136
PREBIOTICS						
Prebiotic (FOS, Raftilose P95, 15 g/day) vs. 4 g/day glucose (placebo)	Randomized, single-blind, placebo-controlled, *cross-over* intervention	Type 2 diabetic patients, n = 20, (9 males and 11 females), age: 56 ± 5.2 for males, 62 ± 4.1 for females; BMI: 29.4 ± 4.22 for males, 27.4 ± 2.72 for females	20 days	ND	No effects on fasting glucose levels and serum lipid concentrations	137
Prebiotic (short-chain FOS, 20 g/day) vs. 20 g/day sucrose (placebo)	Randomized, double-blind, placebo-controlled, *cross-over* intervention	Type 2 diabetic patients, n = 10, (6 males and 4 females), overall age: 57 ± 6.32; overall BMI: 28 ± 3.16	4 weeks	ND	No effects on body weight, fasting plasma glucose, insulin, and lipid levels, insulin binding to erythrocytes and glucose disappearance rate	138
Prebiotic bread and muffins with arabinoxylan (14% AX) vs. conventional bread and muffins (0% AX) as placebo; 4–5 slices of bread and 1–2 of muffins per day	Randomized placebo-controlled, *cross-over* intervention	Type 2 diabetic patients, n = 15, (6 males and 9 females), overall age: 60 ± 7.75, overall BMI: 28.1 ± 3.49	5 weeks	ND	Reduced fasting glycemia, post-oral glucose tolerance test (OGTT) glycemia and insulinemia; No effects on serum lipid levels, body weight and blood pressure	139
Prebiotic (FOS, Raftilose P95, 16 g/day) vs. 16 g/day maltodextrin (placebo)	Randomized, double-blind, placebo-controlled, *cross-over* intervention	Nonalcoholic steatohepatitis males, n = 7, age: 54.6 ± 9.41, BMI: 29.2 ± 6.2	8 weeks	ND	Reduced aspartate aminotransferase; No effects on fasting glucose, insulin and lipid levels	140
Prebiotic bread and powder or bread rolls (15 g AX/day) vs.	Randomized, single-blind,	Patients with impaired glucose tolerance, n = 11 (4 males and 7	6 weeks	ND	Reduced fasting glucose, TG, and apolipoprotein A1;	141

conventional bread and rolls without AX (placebo)	placebo-controlled, *cross-over* intervention	females), overall age: 55.5 ± 6.2, overall BMI: 30.1 ± 5.7			No effects on body weight and fat, leptin, adiponectin, insulin, resistin and NEFA levels	
Prebiotic (FOS, Raftilose P95, 21 g/day) vs. maltodextrin (placebo)	Randomized, double-blind, placebo-controlled, *parallel* intervention	Prebiotic group, n = 21 (4 males and 17 females), age: 41.9 ± 12.7, BMI: 30.4 ± 3.4; Placebo group, n = 18 (3 males and 15 females), age: 38.6 ± 13.0, BMI: 29.8 ± 4.0	12 weeks	ND	Reduced body weight, trunk fat, fat mass, energy intake, GIP; No difference in fasting glucose, insulin, ghrelin, GLP-1, PYY and leptin levels After meal tolerance test (MTT): reduced glycemia, insulin, AUC for ghrelin, AUC for PYY, AUC for leptin, but no difference in GIP level or AUC for GLP-1	90
Prebiotic (inulin and FOS enriched cookies, with ALA) vs. conventional cookies group (control)	Randomized, double-blind, controlled, *parallel* intervention	Prebiotic group, n = 15 (6 males and 9 females), age: 50.6 ± 15.2, BMI: 39.9 ± 6.2 Placebo group, n = 15 (6 males and 9 females), age: 50.8 ± 15.1, BMI: 38.5 ± 7.2	One month	ND	No effects observed on anthropometric measures. TG and LDL decreased in males in treatment group and intake of dietary fiber and soluble fiber increased in both males and females	142
High viscosity polysaccharide (HVP) fiber (6–10 g/ day) vs. inulin (control)	Randomized, double-blind, controlled, *parallel* intervention	HVP group, n = 30 (15 males and 15 females), for males (age: 38.1 ± 7.2, BMI: 29.8 ± 1.2); for females (age: 34.7 ± 10.4, BMI: 30.1 ± 2.5); Inulin group, n = 30 (13 males and 17 females), for males (age: 38.8 ± 7.1, BMI: 30.0 ± 1.5); for females (age: 37.1 ± 10.8; BMI: 31.7 ± 1.8)	15 weeks	ND	Lower body weight and hip circumference, TG, LDL, higher HDL in HVP fiber group compared to the control group	143
Catechin-rich green tea with inulin (650 mL/day) vs. conventional green tea (control)	Controlled, *parallel* intervention	Prebiotic tea group, n = 15 (8 males and 7 females), age: 25.5 ± 5.81, BMI: 26.7 ± 2.71; Control group, n = 15 (8 males and 7 females), age: 27.6 ± 8.13, BMI: 27.3 ± 3.45	6 weeks	ND	Reduced body weight, BMI, hip circumference, fat mass, blood pressure and fasting glucose in the prebiotic tea group	144
GOS (5.5 g/day) vs. maltodextrin (placebo)	Randomized, double-blind, placebo-controlled, *cross-over* intervention	n = 45; 16 males (age: 42.8 ± 12.1, BMI: 30.7 ± 5.3); 29 females (age: 46.4 ± 11.8, BMI: 32.1 ± 6.3)	12 weeks	Higher numbers of *Bifidobacterium* and *Bacteroides*, lower numbers of *C. histolyticum* and *Desulfovibrio* in GOS group	Lower fecal calprotectin level, plasma CRP, TC, TG, insulin, TC: HDL-C ratio in the GOS group	145
Inulin-type fructan s (16 g/day) vs. maltodextrin (placebo)	Randomized, double-blind, placebo-controlled, *parallel* intervention	Prebiotic group, n = 15 (age: 47 ± 9, BMI:36.1 ± 4.1) Placebo group, n = 15 (age: 48 ± 8, BMI: 35.6 ± 4.3)	3 months	Increased numbers of *Bifidobacterium* and *F. prausnitzii*; decreased numbers of *Bacteroides intestinalis*, *B. vulgatus*	No effects on anthropometrics and blood parameters; No clustering of urine metabolites profiles	146

**Subjects' age and BMI were reproduced as mean + / − SD from some of the studies. Unless otherwise specified, subjects were clinically healthy.*

Relative to probiotics, studies with prebiotics have yielded more promising results concerning effects on obesity, with reductions in body weight, fat mass, and hip circumference.[90,143,144] One thing to note is that many studies just examine body weight to assess changes in the obese status, which may not accurately reflect the impact of an intervention on metabolic health.[145] Measures such as body composition, fat mass and fat/lean mass ratios are more precise indicators of obesity.[90,147] Thus, many of the negative outcomes from probiotic/prebiotic trials reporting changes only in body weight may have yielded favorable outcomes if more sensitive measures of obesity were utilized.

The delivery vehicle is another important consideration for pro-, pre- and synbiotics, as it has a bearing on the effective amounts of these ingredients arriving in the gut. This is especially true in case of incorporation into foods for, e.g., dairy products, bread, muffins, or cookies, as unlike in supplements, the biological activity will be influenced by other ingredients in the food product, such as simple sugars and fats. As mentioned earlier, a major deficiency of many studies is a lack of characterization of gut microbiota composition. Only two very recent studies have addressed this issue, and reported quantification of the obese microbiota using FISH[146] or q-PCR.[145] A "bifidogenic" effect was observed with GOS and inulin treatment in overweight and obese subjects, which when taken together with beneficial changes in lipid levels and inflammatory biomarkers indicates a desirable role for these prebiotics in obesity and metabolic syndrome. An in-depth characterization of the gut microbiota in response to pro-, pre-, or synbiotic interventions in future studies using next-generation deep-sequencing methodologies will be key to obtaining a complete picture that will enable us to effectively harness the potential of dietary modulation of gut microbiota to manage metabolic disease.

CONCLUSIONS

As the globalized world faces the modern-day pandemic of obesity, a quest for new effective solutions has become a high research priority. This is primarily because the majority of individuals are unable to sustain the established diet and physical activity regimens aimed at achieving and maintaining a healthy body weight over the long term. The gut microbiota, by virtue of being critically positioned at the site of host nutrient/energy metabolism and by the ability to influence it, has thus gained traction as a novel therapeutic target that holds promise. The inter-relationship between diet, gut microbiota and obesity is highly interactive and dynamic. While dietary caloric intake is the predominant factor in the energy balance equation, we now know that certain gut microbes can harvest "extra" energy from the diet, to the extent of ~150 kcal/day; equivalent to a possible additional body weight of fifteen pounds (6.8 kg) over the course of a year. Aside from caloric content, the composition of the diet is also an important factor underlying diet–microbe interactions relevant to obesity, as exemplified by the microbial dysbiosis in the gut and resulting systemic inflammation in response to a high-fat diet. Gut microbes also impact obesity by their ability to regulate appetite, food intake and satiety: yet another part of the energy balance picture. Probiotics, prebiotics and synbiotics are effective approaches to modulating the gut microbiota composition, and their role is now also being investigated in the context of obesity. While the majority of these interventions did not result in an alteration in body weight, fat mass or BMI, several reported a favorable impact on biomarkers of relevance to obesity and related metabolic conditions, such as plasma lipid levels, insulin sensitivity, plasma adiponectin and fasting glucose levels. Overall, these outcomes suggest that gut microbiota modulation through pre-, pro- or synbiotics may not be sufficient as a stand-alone approach for weight loss in human subjects, but may help alleviate the risk and manage the symptoms of obesity-associated conditions such as cardiovascular disease, insulin resistance and type 2 diabetes. Their effectiveness in mediating both the direct and indirect outcomes may be enhanced by application-specific probiotic strain selection/development and product formulation. The development of effective formulations will also be assisted in future once a consensus emerges on the composition of obesity-related gut microbiota. In conjunction with a healthy diet and lifestyle, pro-, pre- and synbiotics consumed as functional foods or dietary supplements may thus help mold our microbiota to maximize the efficacy of obesity-interventions.

References

1. Harvard Obesity Prevention Source <http://www.hsph.harvard.edu/obesity-prevention-source/obesity-consequences/health-effects/> accessed on 20.05.14.
2. WHO Fact Sheets Obesity and Overweight <http://www.who.int/mediacentre/factsheets/fs311/en/> accessed on 20.05.14.

3. Ogden CL, Carroll MD, Kit BK, Flegal KM. Prevalence of obesity in the United States, 2009–2010. Centers for Disease Control and Prevention National Center for Health Statistics; 2012.
4. Xi B, Liang Y, He T, et al. Secular trends in the prevalence of general and abdominal obesity among Chinese adults, 1993–2009. *Obesity Rev Off J Int Assoc Study Obesity*. 2012;13(3):287–296.
5. Wang Y, Chen HJ, Shaikh S, Mathur P. Is obesity becoming a public health problem in India? Examine the shift from under- to overnutrition problems over time. *Obesity Rev Off J Int Assoc Study Obesity*. 2009;10(4):456–474.
6. Colditz GA. Economic costs of obesity. *Am J Clin Nutr*. 1992;55(2 suppl):503S–507SS.
7. Cawley J, Meyerhoefer C. The medical care costs of obesity: an instrumental variables approach. *J Health Econ*. 2012;31:219–230.
8. Wang YC, McPherson K, Marsh T, Gortmaker SL, Brown M. Health and economic burden of the projected obesity trends in the USA and the UK. *Lancet*. 2011;378(9793):815–825.
9. Razquin C, Marti A, Martinez JA. Evidences on three relevant obesogenes: MC4R, FTO and PPARgamma. Approaches for personalized nutrition. *Mol Nutr Food Res*. 2011;55(1):136–149.
10. Zhang Y, Proenca R, Maffei M, Barone M, Leopold L, Friedman JM. Positional cloning of the mouse obese gene and its human homologue. *Nature*. 1994;372(6505):425–432.
11. Pelleymounter MA, Cullen MJ, Baker MB, et al. Effects of the obese gene product on body weight regulation in ob/ob mice. *Science*. 1995;269(5223):540–543.
12. Strobel A, Issad T, Camoin L, Ozata M, Strosberg AD. A leptin missense mutation associated with hypogonadism and morbid obesity. *Nat Genet*. 1998;18(3):213–215.
13. Martin SS, Qasim A, Reilly MP. Leptin resistance: a possible interface of inflammation and metabolism in obesity-related cardiovascular disease. *J Am College Cardiol*. 2008;52(15):1201–1210.
14. Clement K, Vaisse C, Lahlou N, et al. A mutation in the human leptin receptor gene causes obesity and pituitary dysfunction. *Nature*. 1998;392(6674):398–401.
15. Chen H, Charlat O, Tartaglia LA, et al. Evidence that the diabetes gene encodes the leptin receptor: identification of a mutation in the leptin receptor gene in db/db mice. *Cell*. 1996;84(3):491–495.
16. Rankinen T, Zuberi A, Chagnon YC, et al. The human obesity gene map: the 2005 update. *Obesity*. 2006;14(4):529–644.
17. Qin J, Li R, Raes J, et al. A human gut microbial gene catalogue established by metagenomic sequencing. *Nature*. 2010;464(7285):59–65.
18. Backhed F, Ley RE, Sonnenburg JL, Peterson DA, Gordon JI. Host–bacterial mutualism in the human intestine. *Science*. 2005;307(5717):1915–1920.
19. Ley RE, Backhed F, Turnbaugh P, Lozupone CA, Knight RD, Gordon JI. Obesity alters gut microbial ecology. *Proc Natl Acad Sci USA*. 2005;102(31):11070–11075.
20. Turnbaugh PJ, Ley RE, Mahowald MA, Magrini V, Mardis ER, Gordon JI. An obesity-associated gut microbiome with increased capacity for energy harvest. *Nature*. 2006;444(7122):1027–1031.
21. Turnbaugh PJ, Backhed F, Fulton L, Gordon JI. Diet-induced obesity is linked to marked but reversible alterations in the mouse distal gut microbiome. *Cell Host Microbe*. 2008;3(4):213–223.
22. Fleissner CK, Huebel N, Abd El-Bary MM, Loh G, Klaus S, Blaut M. Absence of intestinal microbiota does not protect mice from diet-induced obesity. *Brit J Nutr*. 2010;104(6):919–929.
23. Zhang C, Zhang M, Wang S, et al. Interactions between gut microbiota, host genetics and diet relevant to development of metabolic syndromes in mice. *ISME J*. 2010;4(2):232–241.
24. Waldram A, Holmes E, Wang Y, et al. Top-down systems biology modeling of host metabotype–microbiome associations in obese rodents. *J Proteome Res*. 2009;8(5):2361–2375.
25. Cani PD, Amar J, Iglesias MA, et al. Metabolic endotoxemia initiates obesity and insulin resistance. *Diabetes*. 2007;56(7):1761–1772.
26. Neyrinck AM, Possemiers S, Verstraete W, De Backer F, Cani PD, Delzenne NM. Dietary modulation of clostridial cluster XIVa gut bacteria (*Roseburia* spp.) by chitin-glucan fiber improves host metabolic alterations induced by high-fat diet in mice. *J Nutr Biochem*. 2012;23(1):51–59.
27. Murphy EF, Cotter PD, Healy S, et al. Composition and energy harvesting capacity of the gut microbiota: relationship to diet, obesity and time in mouse models. *Gut*. 2010;59(12):1635–1642.
28. Murphy EF, Cotter PD, Hogan A, et al. Divergent metabolic outcomes arising from targeted manipulation of the gut microbiota in diet-induced obesity. *Gut*. 2013;62(2):220–226.
29. Serino M, Luche E, Gres S, et al. Metabolic adaptation to a high-fat diet is associated with a change in the gut microbiota. *Gut*. 2012;61(4):543–553.
30. Ley RE, Turnbaugh PJ, Klein S, Gordon JI. Microbial ecology: human gut microbes associated with obesity. *Nature*. 2006;444(7122):1022–1023.
31. Turnbaugh PJ, Hamady M, Yatsunenko T, et al. A core gut microbiome in obese and lean twins. *Nature*. 2009;457(7228):480–484.
32. Le Chatelier E, et al. Richness of human gut microbiome correlates with metabolic markers. *Nature*. 2013;500(7464):541–546.
33. Duncan SH, Lobley GE, Holtrop G, et al. Human colonic microbiota associated with diet, obesity and weight loss. *Int J Obesity*. 2008;32(11):1720–1724.
34. Duncan SH, Belenguer A, Holtrop G, Johnstone AM, Flint HJ, Lobley GE. Reduced dietary intake of carbohydrates by obese subjects results in decreased concentrations of butyrate and butyrate-producing bacteria in feces. *Appl Environ Microbiol*. 2007;73(4):1073–1078.
35. Schwiertz A, Taras D, Schafer K, et al. Microbiota and SCFA in lean and overweight healthy subjects. *Obesity*. 2010;18(1):190–195.
36. Armougom F, Henry M, Vialettes B, Raccah D, Raoult D. Monitoring bacterial community of human gut microbiota reveals an increase in Lactobacillus in obese patients and Methanogens in anorexic patients. *PloS One*. 2009;4(9):e7125.
37. Million M, Maraninchi M, Henry M, et al. Obesity-associated gut microbiota is enriched in Lactobacillus reuteri and depleted in Bifidobacterium animalis and Methanobrevibacter smithii. *Int J Obesity*. 2012;36(6):817–825.
38. Million M, Angelakis E, Maraninchi M, et al. Correlation between body mass index and gut concentrations of Lactobacillus reuteri, Bifidobacterium animalis, Methanobrevibacter smithii and Escherichia coli. *Int J Obesity*. 2013;37(11):1460–1466.

39. Fei N, Zhao L. An opportunistic pathogen isolated from the gut of an obese human causes obesity in germfree mice. *ISME J*. 2013;7 (4):880–884.
40. Qin J, Li Y, Cai Z, et al. A metagenome-wide association study of gut microbiota in type 2 diabetes. *Nature*. 2012;490(7418):55–60.
41. Samuel BS, Gordon JI. A humanized gnotobiotic mouse model of host–archaeal–bacterial mutualism. *Proc Natl Acad Sci U S A*. 2006;103 (26):10011–10016.
42. Zhang H, DiBaise JK, Zuccolo A, et al. Human gut microbiota in obesity and after gastric bypass. *Proc Natl Acad Sci U S Am*. 2009;106 (7):2365–2370.
43. Angelakis E, Armougom F, Million M, Raoult D. The relationship between gut microbiota and weight gain in humans. *Future Microbiol*. 2012;7(1):91–109.
44. Tuohy KM, Costabile A, Fava F. The gut microbiota in obesity and metabolic disease – a novel therapeutic target. *Nutr Ther Metabol*. 2009;27(3):113–133.
45. Backhed F, Ding H, Wang T, et al. The gut microbiota as an environmental factor that regulates fat storage. *Proc Natl Acad Sci U S A*. 2004;101(44):15718–15723.
46. Backhed F, Manchester JK, Semenkovich CF, Gordon JI. Mechanisms underlying the resistance to diet-induced obesity in germ-free mice. *Proc Natl Acad Sci U S A*. 2007;104(3):979–984.
47. Hildebrandt MA, Hoffmann C, Sherrill-Mix SA, et al. High-fat diet determines the composition of the murine gut microbiome independently of obesity. *Gastroenterology*. 2009;137(5):1716–1724:e1–2
48. Vijay-Kumar M, Aitken JD, Carvalho FA, et al. Metabolic syndrome and altered gut microbiota in mice lacking Toll-like receptor 5. *Science*. 2010;328(5975):228–231.
49. Vrieze A, Van Nood E, Holleman F, et al. Transfer of intestinal microbiota from lean donors increases insulin sensitivity in individuals with metabolic syndrome. *Gastroenterology*. 2012;143(4):913–916:e7
50. Hooper LV, Wong MH, Thelin A, Hansson L, Falk PG, Gordon JI. Molecular analysis of commensal host–microbial relationships in the intestine. *Science*. 2001;291(5505):881–884.
51. Stappenbeck TS, Hooper LV, Gordon JI. Developmental regulation of intestinal angiogenesis by indigenous microbes via Paneth cells. *Proc Natl Acad Sci U S A*. 2002;99(24):15451–15455.
52. Kahn BB, Alquier T, Carling D, Hardie DG. AMP-activated protein kinase: ancient energy gauge provides clues to modern understanding of metabolism. *Cell Metab*. 2005;1(1):15–25.
53. Peng L, Li ZR, Green RS, Holzman IR, Lin J. Butyrate enhances the intestinal barrier by facilitating tight junction assembly via activation of AMP-activated protein kinase in Caco-2 cell monolayers. *J Nutr*. 2009;139(9):1619–1625.
54. Gao Z, Yin J, Zhang J, et al. Butyrate improves insulin sensitivity and increases energy expenditure in mice. *Diabetes*. 2009;58 (7):1509–1517.
55. Jumpertz R, Le DS, Turnbaugh PJ, et al. Energy-balance studies reveal associations between gut microbes, caloric load, and nutrient absorption in humans. *Am J Clin Nutr*. 2011;94(1):58–65.
56. McNeil NI. The contribution of the large intestine to energy supplies in man. *Am J Clin Nutr*. 1984;39(2):338–342.
57. Topping DL, Clifton PM. Short-chain fatty acids and human colonic function: roles of resistant starch and nonstarch polysaccharides. *Physiol Rev*. 2001;81(3):1031–1064.
58. Cummings JH. Short chain fatty acids in the human colon. *Gut*. 1981;22(9):763–779.
59. Macfarlane S, Macfarlane GT. Regulation of short-chain fatty acid production. *Proc Nutr Soc*. 2003;62(1):67–72.
60. Macfarlane GT, Gibson GR, Cummings JH. Comparison of fermentation reactions in different regions of the human colon. *J Appl Bacteriol*. 1992;72(1):57–64.
61. Borthakur A, Saksena S, Gill RK, Alrefai WA, Ramaswamy K, Dudeja PK. Regulation of monocarboxylate transporter 1 (MCT1) promoter by butyrate in human intestinal epithelial cells: involvement of NF-kappaB pathway. *J Cell Biochem*. 2008;103(5):1452–1463.
62. Borthakur A, Priyamvada S, Kumar A, et al. A novel nutrient sensing mechanism underlies substrate-induced regulation of monocarboxylate transporter-1. *Am J Physiol Gastrointest Liver Physiol*. 2012;303(10):G1126–G1133.
63. Goncalves P, Catarino T, Gregorio I, Martel F. Inhibition of butyrate uptake by the primary bile salt chenodeoxycholic acid in intestinal epithelial cells. *J Cell Biochem*. 2012;113(9):2937–2947.
64. Conterno L, Fava F, Viola R, Tuohy KM. Obesity and the gut microbiota: does up-regulating colonic fermentation protect against obesity and metabolic disease? *Gene Nutr*. 2011;6(3):241–260.
65. Fava F, Gitau R, Griffin BA, Gibson GR, Tuohy KM, Lovegrove JA. The type and quantity of dietary fat and carbohydrate alter faecal microbiome and short-chain fatty acid excretion in a metabolic syndrome 'at-risk' population. *Int J Obesity*. 2013;37(2):216–223.
66. De Filippo C, Cavalieri D, Di Paola M, et al. Impact of diet in shaping gut microbiota revealed by a comparative study in children from Europe and rural Africa. *Proc Natl Acad Sci U S A*. 2010;107(33):14691–14696.
67. Karaki S, Tazoe H, Hayashi H, et al. Expression of the short-chain fatty acid receptor, GPR43, in the human colon. *J Mol Histol*. 2008;39 (2):135–142.
68. Tazoe H, Otomo Y, Karaki S, et al. Expression of short-chain fatty acid receptor GPR41 in the human colon. *Biomed Res (Tokyo, Japan)*. 2009;30(3):149–156.
69. Stoddart LA, Smith NJ, Milligan G. International Union of Pharmacology. LXXI. Free fatty acid receptors FFA1, -2, and -3: pharmacology and pathophysiological functions. *Pharmacol Rev*. 2008;60(4):405–417.
70. Brown AJ, Goldsworthy SM, Barnes AA, et al. The Orphan G protein–coupled receptors GPR41 and GPR43 are activated by propionate and other short chain carboxylic acids. *J Biol Chem*. 2003;278(13):11312–11319.
71. Le Poul E, Loison C, Struyf S, et al. Functional characterization of human receptors for short chain fatty acids and their role in polymorphonuclear cell activation. *J Biol Chem*. 2003;278(28):25481–25489.
72. Xiong Y, Miyamoto N, Shibata K, et al. Short-chain fatty acids stimulate leptin production in adipocytes through the G protein-coupled receptor GPR41. *Proc Natl Acad Sci U S A*. 2004;101(4):1045–1050.

73. Samuel BS, Shaito A, Motoike T, et al. Effects of the gut microbiota on host adiposity are modulated by the short-chain fatty-acid binding G protein-coupled receptor, Gpr41. *Proc Natl Acad Sci U S A*. 2008;105(43):16767–16772.
74. Dewulf EM, Cani PD, Neyrinck AM, et al. Inulin-type fructans with prebiotic properties counteract GPR43 overexpression and PPARgamma-related adipogenesis in the white adipose tissue of high-fat diet-fed mice. *J Nutr Biochem*. 2011;22(8):712–722.
75. Hong YH, Nishimura Y, Hishikawa D, et al. Acetate and propionate short chain fatty acids stimulate adipogenesis via GPCR43. *Endocrinology*. 2005;146(12):5092–5099.
76. Ge H, Li X, Weiszmann J, et al. Activation of G protein-coupled receptor 43 in adipocytes leads to inhibition of lipolysis and suppression of plasma free fatty acids. *Endocrinology*. 2008;149(9):4519–4526.
77. Bjursell M, Admyre T, Goransson M, et al. Improved glucose control and reduced body fat mass in free fatty acid receptor 2-deficient mice fed a high-fat diet. *Am J Physiol Endocrinol Metabol*. 2011;300(1):E211–E220.
78. Kimura I, Ozawa K, Inoue D, et al. The gut microbiota suppresses insulin-mediated fat accumulation via the short-chain fatty acid receptor GPR43. *Nat Commun*. 2013;4:1829.
79. Kimura I, Inoue D, Maeda T, et al. Short-chain fatty acids and ketones directly regulate sympathetic nervous system via G protein-coupled receptor 41 (GPR41). *Proc Natl Acad Sci U S A*. 2011;108(19):8030–8035.
80. de Graaf C, Blom WA, Smeets PA, Stafleu A, Hendriks HF. Biomarkers of satiation and satiety. *Am J Clin Nutr*. 2004;79(6):946–961.
81. Ford H, Frost G. Glycaemic index, appetite and body weight. *Proc Nutr Soc*. 2010;69(2):199–203.
82. Cherbut C, Ferrier L, Roze C, et al. Short-chain fatty acids modify colonic motility through nerves and polypeptide YY release in the rat. *Am J Physiol*. 1998;275(6 Pt 1):G1415–G1422.
83. Freeland KR, Wolever TM. Acute effects of intravenous and rectal acetate on glucagon-like peptide-1, peptide YY, ghrelin, adiponectin and tumour necrosis factor-alpha. *Brit J Nutr*. 2010;103(3):460–466.
84. Reimer RA, Maurer AD, Eller LK, et al. Satiety hormone and metabolomic response to an intermittent high energy diet differs in rats consuming long-term diets high in protein or prebiotic fiber. *J Proteome Res*. 2012;11(8):4065–4074.
85. Belobrajdic DP, King RA, Christophersen CT, Bird AR. Dietary resistant starch dose-dependently reduces adiposity in obesity-prone and obesity-resistant male rats. *Nutr Metabol*. 2012;9(1):93.
86. Overduin J, Schoterman MH, Calame W, Schonewille AJ, Ten Bruggencate SJ. Dietary galacto-oligosaccharides and calcium: effects on energy intake, fat-pad weight and satiety-related, gastrointestinal hormones in rats. *Brit J Nutr*. 2013;109(7):1338–1348.
87. Cani PD, Dewever C, Delzenne NM. Inulin-type fructans modulate gastrointestinal peptides involved in appetite regulation (glucagon-like peptide-1 and ghrelin) in rats. *Brit J Nutr*. 2004;92(3):521–526.
88. Delzenne NM, Cani PD, Daubioul C, Neyrinck AM. Impact of inulin and oligofructose on gastrointestinal peptides. *Brit J Nutr*. 2005;93 (suppl 1):S157–S161.
89. Cani PD, Neyrinck AM, Maton N, Delzenne NM. Oligofructose promotes satiety in rats fed a high-fat diet: involvement of glucagon-like Peptide-1. *Obes Res*. 2005;13(6):1000–1007.
90. Parnell JA, Reimer RA. Weight loss during oligofructose supplementation is associated with decreased ghrelin and increased peptide YY in overweight and obese adults. *Am J Clin Nutr*. 2009;89(6):1751–1759.
91. Freeland KR, Wilson C, Wolever TM. Adaptation of colonic fermentation and glucagon–like peptide-1 secretion with increased wheat fibre intake for 1 year in hyperinsulinaemic human subjects. *Brit J Nutr*. 2010;103(1):82–90.
92. Tarini J, Wolever TM. The fermentable fibre inulin increases postprandial serum short-chain fatty acids and reduces free-fatty acids and ghrelin in healthy subjects. *Applied physiology, nutrition, and metabolism = Physiologie appliquee, nutrition et metabolisme*. 2010;35(1):9–16.
93. Fukunaga T, Sasaki M, Araki Y, et al. Effects of the soluble fibre pectin on intestinal cell proliferation, fecal short chain fatty acid production and microbial population. *Digestion*. 2003;67(1–2):42–49.
94. Shen L, Keenan MJ, Raggio A, Williams C, Martin RJ. Dietary-resistant starch improves maternal glycemic control in Goto-Kakizaki rat. *Mol Nutr Food Res*. 2011;55(10):1499–1508.
95. Dao TM, Waget A, Klopp P, et al. Resveratrol increases glucose induced GLP-1 secretion in mice: a mechanism which contributes to the glycemic control. *PloS One*. 2011;6(6):e20700.
96. Tachon S, Zhou J, Keenan M, Martin R, Marco ML. The intestinal microbiota in aged mice is modulated by dietary resistant starch and correlated with improvements in host responses. *FEMS Microbiol Ecol*. 2013;83(2):299–309.
97. Duca FA, Swartz TD, Sakar Y, Covasa M. Increased oral detection, but decreased intestinal signaling for fats in mice lacking gut microbiota. *PloS One*. 2012;7(6):e39748.
98. Lin HV, Frassetto A, Kowalik Jr. EJ, et al. Butyrate and propionate protect against diet-induced obesity and regulate gut hormones via free fatty acid receptor 3-independent mechanisms. *PloS One*. 2012;7(4):e35240.
99. Hotamisligil GS, Shargill NS, Spiegelman BM. Adipose expression of tumor necrosis factor-alpha: direct role in obesity-linked insulin resistance. *Science*. 1993;259(5091):87–91.
100. Hotamisligil GS. Inflammation and metabolic disorders. *Nature*. 2006;444(7121):860–867.
101. Cani PD, Possemiers S, Van de Wiele T, et al. Changes in gut microbiota control inflammation in obese mice through a mechanism involving GLP-2-driven improvement of gut permeability. *Gut*. 2009;58(8):1091–1103.
102. Amar J, Chabo C, Waget A, et al. Intestinal mucosal adherence and translocation of commensal bacteria at the early onset of type 2 diabetes: molecular mechanisms and probiotic treatment. *EMBO Mol Med*. 2011;3(9):559–572.
103. Weisberg SP, McCann D, Desai M, Rosenbaum M, Leibel RL, Ferrante Jr. AW. Obesity is associated with macrophage accumulation in adipose tissue. *J Clin Invest*. 2003;112(12):1796–1808.
104. Dalmas E, Clement K, Guerre-Millo M. Defining macrophage phenotype and function in adipose tissue. *Trends Immunol*. 2011;32 (7):307–314.
105. Orr JS, Puglisi MJ, Ellacott KL, Lumeng CN, Wasserman DH, Hasty AH. Toll-like receptor 4 deficiency promotes the alternative activation of adipose tissue macrophages. *Diabetes*. 2012;61(11):2718–2727.
106. Caesar R, Reigstad CS, Backhed HK, et al. Gut-derived lipopolysaccharide augments adipose macrophage accumulation but is not essential for impaired glucose or insulin tolerance in mice. *Gut*. 2012;61(12):1701–1707.

107. Mitic LL, Anderson JM. Molecular architecture of tight junctions. *Ann Rev Physiol*. 1998;60:121–142.
108. Fasano A. Intestinal permeability and its regulation by zonulin: diagnostic and therapeutic implications. *Clin Gastroenterol Hepatol Off Clin Pract J Am Gastroenterol Assoc*. 2012;10(10):1096–1100.
109. Dube PE, Forse CL, Bahrami J, Brubaker PL. The essential role of insulin-like growth factor-1 in the intestinal tropic effects of glucagon-like peptide-2 in mice. *Gastroenterology*. 2006;131(2):589–605.
110. Dube PE, Rowland KJ, Brubaker PL. Glucagon-like peptide-2 activates beta-catenin signaling in the mouse intestinal crypt: role of insulin-like growth factor-I. *Endocrinology*. 2008;149(1):291–301.
111. Tsai CH, Hill M, Asa SL, Brubaker PL, Drucker DJ. Intestinal growth-promoting properties of glucagon-like peptide-2 in mice. *Am J Physiol*. 1997;273(1 Pt 1):E77–E84.
112. Escamilla J, Lane MA, Maitin V. Cell-free supernatants from probiotic *Lactobacillus casei* and *Lactobacillus rhamnosus* GG decrease colon cancer cell invasion in vitro. *Nutr Cancer*. 2012;64(6):871–878.
113. Ewaschuk JB, Diaz H, Meddings L, et al. Secreted bioactive factors from *Bifidobacterium infantis* enhance epithelial cell barrier function. *Am J Physiol Gastro Liver Physiol*. 2008;295(5):G1025–G1034.
114. Cario E, Gerken G, Podolsky DK. Toll-like receptor 2 controls mucosal inflammation by regulating epithelial barrier function. *Gastroenterology*. 2007;132(4):1359–1374.
115. Caricilli AM, Picardi PK, de Abreu LL, et al. Gut microbiota is a key modulator of insulin resistance in TLR 2 knockout mice. *PLoS Biol*. 2011;9(12):e1001212.
116. Davis JE, Braucher DR, Walker-Daniels J, Spurlock ME. Absence of Tlr2 protects against high-fat diet-induced inflammation and results in greater insulin-stimulated glucose transport in cultured adipocytes. *J Nutr Biochem*. 2011;22(2):136–141.
117. Ehses JA, Meier DT, Wueest S, et al. Toll-like receptor 2-deficient mice are protected from insulin resistance and beta cell dysfunction induced by a high-fat diet. *Diabetologia*. 2010;53(8):1795–1806.
118. Chen G, Shaw MH, Kim YG, Nunez G. NOD-like receptors: role in innate immunity and inflammatory disease. *Ann Rev Pathol*. 2009;4:365–398.
119. Rehman A, Sina C, Gavrilova O, et al. Nod2 is essential for temporal development of intestinal microbial communities. *Gut*. 2011;60(10):1354–1362.
120. Macho Fernandez E, Valenti V, Rockel C, et al. Anti-inflammatory capacity of selected lactobacilli in experimental colitis is driven by NOD2-mediated recognition of a specific peptidoglycan-derived muropeptide. *Gut*. 2011;60(8):1050–1059.
121. Martinon F, Burns K, Tschopp J. The inflammasome: a molecular platform triggering activation of inflammatory caspases and processing of proIL-beta. *Mol Cell*. 2002;10(2):417–426.
122. Henao-Mejia J, Elinav E, Jin C, et al. Inflammasome-mediated dysbiosis regulates progression of NAFLD and obesity. *Nature*. 2012;482(7384):179–185.
123. Furusawa Y, et al. Commensal microbe-derived butyrate induces the differentiation of colonic regulatory T cells. *Nature*. 2013;504(7480):446–450.
124. Arpaia N, Campbell C, Fan X, et al. Metabolites produced by commensal bacteria promote peripheral regulatory T-cell generation. *Nature*. 2013;504(7480):451–455.
125. Hill JO, Wyatt HR. Relapse in obesity treatment: biology or behavior? *Am J Clin Nutr*. 1999;69(6):1064–1065.
126. Ayyad C, Andersen T. Long-term efficacy of dietary treatment of obesity: a systematic review of studies published between 1931 and 1999. *Obesity Rev Off J Int Assoc Study Obesity*. 2000;1(2):113–119.
127. Fuller R. Probiotics in human medicine. *Gut*. 1991;32(4):439–442.
128. FAO/WHO. Joint FAO/WHO working group report on drafting guidelines for evaluation of probiotics in food. London, Ontario, Canada; 2002.
129. Kolida S, Gibson GR. Synbiotics in health and disease. *Ann Rev Food Sci Technol*. 2011;2:373–393.
130. Agerholm-Larsen L, Raben A, Haulrik N, Hansen AS, Manders M, Astrup A. Effect of 8 week intake of probiotic milk products on risk factors for cardiovascular diseases. *Eur J Clin Nutr*. 2000;54(4):288–297.
131. Andreasen AS, Larsen N, Pedersen-Skovsgaard T, et al. Effects of *Lactobacillus acidophilus* NCFM on insulin sensitivity and the systemic inflammatory response in human subjects. *Brit J Nutr*. 2010;104(12):1831–1838.
132. Sadrzadeh-Yeganeh H, Elmadfa I, Djazayery A, Jalali M, Heshmat R, Chamary M. The effects of probiotic and conventional yoghurt on lipid profile in women. *Brit J Nutr*. 2010;103(12):1778–1783.
133. Ejtahed HS, Mohtadi-Nia J, Homayouni-Rad A, Niafar M, Asghari-Jafarabadi M, Mofid V. Probiotic yogurt improves antioxidant status in type 2 diabetic patients. *Nutrition*. 2012;28(5):539–543.
134. Leber B, Tripolt NJ, Blattl D, et al. The influence of probiotic supplementation on gut permeability in patients with metabolic syndrome: an open label, randomized pilot study. *Eur J Clin Nutr*. 2012;66(10):1110–1115.
135. Moroti C, Souza Magri LF, de Rezende Costa M, Cavallini DC, Sivieri K. Effect of the consumption of a new symbiotic shake on glycemia and cholesterol levels in elderly people with type 2 diabetes mellitus. *Lipids Health Dis*. 2012;11:29.
136. Alles MS, de Roos NM, Bakx JC, van de Lisdonk E, Zock PL, Hautvast GA. Consumption of fructooligosaccharides does not favorably affect blood glucose and serum lipid concentrations in patients with type 2 diabetes. *Am J Clin Nutr*. 1999;69(1):64–69.
137. Luo J, Van Yperselle M, Rizkalla SW, Rossi F, Bornet FR, Slama G. Chronic consumption of short-chain fructooligosaccharides does not affect basal hepatic glucose production or insulin resistance in type 2 diabetics. *J Nutr*. 2000;130(6):1572–1577.
138. Lu ZX, Walker KZ, Muir JG, O'Dea K. Arabinoxylan fibre improves metabolic control in people with Type II diabetes. *Eur J Clin Nutr*. 2004;58(4):621–628.
139. Daubioul CA, Horsmans Y, Lambert P, Danse E, Delzenne NM. Effects of oligofructose on glucose and lipid metabolism in patients with nonalcoholic steatohepatitis: results of a pilot study. *Eur J Clin Nutr*. 2005;59(5):723–726.
140. Garcia AL, Steiniger J, Reich SC, et al. Arabinoxylan fibre consumption improved glucose metabolism, but did not affect serum adipokines in subjects with impaired glucose tolerance. *Horm Metab Res*. 2006;38(11):761–766.

141. Garcia AL, Otto B, Reich SC, et al. Arabinoxylan consumption decreases postprandial serum glucose, serum insulin and plasma total ghrelin response in subjects with impaired glucose tolerance. *Eur J Clin Nutr*. 2007;61(3):334–341.
142. de Luis DA, de la Fuente B, Izaola O, et al. Double blind randomized clinical trial controlled by placebo with an alpha linoleic acid and prebiotic enriched cookie on risk cardiovascular factor in obese patients. *Nutricion Hospitalaria: Organo Oficial de la Sociedad Espanola de Nutricion Parenteral y Enteral*. 2011;26(4):827–833.
143. Lyon M, Wood S, Pelletier X, Donazzolo Y, Gahler R, Bellisle F. Effects of a 3-month supplementation with a novel soluble highly viscous polysaccharide on anthropometry and blood lipids in nondieting overweight or obese adults. *J Hum Nutr Dietetics Off J Brit Dietetic Assoc*. 2011;24(4):351–359.
144. Yang HY, Yang SC, Chao JC, Chen JR. Beneficial effects of catechin-rich green tea and inulin on the body composition of overweight adults. *Brit J Nutr*. 2012;107(5):749–754.
145. Dewulf EM, Cani PD, Claus SP, et al. Insight into the prebiotic concept: lessons from an exploratory, double blind intervention study with inulin-type fructans in obese women. *Gut*. 2013;62(8):1112–1121.
146. Vulevic J, Juric A, Tzortzis G, Gibson GR. A mixture of trans-galactooligosaccharides reduces markers of metabolic syndrome and modulates the fecal microbiota and immune function of overweight adults. *J Nutr*. 2013;143(3):324–331.
147. Kadooka Y, Sato M, Imaizumi K, et al. Regulation of abdominal adiposity by probiotics (*Lactobacillus gasseri* SBT2055) in adults with obese tendencies in a randomized controlled trial. *Eur J Clin Nutr*. 2010;64(6):636–643.
148. Chapman CM, Gibson GR, Rowland I. Health benefits of probiotics: are mixtures more effective than single strains? *Eur J Nutr*. 2011;50 (1):1–17.
149. Million M, Angelakis E, Paul M, Armougom F, Leibovici L, Raoult D. Comparative meta-analysis of the effect of Lactobacillus species on weight gain in humans and animals. *Microbial Pathogen*. 2012;53(2):100–108.
150. Cani PD, Bibiloni R, Knauf C, et al. Changes in gut microbiota control metabolic endotoxemia-induced inflammation in high-fat diet-induced obesity and diabetes in mice. *Diabetes*. 2008;57(6):1470–1481.
151. Cho I, Yamanishi S, Cox L, et al. Antibiotics in early life alter the murine colonic microbiome and adiposity. *Nature*. 2012;488 (7413):621–626.
152. Gibson GR, Roberfroid MB. Dietary modulation of the human colonic microbiota: introducing the concept of prebiotics. *J Nutr*. 1995;125 (6):1401–1412.
153. Roberfroid M, Gibson GR, Hoyles L, et al. Prebiotic effects: metabolic and health benefits. *Brit J Nutr*. 2010;104(suppl 2):S1–S63.

CHAPTER

12

An Apple a Day Keeps the Doctor Away – Inter-Relationship Between Apple Consumption, the Gut Microbiota and Cardiometabolic Disease Risk Reduction

Athanasios Koutsos and Julie A. Lovegrove*[†]

***Hugh Sinclair Unit of Human Nutrition and Institute for Cardiovascular and Metabolic Research (ICMR), Department of Food and Nutritional Sciences, University of Reading, Reading RG6 6AP, UK; Research and Innovation Centre, Fondazione Edmund Mach, San Michele all'Adige, Trento, Italy [†]Hugh Sinclair Unit of Human Nutrition and Institute for Cardiovascular and Metabolic Research (ICMR), Department of Food and Nutritional Sciences, University of Reading, Reading RG6 6AP, UK**

INTRODUCTION

There is now considerable scientific evidence that a diet rich in fruits and vegetables can improve human health and protect against chronic diseases.[1] Apples are among the most frequently consumed fruits in the world,[2] since they are available the whole year, are inexpensive, convenient to consume and there is, also, the general perception that apples are good for our health. They are consumed as fresh fruit in addition to juices, dried fruit, ciders, concentrates and purees and constitute an important part of the human diet.[3] Several epidemiological studies have linked frequent apple consumption with a reduced risk of cardiovascular disease (CVD), specific cancers and diabetes.[4–7] Moreover, intervention studies in humans and animals support these beneficial effects and relate apple intake with reduced blood lipid levels, weight loss, antioxidant activity, antiinflammatory properties, positive effects on glucose homeostasis, enhanced vascular function, lower blood pressure and beneficial modulation of the human gut microbiota and their metabolic output.[8–17] In this chapter, we will present the evidence concerning the ability of apples to reduce cardiometabolic disease risk. The possible mechanisms of action will be discussed, especially those linked to the human gut microbiota and gut function. The nutritional composition of apples will be addressed initially, with identification of molecular components which are more likely to be responsible for the health effects of apples.

APPLE COMPONENTS

Apples, like all whole plant foods, contain a range of nutrients with different biological activities. These potentially bioactive compounds include polyphenols, vitamins, minerals, lipids, proteins/peptides and carbohydrates, especially complex carbohydrates which are not readily digested in the upper gut, but may survive passage

Diet-Microbe Interactions in the Gut.
DOI: http://dx.doi.org/10.1016/B978-0-12-407825-3.00012-5

through the stomach and small intestine and reach the colon. Non-digested dietary carbohydrate, especially complex plant polysaccharides, are the main energy source of the human gut microbiota which derive energy and carbon by fermenting these fiber compounds, generating a range of end products, most notably the short-chain fatty acids (SCFA), acetate, butyrate and propionate, which become available to the host. SCFA, as will be discussed below, are emerging as powerful mediators of inter-kingdom communication within the human body, regulating diverse physiological functions not only in the gut but systemically, including regulation of food intake (satiety control through incretin production in the gut and modulation of the gut:brain axis), control of gut permeability, regulation of inflammatory processes and immune tolerance, cholesterol synthesis in the liver, fat storage and adipocyte hormonal (leptin) signalling and possibly thermogenesis.[18–20] It is estimated that most of the dietary polyphenolic compounds derived from plant foods are also non-digestible in the upper gut and only become biologically available, and in some cases, biologically active once they have been catabolized by the intestinal microbiota.[21] Moreover, complex polyphenol compounds, for example condensed tannins (proanthocyanidins) in apple, have been shown to inhibit the activity of gastrointestinal digestive enzymes, especially those involved in starch degradation, and can therefore impact on the digestion of complex plant carbohydrates and the amount of these fermentable materials which reach the gut microbiota in the colon. Therefore not only do the individual molecular constituents of apple mediate these health effects, but they may act synergistically, boosting microbial activities in the colon and possibly modulating the microbiome in structure and function.

Simple Carbohydrates

Apple sugar content plays an important role in sensory properties and consumer acceptance.[22] It can vary depending on the variety, exposure to sunlight, temperature and storage period.[22,23] According to the United States Department of Agriculture (USDA), total sugar content is approximately 10.4 g/100 g. A study by Feliciano et al. (2010) focused on 10 traditional and exotic varieties in Portugal reported a slightly higher sugar content of 12 g/100 g.[22] Moreover, total sugar content between 10 apple varieties cultivated in Romania ranged from 9.53 to 12.34%.[23] Fructose is the predominant sugar, (approximately 6 g/100 g). Sucrose and glucose are also abundant (Table 12.1).

Animal model studies suggest that a high fructose diet, mainly as part of sucrose or high-fructose corn syrup (a key additive in processed foods, soft drinks and other beverages), has been associated with insulin resistance, impaired glucose tolerance, hyperlipidemia and hypertension, as reviewed by Elliott et al. in 2002.[24] However,

TABLE 12.1 Composition of Apples (*Malus domestica*), Raw with Skin (USDA National Nutrient Database for Standard Reference)

Nutrients and Units	Value per 100 g
Water (g)	85.56
Energy (kJ)	218
Protein (g)	0.26
Total lipid, fat (g)	0.17
Carbohydrate (g)	13.81
Total dietary fibre (g)	2.40
Sugars, total (g)	10.39
Sucrose (g)	2.07
Glucose (dextrose) (g)	2.43
Fructose (g)	5.90
Lactose (g)	0.00
Maltose (g)	0.00
Galactose (g)	0.00
Starch (g)	0.05

interestingly only the consumption of sweetened fruit juices, but not 100% fruit juices, have been associated with type 2 diabetes in prospective cohort studies.[25–27] This suggests that the naturally occurring fructose and sugar in 100% fruit juices and whole fresh fruits may have a different metabolic impact than fruit juices with added sugar,[27,28] although intervention studies are required to confirm these findings. It has been proposed that bioactive compounds in fruit juices including vitamins, minerals, polyphenols and fiber may counteract the potential adverse effects of sugars.[25]

Vitamins and Minerals

Apples are a good source of vitamin C and E. Moreover they contain lutein, folic acid, potassium and magnesium.[2] Vitamin C and E together with lutein are important antioxidants and may contribute to protective health effects against chronic diseases.[29,30] Folic acid, as well as potassium and magnesium, has been associated with a reduced risk of CVD.[31,32]

Fiber

Apple fiber content is relatively high, approximately 2–3%, compared with other fresh fruits.[2,22] Insoluble fiber represents 50% of the total fiber and includes cellulose and hemicellulose. Soluble fiber, mainly pectin, includes homogalacturonans and rhamnogalacturonans, with homogalacturonans being the main fraction consisting of long chains of α (1–4) linked galacturonic acids. The carboxyl groups in these chains can be esterified by methanol. The degree of methylation (DM) strongly influences the functional and physicochemical properties of pectins.[33–35] Pectin has been associated with improved insulin resistance in rats[36] and cholesterol-lowering effects in humans, with the degree of esterification, molecular weight and structure having an important influence on the degree of effect.[37–39] Moreover, pectin is a gelling agent that affects transit time, gastric emptying and nutrient absorption.[39–43] It is resistant to hydrolysis and, thus, reaches the large intestine where it undergoes fermentation by the gut microbiota.[15,44–46]

Moreover, immature apples contain a high amount of starch. It was reported that immature Granny Smith apples had a starch content of 53.2% on a dry weight basis, whereas Royal Gala had 44%.[47] However, when apples ripen, starch is degraded to sugars and only a few cells contain starch.[48] Mature Marie Menard, a cider apple variety[14] and mature Fuji apples[49] contain 3 and 4% of starch, respectively (dry-weight basis). Starch is the major dietary carbohydrate which can be digested and absorbed by the human gut and represents one of the key dietary energy sources for human metabolism. However despite this, small amounts of undigested starch have been shown to reach the gut microbiota in the colon where it is extensively fermented into SCFA.[50] This is due to the fact that different starches have different degrees of digestibility. Four types of resistant starches (RS) exist, differing in molecular structure or chemical branching, which render them differentially susceptible to the activities of human digestive processes in the stomach and small intestine.[50,51] RS type II is resistant to digestion by virtue of its structured organization into starch granules in fresh plants, including apples. These granules are resistant to digestion in the upper gut and allow starch to enter the colon and become available for fermentation into SCFA.[50,51] Moreover, these RS can have differential effects on the gut microbiota[50,51] and have been shown to mediate health effects in animals and humans, including improvements in markers of metabolic disease.[52–56] However these potential effects may not be of great significance in mature apples due to the low starch and potentially RS levels.

Polyphenols

Many studies suggest an important role for polyphenols present in plant foods in the prevention of diseases related with oxidative stress and inflammation, such as cancer, cardiovascular and neurodegenerative diseases, as has been reviewed extensively by Scalbert et al. (2005).[1] The daily intake of polyphenols has been estimated to be 1 g/day, with two-thirds consisting of flavonoids and one-third of phenolic acids.[57] Their concentration within foods is influenced by genetic and environmental factors, including plant variety, growing season, fruit maturity, geographical location as well as storage and processing.[58] Apples are an important source of polyphenols including flavanols and flavonol glycosides. Many studies have explored the polyphenol content of several apple varieties,[3,58–61] but only the most recent ones have included the determination of oligomeric and polymeric proanthocyanidins, which represent an important class of apple polyphenols.[3,60] Vrhovsek et al. (2004) reported the polyphenol content of eight apple varieties sampled in Trentino, Italy, which represented approximately

88.1% of the total production of dessert apple in Italy with similar levels for southern Europe (France, Spain, Greece and Portugal) and 70.1% of Western Europe in 2002.[60] According to their work, the average content of total polyphenols was 110.2 mg/100 g of fresh fruit. The total polyphenol content, measured by an optimized Folin–Ciocalteu (FC) method (sugars, amino acids and ascorbic acid were removed) and explained by chromatographic analysis, between the different varieties ranged from 66.2 to 211.9 mg/100 g according to the following increasing order: Fuji, Braeburn, Royal Gala, Golden Delicious, Morgenduft, Granny Smith, Red Delicious and Renetta.[60] The most important polyphenol groups in those apples (with the skin) are represented in Figure 12.1. A single serving of apples (150 g) in Italy (and Western Europe) contains as a mean concentration: 101.9 mg (121.7 mg) of flavanols, 13.6 mg (16.2 mg) of hydroxycinnamates, 8.2 mg (8.9 mg) of flavonols, 3.8 mg (4.4 mg) of dihydrochalcones and 0.9 mg (1.2 mg) of anthocyanins.[60] The type and the range of the polyphenolic compounds were similar to a later study of 67 old and new apple varieties grown in Western Europe.[3] The total polyphenol content of apples were also similar with broad values, for apples from Greece,[62] New Zealand[63] and United States.[59] In other varieties, Kevers et al. (2011) reported a mean total phenolic content between 140 and 447 mg [chlorogenic acid equivalents (CAE)/100 g fresh weight (FW)] whereas, Imeh and Khochar (2002) found higher values in the range of 300–535 mg (Gallic Acid Equivalent (GAE)/100 g FW).[64,65] Moreover, Ceymann et al. (2012) examined 104 European apple cultivars, but focused only on 12 individual polyphenols.[61] The differences in the concentration of apple polyphenols among these studies could be due to the different cultivars, area of cultivation, maturity, storage, extraction procedures, analytical techniques and pre- or post-harvest factors.[61]

Apple polyphenols are present in different parts of the fruit. A major source is the skin, which contains all the flavonols and anthocyanins in addition to an important amount of dihydrochalcones.[58,60] Phenolic acids are present in the flesh whereas most of the dihydrochalcones are in the core and the seeds.[66] This distribution in different parts of the fruit also affects polyphenol concentration of apple juice where only small amounts of quercetin glycosides and dihydrochalcones are present. However, polyphenol content is also affected by the technological procedure. The oxidative conditions and the clarification process during the production of clear apple juice reduces the phenolic content.[67] In contrast, the anaerobic conditions and the lack of a clarification step during cloudy apple juice production prevents an important loss of polyphenols.[67]

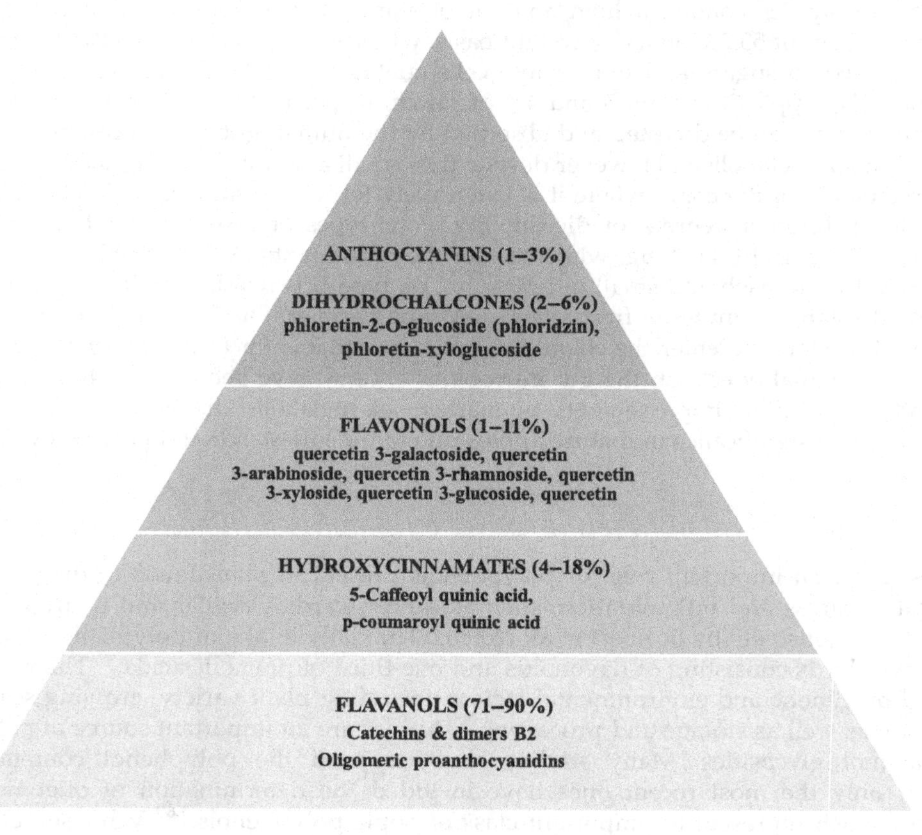

FIGURE 12.1 Apples polyphenol composition. *Data from Vrhovsek et al. (2004).*[60]

Polyphenols Bioavailability

The health effects of polyphenols depend on their intake and bioavailability.[68] They exist in different chemical forms and their structure could influence their absorption in the small intestine and in the colon through their metabolism by the gut microbiota.[57] Many studies have focused on pure polyphenols or plant extracts in order to explore potential health effects, however, a growing number of studies show that the concentration of the parent polyphenols in human plasma is often low.[57,69] Bioconverted and conjugated forms of intact polyphenols could have a more active role than the parent compound.[69] Moreover, food matrix could also influence the absorption, positively or negatively.[70] The possible metabolic routes of polyphenols after consumption are shown in Figure 12.2. Aglycones and a few glycosides can be directly absorbed in the small intestine.[69] However, most of the polyphenols have to be degraded by enzymes. Glycosylated polyphenols require loss of their sugar moiety, by the activity of glycosidases, before being absorbed, whereas polyphenols with more complex sugars reach the colon intact.[69] Furthermore, non-glycosylated polyphenols, such as polymeric flavanols, require cleavage, whereas esterified phenolic acids require hydrolyzation.[69] During the intestinal absorption and in the liver, polyphenols can be subjected to phase II metabolism leading to the formation of glucuronides and/or sulfates.[69] Polyphenols can circulate in the blood or be resecreted into the intestine via bile as a result of the enterohepatic circulation. The latter fraction can be either deconjugated by gut microbes and absorbed in the colon or further metabolized by the gut microbiota.[57,69,71]

The consumption of 1L of cloudy apple juice by 11 healthy ileostomy subjects indicated that 58.3% of the total polyphenols were absorbed or degraded, whereas the remaining were detected in the ileostomy fluid.[72] Procyanidin polymers are the major fraction of polyphenols in cloudy apple juice and only the low-molecular-weight fractions are absorbed [degree of polymerization (DP) < 3].[72] Kahle et al. (2007) showed that although there is a cleavage of the oligomeric procyanidins into smaller units, the major fraction (90%) was not absorbed and thus, reached the colon.[72] Both *in vitro* and animal studies support the theory that polymerization decreases intestinal absorption.[73]

In another study by Hagl et al. (2011), 10 healthy ileostomy subjects consumed 0.7L of apple smoothie containing 60% of cloudy apple juice and 40% apple puree.[70] After 8 h $63 \pm 16.1\%$ of the total polyphenols and D-(−)-quinic acid were detected in the ileostomy bags.[70] Hydrolysis of hydroxycinnamic acids in the small intestine has been observed, including caffeoylquinic acids and p-coumaroylquinic acids, leading to D-(−)quinic acid liberation.[70,72] Other studies, however, suggested that the cleavage and the absorption of chlorogenic acid (5-caffeoylquinic acid) occur mainly in the colon by the action of microbial esterases[57,73,74] Moreover, hydroxycinnamic acids were absorbed to a lesser extent after apple smoothie intake compared to a cloudy apple juice, indicating that the higher quantity of cell wall compounds in apple smoothie affected bioavailability.[70,72]

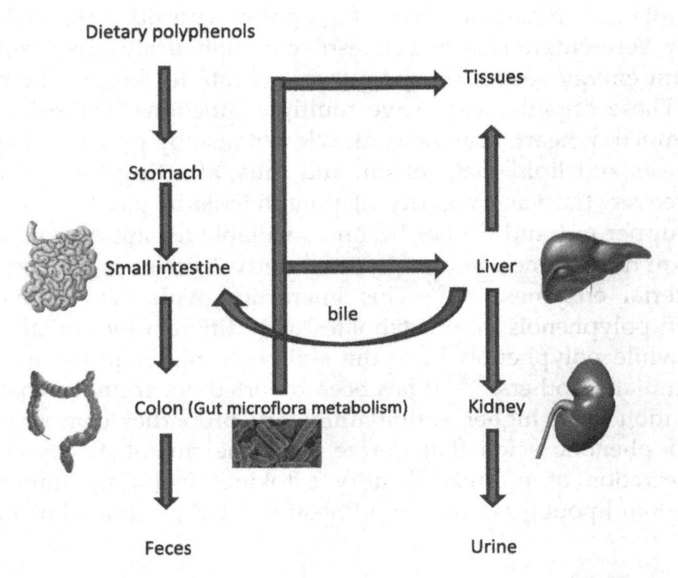

FIGURE 12.2 Possible routes after polyphenol consumption. *Modified from Scalbert and Williamson (2000).*[57]

It has been reported that the bioavailability of quercetin from apples is only 30% relative to quercetin from onions, due to differences in quercetin conjugates.[75] The nature of sugar moiety affects quercetin absorption.[73] Onions have higher amounts of quercetin aglycones and quercetin glucosides which are more bioavailable compared to quercetin monoglycosides and quercetin rutinosides present in apples.[75,76] Other studies reported that quercetin glucosides were more effectively absorbed compared to pure quercetin and that quercetin appeared in plasma only in conjugated forms with high half-lives (reviewed in Ref. 73). The main quercetin glycosides present in the ileostomy bags of human volunteers after the consumption of apple smoothie or cloudy apple juice were 3-O-rhamnoside and 3-O-arabinoside, but with much higher recovery rates after apple smoothie intake compared with the cloudy apple juice.[70,72]

Absorption and metabolism of the major dihydrochalcones in apple products including phloretin-2′-O-(2″-O-xylosyl)glucoside, phloretin-2′-O-glucoside and the aglycone phloretin was explored in nine healthy and five ileostomy subjects after the consumption of 500 mL of apple cider, for 24 h.[77] Phloretin-2′-O-glucuronide was detected in plasma, in addition to ileal fluid and urine, indicating glucuronidation of dihydrochalcones before absorption, mainly in the small intestine.[77] In total, 38.6% of the dihydrochalcones were detected in the ileal fluid. Moreover, phloretin-2′-O-(2″-O-xylosyl) glucoside was present in the ileal fluid, since it is absorbed to a lesser degree compared with phloretin-2′-O-glucoside and thus, reached the large intestine in healthy subjects together with phloretin conjugates (glucuronides and sulfates) and unconjugated phloretin.[77] These results are consistent with Kahle et al. (2007)[70] and Hagl et al. (2011).[72]

THE HUMAN GUT MICROBIOTA

The human gut microbiota is a diverse community of microorganisms, comprising approximately 10^{12} bacterial cells and up to 1000 different species that colonize the gastrointestinal tract.[78,79] *Bacteroidetes* and *Firmicutes* are the main two divisions, representing more than 90% of all the phylotypes, followed by *Proteobacteria*, *Actinobacteria*, *Fusobacteria* and *Verrucomicrobia*.[80] This microbiota plays an important role in human health and well-being by increasing the efficiency of energy harvest through the fermentation of non digestible dietary compounds, modulating immune function, synthesizing vitamins, such as B_{12} and K, and inhibiting the growth of potential pathogens.[81] Bifidobacteria, lactobacilli and butyrate-producing bacteria, such as *Faecalibacterium prausnitzii*, are good examples of health-promoting saccharolytic-type bacteria.[80,81] A potential modulation of the gut microbiota composition towards the beneficial bacteria may decrease risk factors related to cardiovascular and metabolic diseases,[82] including blood lipid levels,[83] inflammation[84] and insulin resistance.[85] In contrast, the prevalence of specific bacteria, such as the *Enterobacteriace* phylum, may be associated with adverse effects including gastrointestinal diseases and toxin production.

Whole plant foods like fruits and vegetables including apples with fiber and polyphenols, have an important role in microbial metabolism. Fermentation of non-digestible carbohydrates, including plant polysaccharides and oligosaccharides, are the main energy source for the gut microbiota leading to the production of SCFAs acetate, propionate and butyrate.[21] These organic acids have multiple functions in the host, not only as an important energy source for intestinal mucosa, heart, brain and muscle but also by playing a significant role in cell function, immune system, thermogenesis and lipid metabolism, and thus, affecting the risk of gastrointestinal disorders, cancers and CVDs.[19,21] Moreover, the vast majority of polyphenols in plant-based foods, including apples, are not readily absorbed in the upper gut and do not become available for absorption across the gut wall until they reach the colon where they are transformed into unglycated derivatives or hydrolyzed into smaller phenolic acids through the action of bacterial enzymes.[57,86,87] This interaction with the gut microbiota is reciprocal, since commensal bacteria transform polyphenols into metabolites with different bioavailabilities, activities and functions, which are better absorbed, while polyphenols have the ability to modulate the gut microbiota, inhibiting some bacterial populations and stimulating others.[88,89] It has been reported, for instance, that quercetin metabolites generated by microbial transformation exert higher antiinflammatory properties compared to the parent compound.[90] Furthermore, dihydroxylated phenolic acids that derive from the microbial degradation of proanthocyanidins, significantly inhibited the secretion of proinflammatory cytokines including tumor necrosis factor α (TNF-α), interleukin-1β (IL-1β), and IL-6 in lipopolysaccharide (LPS)-stimulated peripheral blood mononuclear cells from six healthy volunteers.[91]

Modulation of the Gut Microbiota Composition – Impact of Apples and Apple Components

Aberrant gut microbiota profiles or microbiota metabolic activities have been associated with obesity and the diseases of obesity including diabetes (type 1 and type 2), CVD and non-alcoholic fatty liver disease.[92–100] These microbiota profiles appear to be modifiable through dietary means,[84,101,102] and at least in animals, have been linked to mechanisms regulating mammalian energy metabolism.[96,103,104] Several studies suggest that diet could have short- and long-term effects on human gut microbiota composition and on its activities related to host physiology and metabolic disease risk, not least production of SCFA, bile acid deconjugation and polyphenol metabolism.[20,83,105–107] However, an individual's profile also plays an important role.[106] Eckburg et al. (2005) reported a greater interindividual variation in stool and mucosa bacterial composition compared to intraindividual differences.[79] This interindividual variation might be related with risk factors of chronic diseases; nevertheless, a potential link has to be confirmed.[82] It has been shown that obese people may have an increased ratio of *Firmicutes/Bacteroidetes* compared to lean subjects,[101] however, recent studies have not always confirmed this difference.[108,109]

Polyphenols and fiber can interact with the gut microbiota. As mentioned above, dietary fiber is not absorbed, it is resistant to digestion and, thus, can be fermented by gastrointestinal microbiota.[110–112] However, in order to be defined as a prebiotic, a fermentable fiber must be selective towards a beneficial microbiota composition.[112,113] According to the recent definition by Gibson et al. (2010), a dietary prebiotic is "a selectively fermented ingredient that results in specific changes in the composition and/or activity of the gastrointestinal microbiota, thus conferring benefit(s) upon host health."[114] Polyphenols may also play an important role in the maintenance of gastrointestinal health.[89] Tzounis et al. (2011) have shown that a daily consumption of a high-cocoa flavanol drink (494 mg/day) for 4 weeks could significantly increase bifidobacteria and lactobacilli populations and significantly decrease clostridia counts compared with a low-cocoa flavanol drink (23 mg/day), in a randomized, double-blind, controlled, crossover, human trial of 22 subjects.[115] In a recent randomized, crossover, controlled human intervention study by Queipo-Ortuno et al. (2012) the impact of red wine, dealcoholized red wine and gin consumption on gut microbiota composition was explored in 10 subjects for 20 days.[116] Red wine consumption significant increased the number of *Enterococcus, Bacteroides* and *Prevotella* genera and significantly decreased the *Clostridium* genera and *Clostridium histolyticum* group.[116] Gin consumption was related with increased levels of *Clostridium* and the *Clostridium histolyticum* group.[116] Moreover, both red wine and dealcoholized red wine increased the levels of *Blautia coccoides–Eubacterium rectale* group, *Bifidobacterium, Eggerthella lent,* and *Bacteroides uniformis,* suggesting possible prebiotic benefits of wine polyphenols.[116]

Pectin, an important apple component, is almost completely fermented *in vitro* and, thus, may change the composition of the human gut microbiota.[44–46] However, chemical characteristics of pectins could influence their fermentability properties as has been shown in *in vitro* studies.[33,35,117] Dongowski et al. (1998) showed that low methoxyl (LM) pectins were fermented faster than high methoxyl (HM) pectins.[33] This is in agreement with Olano-Martin et al. (2002).[117] As the authors of these studies indicated, galacturonic acids in LM pectins are exposed more to microbes and their enzymes compared with HM pectins and as a result they can be rapidly fermented. However, in a later *in vitro* study, Gulfi et al. (2006) reported the opposite results and concluded that HM pectins were associated with higher amounts of SCFA and gases compared to LM pectins.[35] Faecal donor variability together with differences in the ratio and the composition of buffer and substrates between these studies may explain the inconsistent results.[35] Focusing on gut microbiota composition, two studies by Salyers et al. (1977) reported that 90 out of 188 cultured strains of several species of *Bacteroides* utilized citrus pectin, whereas only *Eubacterium eligens* was able to utilize pectin among 154 strains of positive anaerobes including species from actinobacteria and *Firmicutes.*[118,119] In contrast, a study by Lopez-Siles et al. (2012) showed that eight out of 10 strains of *Faecalibacterium prausnitzii* utilized apple pectin.[120] Moreover, *Faecalibacterium prausnitzii* strains were able to compete for apple pectin with other known pectin-utilizing species such as *Bacteroides thetaiotaomicron* and *Eubacterium eligens.*[120] *Faecalibacterium prausnitzii* belongs to *Firmicutes* and it is a dominant member of gut microbiota, an important butyrate producer and a potential antiinflammatory agent.[121] Comparing the fermentability properties of pectin and pectic oligosaccharides (POS) Olano-Martin et al. (2002) reported that POS were associated with a higher bifidobacteria number compared to pectin.[117] Moreover, *Bacteroides* and *Clostridium* species have been reported to grow better on pectin than on POS.[117] Similarly, in a recent study by Chen et al. (2013), apple POS prepared by a high-pressure homogenization technology significantly increased bifidobacteria and lactobacilli numbers, while decreasing clostridia and *Bacteroides.*[122] Pectin, on the other hand, significantly increased bifidobacteria, but also clostridia, *Bacteroides* and eubacteria.[122] It seems that pectin is degraded by many intestinal bacteria including *Bacteroides*, eubacteria, clostridia and bifidobacteria,[117,122,123] whereas selectivity towards bifidobacteria has only been confirmed for POS, indicating a potential prebiotic effect.[117,122]

The effects of apple polyphenols on gut microbiota composition are not fully understood. Procyanidins, are not absorbed in the small intestine and, thus, could reach the large intestine.[72] Using an *in vitro* batch culture fermentation system, Bazzocco et al. (2008) reported that apple proanthocyanidins inhibited SCFA production, suggesting a potential reduction of the beneficial saccharolytic fermentation.[124] Levrat et al. (1993) compared the effects of citrus pectin and condensed tannins (proanthocyanidins) intake on cecal fermentations in rats.[125] They reported that the pectin diet was related with a high cecum content and increased levels of volatile fatty acids (VFA), whereas tannins resulted in a reduction of cecal VFA concentrations.[125] Kosmala et al. (2011) reported that although a 5% unprocessed apple pomace intake in rats, containing 61% of fiber and 0.23% of polyphenols was associated with a high cecal SCFA compared to a control diet, removal of the polyphenol fraction reduced the pH and beneficially modified the balance between glycolytic and proteolytic activity of cecal microbiota through a modified ratio of SCFA and branched-chain fatty acids (BCFA).[126] In contrast, Aprikian et al. (2003) indicated that cecal pH in rats was beneficially decreased only after a combined diet of apple pectin (5%) and high-polyphenol freeze-dried apple (10%) compared with the individual diets.[127] However, all of the previous studies did not indicate the specific gut microbiota composition.

Several animal studies have explored the effect of apple products consumption on gut microbiota.[10,15,128] In a controlled rat feeding study, apple pomace extraction juice colloids (5%), which were rich in dietary fiber and polyphenols, increased the population of *Eubacterium rectale cluster, Bacteroidaceae,* caecal content weight and SCFA concentration (mainly acetate and propionate), indicating microbial fermentation of pectin.[10] Furthermore, an alcohol pomace extract rich in insoluble fiber was also related to those beneficial effects, but further increased butyrate concentration and *Eubacterium rectale cluster* compared to the juice colloids and the control.[10] In addition, the same group reported that a 4-week consumption of apple juice, rich in polyphenols and fiber, increased the population of *Lactobacillus, Bifidobacterium,* acetate amount, total SCFA and the excretion of bile acids and neutral sterols, in rats.[128] Acetate production has been associated with pectin fermentation, and polyphenols, such as quercetin-3-glucoside have been linked to increased SCFA concentration.[128] A 14-week chronic feeding study in rats significantly modified microbial composition in the cecum, after whole-apple consumption; with no effects after the intake of apple juice, puree or pomace.[15] The addition of 7% of pectin increased *Clostridium coccoides* group and the expression of genes related to butyryl-coenzyme A (CoA) transferase, which are present in bacteria of the *Clostridium Cluster XIVa, Roseburia-Eubacterium rectale cluster* and *Faecalibacterium prausnitzii.*[15] Moreover, it also resulted in a decrease of *Bacteroides* spp. compared to a control.[15]

In a small human study, eight healthy subjects consumed two apples daily for 2 weeks.[16] Bifidocteria populations were significantly increased, whereas a trend was shown towards increased levels of *Lactobacillus, Streptococcus* and *Enterococcus.*[16] A significant reduction was also observed for lecithinase-positive clostridia, including *C.perfringens.*[16] However, the limitation of this study was the small number of subjects and the use of unreliable culture based microbiological techniques. In a very recent crossover study by Ravn-Haren et al. (2012), 23 healthy subjects consumed whole apples daily (550 g), apple pomace (22 g), clear and cloudy apple juice (500 mL) from the same apple variety, during a 4-week period.[129] Although the intake of whole apples and pomace was associated with a lower pH value in feces, decreased excretion of lithocholic acid (LCA) (including cloudy apple juice consumption) and some differences in denaturing gradient gel electrophoresis (DGGE) profiles of some subjects, a potential change in gut microbiota composition was not confirmed by qPCR.[129] The beneficial effects, mainly from whole apples, could also indicate a synergistic effect of apple polyphenols and fiber, as has been shown for animal studies.[127] Further human studies with a large number of subjects, a longer feeding period and recent metagenomic techniques are necessary.

CARDIOMETABOLIC DISEASE RISK – EPIDEMIOLOGICAL STUDIES

Cardiovascular disease risk factors including dyslipidemia, obesity, hypertension, inflammation and diabetes are some of the biggest health challenges today. Epidemiological data suggest that flavonoid consumption may reduce CVD risk and apple might be a major contributor to these effects.[6,130] In a chronic study by Knekt et al. (2002) of 10,054 Finnish men and women, total mortality was strongly inversely associated with a high flavonoid consumption, with apple showing the strongest correlation (Relative Risk, 95% CI: 0.87; 0.77, 0.99; $P = 0.003$).[6] In Iowa Women's Health Study, 34,489 subjects free of CVD were followed up for 16 years.[130] Flavanones, anthocyanidins and several foods including apples, pears, grapefruit, strawberries, chocolate and wine were associated with reduced risk of coronary heart disease and CVD.[130] In contrast, in a prospective study of 38,445 women by Sesso et al. (2003), flavonoid intake was not strongly related with reduced CVD or vascular events risk after a follow-up of 6.9 years.[131] Apple was associated with a 13–22% reduction, however, this relationship was not statistically significant.[131] Similarly, in an earlier study of 805 elderly men, followed up for 5 years, a high apple consumption (>110 g/d) was inversely associated with coronary disease mortality although similarly the relationship was not statistically significant.[132]

CARDIOMETABOLIC RISK FACTORS

Lipid Metabolism

The effects of apple products or their active components, on blood lipid parameters and other CVD risk factors have been explored in a number of animal studies.[10,128,133–135] Several human intervention studies also support the hypothesis that frequent apple consumption may beneficially modulate blood lipid levels. Only human intervention studies will be presented in this section; animal studies will be discussed in the section Mechanisms Explaining the Potential Lipid Lowering Effects in order to describe the potential underlying mechanisms of action. A summary of the main results of the human trials are presented in Table 12.2.

Total cholesterol (TC) and low-density-lipoprotein cholesterol (LDL-C) were significantly decreased in a dose-dependent manner after the daily consumption of food supplements containing apple polyphenols (0, 300, 600 and 1500 mg), during a 4-week parallel, placebo-control study in 48 healthy, mild hypercholesterolemic subjects.[142] Although the main effects were observed after the highest polyphenol concentration (1500 mg), which is considerably

TABLE 12.2 Effects of Apples and Apple Product Consumption on Blood Lipid Levels in Humans

Author	Product Given – Dose	Subjects (n, sex) – Study Duration	Study Design	Results
Gormley et al. (1977)[136]	3 apples p/d Control group: no more than 3 apples p/w	76 mildly hypercholesterolemic men – 4 months	Parallel, paired according to cholesterol and fruit dietary fiber intake	TC: ↓ 8.1% maximum HDL: ↑
Mayne et al. (1982)[137]	15 g apple fiber (prepared from freeze-dried fresh apples) No control	12 men and women with type 2 diabetes – 7 weeks	Intervention study	TC: → HDL: → TAG: →
Sable-Amplis et al. (1983)[138]	2–3 apples (350–400 g) p/d No control	30 hypercholesterolemic men and women – 1 month	Intervention study	Men: TC ↓ 8.6% Women: TC ↓ 11.7%
Mahalko et al. (1984)[139]	26 or 52 g of dried whole apple added to bread Control: low-fiber white bread	18 men and women with type 2 diabetes – 1 month	Randomized, crossover	TC: ↑ 6.4% (52 g) LDL: ↑ 10.8% (52 g) HDL/TG: → (both)
Mee and Gee (1997)[140]	568 mL of fiber-supplemented juice or filtered juice (control)	25 mildly hypercholesterolemic men – 6 weeks	Randomized, crossover	TC: ↓ 8.2% (supplemented) LDL: ↓ 14.1 (supplemented) HDL: ↓ 8.5 (filtered) TAG: → (both)
Davidson et al. (1998)[141]	720 mL of fiber-supplemented juice [0, 5, 9 or 15 g/ d of gum arabic and pectin (4:ratio)]	85 hypercholesterolemic men and women – 18 weeks	Randomized, double blinded, controlled, parallel	TC: → LDL: → HDL: → TAG: →
Hyson et al. (2000)[8]	340 g apple or 375 mL apple juice p/d	25 healthy men and women – 6 weeks	Randomized, unblinded, crossover	TC: → LDL: → HDL: → TAG: →

(Continued)

TABLE 12.2 (Continued)

Author	Product Given – Dose	Subjects (n, sex) – Study Duration	Study Design	Results
de Oliveira et al. (2003)[11]	300 g apple or 300 g pear or 60 g oat cookies/day	49 hypercholesterolemic, overweight women 35 followed for 12 weeks	Randomized to each different diet	TC: ↓ (oat group) TAG: ↑ (fruit group)
Nagasako-Akazome et al. (2005)[142]	300 or 600 or 1500 mg apple polyphenols supplements/d Control: supplement without polyphenols	48 healthy hypercholesterolemic men and women – 4 weeks	Randomized, double-blinded, placebo-controlled, parallel	TC: ↓ LDL: ↓ HDL: → TAG: → All the effects in the highest polyphenol amount group (900 mg/d)
Avci et al. (2007)[143]	1 apple/d	15 elderly subjects – 1 month	Intervention study	TC: → LDL: → HDL: → TAG: →
Nagasako-Akazome et al. (2007)[12]	600 mg apple or hop bract polyphenols capsules p/d Control: capsules without polyphenols	71 moderately obese male and female subjects – 12 weeks	Randomized, double-blinded, placebo-controlled, parallel	TC: ↓ LDL: ↓ HDL: → TAG: → VFA: ↓ Adiponectin: ↑
Vafa et al. (2011)[144]	300 g apple/d Control group: no apple intake	46 overweight and hyperlipidemic men – 8 weeks	Randomized	TC: → LDL: → HDL: → TAG: ↓ (in control group compared with the apple group) VLDL: ↓ (in control group compared with the apple group) TAG: → (within the apple treatment) VLDL: → (within the apple treatment)
Barth et al. (2012)[145]	750 mL of cloudy apple juice Control: isocaloric beverage without polyphenols	68 overweight men (n = 38 apple juice, n = 30 control) – 4 weeks	Randomized, blinded controlled, parallel	TC: → LDL: → HDL: → TAG: → % total body fat: ↓ Body fat mass: ↓ (only in IL-6-174 C/C variant compared with G-allele carriers)
Chai et al. (2012)[146]	75 g of dried apples or dried plum (comparative control)	160 postmenopausal women – 1 year	Randomized, single blinded, parallel	TC: ↓ (between the treatments) TC: ↓ 13% (within apple treatment) LDL: ↓ 24% (within apple treatment) HDL: → TAG: → TC:HDL: ↓ (within apple treatment) LDL:HDL: ↓ (within apple treatment)
Ravn-Haren et al. (2012)[129]	550 g whole apples, 22 g apple pomace, 500 mL clear or cloudy apple juice Control: period of restricted diet	23 healthy subjects – 4 weeks	Randomized, single blinded, crossover	Treatment resulted in significant effects in TC and LDL. No signifcant changes compared to the control period TC: → LDL: → HDL: → TAG: →

↑: significant increase, ↓ significant decrease, →: no statistical difference, p/d: per day, p/w: per week, TC: total cholesterol, LDL: low-density lipoprotein, VLDL: very-low-density lipoprotein, HDL: high-density lipoprotein, TAG: triacylglycerol, VFA: viscelar fat area.

higher than the level of polyphenols found in an apple, a later study by the same group indicated similar effects with a much lower amount.[12] In particular, a daily consumption of 600 mg of apple polyphenol-containing capsules significantly reduced TC and LDL-C, improved adiponectin levels and visceral fat area (VFA) in a 12-week parallel, placebo-controlled study of 71 overweight subjects.[12] It is well known that adiponectin level, a hormone secreted by adipose tissue, decreases concurrently with an increase in visceral fat and is also associated with insulin resistance.[12] The beneficial effects on VFA and body fat composition were also shown when an apple beverage or a cloudy apple juice was used as an alternative to polyphenol-containing capsules.[145,147] Barth et al. (2012) during a parallel, intervention study in 68 overweight subjects, assessed the effect of a daily consumption of 750 mL cloudy apple juice (n = 38), containing 800 mg of polyphenols, or a control for 4 weeks (n = 30).[145] Although no significant effects on lipid, adipokine and cytokine levels were reported, they indicated an interaction between IL-6-174 G/C polymorphism and body fat reduction, attributed to the cloudy apple juice consumption.[145] In another study, 23 healthy subjects consumed whole apples (550 g), apple pomace (22 g), clear and cloudy apple juices (500 mL) daily, each for 4 weeks in a crossover design including a control period.[129] Treatment resulted in significant effects in TC and LDL-C.[129] However, the results were not statistically significantly different when comparing each treatment with the control period.[129] TC and LDL-C concentrations were significantly increased after clear apple juice consumption compared to whole apple or apple pomace.[129] Health benefits were also observed when 75 g of dried apples or dried plum (comparative control) were consumed daily for 1 year in a parallel study of 160 postmenopausal women.[146] TC concentration was significantly decreased at 6 months compared to the control, whereas within the intervention group TC reduction was achieved as early as 3 months.[146] Two to three apples per day for 1 month significantly decreased TC levels in 30 hypercholesterolemic men and women.[138] In addition, apple pectin has been shown to significantly reduce TC levels.[148] In a meta-analysis by Brown et al. (1999), it has been suggested that 1 g of pectin per day could lower LDL-C by 0.055 mmol/L.[38] In a 6-week crossover study, 25 mildly hypercholesterolemic men consumed 567 mL of a fiber-supplemented unfiltered apple juice per day containing 5 g of apple fiber and 5 g of gum arabic or 567 mL of a control unfiltered apple juice without any fiber.[140] TC and LDL-C were significantly decreased by 8 and 14%, respectively, after the intake of the supplemented apple juice.[140] Although the supplemented apple juice also contained 200 mg of ascorbic acid, the lipid-lowering effect was attributed to the combined effect of apple fiber and gum arabic.[140]

However, some studies did not show any significant beneficial effects with a limited number suggesting some adverse effects. Hyson et al. (2000) showed that daily consumption of 375 mL of apple juice or 340 g of cored whole apple for 6 weeks did not have any effect on lipid levels of 25 normolipidemic subjects.[8] Daily consumption of 300 g of apples for 12 weeks significantly increased triacylglycerol (TAG) levels in a parallel study of 49 hypercholesterolemic women examining the effects of apple, pear or oat cookies, although they also reported a small but significant weight loss of 1.22 kg, after the fruit consumption.[11] A study of 15 older individuals reported that a daily intake of 2 g of apples per kg body weight (approximately 1 apple) for 1 month did not cause any significant differences in lipid levels.[143] Moreover, a more recent parallel study of 46 overweight men by Vafa et al. (2011) showed significant decreased serum TAG concentrations and very-low-density lipoprotein (VLDL) in a control group (no apple consumption) compared with a daily consumption of 300 g of apples for 8 weeks (intervention group).[144] TC and LDL-C levels were also significantly increased after the consumption of 52 g of dried apple (equivalent to 260 g of fresh apple) added to bread compared with a white bread control low in fiber, in a crossover 1-month study of 18 men and women with type 2 diabetes.[139] The lipid-lowering effect of gum arabic-pectin supplement was not observed in a study of 85 hypercholesterolemic subjects consuming 720 mL of apple juice containing 0, 5, 9 and 15 g of gum arabic and pectin (4:1 ratio) for 18 weeks.[141]

In general, there is some evidence to support a 5–8% reduction (approximately 0.5 mmol/L) in total cholesterol after the intake of three apples per day.[2] Moreover, apple juice consumption is associated with no effects or increased levels of TAG.[2] Clear apple juice lacks water-soluble pectin and has a low polyphenol content due to a clarification process.[149] The production of cloudy apple juice does not include clarification and, thus, may maintain an important polyphenol and pectin content.[67,149] Apple pomace, a by-product of juice production, consists of approximately 25% of fresh apple weight, and is a high source of fiber and polyphenols and therefore a valuable material for functional food products.[150,151] Apart from the polyphenol and fiber content, food matrix may also play an important role; whole apples contain intact cell wall components, which could contribute to the effects on lipid and glucose metabolism as will be described below.

Mechanisms Explaining the Potential Lipid Lowering Effects

Inhibition of Enterohepatic Circulation

Bile acids are synthesized in the liver from cholesterol and are conjugated with the amino acids glycine and taurine to form bile salts before secretion via the gall bladder. Following a meal bile salts are released into the small intestine and emulsify lipids to form micelles, facilitating lipid digestion and absorption.[152] Then, bile salts are deconjugated by bacteria, absorbed in the ileum with the majority returned to the liver as bile acids, where they are conjugated and secreted into bile. This continuous process of secretion, absorption and resecretion is known as enterohepatic circulation.[152] The enzyme responsible for the deconjugation of bile acids from their amino acid moiety is bile salt hydrolase (BSH), which has been isolated from several gut bacteria, including species of bifidobacteria and lactobacilli.[153] Once deconjugated, bile acids are less efficacious in absorbing dietary fat leading to less cholesterol absorption.[153,154] Moreover, the increased excretion of bile acids in the feces results in the formation of new bile acids in the liver from circulating cholesterol.[153,154] It has been suggested that polyphenols could bind to cholesterol and/or bile acid and enhance their excretion in the feces.[142] Furthermore, dietary procyanidins, present in high amounts in apples, could lower cholesterol absorption by decreasing micellar cholesterol solubility.[155,156] The concomitant inhibition of the enterohepatic circulation may result in an increased transformation of cholesterol to bile acids and thus, lower cholesterol concentration in the blood.[142] Interestingly, apple pectin is a viscous fiber with the ability to bind bile acids and cholesterol in the upper gut, thus reducing its efficacy in fat absorption.[133] It also can increase its passage into the colon where bile acids may be deconjugated via BSH activity of the gut microbiota, and extensively fermented into SCFA which independently affect physiological activities related to host energy metabolism both within the gut and systemically.

Modulation of Lipid Metabolism

Animal studies have shown that apple polyphenols, including mainly procyanidins, may modulate the activity of hepatic cholesterol 7α-hydroxylase (CYP7A1) and, therefore, lower cholesterol concentration in the liver and increase the excretion of acidic steroids.[156] CYP7A1 is the first and rate-limiting step of bile acid synthesis and thus, of cholesterol degradation.[157] In contrast, the activity of 3-hydroxy-3-methylglutaryl-coenzyme A (HMG-CoA), a cholesterol synthesis rate determining enzyme, significantly decreased after apple polyphenol intake in rats compared to a control.[12,142] Other effects of apple polyphenols related to lipid metabolism include a reduction in leptin levels,[156] adipose tissue weight,[158,159] activation of fatty acid β oxidation in the liver,[160] inhibition of hepatic fatty acid synthesis,[159] reduction of cholesterol esterification and intestinal lipoprotein secretion[161] and lower body weight gain.[10] However, those effects require confirmation in human studies. Pectin may also play an important role in some of these mechanisms including bile acid synthesis, cholesterol synthesis, increased satiety and reduced energy intake, with the degree of esterification, molecular weight and pectin source affecting their efficacy.[38,39,162–164] Moreover, modulation of the gut microbiota composition through pectin fermentation and the production of SCFA may also affect lipid metabolism.[18,19,38] Butyrate plays an important role in colonic function, whereas acetate and propionate have an impact on metabolic processes at a systemic level, and may possess opposing effects on lipid metabolism, as reviewed by Wong et al. (2012).[82] Propionate may inhibit cholesterol synthesis whereas acetate could increase hepatic lipogenesis, however, the results are inconsistent.[82] Moreover, pectin has been shown to increase the expression of monocarboxylic acid transporters responsible for the uptake of SCFA from the gut, thereby up-regulating the flow of SCFA from the colon.[165,166] Other potential effects of the gut microbiota include cholesterol binding to bacterial cell walls and the conversion of cholesterol to coprostanol by bacterial enzymes leading to direct excretion in feces.[167]

Digestive Enzyme Inhibition

Pancreatic lipase is a key enzyme for TAG absorption and many studies have associated polyphenols with lipase inhibition. Sugiyama et al. (2007) showed that oligomeric apple procyanidins had an inhibitory effect on pancreatic lipase activity *in vitro* which was higher than other polyphenolic compounds of an apple extract.[168] These results were also confirmed in both an animal and a human study where oligomeric apple procyanidins inhibited postprandial TAG absorption.[168]

Polyphenol–Pectin Synergistic Effect

The potential lipid-lowering effect of a daily consumption of three apples which has been suggested by human studies, can not be independently attributed to polyphenols or pectin. According to Jensen et al. (2009), 6 g of pectin or 600–1500 mg of apple polyphenols are responsible for significant cholesterol reduction in humans, when tested separately.[2] However, three apples contain much lower amounts of polyphenols and fiber, indicating a potential interaction between the two components. In an animal study Aprikian et al. (2003) reported that a diet including both apple pectin and polyphenols resulted in significant cholesterol reduction compared to a diet with either apple polyphenols or pectin.[10,127,128,134]

Blood Pressure and Vascular Function

Few studies have explored the potential role of apple products on blood pressure and vascular function. A potential decrease of blood pressure and improvement of vascular function may have beneficial effects on cardiovascular health.[169,170] In an acute human study of 30 subjects, Bondonno et al. (2012) found that an apple blend rich in flavonols (360 mg of quercetin) resulted in significantly higher flow-mediated dilatation (a non-invasive technique for the determination of endothelium-dependent vasodilation) and lower systolic blood pressure compared to a control.[17] However, these results are not supported by an earlier crossover chronic study of 30 hypercholesterolemic subjects after a daily consumption of 40 g of lyophilized apples compared to a control.[171] Many other studies have focused on the potential hypotensive effects of quercetin, a flavonol widely found in apples, onions and other food products, as reviewed by Larson et al. (2012).[172] In a double-blind, placebo-controlled, crossover study, a daily consumption of 730 mg of quercetin for 28 days significantly decreased blood pressure in hypertensive subjects (n = 22) but not in prehypertensive volunteers (n = 19).[172] Higher levels of quercetin (1000 mg) and 200 mg of rutin (quercetin rhamnoglucoside) had no effect in 27 normotensive human subjects.[173] In contrast, Egert et al. (2010) found a blood-pressure-lowering effect in obese and overweight prehypertensive volunteers with a much lower amount, 150 mg.[174] However, this was only in subjects homozygous for the apolipoprotein E3 genotype.[174] Apples generally contain a low amount of quercetin, approximately 5 mg/100 g, and thus any potential effects on blood pressure could not be attributed only to the quercetin content.[174] Further chronic human studies are necessary to assess the impact of apple product consumption on vascular function.

Inflammation

Inflammation is a key risk factor for coronary heart disease and metabolic syndrome. A chronic, systemic, low-grade inflammation which includes elevated blood levels of inflammatory molecules, such as acute-phase proteins, proinflammatory cytokines, chemokines and adhesion molecules is associated with several metabolic abnormalities and chronic diseases.[175,176] There is strong evidence that polyphenols could exert antiinflammatory/immunomodulatory properties,[177,178] however, few studies have focused on the effects of apple polyphenols. In an *in vitro* study by Andre et al. (2012) the potential antiinflammatory effect of apple procyanidins from 109 different cultivars was tested using cell-based assays.[179] Cultivars with a high procyanidin content were able to inhibit nuclear factor-kappa B (NF-κB), a transcription factor involved in the expression of inflammatory molecules including TNF-α, IL-1β, IL-6 and IL-8.[179] Jung et al. (2009) reported that procyanidin B1, procyanidin B2 and phloretin from apple juice extracts were related to an antiinflammatory activity *in vitro*.[180] Apple pectin has been also associated with antiinflammatory effects in animal studies by down regulating proinflammatory cytokine expression, including TNF-α, and immunoglobulin production.[181,182] whereas citrus pectin has been reported to block lipopolysaccharide (LPS) signaling pathways in LPS-activated macrophages.[183] Moreover, in an animal study the intake of 7.6% lyophilized apple (Marie Menard cider variety), rich in polyphenols was associated with reduced colonic inflammation in HLA-B27 transgenic rats, which develop spontaneous intestinal inflammation, compared to an apple variety low in polyphenols (Golden Delicious).[14] Studies in humans are scarce. In a cross sectional observational epidemiological study of 8335 US subjects the association between dietary flavonoid intake and serum C-reactive protein (CRP) levels was assessed.[184] CRP is a biomarker of chronic inflammation and a sensitive predictor of CVD.[185] Consumption of apples and vegetables were inversely associated with serum CRP levels, after the appropriate adjustment for covariates ($P < 0.05$).[184] However, in a randomized, controlled, parallel human dietary intervention of 77 subjects, apple consumption was not associated with an antiinflammatory activity.[186] The effects of apple intake on inflammation seem to be inconclusive and are mainly supported by *in vitro* or animal studies.

Gut microbiota composition may also modulate systemic inflammation. LPS, a constituent of Gram negative bacteria triggers the secretion of proinflammatory molecules. Elevated LPS levels in blood circulation, mainly through a high-fat diet, contribute to metabolic endotoxemia[187] which modulate glucose and insulin metabolism and play an important role in the progression and the rupture of the atherosclerotic plaque.[188] Chronic exposure to increased blood plasma LPS concentration may lead to insulin resistance an important mediator in the pathophysiology of the metabolic syndrome.[189] Prebiotic dietary fiber,[187] pectin[190,191] and apple polyphenols[192] may reduce metabolic endotoxemia by improving gut barrier function and reducing intestinal permeability and uptake of LPS. In high-fat fed mice, prebiotic oligofructose reduced endotoxemia, inflammation development and improved insulin secretion by increasing bifidobacteria levels.[187]

Antioxidant Role

Polyphenols, as antioxidants, could protect human cells from oxidative damage and, thus, may reduce the risk of many degenerative diseases.[1] In a study by Drogoudi et al. (2008) the *in vitro* antioxidant activity of apple peel and flesh from specific apple varieties – Fuji, Golden Delicious, Granny Smith, Jonagored, Mutsu, Starkrimson and Fyriki – was explored.[193] In the flesh tissue, Fyriki had the highest antioxidant activity and total polyphenol content, whereas Fuji, Golden Delicious and Granny Smith the lowest values. In the peel tissue, the highest antioxidant activity and polyphenol levels were found in Stakrimson and the lowest in Golden Delicious and Granny Smith. In addition, the authors reported that apple peel contained from 1.5 to 9.2 times higher total antioxidant activity and 1.2 to 3.3 times higher polyphenol content compared with the flesh.[193] Other *in vitro* studies by Khanizadeh et al. (2007) and Wojdylo et al. (2008) have also reported an association between total polyphenols and antioxidant activity whereas van der Sluis et al. (2001) did not show any significant relation.[3,194,195] Apple juices are also an important source of polyphenolic compounds, correlated with antioxidant activity.[196] Seeram et al. (2008) assessed the antioxidative potential of several polyphenol-rich beverages using four different tests Trolox equivalent antioxidant capacity (TEAC), total oxygen radical absorbance capacity (ORAC), free radical scavenging capacity by 2,2-diphenyl-1-picrylhydrazyl (DPPH), and ferric reducing antioxidant power (FRAP).[197] The antioxidant capacity was classified in the following order: pomegranate juice > red wine > concord grape juice > blueberry juice > black cherry juice, acaí juice, cranberry juice > orange juice, iced tea beverages, apple juice.[197] A cloudy apple juice which is richer in procyanidins compared to a clear apple juice was associated with a stronger antioxidant capacity.[67]

LDL-C oxidation is unquestionably an important contributor to atherogenesis. However, although the extent to which LDL-C can be oxidized ex vivo has been extensively investigated, its impact *in vivo* is contentious. The potential inhibition of LDL-C oxidation was tested for six commercial apple juices and the peel, flesh and whole fresh Red Delicious apples.[198] All the tested products inhibited LDL-C oxidation assessed by an *in vitro* copper catalyzed human LDL-C oxidation system. The inhibition of the test products was 21, 34 and 38% for flesh, whole apple and peel, respectively, and 9–34% for the juices.[198] Apple peel extract rich in quercetin also inhibited LDL-C oxidation *in vitro*.[199] Moreover, apple polyphenols have been shown to increase the excretion of cholesterol oxidation products and protect against peroxidation in rats.[13,134]

There are many studies to support an antioxidative potential of apple polyphenols, mainly in apple peel, but also in flesh and juice, yet an *in vitro* antioxidant activity can not prove *in vivo* biological activity in humans. The concentration of a parent polyphenol compound in human plasma is often too low to exert a strong antioxidant effect.[57] Moreover, polyphenols may appear in plasma as bioconverted and conjugated forms with different biological activities which may also alter the antioxidant properties.[69] Further *in vivo* studies are necessary to confirm an antioxidant role. Beneficial acute effects after the consumption of whole apple or apple juice have been reported in human diet trials.[200,201] Pure fruit juices, including apple juice, inhibited reactive oxygen species generation for up to 90 min after their consumption by 10 healthy young men.[200] In addition, 600 g of whole apples also acutely reduced oxidative stress levels in six healthy male volunteers. *In vivo* antioxidant effects were also tested in chronic human studies.[201] Twenty-eight participants consumed 340 g of whole apple or 375 mL of apple juice per day in a crossover intervention study by Hyson et al. (2000).[8] After 6 weeks of apple juice intake, the lag time to copper-induced LDL-C oxidation was significantly increased by 20% and conjugated dienes reduced by 7%. Lag time was not affected by the whole-apple consumption; however, conjugated dienes were also significantly reduced by 5%. The possible reasons for those results are not clear. The authors suggested that the antioxidant components of whole apple might be less readily absorbed or subjected to bioconversion in the gut, compared to apple juice.[8] In a recent human study by Zhao et al. (2013), consumption of one apple per day for 4 weeks resulted in significantly reduced plasma levels of oxidized LDL/beta2-glycoprotein I complex, an important contributor to atherosclerosis.[202] However it is of note that, a potential increase of human plasma antioxidant activity after whole apple or apple juice consumption has been attributed to an increase of uric acid levels due to the fructose content of apple products and not to apple antioxidant polyphenols.[203,204] Moreover, although Lotito et al. (2004b) reported *in vitro* antioxidant effects of apple polyphenols they did not confirm this *in vivo* in human subjects.[205] Other antioxidants such as, ascorbate, uric acid or glutathione are in much higher concentration in the blood compared to polyphenols and thus, exert a higher antioxidant activity.[206,207] However, polyphenols may exert a direct protective antioxidative effect in the gastrointestinal tract where their concentration is higher than plasma, especially after they have been converted by the gut microbiota.[208] These effects in the gut may include decreased lipid peroxidation products, reduced mucosal oxidative stress and inflammation, favoring fermentation of strict anaerobic microbes through the reduction of the redox potential.[209–211]

DIABETES RISK

The prevalence of diabetes is increasing dramatically worldwide and it is expected to rise from 171 million cases in 2000 to 366 million in 2030.[212] Although type 1 diabetes (T1DM), in line with other autoimmune diseases, is on the increase, the current explosion in diabetes incidence appears to be driven by the increase in new cases of type 2 diabetes (T2DM). Fortunately, T2DM appears to be responsive to lifestyle intervention compared with T1DM, especially through modulation of diet, with apples in particular appearing to be a useful means of reducing T2DM modifiable risk factors. Epidemiological studies have observed that apple consumption may be associated with a lower risk of T2DM.[6,7] In a prospective cross-sectional study of 38,018 women, the association between dietary flavonoids and the risk of T2DM, insulin resistance and systemic inflammation was assessed.[7] Total or specific flavonoids did not show any effect. However, the consumption of ≥1 apple per day resulted in a significant 28% reduction of T2DM risk compared with no apple consumption.[7] Dietary intervention studies supported these effects. A 15% unprocessed apple pomace diet, containing both fiber and polyphenols, in Wistar rats for 4 weeks was associated with a significant decrease in serum glucose levels.[213] Similarly, an acute human study of nine subjects by Johnston et al. (2002) indicated that the consumption of a clear or a cloudy apple juice significantly delayed glucose absorption compared with a control.[9] This was supported by changes in gastrointestinal hormones.[9] In contrast, Vaaler et al. (1982) did not show a different postprandial blood glucose response after the consumption of apple, orange or banana compared to equal amounts of glucose, in insulin dependent diabetics.[214] Moreover, Haber et al. (1977) reported that the consumption of apple juice, apple puree or whole apples did not have important differences in plasma glucose levels of 10 normal subjects, however, insulin levels rose to higher concentrations after juice and puree compared to apples.[215] The potential mechanisms by which apple polyphenols could impact on T2DM risk and lower glucose response may be related to a decreased oxidative stress and, therefore, protection of pancreatic β cells from potential damage, inhibition of sugar transport by sodium-glucose linked transporter 1 (SGLT1) or decreased activity of mucosal sucrase and other endogenous digestive enzymes related to carbohydrate metabolism.[9,213,216,217] Phloridzin, phloretin xyloglucoside and quercetin have been considered responsible for these effects[6,9,218–222] and thus, apple polyphenols may be associated with an improved insulin sensitivity by preventing hyperglycemia. In addition apple pectin has been shown to exert beneficial effects on glucose and insulin levels.[36,223] A high methoxylated apple pectin consumption decreased blood glucose and plasma insulin in Zucker fatty rats compared to β-glucan intake and a control.[36] In a human study of 12 patients with T2DM, 15 g of apple fiber per day (17% pectin) significantly reduced fasting plasma glucose and glycosylated haemoglobin over a period of 7 weeks.[137] Thus, pectin, through its viscosity and the potential delay in gastric emptying, the decreased rate of absorption across the intestinal epithelial cells and the potential impact on satiety and weight gain may also contribute to these effects.[36,223,224]

CONCLUSION

There is supporting evidence that apples and apple components upregulate the composition and the activity of the gut microbiota and may improve biomarkers related to cardiometabolic disease. Data from animal studies represent a stronger evidence base compared with more limited data from human studies which are scarce and often inconsistent. Beneficial effects on key CVD risk markers such as blood pressure and vascular function are limited, whereas evidence for antiinflammatory activities and beneficial effects of apples and apple components on glucose and insulin levels has been mainly derived from *in vitro* and animal studies. The strongest effects are related to blood lipid levels where a frequent apple consumption of approximately three apples per day has been reported to reduce TC in humans. Moreover, apple polyphenols possess a relatively high antioxidant activity which may contribute to beneficial effects, although this potential mechanism is less well supported.

It is proposed that polyphenols and fiber are the main contributors to the observed effects, either individually or in a synergistic interaction. However, bioavailability and metabolism of the numerous apple polyphenols is not completely understood. Moreover, limited data are available for the characterization of apple fiber, in particular pectin, in the different apple varieties and the impact of the physicochemical properties and structure on the degree of effect. Suitably powered randomly controlled, human dietary intervention studies with long intervention periods are required to clarify the protective effects of apples and juice consumption. Further *in vivo* animal models are also necessary to explore the potential multiple mechanisms of actions.

References

1. Scalbert A, Manach C, Morand C, Remesy C, Jimenez L. Dietary polyphenols and the prevention of diseases. *Crit Rev Food Sci Nutr.* 2005;45(4):287–306.
2. Jensen EN, Buch-Andersen T, Ravn-Haren G, Dragsted LO. Mini-review: the effects of apples on plasma cholesterol levels and cardiovascular risk – a review of the evidence. *J Hortic Sci Biotechnol.* 2009;34–41.
3. Wojdylo A, Oszmianski J, Laskowski P. Polyphenolic compounds and antioxidant activity of new and old apple varieties. *J Agric Food Chem.* 2008;56(15):6520–6530.
4. Knekt P, Jarvinen R, Reunanen A, Maatela J. Flavonoid intake and coronary mortality in Finland: a cohort study. *Br Med J.* 1996;312 (7029):478–481.
5. Arts ICW, Jacobs DR, Harnack LJ, Gross M, Folsom AR. Dietary catechins in relation to coronary heart disease death among postmenopausal women. *Epidemiology.* 2001;12(6):668–675.
6. Knekt P, Kumpulainen J, Jarvinen R, et al. Flavonoid intake and risk of chronic diseases. *Am J Clin Nutr.* 2002;76(3):560–568.
7. Song YQ, Manson JE, Buring JE, Sesso HD, Liu SM. Associations of dietary flavonoids with risk of type 2 diabetes, and markers of insulin resistance and systemic inflammation in women: a prospective study and cross-sectional analysis. *J Am Coll Nutr.* 2005;24(5):376–384.
8. Hyson D, Studebaker-Hallman D, Davis PA, Gershwin ME. Apple juice consumption reduces plasma low-density lipoprotein oxidation in healthy men and women. *J Med Food.* 2000;3(4):159–166.
9. Johnston KL, Clifford MN, Morgan LM. Possible role for apple juice phenolic, compounds in the acute modification of glucose tolerance and gastrointestinal hormone secretion in humans. *J Sci Food Agric.* 2002;82(15):1800–1805.
10. Sembries S, Dongowski G, Jacobasch G, Mehrlander K, Will F, Dietrich H. Effects of dietary fibre-rich juice colloids from apple pomace extraction juices on intestinal fermentation products and microbiota in rats. *Br J Nutr.* 2003;90(3):607–615.
11. de Oliveira MC, Sichieri R, Moura AS. Weight loss associated with a daily intake of three apples or three pears among overweight women. *Nutrition.* 2003;19(3):253–256.
12. Nagasako-Akazome Y, Kanda T, Ohtake Y, Shimasaki H, Kobayashi T. Apple polyphenols influence cholesterol metabolism in healthy subjects with relatively high body mass index. *J Oleo Sci.* 2007;56(8):417–428.
13. Ogino Y, Osada K, Nakamura S, Ohta Y, Kanda T, Sugano M. Absorption of dietary cholesterol oxidation products and their downstream metabolic effects are reduced by dietary apple polyphenols. *Lipids.* 2007;42(2):151–161.
14. Castagnini C, Luceri C, Toti S, et al. Reduction of colonic inflammation in HLA-B27 transgenic rats by feeding Marie Menard apples, rich in polyphenols. *Br J Nutr.* 2009;102(11):1620–1628.
15. Licht TR, Hansen M, Bergstrom A, et al. Effects of apples and specific apple components on the cecal environment of conventional rats: role of apple pectin. *BMC Microbiol.* 2010;10.
16. Shinohara K, Ohashi Y, Kawasumi K, Terada A, Fujisawa T. Effect of apple intake on fecal microbiota and metabolites in humans. *Anaerobe.* 2010;16(5):510–515.
17. Bondonno CP, Yang X, Croft KD, et al. Flavonoid-rich apples and nitrate-rich spinach augment nitric oxide status and improve endothelial function in healthy men and women: a randomized controlled trial. *Free Radic Biol Med.* 2012;52(1):95–102.
18. Macfarlane S, Macfarlane GT. Regulation of short-chain fatty acid production. *Proc Nutr Soc.* 2003;62(1):67–72.
19. Wong JMW, de Souza R, Kendall CWC, Emam A, Jenkins DJA. Colonic health: Fermentation and short chain fatty acids. *J Clin Gastroenterol.* 2006;40(3):235–243.
20. Conterno L, Fava F, Viola R, Tuohy KM. Obesity and the gut microbiota: does up-regulating colonic fermentation protect against obesity and metabolic disease? *Genes Nutr.* 2011;6(3):241–260.
21. Tuohy KM, Conterno L, Gasperotti M, Viola R. Up-regulating the human intestinal microbiome using whole plant foods, polyphenols, and/or fiber. *J Agric Food Chem.* 2012;60(36):8776–8782.
22. Feliciano RP, Antunes C, Ramos A, et al. Characterization of traditional and exotic apple varieties from Portugal. Part 1—Nutritional, phytochemical and sensory evaluation. *J Funct Foods.* 2010;2(1):35–45.
23. Campeanu G, Neata G, Darjanschi G. Chemical Composition of the Fruits of Several Apple Cultivars Growth as Biological Crop. *Notulae Botanicae Horti Agrobotanici Cluj-Napoca.* 2009;37(2):161–164.
24. Elliott SS, Keim NL, Stern JS, Teff K, Havel PJ. Fructose, weight gain, and the insulin resistance syndrome. *Am J Clin Nutr.* 2002;76 (5):911–922.
25. Schulze MB, Manson JE, Ludwig DS, et al. Sugar-sweetened beverages, weight gain, and incidence of type 2 diabetes in young and middle-aged women. *JAMA.* 2004;292(8):927–934.
26. Palmer JR, Boggs DA, Krishnan S, Hu FB, Singer M, Rosenberg L. Sugar-sweetened beverages and incidence of type 2 diabetes mellitus in African American women. *Arch Intern Med.* 2008;168(14):1487–1492.

27. Eshak ES, Iso H, Mizoue T, Inoue M, Noda M, Tsugane S. Soft drink, 100% fruit juice, and vegetable juice intakes and risk of diabetes mellitus. *Clin Nutr.* 2013;32(2):300–308.
28. Sartorelli DS, Franco LJ, Damiao R, et al. Dietary glycemic load, glycemic index, and refined grains intake are associated with reduced beta-cell function in prediabetic Japanese migrants. *Arq Bras Endocrinol Metabol.* 2009;53(4):429–434.
29. Lee IM, Cook NR, Gaziano JM, et al. Vitamin E in the primary prevention of cardiovascular disease and cancer – The Women's Health Study: a randomized controlled trial. *JAMA.* 2005;294(1):56–65.
30. Ye Z, Song H. Antioxidant vitamins intake and the risk of coronary heart disease: meta-analysis of cohort studies. *Eur J Cardiovasc Prev Rehabil.* 2008;15(1):26–34.
31. Ward M, McNulty H, McPartlin J, Strain JJ, Weir DG, Scott JM. Plasma homocysteine, a risk factor for cardiovascular disease, is lowered by physiological doses of folic acid. *Qjm–Monthly J Assoc Physicians.* 1997;90(8):519–524.
32. Sheehan JP, Seelig MS. Interactions of magnesium and potassium in the pathogenesis of cardiovascular disease. *Magnesium.* 1984;3 (4–6):301–314.
33. Dongowski G, Lorenz A. Unsaturated oligogalacturonic acids are generated by in vitro treatment of pectin with human faecal flora. *Carbohydr Res.* 1998;314(3–4):237–244.
34. Dongowski G, Lorenz A, Proll A. The degree of methylation influences the degradation of pectin in the intestinal tract of rats and in vitro. *J Nutr.* 2002;132(7):1935–1944.
35. Gulfi M, Arrigoni E, Amado R. The chemical characteristics of apple pectin influence its fermentability in vitro. *Lwt–Food Sci Technol.* 2006;39(9):1001–1004.
36. Sanchez D, Muguerza B, Moulay L, Hernandez R, Miguel M, Aleixandre A. Highly methoxylated pectin improves insulin, resistance and other cardiometabolic risk factors in Zucker fatty rats. *J Agric Food Chem.* 2008;56(10):3574–3581.
37. Dongowski G. Influence of pectin structure on the interaction with bile-acids under in-vitro conditions. *Z Lebensm Unters Forsch.* 1995;201 (4):390–398.
38. Brown L, Rosner B, Willett WW, Sacks FM. Cholesterol-lowering effects of dietary fiber: a meta-analysis. *Am J Clin Nutr.* 1999;69(1):30–42.
39. Brouns F, Theuwissen E, Adam A, Bell M, Berger A, Mensink RP. Cholesterol-lowering properties of different pectin types in mildly hyper-cholesterolemic men and women. *Eur J Clin Nutr.* 2012;66(5):591–599.
40. Spiller GA, Chernoff MC, Hill RA, Gates JE, Nassar JJ, Shipley EA. Effect of purified cellulose, pectin, and a low-residue diet on fecal volatile fatty-acids, transit-time, and fecal weight in humans. *Am J Clin Nutr.* 1980;33(4):754–759.
41. Schwartz SE, Levine RA, Singh A, Scheidecker JR, Track NS. Sustained pectin ingestion delays gastric-emptying. *Gastroenterology.* 1982;83 (4):812–817.
42. Tamura M, Nakagawa H, Tsushida T, Hirayama K, Itoh K. Effect of pectin enhancement on plasma quercetin and fecal flora in rutin-supplemented mice. *J Food Sci.* 2007;72(9):S648–S651.
43. Nishijima T, Iwai K, Saito Y, Takida Y, Matsue H. Chronic ingestion of Apple Pectin can enhance the absorption of Quercetin. *J Agric Food Chem.* 2009;57(6):2583–2587.
44. Titgemeyer EC, Bourquin LD, Fahey GC, Garleb KA. Fermentability of various fiber sources by human fecal bacteria in vitro. *Am J Clin Nutr.* 1991;53(6):1418–1424.
45. Barry JL, Hoebler C, Macfarlane GT, et al. Estimation of the fermentability of dietary fiber in-vitro – A European Interlaboratory Study. *Br J Nutr.* 1995;74(3):303–322.
46. Bourquin LD, Titgemeyer EC, Fahey GC. Fermentation of various dietary fiber sources by human fecal bacteria. *Nutr Res.* 1996;16 (7):1119–1131.
47. Stevenson DG, Domoto PA, Jane JL. Structures and functional properties of apple (*Malus domestica* Borkh) fruit starch. *Carbohydr Polym.* 2006;63(3):432–441.
48. Singh N, Inouchi N, Nishinari K. Morphological, structural, thermal, and rheological characteristics of starches separated from apples of different cultivars. *J Agric Food Chem.* 2005;53(26):10193–10199.
49. Bowen JH, Watkins CB. Fruit maturity, carbohydrate and mineral content relationships with watercore in 'Fuji' apples. *Postharvest Biol Technol.* 1997;11(1):31–38.
50. Lesmes U, Beards EJ, Gibson GR, Tuohy KM, Shimoni E. Effects of resistant starch type III polymorphs on human colon microbiota and short chain fatty acids in human gut models. *J Agric Food Chem.* 2008;56(13):5415–5421.
51. Martinez I, Kim J, Duffy PR, Schlegel VL, Walter J. Resistant starches types 2 and 4 have differential effects on the composition of the fecal microbiota in human subjects. *PLOS One.* 2010;5:11.
52. So P-W, Yu W-S, Kuo Y-T, et al. Impact of resistant starch on body fat patterning and central appetite regulation. *PLOS One.* 2007;2:12.
53. Johnston KL, Thomas EL, Bell JD, Frost GS, Robertson MD. Resistant starch improves insulin sensitivity in metabolic syndrome. *Diabet Med.* 2010;27(4):391–397.
54. Shen Q, Zhao L, Tuohy KM. High-level dietary fibre up-regulates colonic fermentation and relative abundance of saccharolytic bacteria within the human faecal microbiota in vitro. *Eur J Nutr.* 2012;51(6):693–705.
55. Conlon MA, Kerr CA, McSweeney CS, et al. Resistant Starches Protect against Colonic DNA Damage and Alter Microbiota and Gene Expression in Rats Fed a Western Diet. *J Nutr.* 2012;142(5):832–840.
56. Belobrajdic DP, King RA, Christophersen CT, Bird AR. Dietary resistant starch dose-dependently reduces adiposity in obesity-prone and obesity-resistant male rats. *Nutr Metab (Lond).* 2012;9:93 10.p.].
57. Scalbert A, Williamson G. Dietary intake and bioavailability of polyphenols. *J Nutr.* 2000;130(8):2073S–2085SS.
58. Tsao R, Yang R, Christopher J, Zhu Y, Zhu HH. Polyphenolic profiles in eight apple cultivars using high-performance liquid chromatography (HPLC). *J Agric Food Chem.* 2003;51(21):6347–6353.
59. Lee KW, Kim YJ, Kim DO, Lee HJ, Lee CY. Major phenolics in apple and their contribution to the total antioxidant capacity. *J Agric Food Chem.* 2003;51(22):6516–6520.
60. Vrhovsek U, Rigo A, Tonon D, Mattivi F. Quantitation of polyphenols in different apple varieties. *J Agric Food Chem.* 2004;52 (21):6532–6538.

61. Ceymann M, Arrigoni E, Schaerer H, Nising AB, Hurrell RF. Identification of apples rich in health-promoting flavan-3-ols and phenolic acids by measuring the polyphenol profile. *J Food Compost Anal*. 2012;26(1−2):128−135.
62. Valavanidis A, Vlachogianni T, Psomas A, Zovoili A, Siatis V. Polyphenolic profile and antioxidant activity of five apple cultivars grown under organic and conventional agricultural practices. *Int J Food Sci Technol*. 2009;44(6):1167−1175.
63. McGhie TK, Hunt M, Barnett LE. Cultivar and growing region determine the antioxidant polyphenolic concentration and composition of apples grown in New Zealand. *J Agric Food Chem*. 2005;53(8):3065−3070.
64. Kevers C, Pincemail J, Tabart J, Defraigne J-O, Dommes J. Influence of cultivar, harvest time, storage conditions, and peeling on the antioxidant capacity and phenolic and ascorbic acid contents of apples and pears. *J Agric Food Chem*. 2011;59(11):6165−6171.
65. Imeh U, Khokhar S. Distribution of conjugated and free phenols in fruits: antioxidant activity and cultivar variations. *J Agric Food Chem*. 2002;50(22):6301−6306.
66. Markowski J, Plocharski W. Determination of phenolic compounds in apples and processed apple products. *J Fruit Ornamental Plant Res*. 2006;14(Suppl. 2):133−142.
67. Oszmianski J, Wolniak M, Wojdylo A, Wawer I. Comparative study of polyphenolic content and antiradical activity of cloudy and clear apple juices. *J Sci Food Agric*. 2007;87(4):573−579.
68. Manach C, Scalbert A, Morand C, Remesy C, Jimenez L. Polyphenols: food sources and bioavailability. *Am J Clin Nutr*. 2004;79 (5):727−747.
69. Possemiers S, Bolca S, Verstraete W, Heyerick A. The intestinal microbiome: A separate organ inside the body with the metabolic potential to influence the bioactivity of botanicals. *Fitoterapia*. 2011;82(1):53−66.
70. Hagl S, Deusser H, Soyalan B, et al. Colonic availability of polyphenols and D-(−)-quinic acid after apple smoothie consumption. *Mol Nutr Food Res*. 2011;55(3):368−377.
71. Kemperman RA, Bolca S, Roger LC, Vaughan EE. Novel approaches for analysing gut microbes and dietary polyphenols: challenges and opportunities. *Microbiol-Sgm*. 2010;156:3224−3231.
72. Kahle K, Huemmer W, Kempf M, Scheppach W, Erk T, Richling E. Polyphenols are intensively metabolized in the human gastrointestinal tract after apple juice consumption. *J Agric Food Chem*. 2007;55(26):10605−10614.
73. Manach C, Williamson G, Morand C, Scalbert A, Remesy C. Bioavailability and bioefficacy of polyphenols in humans. I. Review of 97 bioavailability studies. *Am J Clin Nutr*. 2005;81(1):230S−242SS.
74. Gonthier MP, Verny MA, Besson C, Remesy C, Scalbert A. Chlorogenic acid bioavailability largely depends on its metabolism by the gut microflora in rats. *J Nutr*. 2003;133(6):1853−1859.
75. Hollman PCH, Katan MB. Absorption, metabolism and health effects of dietary flavonoids in man. *Biomed Pharmacother*. 1997;51 (8):305−310.
76. Boyer J, Liu RH. Apple phytochemicals and their health benefits. *Nutr J*. 2004;3(5):1−15.
77. Marks SC, Mullen W, Borges G, Crozier A. Absorption, metabolism, and excretion of cider dihydrochalcones in healthy humans and subjects with an ileostomy. *J Agric Food Chem*. 2009;57(5):2009−2015.
78. Guarner F, Malagelada JR. Gut flora in health and disease. *Lancet*. 2003;361(9356):512−519.
79. Eckburg PB, Bik EM, Bernstein CN, et al. Diversity of the human intestinal microbial flora. *Science*. 2005;308(5728):1635−1638.
80. Robles Alonso V, Guarner F. Linking the gut microbiota to human health. *Br J Nutr*. 2013;109:S21−S26.
81. Wallace TC, Guarner F, Madsen K, et al. Human gut microbiota and its relationship to health and disease. *Nutr Rev*. 2011;69(7):392−403.
82. Wong JM, Esfahani A, Singh N, et al. Gut microbiota, diet and heart disease. *J AOAC Int*. 2012;95(1):24−30.
83. Jones ML, Tomaro-Duchesneau C, Martoni CJ, Prakash S. Cholesterol lowering with bile salt hydrolase-active probiotic bacteria, mechanism of action, clinical evidence, and future direction for heart health applications. *Expert Opin Biol Ther*. 2013;13(5):631−642.
84. Vulevic J, Juric A, Tzortzis G, Gibson GR. A mixture of trans-galactooligosaccharides reduces markers of metabolic syndrome and modulates the fecal microbiota and immune function of overweight adults. *J Nutr*. 2013;143(3):324−331.
85. Delzenne NM, Cani PD. Gut microbiota and the pathogenesis of insulin resistance. *Curr Diab Rep*. 2011;11(3):154−159.
86. Del Rio D, Costa LG, Lean MEJ, Crozier A. Polyphenols and health: what compounds are involved? *Nutr Metab Cardiovasc Dis*. 2010;20(1):1−6.
87. Del Rio D, Rodriguez-Mateos A, Spencer JPE, Tognolini M, Borges G, Crozier A. Dietary (Poly)phenolics in human health: structures, bioavailability, and evidence of protective effects against chronic diseases. *Antioxid Redox Signal*. 2013;18(14):1818−1892.
88. Laparra JM, Sanz Y. Interactions of gut microbiota with functional food components and nutraceuticals. *Pharmacol Res*. 2010;61(3):219−225.
89. Hervert-Hernandez D, Goni I. Dietary polyphenols and human gut microbiota: a review. *Food Rev Int*. 2011;27(2):154−169.
90. Comalada M, Camuesco D, Sierra S, et al. In vivo quercitrin anti-inflammatory effect involves release of quercetin, which inhibits inflammation through down-regulation of the NF-kappa B pathway. *Eur J Immunol*. 2005;35(2):584−592.
91. Monagas M, Khan N, Andres-Lacueva C, et al. Dihydroxylated phenolic acids derived from microbial metabolism reduce lipopolysaccharide-stimulated cytokine secretion by human peripheral blood mononuclear cells. *Br J Nutr*. 2009;102(2):201−206.
92. Dumas ME, Barton RH, Toye A, et al. Metabolic profiling reveals a contribution of gut microbiota to fatty liver phenotype in insulin-resistant mice. *Proc Natl Acad Sci USA*. 2006;103(33):12511−12516.
93. Abu-Shanab A, Quigley EM. The role of the gut microbiota in nonalcoholic fatty liver disease. *Nat Rev Gastroenterol Hepatol*. 2010;7 (12):691−701.
94. Kootte RS, Vrieze A, Holleman F, et al. The therapeutic potential of manipulating gut microbiota in obesity and type 2 diabetes mellitus. *Diabetes Obes Metab*. 2012;14(2):112−120.
95. Wang Z, Klipfell E, Bennett BJ, et al. Gut flora metabolism of phosphatidylcholine promotes cardiovascular disease. *Nature*. 2011;472 (7341):57−U82.
96. Cani PD, Osto M, Geurts L, Everard A. Involvement of gut microbiota in the development of low-grade inflammation and type 2 diabetes associated with obesity. *Gut Microbes*. 2012;3(4):279−288.
97. Qin J, Li Y, Cai Z, et al. A metagenome-wide association study of gut microbiota in type 2 diabetes. *Nature*. 2012;490(7418):55−60.

98. de Goffau MC, Luopajarvi K, Knip M, et al. Fecal microbiota composition differs between children with beta-cell autoimmunity and those without. *Diabetes*. 2013;62(4):1238–1244.
99. Koeth RA, Wang Z, Levison BS, et al. Intestinal microbiota metabolism of l-carnitine, a nutrient in red meat, promotes atherosclerosis. *Nat Med*. 2013.
100. Murri M, Leiva I, Gomez-Zumaquero JM, et al. Gut microbiota in children with type 1 diabetes differs from that in healthy children: a case-control study. *BMC Med*. 2013;11:46.
101. Ley RE, Turnbaugh PJ, Klein S, Gordon JI. Microbial ecology – human gut microbes associated with obesity. *Nature*. 2006;444 (7122):1022–1023.
102. Fava F, Gitau R, Griffin BA, Gibson GR, Tuohy KM, Lovegrove JA. The type and quantity of dietary fat and carbohydrate alter faecal microbiome and short-chain fatty acid excretion in a metabolic syndrome 'at-risk' population. *Int J Obes (Lond)*. 2013;37(2):216–223.
103. Caesar R, Fak F, Backhed F. Effects of gut microbiota on obesity and atherosclerosis via modulation of inflammation and lipid metabolism. *J Intern Med*. 2010;268(4):320–328.
104. Neyrinck AM, Van Hee VF, Piront N, et al. Wheat-derived arabinoxylan oligosaccharides with prebiotic effect increase satietogenic gut peptides and reduce metabolic endotoxemia in diet-induced obese mice. *Nutr Diabetes*. 2012;2:e28.
105. De Filippo C, Cavalieri D, Di Paola M, et al. Impact of diet in shaping gut microbiota revealed by a comparative study in children from Europe and rural Africa. *Proc Natl Acad Sci USA*. 2010;107(33):14691–14696.
106. Flint HJ, Scott KP, Louis P, Duncan SH. The role of the gut microbiota in nutrition and health. *Nat Rev Gastroenterol Hepatol*. 2012;9 (10):577–589.
107. Scott KP, Gratz SW, Sheridan PO, Flint HJ, Duncan SH. The influence of diet on the gut microbiota. *Pharmacol Res*. 2013;69(1).
108. Schwiertz A, Taras D, Schaefer K, et al. Microbiota and SCFA in lean and overweight healthy subjects. *Obesity*. 2010;18(1):190–195.
109. Jumpertz R, Duc Son L, Turnbaugh PJ, et al. Energy-balance studies reveal associations between gut microbes, caloric load, and nutrient absorption in humans. *Am J Clin Nutr*. 2011;94(1):58–65.
110. Mann JI, Cummings JH. Possible implications for health of the different definitions of dietary fibre. *Nutr Metab Cardiovasc Dis*. 2009;19 (3):226–229.
111. Macfarlane S, Macfarlane GT, Cummings JH. Review article: prebiotics in the gastrointestinal tract. *Aliment Pharmacol Ther*. 2006;24(5):701–714.
112. Roberfroid M, Gibson GR, Hoyles L, et al. Prebiotic effects: metabolic and health benefits. *Br J Nutr*. 2010;104:S1–S63.
113. Gibson GR, Roberfroid MB. Dietary modulation of the human colonic microbiota – Introducing the concept of prebiotics. *J Nutr*. 1995;125(6):1401–1412.
114. Gibson GR, Scott KP, Rastall RA, et al. Dietary prebiotics: current status and new definition. *Food Sci Technol Bull Funct Foods*. 2010;7 (1):1–19.
115. Tzounis X, Rodriguez-Mateos A, Vulevic J, Gibson GR, Kwik-Uribe C, Spencer JPE. Prebiotic evaluation of cocoa-derived flavanols in healthy humans by using a randomized, controlled, double-blind, crossover intervention study. *Am J Clin Nutr*. 2011;93(1):62–72.
116. Isabel Queipo-Ortuno M, Boto-Ordonez M, Murri M, et al. Influence of red wine polyphenols and ethanol on the gut microbiota ecology and biochemical biomarkers. *Am J Clin Nutr*. 2012;95(6):1323–1334.
117. Olano-Martin E, Gibson GR, Rastall RA. Comparison of the in vitro bifidogenic properties of pectins and pectic-oligosaccharides. *J Appl Microbiol*. 2002;93(3):505–511.
118. Salyers AA, Vercellotti JR, West SEH, Wilkins TD. Fermentation of mucin and plant polysaccharides by strains of bacteroides from human colon. *Appl Environ Microbiol*. 1977;33(2):319–322.
119. Salyers AA, West SEH, Vercellotti JR, Wilkins TD. Fermentation of mucins and plant polysaccharides by anaerobic bacteria from human colon. *Appl Environ Microbiol*. 1977;34(5):529–533.
120. Lopez-Siles M, Khan TM, Duncan SH, Harmsen HJM, Garcia-Gil LJ, Flint HJ. Cultured representatives of two major phylogroups of human colonic faecalibacterium prausnitzii can utilize pectin, uronic acids, and host-derived substrates for growth. *Appl Environ Microbiol*. 2012;78(2):420–428.
121. Sokol H, Pigneur B, Watterlot L, et al. Faecalibacterium prausnitzii is an anti-inflammatory commensal bacterium identified by gut microbiota analysis of Crohn disease patients. *Proc Natl Acad Sci USA*. 2008;105(43):16731–16736.
122. Chen J, Liang R-h, Liu W, et al. Pectic-oligosaccharides prepared by dynamic high-pressure nnicrofluidization and their in vitro fermentation properties. *Carbohydr Polym*. 2013;91(1):175–182.
123. Dongowski G, Lorenz A, Anger H. Degradation of pectins with different degrees of esterification by *Bacteroides thetaiotaomicron* isolated from human gut flora. *Appl Environ Microbiol*. 2000;66(4):1321–1327.
124. Bazzocco S, Mattila I, Guyot S, Renard CMGC, Aura A-M. Factors affecting the conversion of apple polyphenols to phenolic acids and fruit matrix to short-chain fatty acids by human faecal microbiota in vitro. *Eur J Nutr*. 2008;47(8):442–452.
125. Levrat MA, Texier MDO, Regerat F, Demigne C, Remesy C. Comparison of the effects of condensed tannin and pectin on cecal fermentations and lipid-metabolism in the rat. *Nutr Res*. 1993;13(4):427–433.
126. Kosmala M, Kolodziejczyk K, Zdunczyk Z, Juskiewicz J, Boros D. Chemical composition of natural and polyphenol-free apple pomace and the effect of this dietary ingredient on intestinal fermentation and serum lipid parameters in rats. *J Agric Food Chem*. 2011;59(17):9177–9185.
127. Aprikian O, Duclos V, Guyot S, et al. Apple pectin and a polyphenol-rich apple concentrate are more effective together than separately on cecal fermentations and plasma lipids in rats. *J Nutr*. 2003;133(6):1860–1865.
128. Sembries S, Dongowski G, Mehrlaender K, Will F, Dietrich H. Physiological effects of extraction juices from apple, grape, and red beet pomaces in rats. *J Agric Food Chem*. 2006;54(26):10269–10280.
129. Ravn-Haren G, Dragsted LO, Buch-Andersen T, et al. Intake of whole apples or clear apple juice has contrasting effects on plasma lipids in healthy volunteers. *Eur J Nutr*. 2012.
130. Mink PJ, Scrafford CG, Barraj LM, et al. Flavonoid intake and cardiovascular disease mortality: a prospective study in postmenopausal women. *Am J Clin Nutr*. 2007;85(3):895–909.
131. Sesso HD, Gaziano JM, Liu S, Buring JE. Flavonoid intake and the risk of cardiovascular disease in women. *Am J Clin Nutr*. 2003;77 (6):1400–1408.

132. Hertog MGL, Feskens EJM, Hollman PCH, Katan MB, Kromhout D. Dietary antioxidant flavonoids and risk of coronary heart-disease – the zutphen elderly study. *Lancet*. 1993;342(8878):1007–1011.
133. Aprikian O, Levrat-Verny MA, Besson C, Busserolles J, Remesy C, Demigne C. Apple favourably affects parameters of cholesterol metabolism and of anti-oxidative protection in cholesterol-fed rats. *Food Chem*. 2001;75(4):445–452.
134. Aprikian O, Busserolles J, Manach C, et al. Lyophilized apple counteracts the development of hypercholesterolemia, oxidative stress, and renal dysfunction in obese zucker rats. *J Nutr*. 2002;132(7):1969–1976.
135. Leontowicz H, Gorinstein S, Lojek A, et al. Comparative content of some bioactive compounds in apples, peaches and pears and their influence on lipids and antioxidant capacity in rats. *J Nutr Biochem*. 2002;13(10):603–610.
136. Gormley TR, Kevany J, Egan JP, McFarlane R. Effect of apples on serum cholesterol levels in humans. *Irish J Food Sci Technol*. 1977;1(2):117–128.
137. Mayne PD, McGill AR, Gormley TR, Tomkin GH, Julian TR, Omoore RR. The effect of apple fiber on diabetic control and plasma-lipids. *Ir J Med Sci*. 1982;151(2):36–41.
138. Sableamplis R, Sicart R, Agid R. Further studies on the cholesterol-lowering effect of apple in human – biochemical mechanisms involved. *Nutr Res*. 1983;3(3):325–328.
139. Mahalko JR, Sandstead HH, Johnson LK, et al. Effect of consuming fiber from corn bran, soy hulls, or apple powder on glucose-tolerance and plasma-lipids in type-ii diabetes. *Am J Clin Nutr*. 1984;39(1):25–34.
140. Mee KA, Gee DL. Apple fiber and gum arabic lowers total and low-density lipoprotein cholesterol levels in men with mild hypercholesterolemia. *J Am Diet Assoc*. 1997;97(4):422–424.
141. Davidson MH, Dugan LD, Stocki J, et al. A low-viscosity soluble-fiber fruit juice supplement fails to lower cholesterol in hypercholesterolemic men and women. *J Nutr*. 1998;128(11):1927–1932.
142. Nagasako-Akazome Y, Kanda T, Ikeda M, Shimasaki H. Serum cholesterol-lowering effect of apple polyphenols in healthy subjects. *J Oleo Sci*. 2005;54(3):143–151.
143. Avci A, Atli T, Erguder IB, et al. Effects of apple consumption on plasma and erythrocyte antioxidant parameters in elderly subjects. *Exp Aging Res*. 2007;33(4):429–437.
144. Vafa MR, Haghighatjoo E, Shidfar F, Afshari S, Gohari MR, Ziaee A. Effects of apple consumption on lipid profile of hyperlipidemic and overweight men. *Int j Prev Med*. 2011;2(2):94–100.
145. Barth SW, Koch TCL, Watzl B, Dietrich H, Will F, Bub A. Moderate effects of apple juice consumption on obesity-related markers in obese men: impact of diet–gene interaction on body fat content. *Eur J Nutr*. 2012;51(7):841–850.
146. Chai SC, Hooshmand S, Saadat RL, Payton ME, Brummel-Smith K, Arjmandi BH. Daily apple versus dried plum: impact on cardiovascular disease risk factors in postmenopausal women. *J Acad Nutr Diet*. 2012;112(8):1158–1168.
147. Akazome Y, Kametani N, Kanda T, Shimasaki H, Kobayashi S. Evaluation of safety of excessive intake and efficacy of long-term intake of beverages containing apple polyphenols. *J Oleo Sci*. 2010;59(6):321–338.
148. Jenkins DJA, Reynolds D, Leeds AR, Waller AL, Cummings JH. Hypocholesterolemic action of dietary fiber unrelated to fecal bulking effect. *Am J Clin Nutr*. 1979;32(12):2430–2435.
149. Markowski J, Baron A, Mieszczakowska M, Plocharski W. Chemical composition of french and polish cloudy apple juices. *J Hort Sci Biotechnol*. 2009;68–74.
150. Schieber A, Hilt P, Streker P, Endress HU, Rentschler C, Carle R. A new process for the combined recovery of pectin and phenolic compounds from apple pomace. *Innovative Food Sci Emerg Technol*. 2003;4(1):99–107.
151. O'Shea N, Arendt EK, Gallagher E. Dietary fibre and phytochemical characteristics of fruit and vegetable by-products and their recent applications as novel ingredients in food products. *Innovative Food Sci Emerg Technol*. 2012;16:1–10.
152. Hofmann AF. The continuing importance of bile acids in liver and intestinal disease. *Arch Intern Med*. 1999;159(22):2647–2658.
153. Kumar M, Nagpal R, Kumar R, et al. Cholesterol-lowering probiotics as potential biotherapeutics for metabolic diseases. *Exp Diabetes Res*. 2012.
154. Begley M, Hill C, Gahan CGM. Bile salt hydrolase activity in probiotics. *Appl Environ Microbiol*. 2006;72(3):1729–1738.
155. Ikeda I, Imasato Y, Sasaki E, et al. Tea catechins decrease micellar solubility and intestinal-absorption of cholesterol in rats. *Biochim Biophys Acta*. 1992;1127(2):141–146.
156. Osada K, Funayama M, Fuchi S, et al. Effects of dietary procyanidins and tea polyphenols on adipose tissue mass and fatty acid metabolism in rats on a high fat diet. *J Oleo Sci*. 2006;55(2):79–89.
157. Monte MJ, Marin JJG, Antelo A, Vazquez-Tato J. Bile acids: Chemistry, physiology, and pathophysiology. *World J Gastroenterol*. 2009;15(7):804–816.
158. Nakazato K, Song H, Waga T. Effects of dietary apple polyphenol on adipose tissues weights in wistar rats. *Exp Anim*. 2006;55(4):383–389.
159. Ohta Y, Sami M, Kanda T, Saito K, Osada K, Kato H. Gene expression analysis of the anti-obesity effect by apple polyphenols in rats fed a high fat diet or a normal diet. *J Oleo Sci*. 2006;55(6):305–314.
160. Murase T, Nagasawa A, Suzuki J, Hase T, Tokimitsu I. Beneficial effects of tea catechins on diet-induced obesity: stimulation of lipid catabolism in the liver. *Int J Obes*. 2002;26(11):1459–1464.
161. Vidal R, Hernandez-Vallejo S, Pauquai T, et al. Apple procyanidins decrease cholesterol esterification and lipoprotein secretion in Caco-2/TC7 enterocytes. *J Lipid Res*. 2005;46(2):258–268.
162. GarciaDiez F, GarciaMediavilla V, Bayon JE, Gonzalez-Gallego J. Pectin feeding influences fecal bile acid excretion, hepatic bile acid and cholesterol synthesis and serum cholesterol in rats. *J Nutr*. 1996;126(7):1766–1771.
163. Gonzalez M, Rivas C, Caride B, Lamas MA, Taboada MC. Effects of orange and apple pectin on cholesterol concentration in serum, liver and faeces. *J Physiol Biochem*. 1998;54(2):99–104.
164. Vergara-Jimenez M, Furr H, Fernandez ML. Pectin and psyllium decrease the susceptibility of LDL to oxidation in guinea pigs. *J Nutr Biochem*. 1999;10(2):118–124.

165. Kirat D, Kondo K, Shimada R, Kato S. Dietary pectin up-regulates monocaboxylate transporter 1 in the rat gastrointestinal tract. *Exp Physiol*. 2009;94(4):422–433.
166. Borthakur A, Priyamvada S, Kumar A, et al. A novel nutrient sensing mechanism underlies substrate-induced regulation of monocarboxylate transporter-1. *Am J Physiol Gastrointest Liver Physiol*. 2012;303(10):G1126–G1133.
167. Ooi L-G, Liong M-T. Cholesterol-lowering effects of probiotics and prebiotics: a review of in vivo and in vitro findings. *Int J Mol Sci*. 2010;11(6):2499–2522.
168. Sugiyama H, Akazome Y, Shoji T, et al. Oligomeric procyanidins in apple polyphenol are main active components for inhibition of pancreatic lipase and triglyceride absorption. *J Agric Food Chem*. 2007;55(11):4604–4609.
169. Brown AA, Hu FB. Dietary modulation of endothelial function: implications for cardiovascular disease. *Am J Clin Nutr*. 2001;73 (4):673–686.
170. Turnbull F, Neal B, Algert C, et al. Effects of different blood-pressure-lowering regimens on major cardiovascular events: results of prospectively-designed overviews of randomised trials. *Lancet*. 2003;362(9395):1527–1535.
171. Auclair S, Chironi G, Milenkovic D, et al. The regular consumption of a polyphenol-rich apple does not influence endothelial function: a randomised double-blind trial in hypercholesterolemic adults. *Eur J Clin Nutr*. 2010;64(10):1158–1165.
172. Larson AJ, Symons JD, Jalili T. Therapeutic potential of quercetin to decrease blood pressure: review of efficacy and mechanisms. *Adv Nutr*. 2012;3(1):39–46.
173. Conquer JA, Maiani G, Azzini E, Raguzzini A, Holub BJ. Supplementation with quercetin markedly increases plasma quercetin concentration without effect on selected risk factors for heart disease in healthy subjects. *J Nutr*. 1998;128(3):593–597.
174. Egert S, Boesch-Saadatmandi C, Wolffram S, Rimbach G, Mueller MJ. Serum lipid and blood pressure responses to quercetin vary in overweight patients by apolipoprotein E genotype. *J Nutr*. 2010;140(2):278–284.
175. Calder PC, Ahluwalia N, Brouns F, et al. Dietary factors and low-grade inflammation in relation to overweight and obesity. *Br J Nutr*. 2011;106(S1):S5–S78.
176. Ahluwalia N, Andreeva VA, Kesse-Guyot E, Hercberg S. Dietary patterns, inflammation and the metabolic syndrome. *Diabetes Metab*. 2013;39(2):99–110.
177. Rahman I, Biswas SK, Kirkham PA. Regulation of inflammation and redox signaling by dietary polyphenols. *Biochem Pharmacol*. 2006;72(11):1439–1452.
178. Gonzalez R, Ballester I, Lopez-Posadas R, et al. Effects of flavonoids and other polyphenols on inflammation. *Crit Rev Food Sci Nutr*. 2011;51(4):331–362.
179. Andre CM, Greenwood JM, Walker EG, et al. Anti-inflammatory procyanidins and triterpenes in 109 apple varieties. *J Agric Food Chem*. 2012;60(42):10546–10554.
180. Jung M, Triebel S, Anke T, Richling E, Erkel G. Influence of apple polyphenols on inflammatory gene expression. *Mol Nutr Food Res*. 2009;53(10):1263–1280.
181. Ye MB, Lim BO. Dietary pectin regulates the levels of inflammatory cytokines and immunoglobulins in interleukin-10 knockout mice. *J Agric Food Chem*. 2010;58(21):11281–11286.
182. Sanchez D, Quinones M, Moulay L, Muguerza B, Miguel M, Aleixandre A. Soluble fiber-enriched diets improve inflammation and oxidative stress biomarkers in zucker fatty rats. *Pharmacol Res*. 2011;64(1):31–35.
183. Chen C-H, Sheu M-T, Chen T-F, et al. Suppression of endotoxin-induced proinflammatory responses by citrus pectin through blocking LPS signaling pathways. *Biochem Pharmacol*. 2006;72(8):1001–1009.
184. Chun OK, Chung S-J, Claycombe KJ, Song WO. Serum c-reactive protein concentrations are inversely associated with dietary flavonoid intake in US adults. *J Nutr*. 2008;138(4):753–760.
185. Ridker PM, Hennekens CH, Buring JE, Rifai N. C-reactive protein and other markers of inflammation in the prediction of cardiovascular disease in women. *N Engl J Med*. 2000;342(12):836–843.
186. Freese R, Vaarala O, Turpeinen AM, Mutanen M. No difference in platelet activation or inflammation markers after diets rich or poor in vegetables, berries and apple in healthy subjects. *Eur J Nutr*. 2004;43(3):175–182.
187. Cani PD, Amar J, Iglesias MA, et al. Metabolic endotoxemia initiates obesity and insulin resistance. *Diabetes*. 2007;56(7):1761–1772.
188. Manco M, Putignani L, Bottazzo GF. Gut Microbiota, Lipopolysaccharides, and innate immunity in the pathogenesis of obesity and cardiovascular risk. *Endocr Rev*. 2010;31(6):817–844.
189. Boroni Moreira AP, Salles Texeira TF, Ferreira AB, MdC Gouveia Peluzio, RdC. Goncalves Alfenas. Influence of a high-fat diet on gut microbiota, intestinal permeability and metabolic endotoxaemia. *Br J Nutr*. 2012;108(5):801–809.
190. Shiau SY, Chang GW. Effects of certain dietary-fibers on apparent permeability of the rat intestine. *J Nutr*. 1986;116(2):223–232.
191. Rabbani GH, Teka T, Saha SK, et al. Green banana and pectin improve small intestinal permeability and reduce fluid loss in bangladeshi children with persistent diarrhea. *Dig Dis Sci*. 2004;49(3):475–484.
192. Bergmann H, Rogoll D, Scheppach W, Melcher R, Richling E. The ussing type chamber model to study the intestinal transport and modulation of specific tight-junction genes using a colonic cell line. *Mol Nutr Food Res*. 2009;53(10):1211–1225.
193. Drogoudi PD, Michailidis Z, PantelidiSa G. Peel and flesh antioxidant content and harvest quality characteristics of seven apple cultivars. *Sci Hortic (Amsterdam)*. 2008;115(2):149–153.
194. Khanizadeh S, Tsao R, Rekika D, Yang R, DeEll J. Phenolic composition and antioxidant activity of selected apple genotypes. *J Food Agric Environ*. 2007;5(1):61–66.
195. van der Sluis AA, Dekker M, de Jager A, Jongen WMF. Activity and concentration of polyphenolic antioxidants in apple: effect of cultivar, harvest year, and storage conditions. *J Agric Food Chem*. 2001;49(8):3606–3613.
196. Gliszczynska-Swiglo A, Tyrakowska B. Quality of commercial apple juices evaluated on the basis of the polyphenol content and the TEAC antioxidant activity. *J Food Sci*. 2003;68(5):1844–1849.
197. Seeram NP, Aviram M, Zhang Y, et al. Comparison of antioxidant potency of commonly consumed polyphenol-rich beverages in the united states. *J Agric Food Chem*. 2008;56(4):1415–1422.

198. Pearson DA, Tan CH, German JB, Davis PA, Gershwin ME. Apple juice inhibits human low density lipoprotein oxidation. *Life Sci.* 1999;64(21):1913–1920.
199. Thilakarathna SH, Rupasinghe HPV, Needs PW. Apple peel bioactive rich extracts effectively inhibit in vitro human LDL cholesterol oxidation. *Food Chem.* 2013;138(1):463–470.
200. Ko SH, Choi SW, Ye SK, Cho BL, Kim HS, Chung MH. Comparison of the antioxidant activities of nine different fruits in human plasma. *J Med Food.* 2005;8(1):41–46.
201. Maffei F, Tarozzi A, Carbone F, et al. Relevance of apple consumption for protection against oxidative damage induced by hydrogen peroxide in human lymphocytes. *Br J Nutr.* 2007;97(5):921–927.
202. Zhao S, Bomser J, Joseph EL, DiSilvestro RA. Intakes of apples or apple polyphenols decease plasma values for oxidized low-density lipoprotein/beta(2)-glycoprotein I complex. *J Funct Foods.* 2013;5(1):493–497.
203. Lotito SB, Frei B. The increase in human plasma antioxidant capacity after apple consumption is due to the metabolic effect of fructose on urate, not apple-derived antioxidant flavonoids. *Free Radic Biology Med.* 2004;37(2):251–258.
204. Godycki-Cwirko M, Krol M, Krol B, et al. Uric acid but not apple polyphenols is responsible for the rise of plasma antioxidant activity after apple juice consumption in healthy subjects. *J Am Coll Nutr.* 2010;29(4):397–406.
205. Lotito SB, Frei B. Relevance of apple polyphenols as antioxidants in human plasma: Contrasting in vitro and in vivo effects. *Free Radic Biology Med.* 2004;36(2):201–211.
206. Williams RJ, Spencer JPE, Rice-Evans C. Flavonoids: antioxidants or signalling molecules? *Free Radic Biology Med.* 2004;36(7):838–849.
207. Lotito SB, Frei B. Consumption of flavonoid-rich foods and increased plasma antioxidant capacity in humans: cause, consequence, or epiphenomenon? *Free Radic Biology Med.* 2006;41(12):1727–1746.
208. Bellion P, Hofmann T, Pool-Zobel BL, et al. Antioxidant effectiveness of phenolic apple juice extracts and their gut fermentation products in the human colon carcinoma cell line caco-2. *J Agric Food Chem.* 2008;56(15):6310–6317.
209. Carrasco-Pozo C, Speisky H, Brunser O, Pastene E, Gotteland M, Apple Peel Polyphenols. Protect against gastrointestinal mucosa alterations induced by indomethacin in rats. *J Agric Food Chem.* 2011;59(12):6459–6466.
210. Natella F, Macone A, Ramberti A, et al. Red wine prevents the postprandial increase in plasma cholesterol oxidation products: a pilot study. *Br J Nutr.* 2011;105(12):1718–1723.
211. Soyalan B, Minn J, Schmitz HJ, et al. Apple juice intervention modulates expression of ARE-dependent genes in rat colon and liver. *Eur J Nutr.* 2011;50(2):135–143.
212. Wild S, Roglic G, Green A, Sicree R, King H. Global prevalence of diabetes – estimates for the year 2000 and projections for 2030. *Diabetes Care.* 2004;27(5):1047–1053.
213. Juskiewicz J, Zary-Sikorska E, Zdunczyk Z, Krol B, Jaroslawska J, Jurgonski A. Effect of dietary supplementation with unprocessed and ethanol-extracted apple pomaces on caecal fermentation, antioxidant and blood biomarkers in rats. *Br J Nutr.* 2012;107(8):1138–1146.
214. Vaaler S, Wiseth R, Aagenaes O. Increase in blood glucose in insulin-dependent diabetics after intake of various fruits. *Acta Med Scand.* 1982;212(5):281–283.
215. Haber GB, Murphy D, Heaton KW, Burroughs LF. Depletion and disruption of dietary fiber – effects on satiety, plasma-glucose, and serum-insulin. *Lancet.* 1977;2(8040):679–682.
216. Hanhineva K, Torronen R, Bondia-Pons I, et al. Impact of dietary polyphenols on carbohydrate metabolism. *Int J Mol Sci.* 2010;11(4):1365–1402.
217. Hyson DA. A Comprehensive review of apples and apple components and their relationship to human health. *Adv Nutr.* 2011;2(5):408–420.
218. Ramachandra R, Shetty AK, Salimath PV. Quercetin alleviates activities of intestinal and renal disaccharidases in streptozotocin-induced diabetic rats. *Mol Nutr Food Res.* 2005;49(4):355–360.
219. Masumoto S, Akimoto Y, Oike H, Kobori M. Dietary phloridzin reduces blood glucose levels and reverses sglt1 expression in the small Intestine in streptozotocin-induced diabetic mice. *J Agric Food Chem.* 2009;57(11):4651–4656.
220. Yamaguchi K, Kato M, Suzuki M, et al. Pharmacokinetic and pharmacodynamic modeling of the effect of a sodium–glucose cotransporter inhibitor, phlorizin, on renal glucose transport in rats. *Drug Metab Dispos.* 2011;39(10):1801–1807.
221. Kobori M, Masumoto S, Akimoto Y, Oike H. Phloridzin reduces blood glucose levels and alters hepatic gene expression in normal BALB/c mice. *Food Chem Toxicol.* 2012;50(7):2547–2553.
222. Shirosaki M, Koyama T, Yazawa K. Apple leaf extract as a potential candidate for suppressing postprandial elevation of the blood glucose level. *J Nutr Sci Vitaminol (Tokyo).* 2012;58(1):63–67.
223. Kim M. High-methoxyl pectin has greater enhancing effect on glucose uptake in intestinal perfused rats. *Nutrition.* 2005;21(3):372–377.
224. Slavin J, Green H. Dietary fibre and satiety. *Nutr Bull.* 2007;32(Suppl. 1):32–42.

CHAPTER

13

Whole Plant Foods and Colon Cancer Risk

Emma M. Brown, Ian Rowland†, Nigel G. Ternan*, Philip Allsopp*, Geoff McMullan* and Chris I.R. Gill**

***Northern Ireland Centre for Food and Health, Centre for Molecular Biosciences, University of Ulster, Coleraine, County Londonderry, N. Ireland, UK †Hugh Sinclair Unit of Human Nutrition, Department of Food and Nutritional Sciences, University of Reading, Reading, UK**

INTRODUCTION

"Low dietary intakes of fruits, vegetables, whole grains, or nuts and seeds or a high dietary intake of salt are individually responsible for 1.5% to more than 4% of the global disease burden."[1] This recent study investigated behavioral and dietary risk factors for noncommunicable diseases including cancer, which remains a major cause of mortality worldwide, contributing to approximately 7.6 million deaths in 2008.[2] The World Cancer Research Fund reports[3,4] highlight the importance of specific lifestyle choices and dietary factors to a diverse range of cancers. Overall, however, the message remains clear: nutritionally poor-quality diets, high in saturated fats, high in refined carbohydrates and poor in fruit and vegetables are correlated with increased cancer risk generally. Colorectal cancer (CRC) is the third most common cause of cancer-related mortality globally and has strong associations with diet. There is considerable epidemiological evidence that fruits, and vegetables and whole-grain cereals are associated with reduced risk of CRC. There is also extensive evidence from *in vitro* studies and animal models that fruit and vegetable consumption can modulate biomarkers of DNA damage and that these effects may be potentially chemoprotective, given the likely role that oxidative damage plays in mutation rate and cancer risk. In this chapter, we review the evidence for effects of whole foods on cancer risk in humans with a focus on cereals, brassica and berry fruits.

DIET AND COLORECTAL CANCER

The majority of colorectal malignancies occur as sporadic forms that appear to arise from benign adenomatous polyps, with carcinomas emerging slowly over a period of 10–20 years.[5,6] Epidemiological data indicate that incidence rates and mortality of CRCs are greatly influenced by age, with the majority of cases being detected in individuals over the age of 60.[7] For individuals diagnosed with CRC, it has been determined that the 5-year survival rate is approximately 50–60%.[3] The age-dependent relationship of CRC development is associated with a multi-step oncogenesis process and a number of histological stages, reflecting the generation of genetic errors in somatic cells over time, have been characterized.[8] Conversely, the inheritance of germline mutations may also result in development of neoplasms at an early age. It is estimated that 15% of colon cancers have a strong familial background, with approximately 5% being due to inherited single-gene syndromes such as familial adenomatous polyposis (FAP) and hereditary non-polyposis colorectal cancer (HNPCC).[9]

The geographical distribution of CRC differs significantly across the world with the highest rates in European countries, North America, Australia, New Zealand and Japan.[10] The relatively recent increase in

Diet-Microbe Interactions in the Gut.
DOI: http://dx.doi.org/10.1016/B978-0-12-407825-3.00013-7

CRC incidence in Japan and in urbanized regions of China is thought to be due to the adoption of a more Western lifestyle, and indicates strongly that diet plays a central role in CRC pathogenesis. In several epidemiological studies, a number of dietary factors were indicated to have significant impact on CRC development. Generally, diets rich in saturated animal fat, and red meat (especially processed meat), together with alcohol intake and smoking, were positively associated with colorectal neoplasia.[3,11] However, evidence exists that vegetable consumption and diets that are high in fiber and starch appear to decrease the risk of CRC.[12,13] This notion is supported by a large body of case-control studies, although results from cohort or prospective studies are less convincing.[14,15] Nevertheless, the protective effects of vegetables and grains against colorectal cancer are attributed to the large number of bioactive phytochemicals present within them, comprising mainly plant polyphenolic secondary metabolites and plant structural and storage polysaccharides which make up dietary fiber.[16–19]

In addition, evidence continues to accumulate linking the composition of the adult gut microbiota with a diverse range of health outcomes including obesity, neurological development, CRC and inflammatory bowel disease.[20] It is becoming clear that dietary intake significantly influences the composition of the intestinal microbiota[21–23] and that resultant gut metabotypes[24] may also be important in how individuals respond to diet with respect to the catabolites produced as a consequence of colonic fermentation. Moreover, the gut microbiota has been shown to contribute to host metabolism (lipid, carbohydrate) and to alterations in inflammatory responses, most likely via their fermentation of dietary components. The stability of the gut microbiota is of importance, with evidence to suggest that a lack of diversity within the microbiota may make an individual more susceptible to disease and poorer health outcomes[25] such as colorectal cancer. As previously mentioned, CRC risk is likely to be affected by environmental factors including diet, especially given the constant exposure of the gut mucosa to the microbiome and/or its metabolites in response to dietary intake. The exact role of gut microbiota in the pathogenesis of colorectal cancer mechanisms, however, remains unclear, but a number of theories have been proposed, as comprehensively addressed in a recent review by Zhu et al., 2013.[26] Figure 13.1 shows the possible mechanisms of microbiota involvement in colorectal cancer.

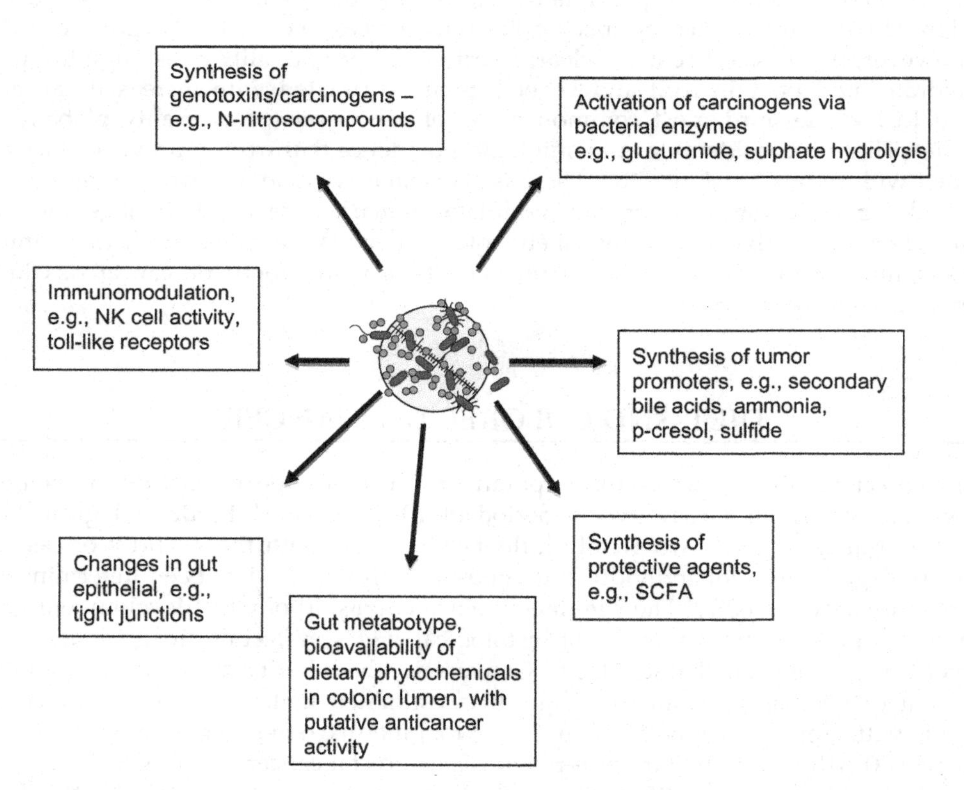

FIGURE 13.1 Possible mechanisms of microbiota involvement in colorectal cancer.

BIOLOGICAL ACTIVITY AND ANTICANCER PROPERTIES OF WHOLE-GRAIN CEREALS

Wheat, rice and maize are the main whole-grain cereals consumed throughout the world, followed by oats, rye and barley. Whole-grain cereals contain diverse components with anticancer potential, which are located mostly in the germ and bran layers that are removed from the endosperm during refining. These bioactives include fiber, resistant starch, phytate, vitamins and a wide range of phenolic compounds including ferulic acid, lignans, alkylresorcinols and flavonoids.[27] The human intestinal microbiota plays an important role in both the bioavailability and biological activity of many of these compounds, releasing smaller phenolic acids from complex polyphenols and fermenting dietary fibers into the short-chain fatty acids, acetate, propionate and butyrate which are then absorbed by the host and mediate important physiological activities not only in the intestine but systemically throughout the human body.

A meta-analysis of case control studies of whole-grain and colorectal cancer or polyps risk reported odds ratios (OR) of <1 in nine out of 10 studies with an overall OR of 0.79 for the highest versus lowest intakes.[28] More recently, Aune et al.[29] conducted a systematic review and meta-analysis of prospective studies, which avoid many of the problems such as recall bias associated with case control studies. Studies from both Europe and the USA were included in the analysis and the whole-grain intake (whole-grain breads, oatmeal, brown rice, high-fiber cereals, porridge) ranged from 61 to 128 g/d. In six studies (7941 cases from 774,806 patients) that were included in the dose−response analysis, relative risk for CRC ranged from 0.73 to 0.89 with summary relative risk (RR) of 0.83 (95% CI 0.78−0.89) for an increase in three servings of whole-grain/d (equivalent to 90 g/d). In addition, there was a linear inverse relationship between CRC risk and dose. The authors also performed a meta-analysis of CRC risk and cereal fiber intake which showed a significant overall RR for high vs. low intake of 0.90 (95% CI 0.83−0.97) and dose−response analysis revealed a summary RR of 0.90 for every 10 g/d increase in intake.[29]

Whilst the epidemiology, both in case control and in prospective studies, strongly suggest that the fiber component of whole-grain plays an important role in its anticancer activity, dietary intervention trials have been largely negative. A systematic review and meta-analysis of five dietary fiber intervention studies (three of which used wheat bran), using as endpoint the recurrence of adenomas two or more years after resection, found no significant effect due to wheat bran or from fiber (RR for adenoma recurrence = 1.04; 95% CI 0.95−1.13),[30] which may reflect the small number of studies, and the lack of homogeneity in the fiber or dose tested including isphagula husk and wheat bran fiber. A subsequent study reported by Lanza et al.,[31] the Polyp Prevention Trial, was a multicentre, randomized controlled trial involving 2079 patients with resected polyps. The experimental diet was high fiber (18 g/1000 kcal), high fruit and vegetable (3.5 servings/1000 kcal) and low fat (20% E) versus a "usual diet" and the endpoint was polyp recurrence at 4 and 8 years. There were no significant effects at either time point, with RR = 1.00 (95% CI 0.90−1.12) and RR = 0.98 (95% CI 0.88−1.09) at 4 and 8 years, respectively.

Resistant starch is a component of whole-grain that has been suggested as an anticancer agent due to its effects on butyrate production in the colon. The short-chain fatty acid, butyrate, is produced by anaerobic microbial fermentation of foodstuffs in the colon and is necessary for colonocyte health.[32] Studies with laboratory animal models of colon cancer are supportive of the efficacy of resistant starch,[33] however, human dietary intervention studies have proved negative. A double-blind, placebo-controlled trial in young familial adenomatous polyposis patients given 30 g resistant starch, with 1- to 7-year follow up,[34] found no effect on total colonic/rectal polyps (RR = 1.05, 95% CI 0.73−1.49). A further study by the same group using patients with Lynch syndrome (hereditary non-polyposis colorectal cancer) given 30 g resistant starch (or digestible starch as a placebo) for up to 4 years, with CRC as endpoint, showed an increased hazard ratio (2.38, 95% CI 0.98−2.55) at 2 years or less, and no effect (HR = 1.09, 95% CI 0.55−2.19) at over 2 years.[35]

In addition to complex polymers, the low-molecular-weight components of whole-grain cereals are receiving increasing attention for their anticancer potential. Some of the most notable phenolic acids include ferulic (FA), caffeic and *p*-coumaric (p-CA) acids, which can comprise up to 200 mg/100 g, and polyphenols such as lignans (0.2−0.6 mg/100 g), alkyresorcinols (12−129 mg/100 g) and flavonoids (30−43 mg/100 g).[27] Most of the data suggesting that such compounds have anticancer activity has come from *in vitro* studies. For example, FA and p-CA inhibit both colon cancer cell proliferation *in vitro* and cell cycle progression at S & G2M phases, and also upregulate several genes involved in cell cycle control.[36] Alkylresorcinols have been shown to inhibit DNA damage induced by hydrogen peroxide and genotoxic fecal extracts in colon cell lines.[37] There is extensive *in vitro* evidence that lignans and their gut microbiota-derived metabolites, enterolactone and enterodiol, not only have antioxidant and antiinflammatory activity but can inhibit cancer cell proliferation, invasion and angiogenesis, and increase apoptosis.[38,39] Epidemiological studies of CRC/adenoma risk and lignan intake have however yielded inconsistent results. Within the Norfolk cohort of European Prospective Investigation into Cancer (EPIC), for example, no association of CRC risk with lignan intake (secoisolariciresinol and matairesinol) was seen in men, but a positive trend (OR = 1.60; 95% CI: 0.96, 2.69) was seen for secoisolaricresinol in women. A case-cohort study of the association between plasma enterolactone concentration and incidence of CRC in a Danish cohort also reported differential effects in men and women.[40] However in this study, in women, for each two-fold increase in enterolactone concentration, a decreased colon risk was seen (IRR = 0.76, 95% CI 0.60–0.96) together with a non-significant decrease in risk of rectal cancer (IRR = 0.83, 0.60–1.14). However in men, a doubling of plasma enterolactone concentration was associated with a higher risk of colon and rectal cancer, significantly in the latter case (IRR = 1.74, CI 1.25–2.44). A recent analysis of lignan intake and adenoma recurrence in the Polyp Prevention Trial found a positive association in women (OR = 2.07, 95% CI 1.22–3.52) but not in men.[41] It is likely that these inconsistent results reflect the diverse sources of lignans in the diet, in addition to the difficulty of accurately estimating dietary lignan intake. A recent study estimated total intake of lignans from food consumption data for adult men and women (19–79 years) from Denmark, Finland, Italy, Sweden, United Kingdom. Utilizing the Dutch lignan database, which includes measures of four lignan precursors – secoisolariciresinol, matairesinol, lariciresinol and pinoresinol – intake ranged from 1 to 2 mg/day with the main contributors to total lignan dietary intake including cereals (rye 458 μg/100 g; wheat μg/100 g), grain products (pasta 16 μg/100 g, vegetables (cabbages 600 μg/100 g), fruit (citrus 112 μg/100 g and berries (334 μg/100 g). Across all the countries intake by gender was comparable, except in Sweden where men tended towards higher consumption of lignans from cereal, and women from fruits and vegetables.[42]

Biomarkers for CRC

Epidemiological studies are suggestive of a link between fruit and vegetable consumption and a reduced risk of colorectal cancer. However, the task of assessing the anticancer effects of food constituents in humans is challenging[43–45] given that, for the most part, intervention studies rely on intermediate biomarkers, such as DNA damage, that are plausibly linked to carcinogenesis[46] in the absence of clinical trials using definitive endpoints such as cancer incidence or mortality. It is well established that the initiating step in CRC development involves exposure to, or uptake of carcinogens, resulting in permanent DNA damage that has a key role in cancer development. Considering the potentially toxic nature of the colon contents, it is not surprising that frequent mutations occur in colonic cells and thus efficient DNA damage detection and repair mechanisms are present to prevent damage becoming a permanent change in the DNA sequence. Despite this, errors in detection and repair mechanisms may occasionally occur and therefore reducing the frequency of DNA damage, or enhancing DNA repair, could be a feasible explanation for the protective effects of fruit and vegetables against the initiation of CRC.

Levels of DNA damage can be determined in tissues and organs either by taking biopsies or by isolating exfoliated epithelial cells such as buccal, bladder or intestinal cells. Systemic DNA damage is usually measured in peripheral blood mononucleated cells (PBMC) as they are easily isolated from blood samples and are believed to reflect DNA damage in other body tissues. Studies of DNA damage can be performed *in vivo*, *ex vivo* as well as *in vitro*. The alkaline comet assay is widely regarded as a valuable tool for genotoxicity testing and biomonitoring of oxidative stress in humans[47] and its applicability to cancer risk studies has been discussed,[48,49] however, inter-laboratory reproducibility still remains an area for improvement. Systemic oxidative DNA damage can also be assessed by measurement, using a variety of chromatographic methods, of the formation of DNA adducts such as 8-hydroxy-2′-deoxyguanosine (8-OHdG). 8-OHdG represents one of the predominant forms of free radical-induced oxidative lesions (as reviewed in Ref. 50). Lipid peroxidation products such as hydroxyeicosatetraenoic acids (HETE) in plasma or urine may also be measured (reviewed in Refs 51,52), although it is interesting to note that there is evidence to suggest that the alkaline comet assay (enzymatic) appears to be more accurate in estimation of low background levels of oxidative damage than are the chromatographic methods.[47]

The use of such biomarkers has benefits due to the non-invasive method of collection: for example, measuring changes in phase II enzymes, carcinogen-induced DNA damage or antioxidant activity in surrogate tissues such as blood lymphocytes after intake of fruit or vegetables is straightforward. However, there are limitations with these methods related to the bioavailability of polyphenols at the site of action, the metabolic breakdown of the parent compounds, and the short half-life of the majority of polyphenols and their metabolites within plasma.

BIOLOGICAL ACTIVITY AND ANTICANCER PROPERTIES OF *BRASSICA* VEGETABLES

In the light of many reports, consumption of *Brassica* vegetables has become associated with reduction of cancer risk at different sites.[53,54] The mechanisms of the chemopreventative effects of cruciferous vegetables (CV) might include alterations to the metabolism of dietary carcinogens due to modulation of phase I and phase II enzymes.[54] Many biologically active compounds found in CV are likely contributors to the alteration of enzyme activities and the overall outcome will reflect the combined effect of these active constituents. For example, glutathione-S-transferases (GSTs) are detoxification enzymes responsible for xenobiotic metabolism and thus they contribute to the increased elimination of potential activated carcinogens from the body. Both isothiocyanates (ITCs) and indoles are responsible for induction of phase II enzymes, however, they have opposite effects on phase I enzymes: indoles are strong inducers of cytochrome P450 enzymes, whereas isothiocyanates usually inhibit these enzymes. Glucosinolates (GLS) and their derivatives have been suggested to prevent oxidative DNA damage and carcinogen-DNA adduct formation, thus obstructing the mutational events that may trigger carcinogenesis. In addition, it has been demonstrated in both cell culture models and animal studies that GLS and their breakdown products inhibit cell proliferation either by induction of apoptosis or via cell cycle arrest.[55]

HUMAN STUDIES

The evidence from a variety of animal studies supports observations from *in vitro* studies that consumption of CV, GLS and their derivatives is associated with a reduction in the number of chemically induced tumors, together with increased activity of detoxifying enzymes and resultant decreases in DNA damage. Anticancer effects of *Brassica* consumption have been also investigated in dietary intervention studies in which intermediate biomarkers were used as the endpoints of CRC (Table 13.1). The application of intermediate biomarkers, including detoxification phase I and II enzymes and oxidative DNA damage, as well as their practical usefulness as an alternative strategy to cancer assessment has been reviewed by Gill and Rowland.[43]

The effect of broccoli and Brussels sprouts intake on metabolism of the heterocyclic amines 2-amino-1-methyl-6-phenylimidazo[4,5-b]pyridine (PhIP), and 2-amino-3,8-dimethylimidazo-[4,5-*f*]quinoxaline (MeIQx) present in cooked meat was investigated by Murray et al.[56] An increase in activity of CYP1A2 (as reflected by caffeine kinetics in saliva) and reduced excretion of intact PhIP and MeIQx in the urine was observed after CV consumption. These results are unsurprising as CYP1A2 is involved in oxidative activation and detoxification of amines. A decrease in intact PhIP and MeIQx in urine indicated an enhanced metabolism of amines, supported by an observed increase in urinary mutagenicity due to the presence of breakdown metabolites. Interestingly, even when the broccoli and Brussels sprouts supplementation period was finished, reduced excretion of intact amines and elevated urinary mutagenicity were sustained, suggesting that other enzyme systems were affected by the CV-rich diet. Studies on the metabolism of PhIP after consumption of broccoli and Brussels sprouts have shown that the major metabolite of PhIP in urine samples is N^2-hydroxy-PhIP-N^2-glucuronide, which is derived from CYP1A2-catalyzed N^2-hydroxylation and subsequent UGT-catalyzed glucuronidation of PhIP.[57] This is consistent with the hypothesis that phase II enzymes take part in detoxification of heterocyclic amines. There is also an indication that the increase in CYP1A2 activity after CV consumption may be affected by polymorphisms within the GSTM1 gene, as GSTM1 null subjects consuming a diet abundant in *Brassica* were observed to have higher CYP1A2 activity compared to individuals lacking the gene deletion.[58] Several other studies have used biomarkers such as caffeine demethylation or hydroxylation of estrone to demonstrate that a diet rich in CV modulates CYP1A2 activity.[59,60]

Studies in humans have demonstrated that consumption of CV contributes to a reduction in cellular susceptibility to carcinogens due to induction of phase II enzymes and detoxification of electrophilic metabolites. As an example, the consumption of Brussels sprouts was shown to induce higher levels of GSTα protein in plasma and amounts of GSTα and GSTπ protein in rectal mucosa.[61,62] Such an increase in both levels and activity of serum GSTα and peripheral-lymphocyte GSTμ was also observed after ingestion of a variety of CV including cabbage, radish sprouts, cauliflower and broccoli. Furthermore, these changes were GSTM1 genotype dependent.[68] The elevated levels of serum GSTα and GST activity were found exclusively in subjects with a GSTM1-null genotype, whereas increased peripheral-lymphocyte GSTμ activity was observed in the GSTM1+ genotype group. It appears that the beneficial effect of CV consumption against colorectal cancer applies mostly to individuals with the GSTM1-null genotype.[69,70] While consumption of CV may induce GST activity and promote detoxification of

TABLE 13.1 Human Studies into the Effects of Cruciferous Vegetable Constituents on Colorectal Cancer

Intervention	Dosage	Exposure Time	Number of Subjects	Effects	Reference
Broccoli, Brussels sprouts	250 g/day	12 days	20 male	CYP1A2 activation; reduced excretion of intact PhIP and MeIQx; UGT-catalyzed glucuronidation	Murray et al., 2001[56] Walters et al., 2004[57]
Broccoli, Cabbage, Cauliflower, Mustard greens	Normal diet, at least once a week	2 years	328: 63 female, 265 male	CYP1A2 activation in GSTM1 null genotype	Probst-Hensch et al., 1998[58]
Radish sprouts, Cauliflower, Broccoli, Cabbage	16 g/day 150 g/day 200 g/day 70 g/day	6 days	40: 21 female, 19 male	CYP1A2 activation; Increased levels and activity of serum GST α (26%, GSTM1 null) and peripheral-lymphocyte GSTμ (18%, GSTM1+)	Lampe et al., 2000[59]
Broccoli	500 g/day	12 days	16: 2 female, 14 male	Enhanced CYP1A2 – mediated by caffeine metabolism	Kall et al., 1996[60]
Brussels sprouts	300 g/day	7 days	10: 5 female, 5 male	Induced levels of GSTα in plasma (male), and GSTα and GSTμ in rectal mucosa	Nijhoff et al., 1995a,b[61,62]
Brussels sprouts	300 g/day	3 weeks 1 week	10 male 10: 5 female, 5 male	Reduced excretion of 8-oxodG Reduced excretion of 8-oxodG (male)	Verhagen et al., 1995[63] Verhagen et al., 1997[64]
Brussels sprouts	300 g/day	6 days	8: 4 female, 4 male	Reduced DNA damage induced by PhIP and H_2O_2 in lymphocytes; decrease in SULT1A1 activity	Hoelzl et al., 2008[65]
Watercress	85 g/day	8 weeks	60: 30 female, 30 male	Reduction in oxidative purine and H_2O_2 induced DNA damage in lymphocytes	Gill et al., 2007[66]
Mixed Cruciferous vegetables (CV) + yoghurt + chlorophyllin tablets	500 g/day CV), Yoghurt, 3.3-oz containers/day based on body weight. 3 × 100 mg sodium copper chlorophyllin tablets/day	Crossover, 2 weeks, high red meat. 2 weeks red meat+ intervention	4 male 4 female	Reduction in DNA damage in rectal tissue Decreased fecal and urine mutagenicity No reduction in DNA damage in lymphocytes	Shaughnessy et al., 2011[67]

chemical carcinogens, conjugation with GST is also the main pathway of ITC metabolism and excretion. Therefore, a slower excretion of ITC in individuals with a GSTM1-null genotype may result in tissue accumulation of ITCs and lead to greater protective effects mediated by a variety of chemoprotective mechanisms.[71]

Another anticancer mechanism associated with CV intake is the reduction of oxidative DNA damage. The urinary excretion rate of 8-oxo-7,8-dihydro-2′deoxyguanosine (8-oxodG) an abundant DNA adduct formed by reactive oxygen species[63,64] can provide an estimate of oxidative DNA damage in the whole body and consumption of Brussels sprouts have been demonstrated to reduce urinary excretion of (8-oxodG). In addition, consumption of Brussels sprouts reduces DNA migration induced by PhIP in lymphocyte samples,[65] an effect explained by the fact that ingestion of Brussels sprouts decreased the level and activity of the enzyme sulfotransferase SULT1A1, which is thought to play a role in the bioactivation of PhIP. Furthermore, endogenous formation of oxidized purines and pyrimidines, and hydrogen peroxide-induced DNA damage in lymphocytes, was significantly reduced post-intervention. Watercress supplementation has also resulted in a reduction in oxidized purines and of hydrogen peroxide-induced DNA damage in lymphocytes,[66] an observation suggested to be related to changes in antioxidant status. Interestingly, a recent – albeit small – study reported that intervention with cruciferous vegetables, yogurt, and chlorophyllin tablets in a group (n = 8) consuming red meat with a high cooking temperature (resulting in increased levels of heterocyclic amines), significantly reduced DNA damage in rectal tissue.[67]

BIOLOGICAL ACTIVITY AND ANTICANCER PROPERTIES OF BERRY FRUITS

The polyphenol content and composition in berries is influenced by genetic and environmental factors such as species and variety, cultivation methods, weather, ripeness at and time of harvesting and duration and conditions of storage.[72–78] The content and composition of the polyphenols varies with the species, with berries of the same genus exhibiting characteristic polyphenol profiles. In the *Vaccinum* genus, for example, the relative amounts of anthocyanins, flavonols, HCA derivatives and proanthocyanidins differ between bilberry, cranberry and lingonberry.[79,80] However in blueberries (V. *angustifolium* × *corymbosum* and *V. corymbosum*), and huckleberry (*V. parvifolium*) HCA derivatives predominate.[81] This can also be illustrated using the Rosaceae family which contains multiple genera including *Rubus*, *Fragaria* and *Sorbus*. Berries from the *Rubus* genus (cloudberry, raspberry and blackberry) primarily contain ellagitannins with differing levels of anthocyanins, while strawberries (*Fragaria* genus) principally contain anthocyanins closely followed by ellagitannins and also proanthocyanidins. However, fruits from the *Sorbus* genus, such as rowanberry, consist chiefly of hydroxycinnamic acid derivatives, with lesser amounts of flavonols and smaller amounts of anthocyanins.[79] In the Grossulariaceae family fruits from the *Ribes* genus including gooseberries, and black- and redcurrants, mainly comprise anthocyanins.[79] As with previous examples from cruciferous vegetables, the evidence for an anticancer impact of fruit, in this case specifically berries, relies predominantly on the use of surrogate markers such as DNA damage, phase II enzyme activity, or DNA adduct formation (Table 13.2). Moreover there are emerging studies in patients with colorectal cancer, in which investigations of tumor tissue both before and after treatment with fruits have been undertaken.

As mentioned above, GSTs are detoxification enzymes responsible for xenobiotic metabolism and thus the increased elimination of potential activated carcinogens from the body. In humans, low GST activity has been associated with increased risk of CRC or incidence of tumors in the colonic mucosa compared to individuals with a higher level of GST activity.[88] Studies have shown that, in humans, dietary supplementation with berry polyphenol-rich drinks have had differing effects on GST activity (Table 13.2). Over a short supplementation period (2 weeks) no change in GST activity was reported,[82,84] but over a longer supplementation period it was observed that GST activity was increased.[84] This difference may be due either to a time delay effect, or possibly because an accumulation of berry polyphenols is required to produce an effect: it was observed that a decrease in levels of human GST protein (hGSTP1) at the start of supplementation was followed by a significant increase in hGSTP1 levels towards the end of the supplementation period.[84]

The effect of berry juice supplementation on antioxidant status has also been investigated by measuring oxidative DNA damage in lymphocytes with additional measurements of blood antioxidant activity giving an indication of the rate of integrated DNA damage within the body (Table 13.2). While measurements of the oxidative potential of blood were not conclusive, an overall decrease in oxidative potential after supplementation with berry juices was observed.[82–84] This may, however, be due to the vitamin C content of berry juice.[84] With polyphenol-rich berry juices, a decrease in oxidized bases, but no change to the frequency of strand breakages, was reported.[83] Interestingly, an inverse correlation was observed between the decreased levels of oxidized bases and increases in GST levels in human lymphocytes over a longer treatment period. Thus, increased levels of detoxification as a result of supplementation may lead to decreased levels of DNA damage. DNA damage in lymphocytes following treatment with berry polyphenols may be important, as the work of Pool-Zobel et al.[46] has demonstrated a correlation between levels of DNA damage in blood lymphocytes and levels of DNA damage in rectal cells. That said, however, measuring the levels of DNA damage in lymphocytes could underestimate the protective effect of berry polyphenols on cells of the colorectum, as these compounds are relatively poorly absorbed from the gut.[46] For example, following consumption of berries, the level of phenolics present in the colonic lumen is likely to be greater than that in systemic circulation, since mmol quantities of various phenolics may be detected in fecal water.[89,90] Moreover, some evidence suggests that these simple phenolics may exert anticancer effects.[91,92]

In pilot studies, supplementation of CRC patients with freeze-dried black raspberry (BRB) suppositories and/or oral BRB slurry or an anthocyanin-rich bilberry extract appeared to have a protective effect.[85–87] Over a

TABLE 13.2 Effect of Supplementation of Berry Extracts/Components on a Range of Biomarkers in Healthy Volunteers or in Colorectal Cancer Patients

Intervention	Dosage	Exposure Time	Number of Subjects	Effects	Reference
Polyphenol-rich juices: A = aronia, blueberry & boysenberry B = green tea, apricot & lime	300 mL/d	10 weeks low polyphenol diet 2 weeks A/B followed by washout & 2 weeks B/A	27 healthy men non-smokers	↓oxidative potential of plasma using TBARS measurement only No effect on strand breaks After 2nd supplementation period with juice A or B↓in oxidized bases observed Both A & B↑IL-2 secretion	Bub et al., 2003[82]
Polyphenol-rich juices: A = aronia, blueberry & boysenberry B = green tea, apricot & lime	300 mL/d	10 weeks low polyphenol diet 2 weeks A/B followed by washout & 2 weeks B/A 24 h incubation *in vitro*	24 healthy men non-smokers	↓ hGSTP1 protein levels after weeks 4 & 6 but↑levels at week 8 ↑ hGSTP1 levels correlate with↓levels of DNA damage No change in GST genes was observed in leukocytes *in vitro* but upregulation of cyp P450 enzymes and sulphotrans-ferase enzymes was observed for B but not A	Hoffmann et al., 2006[83]
Commercially available cranberry juice (CJ)	750 mL/d	2 weeks of cranberry or placebo juice	20 healthy women (18 non-smokers)	↑ with CJ compared to placebo No difference in GSH-Px, CAT or SOD activity No effect on 8-OHdG levels No effect on endogenous or induced DNA damage in lymphocytes	Duthie et al., 2006[84]
Black raspberry (BRB – freeze dried)	60 g BRB daily (3 × 20 g in 100 mL water)	2–4 weeks	25 CRC patients	↓ staining of proliferative cells in tumor tissue but not normal tissue ↑ staining of apoptotic cells in tumor but not normal tissue ↓ staining of blood vessels in both normal and tumor tissue after berry treatment	Wang et al., 2007[85]
Mirtocyan (anthocyanin rich bilberry extract)	1.4 g/d 2.8 g/d 5.6 g/d	7 days	25 CRC patients	↓ proliferation by 7% ↓ circulating IGF-1 concentrations 179 ng/g of tumor at highest dose	Thomasset et al., 2009[86]

(Continued)

TABLE 13.2 (Continued)

Intervention	Dosage	Exposure Time	Number of Subjects	Effects	Reference
Black raspberry (BRB – freeze dried)	60 g BRB (20 g/3 × /day) + 2 × 700 mg BRB Sup or oral placebo + 2 × 700 mg BRB Sup	9 months	14 FAP patients previously undergone colectomy	↓ rectal polyp by 59% in patients with oral and suppository treatment ↓ rectal polyp by 36% in patients treated with suppository only (compared with patient polyp numbers at baseline)	Stoner et al., 2008[87]

↓ = decrease. ↑ = increase. Sup = suppository. NM = not mentioned. ICH = immunohistochemical staining.

short period (2–4 weeks) orally supplemented BRB extract was able to significantly reduce proliferation in colon tumors, but crucially not in normal colonic mucosa. At the same time there was a non-significant increase in staining of apoptotic cells from tumors but not from normal mucosa, and a non-significant decrease in blood vessel staining from both normal and tumor tissue, indicating that tumor growth may be decreased via lessened blood supply and increased apoptosis.[85] With bilberry powder supplementation, a 7% decrease in tumor proliferation was observed and accumulation of anthocyanins in the tumor tissue to a level of 179 ng/g showed uptake of the treatment in the desired tissue over the short supplementation period.[86] A longer-term study (9 months) with BRB oral slurry, or BRB suppository in addition to oral treatment, with FAP patients after colectomy showed a reduction in the number of rectal polyps compared to the baseline.[87] A greater reduction was observed for both berry treatments compared to the berry suppository alone, which could be due to a number of different factors such as total increased concentration in tissues, products from digestive degradation in the upper gastrointestinal tract having a higher activity than the parent compounds, or the differences in absorption between anthocyanins in solution compared to the solid format. Moreover, recent evidence suggests that dietary flavonoids such as tiliroside and d(−)-epigallocatechin-3-gallate can inhibit carbohydrate digestion,[93–95] potentially increasing the amount of fermentable carbohydrate reaching the colon. This could, in turn, modulate the putative protective effects associated with fiber including an increase in butyrate formation, or bile acid binding activities of polyphenols/fiber, thus increasing fecal excretion of bile/cholesterol. Indeed, there is some evidence to suggest that the polyphenols themselves can act in a prebiotic manner and could further contribute to fermentation in the colon. Further evidence from studies with CRC patients suggests that consumption of black raspberries decreases proliferation markers in colon tissue, demethylates tumor suppressor genes and modulates other biomarkers of tumor development in the human colon and rectum[96] and beneficially modulate inflammatory markers such as IL-8[97] and that these effects may be mediated by inhibition of DNA methyltransferase 1 (DNMT1).[98]

CONCLUSION

There is considerable epidemiological evidence that whole plant foods, fruits, vegetables and whole-grain cereals are associated with reduced risk of CRC. There is also extensive evidence that fruit and vegetable consumption can modulate biomarkers of DNA damage. *In vitro* studies and animal models suggest that these effects may be potentially chemo-protective due to the likely role that oxidative damage plays in mutation rate and cancer risk. The dietary intervention trials with exemplar fruits and vegetables such as brassicas and berries are generally consistent and indicate a capacity to significantly decrease oxidative damage to DNA, however they in themselves represent only weak evidence for anti-carcinogenicity as they rely on surrogate putative markers for cancer risk. To understand the protective effects of fruit and vegetables consumption on colorectal cancer risk, future studies will need to be well controlled, using suitable and clinically relevant endpoints to support the tentative emerging evidence for a protective effect. An understanding of the comparative effectiveness of whole foods and their respective dietary phytochemicals is only relevant if we improve our understanding of which components survive and are present *in situ*, and which metabolites are formed.[99] In view of the impact that an individual's gut microbiota may have, with respect to bioavailability and ultimately the bioactivity of dietary phenols, the importance of establishing an individual's (poly)phenol metabolizing phenotype or gut metabotype will become of increasing relevance in studies of colon cancer.[24,100] Moreover, fermentable carbohydrates (fiber) present in whole foods might act in synergy with the mixture of (poly)phenols present in berries and contribute to their overall anti-genotoxic activities, as these fermentable dietary fibers can beneficially modulate both the composition and metabolic output of the human gut microbiota.[101]

References

1. Ezzati M, Riboli E. Behavioural and dietary risk factors for noncommunicable diseases. *N Engl J Med*. 2013;369(10):954–964:5.
2. Ferlay J, Shin HR, Bray F, Forman D, Mathers C, Parkin DM. Estimates of worldwide burden of cancer in 2008: GLOBOCAN 2008. *Int J Cancer*. 2010;127(12):2893–2917.
3. WCRF. *Food, Nutrition and the Prevention of Cancer: A Global Perspective*. Washington, DC: World Cancer Research Fund/American Institute for Cancer Research; 1997.
4. Wiseman M. The second World Cancer Research Fund/American Institute for Cancer Research expert report. Food, nutrition, physical activity, and the prevention of cancer: a global perspective. *Proc Nutr Soc*. 2008;67(3):253–256.
5. Kinzler KW, Vogelstein B. Lessons from hereditary colorectal cancer. *Cell*. 1996;87:159–170.
6. Peipins LA, Sandler RS. Epidemiology of colorectal adenomas. *Epidemiol Rev*. 1994;16:273–297.
7. Farley J, Bray F, Pisani P, Parkin DM. *GLOBOCAN 2002: Cancer Incidence, Mortality and Prevalence Worldwide*. IARC CancerBase **No. 5**. Lyon, France: IARC Press; 2004.
8. Fearon ER, Vogelstein B. A genetic model for colorectal tumorigenesis. *Cell*. 1990;61:759–767.
9. Johns LE, Houlston RS. A systematic review and meta-analysis of familial colorectal cancer risk. *Am J Gastroenterol*. 2001;96:2992–3003.
10. Jass JR. Lower gastrointestinal tract. In: Alison MR, ed. *The Cancer Handbook*. John Wiley & Sons; 2005:545–561. on-line edition.
11. World Health Organization. Colorectal cancer. In: Stewart BW, Kleihues P, eds. *World Cancer Report*. Lyon, France: IARC Press; 2003:198–202.
12. COMA. *Nutritional Aspects of the Development of Cancer: Report of the Working Group on Diet and Cancer of the Committee on Medical Aspects of Food and Nutrition Policy*. London: The Stationery Office; 1998.
13. Bingham SA, Day NE, Luben R, et al. Dietary fibre in food and protection against colorectal cancer in the European Prospective Investigation into Cancer and Nutrition (EPIC): an observational study. *Lancet*. 2003;361:1496–1501.
14. Bingham SA. Diet and colorectal cancer prevention. *Biochem Soc Trans*. 2000;28:2–6.
15. Dove-Edwin I, Thomas HJ. Review article: the prevention of colorectal cancer. *Aliment Pharmacol Ther*. 2001;15:323–336.
16. Levi F, Pasche C, Lucchini F, La Vecchia C. Selected micronutrients and colorectal cancer. a case-control study from the canton of Vaud, Switzerland. *Eur J Cancer*. 2000;36:2115–2119.
17. Voorrips LE, Goldbohm RA, van Poppel G, Sturmans F, Hermus RJ, van den Brandt PA. Vegetable and fruit consumption and risks of colon and rectal cancer in a prospective cohort study: the Netherlands Cohort Study on Diet and Cancer. *Am J Epidemiol*. 2000;152:1081–1092.
18. Verhoeven DT, Goldbohm RA, van Poppel G, Verhagen H, van den Brandt PA. Epidemiological studies on brassica vegetables and cancer risk. *Cancer Epidemiol Biomarkers Prev*. 1996;5:733–748.
19. Erlund I, Freese R, Marniemi J, Hakala P, Alfthan G. Bioavailability of quercetin from berries and the diet. *Nutr Cancer*. 2006;54(1):13–17: Review.
20. Arumugam M, Raes J, Pelletier E, et al. Enterotypes of the human gut microbiome. *Nature*. 2011;473(7346):174–180.
21. De Filippo C, Cavalieri D, Di Paola M, et al. Impact of diet in shaping gut microbiota revealed by a comparative study in children from Europe and rural Africa. *Proc Natl Acad Sci USA*. 2010;107(33):14691–14696.
22. Tuohy KM, Conterno L, Gasperotti M, Viola R. Up-regulating the human intestinal microbiome using whole plant foods, polyphenols, and/or fibre. *J Agric Food Chem*. 2010;60:8776–8782.
23. Claesson MJ, Jeffery IB, Conde S, et al. Gut microbiota composition correlates with diet and health in the elderly. *Nature*. 2012;488 (7410):178–184.
24. Bolca S, Van de Wiele T, Possemiers S. Gut metabotypes govern health effects of dietary polyphenols. *Curr Opin Biotechnol*. 2013;24 (2):220–225.
25. Lozupone CA, Stombaugh JI, Gordon JI, Jansson JK, Knight R. Diversity, stability and resilience of the human gut microbiota. *Nature*. 2012;489(7415):220–230.
26. Zhu Q, Gao R, Wu W, Qin H. The role of gut microbiota in the pathogenesis of colorectal cancer. *Tumour Biol*. 2013;34(3):1285–1300.
27. Fardet A. New hypotheses for the health-protective mechanisms of whole-grain cereals: what is beyond fibre? *Nutr Res Rev*. 2010;23:65–134.
28. Jacobs Jr. DR, Marquart L, Slavin J, Kushi LH. Whole-grain intake and cancer: an expanded review and meta-analysis. *Nutr Cancer*. 1998;30:85–96.
29. Aune D, Chan DSM, Lau R, et al. Dietary fibre, whole grains, and risk of colorectal cancer: systematic review and dose–response meta-analysis of prospective studies. *BMJ*. 2011;343:d6617.
30. Asano TK, McLeod RS. Dietary fibre for the prevention of colorectal adenomas and carcinomas. *Cochrane Database Sys Rev*. 2002; CD003430.
31. Lanza E, Yu B, Murphy G, et al. The polyp prevention trial continued follow-up study: no effect of a low-fat, high-fibre, high-fruit, and -vegetable diet on adenoma recurrence eight years after randomization. *Cancer Epidemiol Biomarkers Prev*. 2007;16:1745–1752.
32. Astbury SM, Corfe BM. Uptake and metabolism of the short-chain fatty acid butyrate, a critical review of the literature. *Curr Drug Metab*. 2012;13(6):815–821.

33. Le Leu RK, Brown IL, Hu Y, Esterman A, Young GP. Suppression of azoxymethane-induced colon cancer development in rats by dietary resistant starch. *Cancer Biol Ther*. 2007;6:1621–1626.
34. Burn J, Bishop DT, Chapman PD, et al. A randomized placebo-controlled prevention trial of aspirin and/or resistant starch in young people with familial adenomatous polyposis. *Cancer Prev Res*. 2011;4(5):655–665.
35. Mathers JC, Movahedi M, Macrae F, et al. Long-term effect of resistant starch on cancer risk in carriers of hereditary colorectal cancer: an analysis from the CAPP2 randomised controlled trial. *Lancet Onc*. 2012;13:1242–1249.
36. Janicke B, Hegardt C, Krogh M, et al. The antiproliferative effect of dietary fibre phenolic compounds ferulic acid and *p*-coumaric acid on the cell cycle of Caco-2 cells. *Nutr Cancer*. 2011;63(4):611–622.
37. Parikka K, Rowland IR, Welch RW, Wähälä K. *In vitro* antioxidant activity and antigenotoxicity of 5-n-alkylresorcinols. *J Agric Food Chem*. 2006;54(5):1646–1650.
38. Qu H, Madl RL, Takemoto DJ, Baybutt RC, Wang W. Lignans are involved in the antitumor activity of wheat bran in colon cancer SW480 cells. *J Nutr*. 2005;135:598–602.
39. Danbara N, Yuri T, Tsujita-Kyutoku M, Tsukamoto R, Uehara N, Tsubura A. Enterolactone induces apoptosis and inhibits growth of Colo 201 human colon cancer cells both *in vitro* and *in vivo*. *Anticancer Res*. 2005;25:2269–2276.
40. Johnsen NF, Olsen A, Thomsen BL, et al. Plasma enterolactone and risk of colon and rectal cancer in a case-cohort study of Danish men and women. *Cancer Causes Control*. 2010;21:153–162.
41. Bobe G, Murphy G, Albert PS, et al. Dietary lignan and proanthocyanidin consumption and colorectal adenoma recurrence in the Polyp Prevention Trial. *Int J Cancer*. 2012;130(7):1649–1659.
42. Tetens I, Turrinim A, Tapanainen H, Phytohealth WP1 working group, et al. Dietary intake and main sources of plant lignans in five European countries. *Food Nutr Res*. 2013;57. Available from: http://dx.doi.org/doi:10.3402/fnr.v57i0.19805.
43. Gill CIR, Rowland IR. Diet and cancer: assessing the risk. *Br J Nutr*. 2002;88(Suppl. 1):S73–S87.
44. Spencer JP, Abd El Mohsen MM, Minihane AM, Mathers JC. Biomarkers of the intake of dietary polyphenols: strengths, limitations and application in nutrition research. *Br J Nutr*. 2008;99(1):12–22.
45. Del Rio D, Rodriguez-Mateos A, Spencer JP, Tognolini M, Borges G, Crozier A. Dietary (poly)phenolics in human health: structures, bioavailability, and evidence of protective effects against chronic diseases. *Antioxid Redox Signal*. 2013;18(14):1818–1892.
46. Pool-Zobel BL, Dornacher I, Lambertz R, Knoll M, Seitz HK. Genetic damage and repair in human rectal cells for biomonitoring: sex differences, effects of alcohol exposure, and susceptibilities in comparison to peripheral blood lymphocytes. *Mutat Res*. 2004;551 (1–2):127–134.
47. Collins AR. Measuring oxidative damage to DNA and its repair with the comet assay. *Biochim Biophys Acta*. 2014;1840(2):794–800.
48. McKenna DJ, McKeown SR, McKelvey-Martin VJ. Potential use of the comet assay in the clinical management of cancer. *Mutagenesis*. 2008;23(3):183–190.
49. Clague J, Shao L, Lin J, et al. Sensitivity to NNKOAc is associated with renal cancer risk. *Carcinogenesis*. 2009;30(4):706–710.
50. Valavanidis A, Vlachogianni T, Fiotakis C. 8-hydroxy-2′ -deoxyguanosine (8-OHdG): a critical biomarker of oxidative stress and carcinogenesis. *J Environ Sci Health C Environ Carcinog Ecotoxicol Rev*. 2009;27:120–139.
51. Niki E. Biomarkers of lipid peroxidation in clinical material. *Biochim Biophys Acta*. 2014;1840(2):809–817.
52. Yoshida Y, Kodai S, Takemura S, Minamiyama Y, Niki E. Simultaneous measurement of F2-isoprostane, hydroxyoctadecadienoic acid, hydroxyeicosatetraenoic acid, and hydroxycholesterols from physiological samples. *Anal Biochem*. 2008;379(1):105–115.
53. Beecher CW. Cancer preventive properties of varieties of *Brassica oleracea*: a review. *Am J Clin Nutr*. 1994;59:1166S–1170S.
54. Verhoeven DT, Verhagen H, Goldbohm RA, van den Brandt PA, van Poppel G. A review of mechanisms underlying anticarcinogenicity by brassica vegetables. *Chem Biol Interact*. 1997;103:79–129.
55. Higdon JV, Delage B, Williams DE, Dashwood RH. Cruciferous vegetables and human cancer risk: epidemiologic evidence and mechanistic basis. *Pharmacological Res*. 2007;55:224–236.
56. Murray S, Lake BG, Gray S, et al. Effect of cruciferous vegetable consumption on heterocyclic aromatic amine metabolism in man. *Carcinogenesis*. 2001;22:1413–1420.
57. Walters DG, Young PJ, Agus C, et al. Cruciferous vegetable consumption alters the metabolism of the dietary carcinogen 2-amino-1-methyl-6-phenylimidazo[4,5-b]pyridine (PhIP) in humans. *Carcinogenesis*. 2004;25:1659–1669.
58. Probst-Hensch NM, Tannenbaum SR, Chan KK, Coetzee GA, Ross RK, Yu MC. Absence of the glutathione S-transferase M1 gene increases cytochrome P4501A2 activity among frequent consumers of cruciferous vegetables in a Caucasian population. *Cancer Epidemiol Biomarkers Prev*. 1998;7:635–638.
59. Lampe JW, King IB, Li S, et al. Brassica vegetables increase and apiaceous vegetables decrease cytochrome P450 1A2 activity in humans: changes in caffeine metabolite ratios in response to controlled vegetable diets. *Carcinogenesis*. 2000;21:1157–1162.
60. Kall MA, Vang O, Clausen J. Effects of dietary broccoli on human in vivo drug metabolizing enzymes: evaluation of caffeine, oestrone and chlorzoxazone metabolism. *Carcinogenesis*. 1996;17:793–799.
61. Nijhoff WA, Grubben M, Nagengast FM, et al. Effects of consumption of Brussels-sprouts on intestinal and lymphocytic Glutathione S-Transferases in humans. *Carcinogenesis*. 1995;16:2125–2128.
62. Nijhoff WA, Mulder TPJ, Verhagen H, Vanpoppel G, Peters WHM. Effects of consumption of Brussels-sprouts on plasma and urinary Glutathione-S-Transferase class-alpha and class-pi in humans. *Carcinogenesis*. 1995;16:955–957.
63. Verhagen H, Poulsen HE, Loft S, Vanpoppel G, Willems MI, Vanbladeren PJ. Reduction of oxidative DNA-damage in humans by Brussels sprouts. *Carcinogenesis*. 1995;16:969–970.
64. Verhagen H, de Vries A, Nijhoff WA, et al. Effect of Brussels sprouts on oxidative DNA-damage in man. *Cancer Lett*. 1997;114:127–130.
65. Hoelzl C, Glatt H, Meinl W, et al. Consumption of Brussels sprouts protects peripheral human lymphocytes against 2-amino-1-methyl-6-phenylimidazo[4,5-b]pyridine (PhIP) and oxidative DNA-damage: results of a controlled human intervention trial. *Mol Nutr Food Res*. 2008;52:330–341.
66. Gill CIR, Haldar S, Boyd LA, et al. Watercress supplementation in diet reduces lymphocyte DNA damage and alters blood antioxidant status in healthy adults. *Am J Clin Nutr*. 2007;85:504–510.

67. Shaughnessy DT, Gangarosa LM, Schliebe B, et al. Inhibition of fried meat-induced colorectal DNA damage and altered systemic genotoxicity in humans by crucifera, chlorophyllin, and yogurt. *PLoS One*. 2011;6(4):e18707.
68. Lampe JW, Chen C, Li S, et al. Modulation of human glutathione S-transferases by botanically defined vegetable diets. *Cancer Epidemiol Biomarkers Prev*. 2000;9:787–793.
69. Seow A, Yuan JM, Sun CL, Van Den Berg D, Lee HP, Yu MC. Dietary isothiocyanates, glutathione S-transferase polymorphisms and colorectal cancer risk in the Singapore Chinese Health Study. *Carcinogenesis*. 2002;23:2055–2061.
70. Lin HJ, Probst-Hensch NM, Louie AD, et al. Glutathione transferase null genotype, broccoli, and lower prevalence of colorectal adenomas. *Cancer Epidemiol Biomarkers Prev*. 1998;7(8):647–652.
71. Lynn A, Collins A, Fuller Z, Hillman K, Ratcliffe B. Cruciferous vegetables and colo-rectal cancer. *Proc Nutr Soc*. 2006;65:135–144.
72. Häkkinen SH, Kärenlampim SO, Heinonen IM, Mykkänen HM, Törrönen AR. Content of the flavonols quercetin, myricetin, and kaempferol in 25 edible berries. *J Agric Food Chem*. 1999;47(6):2274–2279.
73. Deighton N, Brennan R, Finn C, Davies HV. Antioxidant properties of domesticated and wild Rubus species. *J Sci Food Agric*. 2000;80:1307–1313.
74. Häkkinen SH, Kärenlampi SO, Mykkänen HM, Törrönen AR. Influence of domestic processing and storage on flavonol contents in berries. *J Agric Food Chem*. 2000;48(7):2960–2965.
75. Connor AM, Luby JJ, Hancock JF, Berkheimer S, Hanson EJ. Changes in fruit antioxidant activity among blueberry cultivars during cold-temperature storage. *J Agric Food Chem*. 2002;50(4):893–898.
76. Anttonen MJ, Karjalainen RO. Environmental and genetic variation of phenolic compounds in red raspberry. *J Food Compost Anal*. 2005;18:759–769.
77. Skupień K, Oszmiański J. Influence of titanium treatment on antioxidants content and antioxidant activity of strawberries. *Acta Sci Pol Technol Aliment*. 2007;6(4):83–93.
78. Castrejón DR, Eichholz I, Rohn S, Kroh LW, Huyskens-Keil S. Phenolic profile and antioxidant activity of highbush blueberry (*Vaccinium corymbosum* L.) during fruit maturation and ripening. *Food Chem*. 2008;109:564–572.
79. Kähkönen MP, Hopia AI, Heinonen M. Berry phenolics and their antioxidant activity. *J Agric Food Chem*. 2001;49(8):4076–4082.
80. Ek S, Kartimo H, Mattila S, Tolonen A. Characterization of phenolic compounds from lingonberry (*Vaccinium vitis*-idaea). *J Agric Food Chem*. 2006;54(26):pp. 9834–9342
81. Taruscio TG, Barney DL, Exon J. Content and profile of flavanoid and phenolic acid compounds in conjunction with the antioxidant capacity for a variety of northwest *Vaccinium* berries. *J Agric Food Chem*. 2004;52(10):3169–3176.
82. Bub A, Watz LB, Blockhaus M, et al. Fruit juice consumption modulates antioxidative status, immune status and DNA damage. *J Nutr Biochem*. 2003;14(2):90–98.
83. Hofmann T, Liegibel U, Winterhalter P, Bub A, Rechkemmer G, Pool-Zobel BL. Intervention with polyphenol-rich fruit juices results in an elevation of glutathione S-transferase P1 (hGSTP1) protein expression in human leucocytes of healthy volunteers. *Mol Nutr Food Res*. 2006;50(12):1191–1200.
84. Duthie SJ, Jenkinson AM, Crozier A, et al. The effects of cranberry juice consumption on antioxidant status and biomarkers relating to heart disease and cancer in healthy human volunteers. *Eur J Nutr*. 2006;45(2):113–122.
85. Wang LS, Sardo C, Henry C, et al. Effect of freeze-dried black raspberries on human colorectal cancer lesions:American Association for Cancer Research Special Conference in Cancer Research *Adv Colon Cancer Res*. 2007;B31.
86. Thomasset S, Berry DP, Cai H, et al. Pilot study of oral anthocyanins for colorectal cancer chemoprevention. *Cancer Prev Res (Phila Pa)*. 2009;2(7):625–633.
87. Stoner G, Hasson H, Sardo C, et al. Regression of rectal polyps in familial adenomatous polyposis patients with freeze-dried black raspberries. *Cancer Prev Res*. 2008;1(7 Suppl.):PR-14.
88. Grubben MJ, Nagengast FM, Katan MB, Peters WH. The glutathione biotransformation system and colorectal cancer risk in humans. *Scand J Gastroenterol Suppl*. 2001;234:68–76.
89. Gill CI, McDougall GJ, Glidewell S, et al. Profiling of phenols in human fecal water after raspberry supplementation. *J Agric Food Chem*. 2010;58(19):10389–10395.
90. Gonzalez-Barrio R, Borges G, Mullen W, Crozier A. Bioavailability of raspberry anthocyanins and ellagitannins following consumption of raspberries by healthy humans and subjects with an ileostomy. *J Agric Food Chem*. 2010;58:3933–3939.
91. Brown EM, Gill CI, McDougall GJ, Stewart D. Mechanisms underlying the anti-proliferative effects of berry components in in vitro models of colon cancer. *Curr Pharm Biotechnol*. 2012;13(1):200–209.
92. Kropat C, Mueller D, Boettler U, et al. Modulation of Nrf2-dependent gene transcription by bilberry anthocyanins *in vivo*. *Mol Nutr Food Res*. 2013;57(3):545–550.
93. Goto T, Horita M, Nagai H, et al. Tiliroside, a glycosidic flavonoid, inhibits carbohydrate digestion and glucose absorption in the gastrointestinal tract. *Mol Nutr Food Res*. 2012;56(3):435–445.
94. Forester SC, Gu Y, Lambert JD. Inhibition of starch digestion by the green tea polyphenol, (−)-epigallocatechin-3-gallate. *Mol Nutr Food Res*. 2012;56(11):1647–1654.
95. Williamson G. Possible effects of dietary polyphenols on sugar absorption and digestion. *Mol Nutr Food Res*. 2013;57(1):48–57.
96. Wang LS, Arnold M, Huang YW, et al. Modulation of genetic and epigenetic biomarkers of colorectal cancer in humans by black raspberries: a phase I pilot study. *Clin Cancer Res*. 2011;17(3):598–610.
97. Marcel V, Petit I, Murray-Zmijewski F, et al. Diverse p63 and p73 isoforms regulate Δ133p53 expression through modulation of the internal TP53 promoter activity. *Cell Death Differ*. 2012;19(5):816–826.
98. Wang LS, Kuo CT, Cho SJ, et al. Black raspberry-derived anthocyanins demethylate tumor suppressor genes through the inhibition of DNMT1 and DNMT3B in colon cancer cells. *Nutr Cancer*. 2013;65(1):118–125.

99. Del Rio D, Borges G, Crozier A. Berry flavonoids and phenolics: bioavailability and evidence of protective effects. *Br J Nutr*. 2010;104 (Suppl 3):S67–S90.

100. van Duynhoven J, Vaughan EE, Jacobs DM, et al. Metabolic fate of polyphenols in the human superorganism. *Proc Natl Acad Sci USA* 2011;108:4531–4538.

101. Fava F, Gitau R, Griffin BA, Gibson GR, Tuohy KM, Lovegrove JA. The type and quantity of dietary fat and carbohydrate alter faecal microbiome and short-chain fatty acid excretion in a metabolic syndrome 'at-risk' population. *Int J Obes (Lond)*. 2013;37(2):216–223.

CHAPTER

14

Population Level Divergence from the Mediterranean Diet and the Risk of Cancer and Metabolic Disease

George Pounis, Marialaura Bonaccio*, Kieran M. Tuohy†, Maria Benedetta Donati*, Giovanni de Gaetano* and Licia Iacoviello**

***IRCCS Istituto Neurologico Mediterraneo Neuromed, Pozzilli, Isernia, Italy †Department of Food Quality and Nutrition, Research and Innovation Centre, Fondazione Edmund Mach, San Michele all'Adige, Trento, Italy**

MEDITERRANEAN DIET AS THE TRADITIONAL DIET OF SOUTHERN EUROPE

Historical Overview

In the 1990s the MeD as nutritional guidance gained widespread recognition following 50 years of study by nutirtionists.[1–5] However, it is the fifth decade of life which should be defined as the time point when the first health benefits become apparent.[6,7] From the first epidemiological observations[8] of the Seven Countries Study[9] to the recent identification of MeD as an "Intangible Cultural Heritage of Humanity" by the United Nations Educational, Scientific, and Cultural Organization (UNESCO)[10] in 2010, much scientific effort has been undertaken to prove the important health aspects of this traditional dietary pattern.

The independence of *Homo sapiens* from their local environmental availability of food is a relatively recent occurrence. The transplantation of plants and animals across regions and the development of transportation and food preservation technologies have enormously increased the options of humans in choosing their food. Moreover, the enormous advances in crop yield as a result of the green revolution in fertilizers and pesticides of the mid-twentieth century, have meant that at least in wealthy countries, food is cheap or at least energy dense, highly processed and mass produced foods are readily available to even to poorest of citizens in the Western world. Indeed, the origins of the MeD should be placed before the ability of people to transfer their eating habits. This is why the beneficial effects of healthy eating were initially recognized in Southern Europe by the famous nutritionist Ancel Keys.[9,11,12] The food type and availability along the coast of the Mediterranean Sea most notably in countries like Greece, Italy, Spain, Southern France and Morocco, evolved together with other cultural and lifestyle factors, and constitutes one of the main progenitors of the very traditional Mediterranean lifestyle. Regardless the presumably very ancient origins of this healthy way of living, its "scientific birth" are placed just in the last century.

The Seven Countries Study revealed that Cretan men had very low death rates from heart disease, despite the moderate to high intake of fat. Indeed, the Cretan diet was similar to other traditional MeDs offering most of the health benefits.[12] Following this, the World Health Organization (WHO) in 1961 tracked that overall life expectancy at the age of 45 years was higher in the Greek population than in any other national group examined at that time.[7] Since then other work has shown that some of the same elements which comprise the MeD are also major components of other traditional diets associated with long life and low incidence of non-communicable, chronic diseases especially heart disease and cancer. A good example is the association between the traditional

Diet-Microbe Interactions in the Gut.
DOI: http://dx.doi.org/10.1016/B978-0-12-407825-3.00014-9

Okinawan diet, where life-long low-calorie intake, high consumption of vegetables and complex carbohydrates and fish, and low intake of red meat, fat, animal protein and refined carbohydrates has been proposed as the main environmental contributor to the low incidence of CVD and cancer, and longevity in Okinawans. Okinawans are now the longest lived human populations, with a life expectancy 1.5 years longer than mainland Japanese and 4.9 years longer than the average in the United States of America (US).[13] More recently, many well-organized Health and Nutrition surveys like NHANES[13–15] in the US and randomized trials like the Lyon Diet Heart Study[16] in Europe, have confirmed the beneficial effect of many components of the MeD pattern.

This weight of evidence, which is not present for other healthy dietary patterns, either traditional or subsequently recommended by different regulatory agencies, resulted in UNESCO inscribing the "Mediterranean Diet" in the List of the Intangible Heritage of Humanity[10] in November 2010.

Expression of Culture and Lifestyle – UNESCO's Recognition

Although the foods commonly consumed in the countries of the Mediterranean are thought to be the main contributor to the health effects of the MeD diet, they may not be the only lifestyle factors capable of impacting on health. Indeed, the MeD is recognized to be more than just the prevailing foods of Mediterranean countries like Spain, Greece, Italy and Morocco. Indeed, UNESCO stated in their proposal that:

> The MeD constitutes a set of skills, knowledge, practices and traditions ranging from the landscape to the table, including the crops, harvesting, fishing, conservation, processing, preparation and, particularly, consumption of food...

Adopting the healthy lifestyle includes the consideration of some cultural aspects as essentials.[3,5,17,18] Socialization is one of them, since sharing food in the company of family and friends around the table represents the social support, which would positively affect healthy food behaviors. Social interaction is fully promoted by the adherence to this way of eating since many foods were part of social customs and festive events. Of course eating in company or as a family unit is not restricted to the Mediterranean countries but it does emphasize the important role relaxation, or rather stress, may have on chronic disease risk. Recent scientific studies highlight stress as a disease mechanism itself, and stress may in fact mediate some of its pathological effects through the gut. Even more amazing, are the observations that gut bacteria can influence stress, mood, depression and cognitive function through metabolic and/or immune interactions linked to diet. In addition, moderation as a Mediterranean philosophy should be applied on serving sizes in order to adopt this pattern. Furthermore, the development of Mediterranean culinary activities is basic for the social reproduction of the identity of each particular culture. However, a clear distinction must be drawn between the MeD and Mediterranean "cuisine," and that over indulgence even on good food is not beneficial for health. Indeed, a low-calorie aspect of the traditional MeD diet is an important aspect to remember when discussing the relative health benefits of dietary habits from around the world, especially since calorie restriction is thought to contribute positively to long life and reduced risk of chronic non-communicable diseases. Moderate physical activity, adequate rest and consumption of foods on the right season constitute and right amounts could summarize the three main cultural aspects of the MeD.

Scientific Definition and Description

WC Willet et al. at Harvard University's School of Public Health defined the MeD[5] as the "...abundance of plant foods such as fruit, vegetables, breads, cereals, potatoes, beans, nuts and seeds; minimally processed, seasonally fresh and locally grown foods; fresh fruits as daily dessert, with sweets containing concentrated sugars or honey consumed a few times per week; olive oil as principal source of fat; dairy products (principally cheese and yogurt) consumed daily in low to moderate amounts; fish and poultry consumed in low to moderate amounts; zero to four eggs consumed weekly; red meat consumed in low amounts; and wine consumed in low to moderate amounts, normally with meals..."

This definition was based on a food pyramid, which reflecting the dietary traditions of Crete and Southern Italy in the early 1960s. This was also the first attempt to fit dietary habits into a pattern, which could be used as a useful tool in public health promotion. The message derived from such a definition was clear and strong: if you get back to the traditional diet of these populations you may gain health benefits. In addition, all dietary recommendations derived from this definition were visualized using the shape of a pyramid. This gave the reader the ability to categorize foods according to their frequency of consumption from frequently at base of the pyramid, to rarely consumed at the top of the pyramid.

Years after the first definition of the MeD pyramid, updates were made by the Scientific Advisory Board of the 15th Anniversary MeD Conference in November 2008.[19] The changes concerned gathering plant foods (fruits, vegetables, grains, nuts, legumes, seeds, olives and olive oil) in a single group to stress their health benefits. In addition, herbs and spices were new items inserted in the MeD pyramid, due to their health benefits and taste. Furthermore, herbs and spices play a key role in the national identities of various Mediterranean cuisines. Finally, the scientific committee decided to change the frequency of consumption of fish and shellfish within the pyramid, recognizing the benefits of eating fish and shellfish at least two times per week.

THE EVIDENCE-BASED HEALTH PROTECTION BY MEDITERRANEAN DIET

First Epidemiological Evidence

The Seven Countries Study,[9,12,20,21] a longitudinal study which started in 1958 in the former Yugoslavia and went on to recruit a total of 12,763 men, 40–59 years of age, in seven countries, the United States, Japan and countries from Northern and Southern Europe. The follow-up period was extended up to 50 years while one of the first findings was that Cretan men had exceptionally low death rates from heart disease, despite moderate to high intake of fat. This ecological indication was firstly attributed to the traditional dietary pattern of this population.

Besides this early conclusion, the major results about the Mediterranean pattern and its relation to clinical endpoints were published much later. Keys et al. in 1986, using data from the same study after 15 years of follow-up, indicated that the 15 cohorts differed in average diets.[12] Death rates were related positively to average percentage of dietary energy from saturated fatty acids, negatively related to dietary energy percentage from monounsaturated fatty acids, and were unrelated to dietary energy percentage from polyunsaturated fatty acids, proteins, carbohydrates and alcohol. All death rates were negatively related to the ratio of monounsaturated to saturated fatty acids.

Trichopoulou et al. (1995) rather than looking at individual nutrients considered adherence to the overall diet and its impact on overall survival.[22] More accurately, 182 elderly residents of three rural Greek villages were followed and their diet was evaluated through a validated extensive semi-quantitative questionnaire on food intake. Results showed that one unit increase in a dietary score, constructed a priori on the basis of eight component characteristics of the traditional diet in the Mediterranean region, was associated with a significant 17% reduction in overall mortality (95% confidence interval 1% to 31%).

Level of Adherence: Is there any Measure?

The study of dietary patterns has recently been proposed as the best way to evaluate dietary habits.[23] In this context the understanding of the level of adherence to a dietary pattern such as MeD gains an important public health meaning. Traditionally, nutrition epidemiology evaluated the association of foods and dietary components individually with a particular illness. More recently research efforts have focused on evaluating total adherence of populations to MeD.

Scoring systems are the most popular methodological approach used to classify populations according to their diet adherence.[23–26] In most dietary scores a higher rate corresponds to higher adherence to a dietary pattern such as the MeD. For example, in 2003 the Greek EPIC cohort following 22,043 subjects for 44 months of follow-up illustrated that an increase of 2 units in a 10-point scale score was associated with a 25% reduction in all-cause mortality.[27] Similar findings were discovered up by the Seven Countries Study in 2004 when the Mediterranean Adequacy Index (MAI) was constructed.[28,29] Fidanza et al. computed the MAI of random samples of men in the 16 Cohorts of this study. MAI showed a strong inverse association with 25-year death rates from coronary heart disease.

Recently an Italian research group published their work on assessing the adherence to the MeD using an ad hoc dietary score for the Italian population.[30] The score was based on the intake of 11 items: high intakes of six typical Mediterranean foods (pasta; typical Mediterranean vegetables such as raw tomatoes, leafy vegetables, onion, and garlic, salad, and fruiting vegetables; fruit; legumes; olive oil; and fish); low intakes of four non-Mediterranean foods (soft drinks, butter, red meat and potatoes); and also alcohol. In a sample of 40,681 subjects followed for 7.9 years, they compared the associations between stroke risk and adherence to four a-priori-defined dietary indices: the Healthy Eating Index 2005[31] developed to test adherence to a healthy diet for the Dietary

Approaches to Stop Hypertension[32] (DASH), study in the US; the Mediterranean Diet Score, designed to estimate adherence to the Greek variant of the Mediterranean diet;[27] and the Italian Mediterranean Index specifically developed to better estimate the adherence to the Italian Mediterranean diet.[30] Although all these dietary patterns tended to be inversely associated with at least one type of stroke, the Italian Mediterranean Index performed somewhat better than the others, having significant inverse associations with all types of stroke and ischemic stroke and also a tendency to be inversely associated with hemorrhagic stroke.

The Moli-Sani Experience

The evaluation of the level of adherence to the MeD in an Italian Mediterranean region was also evaluated through the Moli-sani Project conducted at the Catholic University of Campobasso. The Moli-sani Project is a prospective, population-based cohort study that in the period 2005–2010 recruited 25,000 citizens from the southern Italian region of Molise with the aim of investigating the importance of genetic–environment interactions to the onset of cardiovascular and tumor diseases, the two major killers affecting Western societies.[33]

The study was entirely paperless and involved for several years over 60 researchers and trained personnel who collected a very large amount of information. Beyond building up one of the biggest biological banks of Europe storing about one million biological samples, the study intensely investigated the dietary habits and lifestyle of participants. Dietary information was obtained using the EPIC food frequency questionnaire; allowing a complete and reliable picture of the eating habits of the population sample. The Molise region is Mediterranean by geography, tradition and culture. Nevertheless, adherence to Mediterranean diet was found to decrease over time and across all age groups. According to earlier data recorded in the northern part of the region, the elderly were those sticking the most to the Mediterranean pattern while the youngest groups had dramatically decreased their adherence to this eating model.[34] This decline was not restricted only to young people; the population as a whole have experienced a huge drop from the traditional trends registered in the southern areas of Italy about 50 years ago. A comparison with data collected in the 1960s within the southern Italian populations of Nicotera (Calabria) and Pollica (Campania) highlighted that the highest percentile of adherence recorded in the Moli-sani cohort was equal to the lowest percentile of adherence in both Pollica and Nicotera studies.[34] This means that during the last five decades people have been progressively shifting from the traditional Mediterranean dietary pattern towards other eating models more similar to the so-called Western dietary patterns. Yet the rate of drift away from the Mediterranean diet appears to be not equal for all individuals. Indeed, this drift from the MeD is worse for disadvantaged people or those on low income. As observed in other regions of the world, less healthy diets are mainly adopted by economically and socially disadvantaged groups and cluster with other unhealthy lifestyles, such as smoking and lack of exercise, leading to a clear social gradient in mortality and morbidity.[35] Recent data from the Moli-sani study has focussed mainly on the role of income showing that people with lower household income stick less to the Mediterranean diet than those having higher household income, independently from other socioeconomic factors such as education.[36] These results were also supported by previous studies arguing that food cost may be one of the mechanisms by which less affluent subjects end up with having lower-quality diets.[37] Indeed, there is growing evidence that eating in a Mediterranean way, or at least getting healthy food, is more expensive than nutrient-poor food, commonly labeled as "junk food."[38,39] Such observations relate an important social commentary on how the food environment and access to healthy foods have changed over the past 50–60 years. It is interesting to note that in other parts of the world, not only have elderly people been shown to consume more "healthy" foods such as fruit, vegetables dairy products and whole-grain cereals compared to younger individuals, but that adherence to these more healthy dietary habits appears to be determined by consumption of these foods as children.[40–42] This has two important implications: (1) that it is likely to be at a young age when population-based nutritional advice or interventions have their greatest chance of success; and (2) that we face a veritable time bomb of chronic non-communicable diseases due to low adherence of younger generations to the MeD and other healthy dietary patterns loosely based on MeD.

An interesting aspect of the results from the Moli-sani study was that the range of household income was quite homogenous, the "poorest" being those earning less than 10,000 Euros net per year while the medium-high category was represented by people with family incomes less than 40,000 Euros per year.[36] The relative economic homogeneity of the sample led authors to conclude that also small gaps in the household income could make the difference in the eating choices of individuals. One of the most accredited hypotheses advanced by the Moli-sani research team is that the economic and financial crisis affecting contemporary societies in recent years might turn out to have a tremendous role in contributing to the sudden shift from healthy-eating patterns recorded in very recent years. Preliminary data obtained from the Moli-sani study indicated that the drop from the highest category of adherence to Mediterranean diet started in the period 2007–2008, the dramatic 2-year period in which the economic crisis just started or at least became apparent. It is likely also that such economic factors are exacerbated by urban compared to rural living, as access or perceived access to fresh fruit and vegetables may be diminished for low income families living in urban centres compared to ready access to nutrient poor "junk food."[43–46] In the light of these data, advice for a return to the traditional Mediterranean diet can not any longer be the only task of the public health agenda; cost and availability of the foods which make up the MeD must also be addressed. Experts should seriously take into account that being Mediterranean today is something dealing with the deepest changes affecting contemporary societies; the impoverishment of the less wealthy groups should be considered as a major risk factor for the practice of healthy lifestyles and consequent health outcomes both at the individual level and in terms of national economic health.

Among the wide panel of factors influencing food choices, *nutrition knowledge* is drawing increasing attention. Based on the assumption that to perform a healthy behavior one has to know what the healthy behavior is, the Moli-sani project has provided interesting hints on this novel aspect involved in the dietary choices of individuals. Investigation on a representative subsample of the cohort has shown that people with higher health-related awareness have greater odds of being in the top category of adherence to Mediterranean diet as compared to those having poor knowledge of health-related topics, independently from other possible confounding factors such as education.[47] Similar results were obtained when the association between mass media exposure and dietary patterns was considered, with people highly exposed to the information delivered by the media having healthier diets.[48]

Mediterranean Diet – From Epidemiology to Clinical Trials

Following the epidemiological observational data confirming the health-protective effects of MeD, evidence from large randomized clinical trials has been forthcoming, solidifying support for the MeD as a healthy diet. The Lyon Diet Heart Study was a trial focusing on secondary prevention of CVD aimed at evaluating whether the Mediterranean diet and lifestyle could reduce the risk of a secondary CVD event.[16,49–52] A first intermediate analysis of data showed a huge protective effect of MeD after 27 months of follow-up. In particular, 275 survivors from an acute coronary event were randomly stratified in two groups, between March 1988 and 1992. The intervention group was administered with a MeD dietary plan while the control group received no specific dietary guidelines. After 46 months of follow-up, the intervention group showed a significantly lower rate of the main cardiovascular outcomes compared to the control group. In addition, the alpha-linolenic acid status was negatively associated with a secondary CVD event. On the contrary, no significant association was observed for other n-3 fatty acids and cardiovascular outcome rates. One of the collateral observations of this trial was that it would be easier for a medical doctor to prescribe patients a drug than to convince them to adhere to a healthy diet and lifestyle. For this reason many doctors abandon the effort of changing patient's dietary habits. On the other hand, it was also observed that only frequent and well-established communication of patients with their medical doctors could help to improve dietary habits.

There are two other large-scale trials indicating health benefits of lifestyle modifications.[53–55] Firstly, the Dietary Approaches to Stop Hypertension (DASH) Trials, where 459 participants after a 3-week run-in period of controlled diet, were randomized to one of the following interventions for 8 weeks: (1) the average American diet (control); (2) a diet rich in fruit and vegetables; or (3) the DASH diet, consisting of fruits, vegetables and low-fat dairy products; included whole grains, poultry, fish and nuts; and was lower in red meat, sweets and sugar-sweetened beverages – dietary aspects not too far from a MeD. The DASH diet was effective both in men and women and in hypertensive and non-hypertensive participants, in lowering both systolic and diastolic blood pressures significantly. The most striking result was the drop in blood pressure seen in the hypertensive population which was comparable to drug therapy. Moreover, the DASH pattern lowered total cholesterol and low-density lipoproteins. A second DASH sodium trial followed testing the effect on blood pressure by combining with the DASH diet restricted sodium intake. This combination lowered blood pressure in all subgroups studied, including non-hypertensive individuals. Unfortunately, very few US adults, even those with elevated BP, currently eat a diet consistent with DASH dietary patterns.

The Dart project was another dietary intervention program focusing on changing lifestyles and consumption of fish, one of the most important health components of MeD, in patients with an acute coronary syndrome.[56] In this trial 2033 men aged <70 years were randomly assigned to two groups. The first group received a recommendation to eat 0.33 g of fish daily and the second 0.10 g. Patients were followed for 2 years and main outcomes were a secondary cardiovascular event or death. Analysis of data showed that patients in the first group had lower cardiovascular death rates compared to the other group. In addition, a protective effect was also observed for blood pressure.

Efforts were also made to evaluate whether the supplementary intake of nutrients included in the MeD could also be a health-protective policy. The most effective trial in establishing such information was the GISSI-Prevenzione trial[57–59] which was focused on fish-derived n-3 fatty acids supplementation. The trial was conducted on 11,324 patients who had a first acute coronary syndrome and followed for a period of 3.5 years. The patients were randomly stratified in four groups: the first receiving 1 g of n-3 fatty acids per day; the second a supplement of 300 mg vitamin E per day; the third both supplements; and the fourth only dietary guidelines. The data analysis revealed that the first group receiving only 1 g n-3 fatty acids supplementary had a lower risk of secondary CVD.

A very recent dietary multicenter trial in Spain assessed MeD for the primary prevention of cardiovascular events.[60] Seven-thousand four-hundred and forty-seven participants at high cardiovascular risk, but cardiovascular disease (CVD)-free at enrollment were randomly assigned to one of three diets: a MeD supplemented with extra-virgin olive oil; a MeD supplemented with mixed nuts, or a control diet (advice to reduce dietary fat). The trial was stopped after a median follow-up of 4.8 years. The two MeD groups had good adherence to the intervention, according to self-reported intake and biomarker analyses. Results of this work supported that the MeD supplemented with extra-virgin olive oil or nuts could reduce the incidence of major cardiovascular events.

MEDITERRANEAN DIET AS A HEALTH PROTECTION MODEL

Preventing Cardio-Metabolic Disease

The healthy value of the MeD has from one side been attributed to its cardio-protective effects. The number of publications on this topic was large enough to allow a meta-analysis published by Sofi et al. in 2008.[24] The studies included prospectively analyzed the association between adherence to MeD, mortality and incidence of diseases; 12 studies, with a total of 1,574,299 subjects followed for a time ranging from three to 18 years were included. Data evaluating cardiovascular mortality in relation to MeD using dietary scores came from four studies including a total of 404,491 subjects and 3876 fatal events. Results supported the association of a two-point increase in the score indexes was associated with 9% reduction in mortality from CVDs.

Several components of MeD have already been appreciated for their effects in preventing CVD. In particular, they could be summarized as food, nutrients, fat content of the diet and energy balance.[61] Randomized, controlled trials and prospective cohort studies had already pointed out the cardiovascular protective effects of several specific foods including fruit and vegetables, whole grains, fish, nuts, chocolate and moderate consumption of alcoholic drink. The main advantage of this dietary methodology is that assessing a given food will more likely represent the synergy of composite effects and interactions of multiple factors, including carbohydrate quality, fiber content, specific fatty acids and proteins, preparation methods, food structure, and bioavailability of inherent micronutrients and phytochemicals.[61–63] Main protective food groups include fruit and vegetables, whole grains, fish, nuts, chocolate and moderate consumption of alcoholic drinks. Trials focusing on vegetables and fruits established effects on lowering of blood pressure, lipid levels, insulin resistance and inflammation and on weight control.[64–69] Benefits have been attributed to a complex set of micronutrients, phytochemicals, and fibers included; to the enhanced bioavailability of these nutrients in their natural form; and to the replacement of less healthful foods in the diet. In addition, whole-grain consumption has been promoted for improving glucose–insulin homeostasis and endothelial function, reducing inflammation and leading to weight loss.[70–72] The fiber content of such food groups has been proposed for these effects. The FLAVURS study at the University of Reading in the UK has very recently shown that increased consumption of flavonoid-rich fruit and vegetables (increasing up to six portions per day for 6 weeks) improves microvascular reactivity and inflammatory status in men at risk of CVD in a single-blind, dose-dependent, parallel, randomized and controlled dietary intervention (n = 174). A second diet with fruit and vegetables but lower flavonoid content mitigated increases in vascular stiffness and reductions in nitric oxide observed in the control group, supporting the advised increased consumption of fruit and vegetables, regardless of flavonoid content, for reducing the risk of CVD.[73]

Fish and seafood is a major cardio-protective food group, which has been much studied for its association with CVD prevention and is a characteristic food of the MeD. In particular, the high n-3 fatty acids content, such as eicosapentaenoic acid (EPA) (20:5 n-3) and docosahexaenoic acid (22:6 n-3) (DHA), has been recognized for their anti-inflammatory, antiarrhythmic and antithrombotic properties; they also inhibit synthesis of cytokines and mitogens, stimulate endothelial-derived nitric oxide and inhibit atherosclerosis in general.[74] Indeed, regular fish consumption is associated with lower incidence of CVD and ischemic stroke, especially risk of cardiac death, among generally healthy populations.[75,76]

Nuts are another food group promoted for preventing CVD and commonly consumed as part of the MeD. In particular, cardiovascular benefits of moderate nut intake have been observed lowering risk factors in trials and reducing CVD risk in prospective cohort studies.[77] Nuts contain several macro- and micronutrients which could be responsible for the aforementioned effects. These include unsaturated fatty acids, plant proteins, folate, fiber and antioxidant vitamins and phytochemicals.

Alcoholic drinks, especially wine, have been extensively studied for their possible benefits on CVD prevention.[78,79] A number of metanalyses have recently been published to establish accurate evidence.[80–83] Di Castelnuovo et al. in 2002 performed a meta-analysis of 26 studies on the relationship between wine or beer consumption and vascular risk.[83] From 13 studies involving 209,418 persons, wine intake was associated with reduced relative risk of vascular disease compared with non-drinkers. From 10 studies involving 176,042 persons a J-shaped relationship between different amounts of wine intake and vascular risk became apparent. Similar observations were derived for another metanalysis of 15 studies including 116,702 subjects. Corrao et al. in 2004 reported threshold values of ischemic and hemorrhagic strokes.[72] For coronary heart disease, a J-shaped relation was observed with a minimum relative risk at 20 g/day, a significant protective effect up to 72 g/day, and a significant increased risk at 89 g/day.

Chocolate is mostly not included in "healthy food"; however, it has recently received much attention, since it is an important dietary source of polyphenols. Recent studies have tried to validate the role of chocolate and a large number of experimental studies indicated beneficial effects of polyphenols in preventing atherosclerosis progression.[84–86] Di Giuseppe et al. in 2008 analyzed cross-sectional data of the Moli-sani study and indicated a J-shaped relationship between dark chocolate consumption and serum levels of C-reactive protein (CRP), an anti-inflammatory marker.[86] Consumers of up to 1 serving (20 g) of dark chocolate every 3 days had serum CRP concentrations that were significantly lower than non-consumers or higher consumers.

It is of interest that dietary interventions that included daily intake of low-fat dairy foods significantly lowered BP, lipid levels, and insulin resistance and improved endothelial function, independent of changes in weight.[54,66,87] Dairy products in general have been proposed as MeD components. However, the active constituents for the cardio-metabolic benefits were not evident. The lower caloric and saturated fatty acids content could be proposed as nutritional advantages compared to whole-fat dairy products. This might be one the major reasons that low-fat dairy products have been promoted through the dietary guidelines for a healthy life by many different organizations.[88,89] A recent study in Sweden examined the relative impact of the type of dairy product on CVD risk with surprising results.[90] In 26,445 individuals of the Malmö Diet and Cancer cohort without history of CVD, overall dairy intake was associated with a significant inverse relationship with CVD risk. Among the specified dairy products [with milk separated into high or low fat, or fermented (yoghurt or cultured sour milk)] this increased relationship with CVD risk was only observed for fermented dairy products. There also appeared to be an important sex effect, with cheese decreasing CVD risk only in women. This study not only challenges long-held beliefs in some parts of the world that dairy intake contributes to CVD risk but also highlights the important role of microorganisms and the "culture" of fermentation, both microbiological and anthropological, in the observed health effects of dairy products. Milk fermentation is a key cultural aspect of the Mediterranean region and fermented dairy products are core components of the MeD.

Protection against Different Types of Cancer

MeD as a health protection model has been appreciated for reducing the risk of developing different types of cancer.[27,91] Referring to the meta-analysis published by Sofi et al. in 2008, six studies comprising 521,366 subjects and 10,929 events of cancer mortality were meta-analyzed.[24] MeD was associated with reduced cancer incidence and mortality while a two-point increase in dietary indices was responsible for a 6% reduction in incidence of or mortality from cancer.

Lung cancer is one of the most common neoplasms and a leading cause of death with smoking considered the major risk factor. However, there is evidence suggesting that a diet high in vegetables and fruits such as MeD could be protective against lung cancer.[92–94] Antioxidant vitamins and phytochemicals could be proposed as the main nutrients underlining this association.[95–99] Vitamin E and carotenoids are probably responsible for some of this effect when sourced from whole foods. On the contrary, the same nutrients consumed through supplements have not proven to be effective in the prevention and treatment of lung cancer.[100–102]

Another type of cancer common among women all over the world, breast cancer, seemed to be partially prevented by diets high in mono-unsaturated fat and high in vegetables and fruits like MeD.[103] Tomato intake especially has been appreciated for its effect in lowering risk of prostate cancer in men.[104–106] In addition, recent literature revealed that following a healthy lifestyle[107] could prevent a high percentage of bowel cancer cases. This approximately MeD lifestyle could be characterized by controlling body weight, being physically active and adhering to a diet including vegetables and fibers, while consuming a large amount of red meat and alcohol may increase the risk.

Considering individually antioxidant nutrients included in MeD, recent studies reported that β-carotene deficiency is associated with increasing risk for lung, bladder and laryngeal cancer.[97,98,108] On the contrary, vitamin C as a nutrient included in fruit and vegetables seemed to be protective against esophageal and stomach cancer.[78] Vitamin E, included in foods such as olive oil and nuts has also proven to exert beneficial effects in cancers of the digestive system.[109–111]

In addition, several findings on the association of selenium intake with the prevention of several types of cancer are available. Whole grains, pasta, nuts, seeds, legumes, potatoes, poultry and fish could be proposed as the major sources of selenium.[112] In 1988, Willett summarized the first results on the possible anti-cancer role of serum selenium levels.[113] In the following years, prospective and intervention human studies in support of these suggestions were performed. Selenium appears to reduce the risk of esophageal, gastric-cardia and lung cancer.[114,115] In addition, trials with selenium showed benefit in reducing the risk of cancer incidence and mortality in all cancers, and specifically in liver, prostate, colo-rectal and lung cancers.[114]

Moreover, fiber intake coming from whole grains seems to reduce the risk of dietary-related cancer; although the mechanisms are still unknown, it is well known that dietary fiber reduces intestinal transit times and increases colonic fermentation.[116,117] The association appears to be strongest for intestinal cancers where reduced exposure to carcinogens or cancer-promoting agents in the lumen contents has been considered a potential mechanism of action. However, microbial fermentation of dietary fiber or carbohydrates which escape digestion in the upper gut leads to the protection of a range of biologically active compounds, not least the short-chain fatty acids (acetate, propionate and butyrate), which have been shown to impact not only on the intestinal cell cycle and to induce apoptosis in intestinal cancer cells, but are now known to act systemically in other body tissues and on the immune system, an important pathological determinant in all cancers.[118–120] On the contrary, high alcohol consumption is associated with an increased risk in a number of cancers. Indeed, 3.5% of cancer deaths worldwide are attributable to consumption of alcohol.[121] The most common types of cancer associated with alcoholic drinking include breast cancer in women, cancer of the mouth, esophagus, pharynx and larynx, colorectal cancer, liver and stomach.[122–124]

MEDITERRANEAN FOOD CONSUMPTION AND HUMAN GUT MICROBIOTA

Mediterranean Food Intake and Impact on Human Gut Microbiota

The human gut microbiota has recently been recognized as an important player in human health, not just in the intestine but systemically.[125–127] Indeed, it is becoming apparent that the gut microbiota impacts greatly on both metabolism and immune function, two key physiological processes involved in chronic non-communicable diseases such as cancer and CVD.[128] As discussed elsewhere in this volume, the gut microbiota contributes to obesity and metabolic syndrome, immune function and autoimmunity, and even impacts on brain development and function. A key modulator of these activities is diet, with dietary fiber or at least carbohydrates which escape digestion in the upper gut and reach the colon, wherein resides the vast majority of the gut microbiota, having a powerful effect on the ecology and metabolic output of the gut microbiota. Indeed, the availability of dietary carbohydrate in the colon determines whether the microbiota as a community carries out saccharolytic fermentation, leading to the production of SCFA[129,130] (mainly acetate, propionate and butyrate) or amino acid fermentation (leading to the production of toxic compounds including phenols, indoles, biogenic amines, p-cresol and ammonia, for example). This occurs because carbohydrate fermentation is energetically more favorable than amino acid fermentation, and therefore the quantity of carbohydrate reaching the colon is a key determinant of whether beneficial or toxic metabolites are produced by the colonic bacteria. The beneficial health effects of SCFA have recently been reviewed,[118,131] but at least in part stem from their ability to act as ligands for GPR41 and 43,[132] two important cell signaling receptors and also the ability of propionate and butyrate to act as histone deacetylase (HDAC) inhibitors,[133] an important epigenetic process involved in the regulation of host physiology including energy storage and thermogenesis, apoptosis and programmed cell death, and inflammation. Of the dietary fibers, a certain class of molecule has been identified which appears to be associated with increased human health via microbiota modulation. A *prebiotic* is a selectively fermented ingredient that results in specific changes in the composition and/or activity of the gastrointestinal microbiota, thus conferring benefit(s) upon host health. Prebiotics, which are common in whole plant foods including fruit, vegetables and whole-grain cereals, selectively stimulate the relative abundance of bacteria seen as "health promoting," most commonly the bifidobacteria, which in turn leads to improved health effects. Bifidobacteria have been shown to improve immune function, produce folate, contribute to SCFA production and cross-feeding, produce metabolites capable of impacting on cognitive function, carry out biohydrogenation of fatty acids contributing to stores of beneficial fats in the body and inhibit important intestinal pathogens.

The concentration and probably, chemical diversity of plant polyphenolic compounds coming in contact with the gut microbiota has also recently been appreciated as an important dietary modulator of both gut microbiota relative composition and metabolic output.[134] It is estimated that in the region of 95% of plant polyphenols[135] consumed daily persist until the colon, where they are free to interact with the gut microbiota. Different classes of plant polyphenolic compounds have been shown to exert antibacterial activities, including inhibition of the Gram-negative enterobacteria which includes many gastrointestinal pathogens,[136] while others have been shown to stimulate bifidobacteria[137] in particular, bacteria linked to improved human health and selected for both by human breast milk and fermentable carbohydrates especially the prebiotics.[138] In addition, complex plant poyphenols, such as condensed tannins present in many fruit and beverages like tea, bind bile acids,[139] increasing their passage from small to large intestine. This may have important implications for the enterohepatic circulation of bile acids and recent studies are showing that this process might not be as straightforward as previously thought. The conventional thinking is that bile acids reaching the colon are deconjugated by the gut microbiota and excreted in feces,[140,141] driving conversion of plasma cholesterol into bile acids in the liver where the bile acid pool is under tight control. Recent studies showing that bacteria in the gut, in this case in ingested probiotic strain *Lactobacillus reuteri* NCIMB 30242 which possesses the bile salt hydrolase, increase plasma deconjugated bile acids and reduce intestinal absorption of cholesterol and other sterols, indicates that microbial deconjugation within the colon may have direct effects on important markers of disease in this case CVD.[142] Such activities are likely to be mediated through bile salt receptors such as FXR-alpha, known to regulate fat, bile salt and glucose homeostasis and also TGR-5, an important cell signaling receptor on many immune cells.[143,144] Similarly, the concentration of bile reaching the colon can in itself modulate the composition of the gut microbiota as certain bacteria are sensitive to bile. Moreover, microbial end products of choline[145] and L-carnitine,[146] commonly components of red meat, have been shown to reduce the size of the bile acid pool which has important consequences for reverse cholesterol transport and foam cell formation, key processes in artherosclerosis.[147] However, which bacteria are involved, the relative potency of their bile salt hydrolases or impact on profiles of deconjugated bile acids returning to the liver or acting on FXR-alpha and other bile salt receptors within the gut remains to be determined.

Plant polyphenols themselves are modulated by the gut microbiota. Gut bacteria play an important hydrolytic role releasing polyphenols from complex food matrices (especially complex polysaccharides complexing certain polyphenols in whole plant foods) and in reducing chemical complexity increasing polyphenol bioavailability. They also impact on the bioactivity of plant polyphenols, as discussed in more detail elsewhere in this volume, where they can either reduce or increase the biological activity of plant polyphenols and thus regulate their ability to mediate health-promoting activities linked to consumption of whole plant foods.[148,149] Indeed, it is becoming apparent that, as we all possess a unique collection of microorganisms within our gut microbiota, different individuals present with different profiles of polyphenol catabolites upon ingestion of polyphenol-rich foods. The existence of these "metabotypes"[150] has important implications for both the relative efficacy of different fruit and vegetables in protecting against chronic non-communicable diseases, and in our ability to discern such health effects as individual polyphenols may be broken down to different catabolites with different biological activities in different individuals, compounding our ability to achieve statistically significant changes in disease biomarkers in intervention studies given that some of these effects may be subtle at best.

Gut bacteria also contain other enzymatic activities, not least linoleic dehydrogenase which converts alpha-linoleic acid to conjugated linoleic acid (CLA),[151] a beneficial fat linked to protection from CVD and certain cancers through improved immune function. Wall et al. (2010) have also shown that feeding alpha-linoleic acid together with the probiotic strain *Bifidobacterium breve* NCIMB 702258, a common intestinal commensal microorganism, increases concentrations of EPA in the liver and DHA in the brain, indicating that intestinal bacteria have the potential to modify the health-promoting ability of dietary lipids. We have recently shown that both the type and quantity of both dietary carbohydrate (irrespective of digestability) and fat (saturated fat, monounsaturated and polyunsaturated fats) impact on the relative abundance of important groups of gut bacteria and also have implications for SCFA production and/or absorption from the gut.[152]

Other compounds produced by the gut microbiota have been shown to impact on host physiology such as gamma-aminobutyrate (GABA) an important neurotransmitter and immune modulator which has also been shown to improve insulin sensitivity,[153] and important components of one-carbon metabolism such as betaine[154] and folic acid,[155] both protective against CVD. Conversely, and depending on the availability of carbohydrate to support saccharolytic rather than amino acid fermentation in the colon, the gut microbiota can lead to the formation of genotoxins, carcinogens and toxicants. A good example is the recent confirmation that trimethylamine produced by the gut microbiota from dietary choline or L-carnitine, leads to atherosclerotic plaque formation by increased foam cell formation, reduced reverse cholesterol transport in macrophages and inflammation.[145,146] This does not appear to happen in patients following a vegan diet or long-term vegetarians, in germ-free animals or where the gut microbiota has been removed using antibiotics.[146] This dichotomous role of the gut microbiota in CVD is a very good example of how diet can shape not only the composition of the gut microbiota but their metabolic and immunological interactions, with clear pathological consequences. It also fits very neatly with what we know about healthy diets or adherence to the MeD from nutritional interventions and epidemiological studies going back decades. It proposes the gut microbiota or diet:microbe interactions as a molecular basis for some of the pathophysiological processes linked particularly to CVD and cancer, and may with further study prove to be at the base of the MeD and MeD-derived-health diet pyramids.[156] Figure 14.1, adapted from Tuohy et al.,[156] schematically represents this overlap between foods and diets recognized to support human health and protect against chronic disease and those likely to support an ecologically diverse, stable and largely carbohydrate-fermenting gut microbiota.

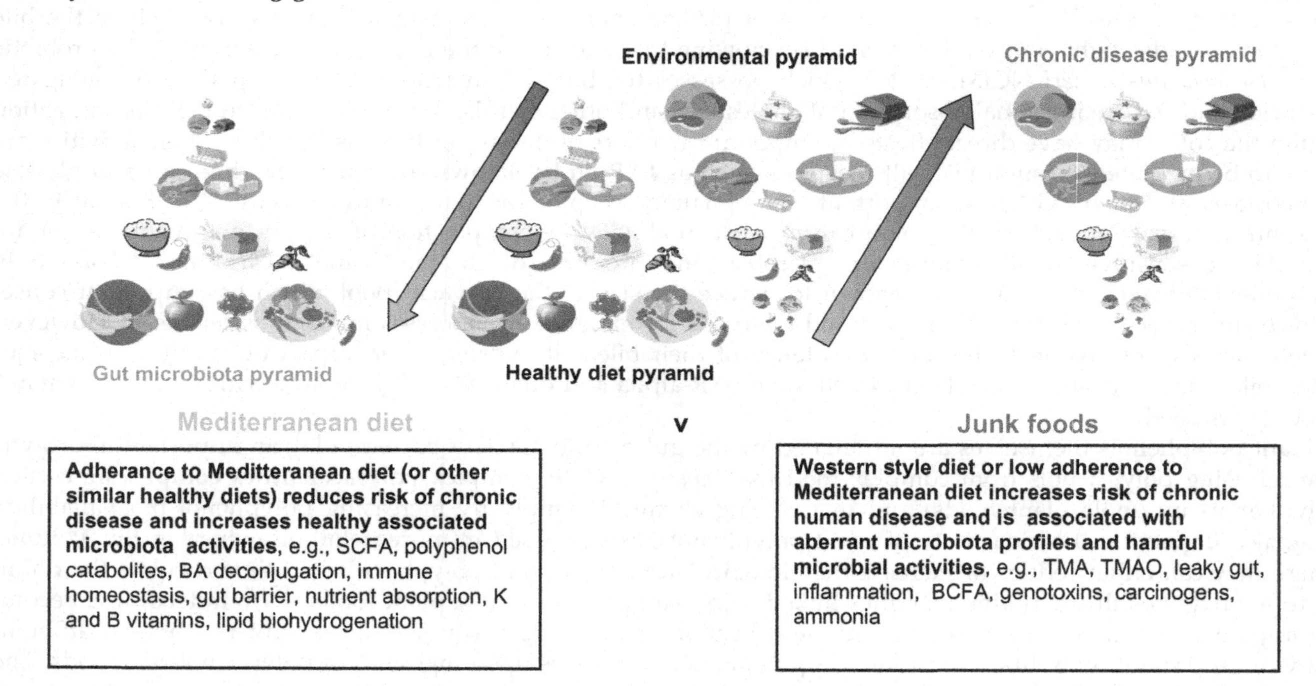

FIGURE 14.1 Adherance to Mediterranean diet (or other healthy-eating pyramids based on the MeD) reduces risk of chronic disease and increases healthy microbiota activities, e.g., SCFA, polyphenol catabolites, BA deconjugation, immune homeostasis, gut barrier, nutrient absorption, K and B vitamins, lipid biohydrogenation. These beneficial activities of the human gut microbiota have been demonstrated to be induced by foods or more commonly, food ingredients/components common in the MeD, such as fermentable fibers, prebiotics and polyphenols, or produced by probiotic microorganisms. Recent evidence combining gut microbiology with traditional nutritional intervention studies suggests that microbiota modulation by foods like fruit, vegetables, and red wine may contribute significantly to the health benefits of the Mediterranean dietary pattern. On the contary, following the Western diet or low adherence to Mediterranean diet increases risk of chronic human disease and is associated with aberrant microbiota profiles and harmful activities, e.g., TMA, TMAO, leaky gut, inflammation, branched chain fatty acids, 2° BA, genotoxins, carcinogens. This dichotomous nature of how diet modulates the gut microbiome and subsequently, human health, is particularly apparent in the case of CVD. *Tuohy et al. (2014).*[156]

Concluding Remarks

Although in vitro fermentation studies and animal models have been extensively used to examine the impact of whole plant foods on the gut microbiota, many of these studies are relatively old and did not employ more recently developed culture-independent molecular microbiological tools or high-resolution metabolite profiling and thus their contribution to understanding how these foods regulate the microbial ecology of the intestine has been limited. Many studies exist showing that isolated fractions from plant-based foods, both fermentable fibers and prebiotics on the one hand, and plant polyphenols on the other, modulate the gut microbiota or require microbial bioconversion for bioavailability and bioactivity. Conversely, only on a few occasions has the impact of whole plant foods or other foods and beverages characteristic of the MeD on both gut microbiota and human health related biomarkers been studied in humans. Costabile et al. (2008) showed that ingestion of a 100% whole-grain wheat breakfast cereal resulted in a significant and selective increase in bifidobacteria and lactobacilli in healthy individuals, and concomitantly gave a reduction in plasma cholesterol in those with highest starting cholesterol levels. More recently, we have shown that whole-grain oat ingestion, again as part of a breakfast cereal, both selectively and beneficially modulates the human gut microbiota and lowers both total and LDL-cholesterol in subjects at risk of the metabolic syndrome. Others have shown that ingestion of whole grains compared to refined grains, apples, blueberry drink, chickpea, coffee and kiwifruit all can mediate a beneficial modulation of the gut microbiota or their metabolic output.[157–162] Many of these studies were relatively small, did not use up-to-date microbiological tools and others did not measure health biomarkers during interventions making it difficult to draw conclusions about either the microbiota modulatory effects of these foods or their ability to impact on human health. On the contrary, a number of recent, well-controlled human dietary interventions with beverages either derived from high polyphenol foods such as chocolate or fermented beverages characteristic of the MeD, have been shown to both modulate the gut microbiota in a beneficial manner and impact on important health biomarkers, in particular, markers of CVD risk. Both red wine and high-flavanol chocolate drinks have been shown to increase the relative abundance of bifidobacteria and butyrate-producing bacteria within the gut microbiota of human volunteers and concomitantly, improve plasma lipid profiles and markers of inflammation.[137,163,164] However, it is fair to say that many foods, especially fruits and vegetables, in animal models both modulate gut microbiota composition and impact favorably on physiological processes linked to chronic non-communicable diseases. There is currently a real need to prove the "cause and effect" relationship between ingestion of different fruits and vegetables in particular, gut microbiota modulation and recognized biomarkers of health, to fully establish the role of the gut microbiota in MeD and similar dietary strategies based on high intake of whole plant foods.

References

1. Ferro-Luzzi A, Branca F. Mediterranean diet, Italian-style: prototype of a healthy diet. *Am J Clin Nutr.* 1995;61(6 suppl):1338S–1345S.
2. Nestle M. Mediterranean diets: historical and research overview. *Am J Clin Nutr.* 1995;61(6 suppl):1313S–1320S.
3. Trichopoulou A, Lagiou P. Healthy traditional Mediterranean diet: an expression of culture, history, and lifestyle. *Nutr Rev.* 1997;55 (11 Pt 1):383–389.
4. Ulbricht TL, Southgate DA. Coronary heart disease: seven dietary factors. *Lancet.* 1991;338(8773):985–992.
5. Willett WC, Sacks F, Trichopoulou A, et al. Mediterranean diet pyramid: a cultural model for healthy eating. *Am J Clin Nutr.* 1995;61(6 suppl):1402S–1406S.
6. The National Diet-Heart Study Final Report. *Circulation* 1968;37(3 suppl):I1–428.
7. WHO, FAO. *Food and Health Indicators in Europe: Nutrition and Health, 1961–1990 (computer program).* Copenhagen: World Health Organization Regional Office for Europe; 1993.
8. Keys A, Taylor H, Blackburn H, Brozek J, Anderson JT, Simonson E. Coronary heart disease among minnesota business and professional men followed fifteen years. *Circulation.* 1963;28:381–395.
9. Keys A. *Seven Countries: A Multivariate Analysis of Death and Coronary Heart Disease. Massachusetts: Harvard University Press Cambridge*; 1980.
10. UNESCO. *Representative List of the Intangible Cultural Heritage of Humanity.* 2010.
11. Keys A. Wine, garlic, and CHD in seven countries. *Lancet.* 1980;1(8160):145–146.
12. Keys A, Menotti A, Karvonen MJ, et al. The diet and 15-year death rate in the seven countries study. *Am J Epidemiol.* 1986;124(6):903–915.
13. Willcox BJ, Willcox DC, Todoriki H, et al. Caloric restriction, the traditional Okinawan diet, and healthy aging: the diet of the world's longest-lived people and its potential impact on morbidity and life span. *Ann N Y Acad Sci.* 2007;1114:434–455.
14. Kerver JM, Yang EJ, Bianchi L, Song WO. Dietary patterns associated with risk factors for cardiovascular disease in healthy US adults. *Am J Clin Nutr.* 2003;78(6):1103–1110.
15. Carter SJ, Roberts M, Salter J, Eaton CB. Relationship between Mediterranean Diet Score and atherothrombotic risk: findings from the Third National Health and Nutrition Examination Survey (NHANES III), 1988–1994. *Atherosclerosis.* 2010;210:630–636.
16. de Lorgeril M, Salen P, Martin JL, Monjaud I, Delaye J, Mamelle N. Mediterranean diet, traditional risk factors, and the rate of cardiovascular complications after myocardial infarction: final report of the Lyon Diet Heart Study. *Circulation.* 1999;99(6):779–785.
17. Ferrari R, Rapezzi C. The Mediterranean diet: a cultural journey. *Lancet.* 2011;377(9779):1730–1731.
18. Guillaume D. The Mediterranean diet: a cultural journey. *Lancet.* 2011;378(9793):766–767. [author reply 7].

19. Oldways HSoPH, European Office of the World Health Organization. Mediterranean Diet Pyramid. 2008.
20. Kromhout D, Keys A, Aravanis C, et al. Food consumption patterns in the 1960s in seven countries. *Am J Clin Nutr*. 1989;49(5):889–894.
21. Menotti A, Conti S, Corradini P, Giampaoli S, Rumi A, Signoretti P. Incidence and prediction of coronary heart disease in the Italian cohorts of the Seven Countries Study. 10 year experience (author's transl). *G Ital Cardiol*. 1980;10(7):792–806.
22. Trichopoulou A, Kouris-Blazos A, Wahlqvist ML, et al. Diet and overall survival in elderly people. *BMJ*. 1995;311(7018):1457–1460.
23. Hu FB. Dietary pattern analysis: a new direction in nutritional epidemiology. *Curr Opin Lipidol*. 2002;13(1):3–9.
24. Sofi F, Cesari F, Abbate R, Gensini GF, Casini A. Adherence to Mediterranean diet and health status: meta-analysis. *BMJ*. 2008;337:a1344.
25. Bach A, Serra-Majem L, Carrasco JL, et al. The use of indexes evaluating the adherence to the Mediterranean diet in epidemiological studies: a review. *Public Health Nutr*. 2006;9(1A):132–146.
26. Willett WC. The Mediterranean diet: science and practice. *Public Health Nutr*. 2006;9(1A):105–110.
27. Trichopoulou A, Costacou T, Bamia C, Trichopoulos D. Adherence to a Mediterranean diet and survival in a Greek population. *N Engl J Med*. 2003;348(26):2599–2608.
28. Alberti-Fidanza A, Fidanza F. Mediterranean adequacy index of Italian diets. *Public Health Nutr*. 2004;7(7):937–941.
29. Fidanza F, Alberti A, Lanti M, Menotti A. Mediterranean Adequacy Index: correlation with 25-year mortality from coronary heart disease in the Seven Countries Study. *Nutr Metab Cardiovasc Dis*. 2004;14(5):254–258.
30. Agnoli C, Krogh V, Grioni S, et al. A priori-defined dietary patterns are associated with reduced risk of stroke in a large Italian cohort. *J Nutr*. 2011;141(8):1552–1558.
31. Guenther PM, Reedy J, Krebs-Smith SM. Development of the Healthy Eating Index–2005. *J Am Diet Assoc*. 2008;108(11):1896–1901.
32. Appel LJ, Moore TJ, Obarzanek E, et al. A clinical trial of the effects of dietary patterns on blood pressure. DASH Collaborative Research Group. *N Engl J Med*. 1997;336(16):1117–1124.
33. Iacoviello L, Bonanni A, Costanzo S, , et al.on Behalf of the Moli-sani Project Investigators The Moli-Sani Project, a randomized, prospective cohort study in the Molise region in Italy, design, rationale and objectives. *Ital J Public Health*. 2007;4:110–118.
34. di Giuseppe R, Bonanni A, Olivieri M, et al. Adherence to Mediterranean diet and anthropometric and metabolic parameters in an observational study in the 'Alto Molise' region: the MOLI-SAL project. *Nutr Metab Cardiovasc Dis*. 2008;18(6):415–421.
35. Leon-Munoz LM, Guallar-Castillon P, Graciani A, et al. Adherence to the Mediterranean diet pattern has declined in Spanish adults. *J Nutr*. 2012;142(10):1843–1850.
36. Bonaccio M, Bonanni AE, Di Castelnuovo A, et al. Low income is associated with poor adherence to a Mediterranean diet and a higher prevalence of obesity: cross-sectional results from the Moli-sani study. *BMJ Open*. 2012;2:6.
37. Drewnowski A. Obesity, diets, and social inequalities. *Nutr Rev*. 2009;67(suppl 1):S36–S39.
38. Andrieu E, Darmon N, Drewnowski A. Low-cost diets: more energy, fewer nutrients. *Eur J Clin Nutr*. 2006;60(3):434–436.
39. Lopez CN, Martinez-Gonzalez MA, Sanchez-Villegas A, Alonso A, Pimenta AM, Bes-Rastrollo M. Costs of Mediterranean and western dietary patterns in a Spanish cohort and their relationship with prospective weight change. *J Epidemiol Community Health*. 2009;63(11):920–927.
40. van der Pols JC, Gunnell D, Williams GM, Holly JM, Bain C, Martin RM. Childhood dairy and calcium intake and cardiovascular mortality in adulthood: 65-year follow-up of the Boyd Orr cohort. *Heart*. 2009;95(19):1600–1606.
41. Maynard M, Gunnell D, Ness AR, Abraham L, Bates CJ, Blane D. What influences diet in early old age? Prospective and cross-sectional analyses of the Boyd Orr cohort. *Eur J Public Health*. 2006;16(3):316–324.
42. Ness AR, Maynard M, Frankel S, et al. Diet in childhood and adult cardiovascular and all cause mortality: the Boyd Orr cohort. *Heart*. 2005;91(7):894–898.
43. Levin KA. Urban-rural differences in adolescent eating behaviour: a multilevel cross-sectional study of 15-year-olds in Scotland. *Public Health Nutr*. 2013;:1–10.
44. Gould AC, Apparicio P, Cloutier MS. Classifying neighbourhoods by level of access to stores selling fresh fruit and vegetables and groceries: identifying problematic areas in the city of Gatineau, Quebec. *Can J Public Health*. 2012;103(6):e433–e437.
45. Caspi CE, Kawachi I, Subramanian SV, Adamkiewicz G, Sorensen G. The relationship between diet and perceived and objective access to supermarkets among low-income housing residents. *Soc Sci Med*. 2012;75(7):1254–1262.
46. Cohen DA, Sturm R, Scott M, Farley TA, Bluthenthal R. Not enough fruit and vegetables or too many cookies, candies, salty snacks, and soft drinks? *Public Health Rep*. 2010;125(1):88–95.
47. Bonaccio M, Di Castelnuovo A, Costanzo S, et al. Nutrition knowledge is associated with higher adherence to Mediterranean diet and lower prevalence of obesity. Results from the Moli-sani study. *Appetite*. 2013;68:139–146.
48. Bonaccio M, Di Castelnuovo A, Costanzo S, et al. Mass media information and adherence to Mediterranean diet: results from the Moli-sani study. *Int J Public Health*. 2012;57(3):589–597.
49. de Lorgeril M, Salen P. The Mediterranean diet in secondary prevention of coronary heart disease. *Clin Invest Med*. 2006;29(3):154–158.
50. de Lorgeril M, Salen P. Dietary prevention of coronary heart disease: the Lyon diet heart study and after. *World Rev Nutr Diet*. 2005;95:103–114.
51. de Lorgeril M, Salen P, Martin JL, Monjaud I, Boucher P, Mamelle N. Mediterranean dietary pattern in a randomized trial: prolonged survival and possible reduced cancer rate. *Arch Intern Med*. 1998;158(11):1181–1187.
52. Kris-Etherton P, Eckel RH, Howard BV, , et al.AHA Science Advisory, Lyon Diet Heart Study Benefits of a Mediterranean-style, National Cholesterol Education Program/American Heart Association Step I Dietary Pattern on Cardiovascular Disease. *Circulation*. 2001;103(13):1823–1825.
53. Appel LJ, Sacks FM, Carey VJ, et al. Effects of protein, monounsaturated fat, and carbohydrate intake on blood pressure and serum lipids: results of the OmniHeart randomized trial. *JAMA*. 2005;294(19):2455–2464.
54. Miller III ER, Erlinger TP, Appel LJ. The effects of macronutrients on blood pressure and lipids: an overview of the DASH and OmniHeart trials. *Curr Atheroscler Rep*. 2006;8(6):460–465.
55. Swain JF, McCarron PB, Hamilton EF, Sacks FM, Appel LJ. Characteristics of the diet patterns tested in the optimal macronutrient intake trial to prevent heart disease (OmniHeart): options for a heart-healthy diet. *J Am Diet Assoc*. 2008;108(2):257–265.
56. Burr ML, Fehily AM, Gilbert JF, et al. Effects of changes in fat, fish, and fibre intakes on death and myocardial reinfarction: diet and reinfarction trial (DART). *Lancet*. 1989;2(8666):757–761.

57. Dietary supplementation with n-3 polyunsaturated fatty acids and vitamin E after myocardial infarction: results of the GISSI-Prevenzione trial. Gruppo Italiano per lo Studio della Sopravvivenza nell'Infarto miocardico. *Lancet.* 1999;354(9177):447–455.
58. Marchioli R, Barzi F, Bomba E, et al. Early protection against sudden death by n-3 polyunsaturated fatty acids after myocardial infarction: time-course analysis of the results of the Gruppo Italiano per lo Studio della Sopravvivenza nell'Infarto Miocardico (GISSI)-Prevenzione. *Circulation.* 2002;105(16):1897–1903.
59. Marchioli R, Marfisi RM, Borrelli G, et al. Efficacy of n-3 polyunsaturated fatty acids according to clinical characteristics of patients with recent myocardial infarction: insights from the GISSI-Prevenzione trial. *J Cardiovasc Med (Hagerstown).* 2007;8(suppl 1):S34–S37.
60. Estruch R, Ros E, Salas-Salvado J, et al. Primary prevention of cardiovascular disease with a Mediterranean diet. *N Engl J Med.* 2013;368 (14):1279–1290.
61. Mozaffarian D, Appel LJ, Van Horn L. Components of a cardioprotective diet: new insights. *Circulation.* 2011;123(24):2870–2891.
62. Jacobs Jr DR, Steffen LM. Nutrients, foods, and dietary patterns as exposures in research: a framework for food synergy. *Am J Clin Nutr.* 2003;78(3 suppl):508S–513S.
63. Mozaffarian D, Ludwig DS. Dietary guidelines in the 21st century—a time for food. *JAMA.* 2010;304(6):681–682.
64. Dauchet L, Amouyel P, Dallongeville J. Fruit and vegetable consumption and risk of stroke: a meta-analysis of cohort studies. *Neurology.* 2005;65(8):1193–1197.
65. Dauchet L, Amouyel P, Hercberg S, Dallongeville J. Fruit and vegetable consumption and risk of coronary heart disease: a meta-analysis of cohort studies. *J Nutr.* 2006;136(10):2588–2593.
66. Elmer PJ, Obarzanek E, Vollmer WM, et al. Effects of comprehensive lifestyle modification on diet, weight, physical fitness, and blood pressure control: 18-month results of a randomized trial. *Ann Intern Med.* 2006;144(7):485–495.
67. He FJ, Nowson CA, Lucas M, MacGregor GA. Increased consumption of fruit and vegetables is related to a reduced risk of coronary heart disease: meta-analysis of cohort studies. *J Hum Hypertens.* 2007;21(9):717–728.
68. McCall DO, McGartland CP, McKinley MC, et al. Dietary intake of fruits and vegetables improves microvascular function in hypertensive subjects in a dose-dependent manner. *Circulation.* 2009;119(16):2153–2160.
69. Svendsen M, Blomhoff R, Holme I, Tonstad S. The effect of an increased intake of vegetables and fruit on weight loss, blood pressure and antioxidant defense in subjects with sleep related breathing disorders. *Eur J Clin Nutr.* 2007;61(11):1301–1311.
70. Anderson JW, Randles KM, Kendall CW, Jenkins DJ. Carbohydrate and fiber recommendations for individuals with diabetes: a quantitative assessment and meta-analysis of the evidence. *J Am Coll Nutr.* 2004;23(1):5–17.
71. Ludwig DS. The glycemic index: physiological mechanisms relating to obesity, diabetes, and cardiovascular disease. *JAMA.* 2002;287 (18):2414–2423.
72. Whelton SP, Hyre AD, Pedersen B, Yi Y, Whelton PK, He J. Effect of dietary fiber intake on blood pressure: a meta-analysis of randomized, controlled clinical trials. *J Hypertens.* 2005;23(3):475–481.
73. Macready AL, George TW, Chong MF, et al. Flavonoid-rich fruit and vegetables improve microvascular reactivity and inflammatory status in men at risk of cardiovascular disease—FLAVURS: a randomized controlled trial. *Am J Clin Nutr.* 2014;99(3):479–489.
74. Connor WE. Importance of n-3 fatty acids in health and disease. *Am J Clin Nutr.* 2000;71(1 suppl):171S–175S.
75. He K, Song Y, Daviglus ML, et al. Fish consumption and incidence of stroke: a meta-analysis of cohort studies. *Stroke.* 2004;35 (7):1538–1542.
76. Wang C, Harris WS, Chung M, et al. n-3 Fatty acids from fish or fish-oil supplements, but not alpha-linolenic acid, benefit cardiovascular disease outcomes in primary- and secondary-prevention studies: a systematic review. *Am J Clin Nutr.* 2006;84(1):5–17.
77. Kris-Etherton PM, Hu FB, Ros E, Sabate J. The role of tree nuts and peanuts in the prevention of coronary heart disease: multiple potential mechanisms. *J Nutr.* 2008;138(9):1746S–1751S.
78. Di Castelnuovo A, Costanzo S, Bagnardi V, Donati MB, Iacoviello L, de Gaetano G. Alcohol dosing and total mortality in men and women: an updated meta-analysis of 34 prospective studies. *Arch Intern Med.* 2006;166(22):2437–2445.
79. Costanzo S, Di Castelnuovo A, Donati MB, Iacoviello L, de Gaetano G. Cardiovascular and overall mortality risk in relation to alcohol consumption in patients with cardiovascular disease. *Circulation.* 2010;121(17):1951–1959.
80. Corrao G, Bagnardi V, Zambon A, La Vecchia C. A meta-analysis of alcohol consumption and the risk of 15 diseases. *Prev Med.* 2004;38 (5):613–619.
81. Costanzo S, Di Castelnuovo A, Donati MB, Iacoviello L, de Gaetano G. Wine, beer or spirit drinking in relation to fatal and non-fatal cardiovascular events: a meta-analysis. *Eur J Epidemiol.* 2011;26(11):833–850.
82. de Gaetano G, Di Castelnuovo A, Rotondo S, Iacoviello L, Donati MB. A meta-analysis of studies on wine and beer and cardiovascular disease. *Pathophysiol Haemost Thromb.* 2002;32(5–6):353–355.
83. Di Castelnuovo A, Rotondo S, Iacoviello L, Donati MB, De Gaetano G. Meta-analysis of wine and beer consumption in relation to vascular risk. *Circulation.* 2002;105(24):2836–2844.
84. Buitrago-Lopez A, Sanderson J, Johnson L, et al. Chocolate consumption and cardiometabolic disorders: systematic review and meta-analysis. *BMJ.* 2011;343:d4488.
85. Di Castelnuovo A, di Giuseppe R, Iacoviello L, de Gaetano G. Consumption of cocoa, tea and coffee and risk of cardiovascular disease. *Eur J Intern Med.* 2012;23(1):15–25.
86. di Giuseppe R, Di Castelnuovo A, Centritto F, et al. Regular consumption of dark chocolate is associated with low serum concentrations of C-reactive protein in a healthy Italian population. *J Nutr.* 2008;138(10):1939–1945.
87. Ard JD, Grambow SC, Liu D, et al. The effect of the PREMIER interventions on insulin sensitivity. *Diabetes Care.* 2004;27(2):340–347.
88. WHO. *Diet, Nutrition and the Prevention of Chronic Disease. WHO Technical Report series 916.* Geneva: WHO; 2003.
89. US Department of Agriculture UDoHaHS. :Home and Garden Bulletin No. 232 *Nutrition and Your Health: Dietary Guidelines for Americans.* 5th ed. Washington, DC: US Government Printing Office; 2000.
90. Sonestedt E, Wirfalt E, Wallstrom P, Gullberg B, Orho-Melander M, Hedblad B. Dairy products and its association with incidence of cardiovascular disease: the Malmo diet and cancer cohort. *Eur J Epidemiol.* 2011;26(8):609–618.
91. Sofi F, Macchi C, Abbate R, Gensini GF, Casini A. Mediterranean diet and health. *Biofactors.* 2013;39(4):335–342.

92. Block G, Patterson B, Subar A. Fruit, vegetables, and cancer prevention: a review of the epidemiological evidence. *Nutr Cancer*. 1992;18(1):1–29.
93. Feskanich D, Ziegler RG, Michaud DS, et al. Prospective study of fruit and vegetable consumption and risk of lung cancer among men and women. *J Natl Cancer Inst*. 2000;92(22):1812–1823.
94. Riboli E, Norat T. Epidemiologic evidence of the protective effect of fruit and vegetables on cancer risk. *Am J Clin Nutr*. 2003;78(3 suppl):559S–569S.
95. Block G. Vitamin C and cancer prevention: the epidemiologic evidence. *Am J Clin Nutr*. 1991;53(1 suppl):270S–282S.
96. Knekt P. Vitamin E and smoking and the risk of lung cancer. *Ann N Y Acad Sci*. 1993;686:280–287 [discussion 7–8]
97. Menkes MS, Comstock GW, Vuilleumier JP, Helsing KJ, Rider AA, Brookmeyer R. Serum beta-carotene, vitamins A and E, selenium, and the risk of lung cancer. *N Engl J Med*. 1986;315(20):1250–1254.
98. Omenn GS, Goodman GE, Thornquist MD, et al. Effects of a combination of beta carotene and vitamin A on lung cancer and cardiovascular disease. *N Engl J Med*. 1996;334(18):1150–1155.
99. Willett W. Vitamin A and selenium intake in relation to human cancer risk. *Princess Takamatsu Symp*. 1985;16:237–245.
100. Albanes D, Heinonen OP, Taylor PR, et al. Alpha-Tocopherol and beta-carotene supplements and lung cancer incidence in the alpha-tocopherol, beta-carotene cancer prevention study: effects of base-line characteristics and study compliance. *J Natl Cancer Inst*. 1996;88(21):1560–1570.
101. Bardia A, Tleyjeh IM, Cerhan JR, et al. Efficacy of antioxidant supplementation in reducing primary cancer incidence and mortality: systematic review and meta-analysis. *Mayo Clin Proc*. 2008;83(1):23–34.
102. Bjelakovic G, Nikolova D, Gluud LL, Simonetti RG, Gluud C. Mortality in randomized trials of antioxidant supplements for primary and secondary prevention: systematic review and meta-analysis. *JAMA*. 2007;297(8):842–857.
103. Gandini S, Merzenich H, Robertson C, Boyle P. Meta-analysis of studies on breast cancer risk and diet: the role of fruit and vegetable consumption and the intake of associated micronutrients. *Eur J Cancer*. 2000;36(5):636–646.
104. Etminan M, Takkouche B, Caamano-Isorna F. The role of tomato products and lycopene in the prevention of prostate cancer: a meta-analysis of observational studies. *Cancer Epidemiol Biomarkers Prev*. 2004;13(3):340–345.
105. Giovannucci E, Rimm EB, Liu Y, Stampfer MJ, Willett WC. A prospective study of tomato products, lycopene, and prostate cancer risk. *J Natl Cancer Inst*. 2002;94(5):391–398.
106. Kirsh VA, Mayne ST, Peters U, et al. A prospective study of lycopene and tomato product intake and risk of prostate cancer. *Cancer Epidemiol Biomarkers Prev*. 2006;15(1):92–98.
107. Kirkegaard H, Johnsen NF, Christensen J, Frederiksen K, Overvad K, Tjonneland A. Association of adherence to lifestyle recommendations and risk of colorectal cancer: a prospective Danish cohort study. *BMJ*. 2010;341:c5504.
108. Tavani A, La Vecchia C. Fruit and vegetable consumption and cancer risk in a Mediterranean population. *Am J Clin Nutr*. 1995;61(6 suppl):1374S–1377S.
109. Braga C, La Vecchia C, Franceschi S, et al. Olive oil, other seasoning fats, and the risk of colorectal carcinoma. *Cancer*. 1998;82(3):448–453.
110. La Vecchia C. Association between Mediterranean dietary patterns and cancer risk. *Nutr Rev*. 2009;67(suppl 1):S126–S129.
111. La Vecchia C, Negri E, Decarli A, D'Avanzo B, Franceschi S. A case-control study of diet and gastric cancer in northern Italy. *Int J Cancer*. 1987;40(4):484–489.
112. Pounis G, Costanzo S, di Giuseppe R, et al. Consumption of healthy foods at different content of antioxidant vitamins and phytochemicals and metabolic risk factors for cardiovascular disease in men and women of the Moli-sani study. *Eur J Clin Nutr*. 2013;67(2):207–213.
113. Willett WC, Stampfer MJ. Selenium and cancer. *BMJ*. 1988;297(6648):573–574.
114. Rayman MP. Selenium in cancer prevention: a review of the evidence and mechanism of action. *Proc Nutr Soc*. 2005;64(4):527–542.
115. Rayman MP. The importance of selenium to human health. *Lancet*. 2000;356(9225):233–241.
116. Fuchs CS, Giovannucci EL, Colditz GA, et al. Dietary fiber and the risk of colorectal cancer and adenoma in women. *N Engl J Med*. 1999;340(3):169–176.
117. Negri E, Franceschi S, Parpinel M, La Vecchia C. Fiber intake and risk of colorectal cancer. *Cancer Epidemiol Biomarkers Prev*. 1998;7(8):667–671.
118. Conterno L, Fava F, Viola R, Tuohy KM. Obesity and the gut microbiota: does up-regulating colonic fermentation protect against obesity and metabolic disease? *Genes Nutr*. 2011;6(3):241–260.
119. Arpaia N, Campbell C, Fan X, et al. Metabolites produced by commensal bacteria promote peripheral regulatory T-cell generation. *Nature*. 2013;504(7480):451–455.
120. Bindels LB, Porporato P, Dewulf EM, et al. Gut microbiota-derived propionate reduces cancer cell proliferation in the liver. *Br J Cancer*. 2012;107(8):1337–1344.
121. Boffetta P, Hashibe M, La Vecchia C, Zatonski W, Rehm J. The burden of cancer attributable to alcohol drinking. *Int J Cancer*. 2006;119(4):884–887.
122. Bagnardi V, Blangiardo M, La Vecchia C, Corrao G. A meta-analysis of alcohol drinking and cancer risk. *Br J Cancer*. 2001;85(11):1700–1705.
123. Longnecker MP, Orza MJ, Adams ME, Vioque J, Chalmers TC. A meta-analysis of alcoholic beverage consumption in relation to risk of colorectal cancer. *Cancer Causes Control*. 1990;1(1):59–68.
124. Moskal A, Norat T, Ferrari P, Riboli E. Alcohol intake and colorectal cancer risk: a dose–response meta-analysis of published cohort studies. *Int J Cancer*. 2007;120(3):664–671.
125. Salonen A, de Vos WM. Impact of diet on human intestinal microbiota and health. *Annu Rev Food Sci Technol*. 2014;5:239–262.
126. Holmes E, Kinross J, Gibson GR, et al. Therapeutic modulation of microbiota–host metabolic interactions. *Sci Transl Med*. 2012;4(137):137rv6.
127. Wallace TC, Guarner F, Madsen K, et al. Human gut microbiota and its relationship to health and disease. *Nutr Rev*. 2011;69(7):392–403.
128. Iebba V, Nicoletti M, Schippa S. Gut microbiota and the immune system: an intimate partnership in health and disease. *Int J Immunopathol Pharmacol*. 2012;25(4):823–833.
129. Macfarlane S, Macfarlane GT. Regulation of short-chain fatty acid production. *Proc Nutr Soc*. 2003;62(1):67–72.
130. Macfarlane GT, Macfarlane S. Bacteria, colonic fermentation, and gastrointestinal health. *J AOAC Int*. 2012;95(1):50–60.

131. Macfarlane GT, Macfarlane S. Fermentation in the human large intestine: its physiologic consequences and the potential contribution of prebiotics. *J Clin Gastroenterol*. 2011;45(suppl):S120–S127.
132. Sleeth ML, Thompson EL, Ford HE, Zac-Varghese SE, Frost G. Free fatty acid receptor 2 and nutrient sensing: a proposed role for fibre, fermentable carbohydrates and short-chain fatty acids in appetite regulation. *Nutr Res Rev*. 2010;23(1):135–145.
133. Astbury SM, Corfe BM. Uptake and metabolism of the short-chain fatty acid butyrate, a critical review of the literature. *Curr Drug Metab*. 2012;13(6):815–821.
134. Fava F, Lovegrove JA, Gitau R, Jackson KG, Tuohy KM. The gut microbiota and lipid metabolism: implications for human health and coronary heart disease. *Curr Med Chem*. 2006;13(25):3005–3021.
135. Clifford MN. Diet-derived phenols in plasma and tissues and their implications for health. *Planta Med*. 2004;70(12):1103–1114.
136. Daglia M. Polyphenols as antimicrobial agents. *Curr Opin Biotechnol*. 2012;23(2):174–181.
137. Clemente-Postigo M, Queipo-Ortuno MI, Boto-Ordonez M, et al. Effect of acute and chronic red wine consumption on lipopolysaccharide concentrations. *Am J Clin Nutr*. 2013;97(5):1053–1061.
138. Roberfroid M, Gibson GR, Hoyles L, et al. Prebiotic effects: metabolic and health benefits. *Br J Nutr*. 2010;104(suppl 2):S1–63.
139. Aprikian O, Duclos V, Guyot S, et al. Apple pectin and a polyphenol-rich apple concentrate are more effective together than separately on cecal fermentations and plasma lipids in rats. *J Nutr*. 2003;133(6):1860–1865.
140. Moore RB, Crane CA, Frantz Jr ID. Effect of cholestyramine on the fecal excretion of intravenously administered cholesterol-4-14C and its degradation products in a hypercholesterolemic patient. *J Clin Invest*. 1968;47(7):1664–1671.
141. Pereira DI, Gibson GR. Effects of consumption of probiotics and prebiotics on serum lipid levels in humans. *Crit Rev Biochem Mol Biol*. 2002;37(4):259–281.
142. Jones ML, Martoni CJ, Prakash S. Cholesterol lowering and inhibition of sterol absorption by Lactobacillus reuteri NCIMB 30242: a randomized controlled trial. *Eur J Clin Nutr*. 2012;66(11):1234–1241.
143. Sayin SI, Wahlstrom A, Felin J, et al. Gut microbiota regulates bile acid metabolism by reducing the levels of tauro-beta-muricholic acid, a naturally occurring FXR antagonist. *Cell Metab*. 2013;17(2):225–235.
144. Stanimirov B, Stankov K, Mikov M. Pleiotropic functions of bile acids mediated by the farnesoid X receptor. *Acta Gastroenterol Belg*. 2012;75(4):389–398.
145. Wang Z, Klipfell E, Bennett BJ, et al. Gut flora metabolism of phosphatidylcholine promotes cardiovascular disease. *Nature*. 2011;472 (7341):57–63.
146. Koeth RA, Wang Z, Levison BS, et al. Intestinal microbiota metabolism of L-carnitine, a nutrient in red meat, promotes atherosclerosis. *Nat Med*. 2013;19(5):576–585.
147. Li F, Jiang C, Krausz KW, et al. Microbiome remodelling leads to inhibition of intestinal farnesoid X receptor signalling and decreased obesity. *Nat Commun*. 2013;4:2384.
148. Tuohy KM, Conterno L, Gasperotti M, Viola R. Up-regulating the human intestinal microbiome using whole plant foods, polyphenols, and/or fiber. *J Agric Food Chem*. 2012;60(36):8776–8782.
149. Del Rio D, Rodriguez-Mateos A, Spencer JP, Tognolini M, Borges G, Crozier A. Dietary (poly)phenolics in human health: structures, bioavailability, and evidence of protective effects against chronic diseases. *Antioxid Redox Signal*. 2013;18(14):1818–1892.
150. Bolca S, Van de Wiele T, Possemiers S. Gut metabotypes govern health effects of dietary polyphenols. *Curr Opin Biotechnol*. 2013;24 (2):220–225.
151. Wall R, Ross RP, Shanahan F, et al. Metabolic activity of the enteric microbiota influences the fatty acid composition of murine and porcine liver and adipose tissues. *Am J Clin Nutr*. 2009;89(5):1393–1401.
152. Fava F, Gitau R, Griffin BA, Gibson GR, Tuohy KM, Lovegrove JA. The type and quantity of dietary fat and carbohydrate alter faecal microbiome and short-chain fatty acid excretion in a metabolic syndrome 'at-risk' population. *Int J Obes (Lond)*. 2013;37(2):216–223.
153. Tian J, Dang HN, Yong J, et al. Oral treatment with gamma-aminobutyric acid improves glucose tolerance and insulin sensitivity by inhibiting inflammation in high fat diet-fed mice. *PLoS One*. 2011;6(9):e25338.
154. Tang WH, Wang Z, Levison BS, et al. Intestinal microbial metabolism of phosphatidylcholine and cardiovascular risk. *N Engl J Med*. 2013;368(17):1575–1584.
155. D'Aimmo MR, Mattarelli P, Biavati B, Carlsson NG, Andlid T. The potential of bifidobacteria as a source of natural folate. *J Appl Microbiol*. 2012;112(5):975–984.
156. Tuohy KM, Fava F, Viola R. 'The way to a man's heart is through his gut microbiota' – dietary pro- and prebiotics for the management of cardiovascular risk. *Proc Nutr Soc*. 2014;:1–14.
157. Langkamp-Henken B, Nieves Jr C, Culpepper T, et al. Fecal lactic acid bacteria increased in adolescents randomized to whole-grain but not refined-grain foods, whereas inflammatory cytokine production decreased equally with both interventions. *J Nutr*. 2012;142 (11):2025–2032.
158. Shinohara K, Ohashi Y, Kawasumi K, Terada A, Fujisawa T. Effect of apple intake on fecal microbiota and metabolites in humans. *Anaerobe*. 2010;16(5):510–515.
159. Vendrame S, Guglielmetti S, Riso P, Arioli S, Klimis-Zacas D, Porrini M. Six-week consumption of a wild blueberry powder drink increases bifidobacteria in the human gut. *J Agric Food Chem*. 2011;59(24):12815–12820.
160. Fernando WM, Hill JE, Zello GA, Tyler RT, Dahl WJ, Van Kessel AG. Diets supplemented with chickpea or its main oligosaccharide component raffinose modify faecal microbial composition in healthy adults. *Benef Microbes*. 2010;1(2):197–207.
161. Jaquet M, Rochat I, Moulin J, Cavin C, Bibiloni R. Impact of coffee consumption on the gut microbiota: a human volunteer study. *Int J Food Microbiol*. 2009;130(2):117–121.
162. Lee YK, Low KY, Siah K, Drummond LM, Gwee KA. Kiwifruit (Actinidia deliciosa) changes intestinal microbial profile. *Microb Ecol Health Dis*. 2012;:23.
163. Queipo-Ortuno MI, Boto-Ordonez M, Murri M, et al. Influence of red wine polyphenols and ethanol on the gut microbiota ecology and biochemical biomarkers. *Am J Clin Nutr*. 2012;95(6):1323–1334.
164. Tzounis X, Rodriguez-Mateos A, Vulevic J, Gibson GR, Kwik-Uribe C, Spencer JP. Prebiotic evaluation of cocoa-derived flavanols in healthy humans by using a randomized, controlled, double-blind, crossover intervention study. *Am J Clin Nutr*. 2011;93(1):62–72.

CHAPTER

15

Diet and the Gut Microbiota – How the Gut: Brain Axis Impacts on Autism

Kieran M. Tuohy, Paola Venuti†, Simone Cuva†, Cesare Furlanello‡, Mattia Gasperotti*, Andrea Mancini*, Florencia Ceppa*, Duccio Cavalieri§, Carlotta de Filippo*, Urska Vrhovsek*, Pedro Mena¶, Daniele Del Rio¶ and Francesca Fava**

***Department of Food Quality and Nutrition, Research and Innovation Centre, Fondazione Edmund Mach, San Michele all'Adige, Trento, Italy †Department of Psychology and Cognitive Sciences, University of Trento, Rovereto, Trento, Italy ††MPBA/Center for Information and Communication Technology, Fondazione Bruno Kessler, Trento, Italy §Department of Computaitonal Biology, Research and Innovation Centre, Fondazione Edmund Mach, San Michele all'Adige, Trento, Italy ¶LS9 Bioactives and Health Interlab Group, Department of Food Science, University of Parma, Parma, Italy**

BACKGROUND

Autism spectrum disorders (ASDs) are a group of neurodevelopmental conditions, with an onset prior to 3 years of age, characterized by qualitative impairments in social interaction and communication, and by the presence of restricted interests and of repetitive and stereotyped behaviors.[1] Until a few decades ago, ASDs were considered quite rare, but have increased 13-fold in the UK since the 1980s and recent prevalence rates considering the broader spectrum in the US are now estimated at 1 in 88 children.[2,3] These shocking statistics combined with the quality of life destruction and socioeconomic implications of a lost generation paints a frightening picture of an emergent epidemic with devastating consequences both for ASD-afflicted families and for society at large.[4] Valid biomarkers or biological tests for ASD are not yet available and there are few effective therapies, compounding medical treatment of ASD children.[5] That of ASD, therefore, remains a diagnosis that is defined completely on the basis of behavior, through a detailed developmental history, parental descriptions of a child's everyday behavior, and direct assessment of the child's social interaction style and communicative and intellectual function.[6] However, our understanding of ASD phenotypic pathogeneses is made all the more complicated by another important issue; in a nutshell, ASDs are far from being homogeneous in their onset and manifestations: 15–30% of children with ASD show a period of stasis of development and even a frank loss of skills, most commonly speech.[6,7] Moreover, beyond the core symptoms, these children often manifest an array of other associated features such as sensorimotor abnormalities (albeit of uncertain origin, see Ref.[6]), poor muscle tone and motor deficits[8,9] and abnormal cognitive profiles.[10,11] Also comorbidities are common, including attention-deficit hyperactivity disorder (ADHD) and anxiety, as well as sleeping,[12,13] gastrointestinal conditions and eating problems.[14,15] This last point is of particular interest since feeding might represent a bridge between environmental and neurobiological factors and thus maybe play a role in the pathways leading to the disease phenotype. It is known both from clinical experience and from the literature that autistic children tend to have a particular relation with feeding and dietary attitudes. There have been reports of atypical feeding behavior, such as sensitivity to food texture and selective preferences for particular food.[16–18] The association between limited repertoire and

Diet-Microbe Interactions in the Gut.
DOI: http://dx.doi.org/10.1016/B978-0-12-407825-3.00015-0

nutrient inadequacy suggests that a very limited diet may put any child at risk for nutritional deficiencies and sub-optimal development, including brain development.[19] Worthy of note are also the findings regarding a gluten–milk-free diet,[20] which even if not consistent often pushes parents to try modifying the dietary intake of their children.

Strong evidence supports a genetic contribution to disease risk in autism and related conditions. The prevalence of autism is four-times greater in boys than in girls,[21] frequently there are abnormalities in head circumference;[22] approximately 50% of ASD individuals have an IQ in the intellectual disability range[23] and frequently autism is associated with other medical conditions, such as epilepsy and epileptiform electroencephalography (EEG) abnormalities[24] or genetic syndromes.[25] Twin studies and molecular genetics research have found that ASD is comprised of many different syndromes with common symptoms and presentation and that genetic predisposition may derive from multiple genotypic determinants.[26,27] However, the many different genes implicated in ASD risk encode a variety of different proteins involved in various physiological processes, including brain development and function, neurotransmitter receptors or transporters, cell adhesion/barrier function proteins, immune-related proteins, proteins involved in cholesterol metabolism or transport, and proteins impacting on mitochondrial function.[26–29] Further not all individuals carrying these particular mutations develop ASD. These facts, and a recent identical fraternal twin study which highlighted that environmental factors might explain about 55% of their relative risk of developing ASD,[30] highlights a growing awareness that environmental factors provide the selective pressure for ASD-type symptoms to emerge from various and sundry predisposing genetic abnormalities, combining to cause overt disease.

Although the pathophysiology of the disease remains unknown, a number of metabolic pathways appear to be modulated in ASD and have been highlighted as possible candidate disease mechanisms, which under given genetic susceptibility may alter neurological development in early childhood.[31–33] Moreover, the recent appreciation of the core contribution of the human microbiome to host metabolic processes and in processing of dietary compounds has raised interest in how microbial activities might impact on ASD.[34,35] Gastrointestinal symptoms have long indicated that gut bacteria might play some role in ASD pathophysiology[14] and indeed various studies have shown that the gut microbiota is altered in ASD,[36] although there is little agreement in the literature as to which bacteria might be involved.[37] There have also been recent insights into the functioning of the gut:brain axis showing that gut bacteria or their metabolic end-products can have surprising physiological activities not just in well-recognized conduits between gut and brain like satiety control and intestinal transit, but also in mood, cognition, behavior, depression and brain development.[38,39]

Diet has an important role in shaping mammalian metabolic flux including the flux of neurochemicals and also in shaping the gut microbiota and their activities.[40–42] Fibers and prebiotics in particular, support a beneficial, saccharolytic gut microbiota characterized by increased relative abundance of bifidobacteria and lactobacilli and short-chain fatty acids (SCFA) production.[43,44] Bifidobacteria and lactobacilli, important members of the beneficial gut microbiota, mediate host immune function,[45] mucosal integrity[46] and produce bioactive compounds including SCFA,[41] beneficial fatty acids,[47,48] folate,[49] and γ-aminobutyric acid (GABA).[50] Strains of both genera are commonly used as probiotics or are present in traditional fermented foods especially dairy products.[44,51] Polyphenols derived from whole plant foods like fruit, cereal grains and vegetables also impact on gut microbiota composition, immune function, act as antioxidants and protect against inflammation in the brain and improve blood–brain barrier (BBB) function[52,53] and may also have thus-far unrecognized effects on brain metabolism or gene expression. Below, and in Figure 15.1, we discuss how different metabolic processes involved in the gut:brain axis appear to be modulated by the activities of the gut microbiome and its interaction with diet and suggest how it may be possible to modulate these processes through dietary means. Although, environmental factors are recognized to play an important role in ASD, there is little conclusive evidence linking diet with either disease onset or progression. However, recent investigations examining how diet shapes the gut/microbiome:brain axis may offer new insights into environmental contributions to disease mechanisms and raise the possibility of at least improving certain ASD symptoms through diet in the near future.

GUT MICROBIOTA AND ASD

The gut:brain axis as a communication highway between the external environment and human brain has important "internal" turnpikes in the shape of the human gut microbiota through which many nutrients and ingested chemicals must pass before being converted into biologically available and active intermediates which

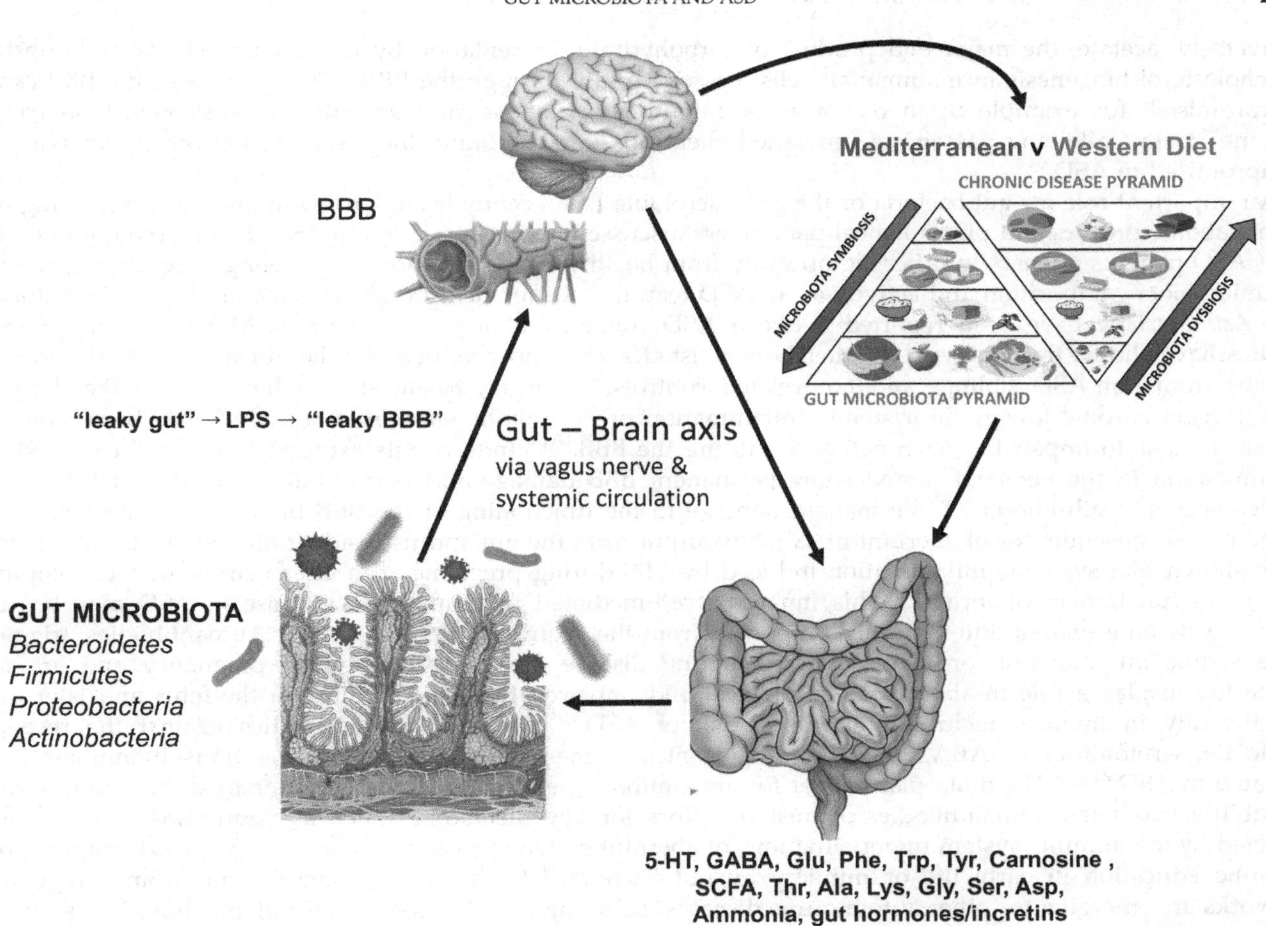

FIGURE 15.1 This schematic represents the putative modulatory capability of diet, especially dietary extremes, represented by the Mediterranean-style diet, rich in components which support the gut microbiota, and the Western-style diet, with high fat and animal protein which mediate a detrimental impact on the gut microbiota and their metabolic output, on the gut:brain axis. Key intermediaries in this communication highway are the metabolites produced/regulated by the gut microbiota [e.g., serotonin/5-hydroxytryptamine (5-HT), gamma-amino butyric acid (GABA), glutamate (Glu), Phenylalanine (Phe), Tryptophan (Trp), tyrosine (Tyr), Carnosine, short-chain fatty acids (SCFA), Threonine (Thr), Alanine (Ala), Lysine (Lys), Glycine (Gly), Serine (Ser), aspartic acid (Asp), Ammonia, gut hormone/incretin production], microbial regulation of barrier integrity (both within the gut and at the blood brain barrier, BBB), and nutrient absorption at the mucosal surface, which is itself controlled in part by gut bacteria. These activities have the potential to impact on neurotransmission, inflammation, redox potential and epigenetic regulation of developmental processes, as well as neurological function, probably through the vagus nerve or in certain cases, may mediate direct effects on the brain.

are then absorbed and directed systemically through the hepatic portal vein. Many chemicals important in brain chemistry are also produced in the intestine by the gut microbiota including tryptophan, dopamine, serotonin, GABA, β-hydroxybutyrate, choline, taurine, acetate, succinate, lactate, acetyl-coA, creatinine, betaine, glutamate, glutamine, p-cresol, trans-indolylacryloylglycine, fatty acids and hippurate.[54,55] Others may be regulated by bacteria (e.g., during digestion and fermentation) or are bacterial components like the Gram-negative bacterial cell wall building block lipopolysaccharide (LPS), which dose-dependently reduces serotonin uptake by human intestinal cells[56] and triggers inflammation both in the periphery and brain impacting on brain function.[57] The neurotransmitter GABA has recently been suggested to play a role in ASD neuronal development,[58] especially considering its switch from stimulator to inhibitor of neurotransmission in infancy.[59] Another neurotransmitter, serotonin [5-hydroxytryptamine (5-HT], has also been suspected to play a role in ASD.[60] Both serotonin and GABA levels are altered in the blood of ASD subjects.[61] Interestingly, both of these neurotransmitters derive from amino acid catabolism, tryptophan and glutamate, respectively. Intestinally derived metabolites, however, may or may not directly impact on neurological development in the brain depending on whether they can pass through the blood–brain barrier (BBB). For example, bacterial GABA produced in the gut may act locally on the enteric nervous system or alter plasma GABA concentrations, but in health will not pass through the BBB.[62]

Conversely, acetate, the major end product of carbohydrate fermentation by the gut microbiota and substrate for cholesterol biogenesis in mammalian cells, passes rapidly through the BBB.[63] None the less, the BBB can be compromised, for example upon oxidative or inflammatory stress, and as with the gastrointestinal barrier, become "leaky" allowing passage of unwanted chemicals into the brain. Interestingly, the BBB is known to be compromised in ASD.[64]

An important role for gut bacteria or the gut microbiota has recently been defined in processes impacting both on metabolite profiles and physiological parameters discussed above in relation to ASD. Firstly, the gut microbiota of ASD patients appears to differ significantly from healthy controls or non-ASD siblings, identifying aberrant gut microbiota composition and activity as an ASD feature.[37] Differences in *Bacteroidetes*, *Firmicutes*, *Proteobacteria* and *Actinobacteria* have been reported between ASD cohorts and non-ASD controls. Moreover, metabonomic studies have shown that many of the metabolites listed above and produced by the gut microbiota differentiate autistic from non-ASD siblings and non-related controls.[34] Further, recent studies have shown that bacterial LPS triggers chronic low-grade systemic inflammation or "metabolic endotoxemia" which has been shown in animal models to impair barrier function, including the BBB.[65,66] Indeed, rats exposed to LPS-induced systemic inflammation in the neonatal period show permanent BBB damage and permeability and altered behavior in adolescence and adulthood.[67,68] Permanent damage to the functioning of the BBB in ASD will only exacerbate pathological consequences of aberrant metabolite output from the gut and its resident microbiota. Similar studies have shown that systemic inflammation induced by LPS during pregnancy can alter neurological development and brain function in offspring.[69] This innate or cell-mediated inflammatory response in ASD may be compounded by an apparent autoimmune component from the acquired immune system. Autoantibodies, triggered by systemic inflammation or maternal autoimmune disease are generated during pregnancy and are now suspected to play a role in abnormal neurological and impaired BBB development in the fetus and later brain functionality in infancy, including increased risk of ASD.[70] Similarly, autoantibodies against the receptors for folate, serotonin and GABA, as well as important immune-related enzymes such as transglutaminase 2, are elevated in ASD.[71–73] The molecular trigger for autoantibody generation is poorly understood, but the intriguing possibility exists that autoantibodies against receptors for key metabolites such as neurotransmitters may be directed by the immune system under situations of aberrant metabolite concentrations in the blood, inappropriate immune education in early life or mimicry by gut bacteria.[74] Such "metabolome-inflammosome" regulatory networks are emerging for other autoimmune diseases including type 1 diabetes and inflammatory bowel disease (IBD), and appear to be closely linked to the gut microbiota.[75,76]

AMINO ACID METABOLISM

Amino acid (AA) metabolism plays an important role in biosynthesis of metabolites involved in neurotransmission and has long been suspected to play a role in ASD.[77] A number of studies have reported raised glutamate to glutamine (Glu:Gln) ratio in the blood of ASD patients.[78,79] Aldred et al.,[78] comparing children with ASD or Asperger's syndrome and their parents, found that patients and their families all had higher than normal (age-matched health controls) plasma concentrations of glutamic acid, phenylaline, asparagine, tyrosine, alanine and lysine, and lower concentrations of glutamine. AA transport and subsequent production of neurotransmitters (e.g., serotonin from tryptophan and the catecholamines dopamine and norepinephrine from tyrosine) has been found to be altered in fibroblasts taken from individuals with different physiological conditions including schizophrenia, bipolar disorder and autism and in the case of tryptophan and alanine, ADAH.[80–83] Fibroblasts use the same AA transporters as the brain and altered AA transport in fibroblasts ex vivo is thought therefore to reflect altered BBB transport of these AA *in vivo*.[83] Ming et al.[84] using MS-based strategies to profile urinary metabolites in 48 ASD children and 53 age-matched controls, found 82 metabolites altered in the ASD samples. Specifically levels of AA (including glycine, serine, threonine, alanine, histidine, glutamyl amino acids), and organic acid and bile derivative taurine, and carnosine were all lower in ASD urines. They also observed altered profiles of gut microbiota metabolites and markers of oxidative stress.

Conversion of glutamate to glutamine is the main mode of ammonia waste disposal in the brain and critical to avoid ammonia toxicity and excess glutamate build-up in synapses to minimize excitotoxicity.[85] Raised Glu:Gln ratio in the blood therefore may indicate altered ammonia detoxification and glutamate cycling in the brain of ASD patients which can impact on behavior.[86] This ammonia toxicity could only be exacerbated by low-fiber, high-protein diets leading to increased systemic contribution of ammonia derived from amino acid fermentation by a proteolytic microbiota in the gut.[87] Microbial catabolism of dietary AA can impact on AA availability and systematic circulation in mammals or result in the production of biologically active compounds such as SCFA, branched-chain fatty acids and biogenic amines.[88] As in hepatic encephalopathy, prebiotic (specifically lactulose) modulation of the gut microbiota and bifidogenesis may act as a sink for colonically produced ammonia and relieve the neurological symptoms of ammonia toxicity.[89] The dipeptide L-carnosine is a suspected neuroprotectant and may enhance frontal lobe function. A single human feeding study found that 800 mg L-carnosine daily for 8 weeks in 31 ASD children brought on improvements in behavior measures (the Gilliam Autism Rating Scale; total score, behavior, socialization and communication subscales and the Receptive One-Word Picture vocabulary test), compared to a placebo treatment.[90] However, although promising, these studies have not since been repeated, were conducted only in small populations and the validity of some of the behavioral measures used may be restricted to certain conditions within ASD such as Asperger's syndrome.

AA levels in blood and urine will be affected by many factors including absorption of AA from digested food, AA and protein degradation, host protein secretion and excretion in feces. Alterations in the relative proportions of these AA are also likely to have knock-on effects on the products of metabolic pathways in which these AA are involved, including the production or relative ratios of different neurotransmitters. Most AA derive from the diet or are produced endogenously, but the gut microbiota too can impact both on the recovery of dietary AA and on production or catabolism of AA.[91,92] Recent studies have also shown that ingested L-arginine or L-glutamine can modify the ability of gut bacteria to utilize other dietary AA.[93] However, as discussed below, we know little about the factors governing amino acid metabolism by the gut microbiota and how it can be modified by other dietary components or the extent to which AA synthesized by the microbiota contribute to human nutrition. Similarly, β-alanine a common AA in meat inhibits the uptake of GABA by intestinal cells such as Caco2.[94] Interestingly, β-alanine could also be formed *in situ* in the gut upon reaction of propionate and ammonia produced by *Candida albicans*, although these compounds are also produced by many other microorganisms in the gut.[95] Shaw[96] reported the occurrence of 3-(3-hydroxyphenyl)-3-hydroxypropionic acid (HPHPA) a rare metabolite of m-tyrosine produced by certain clostridia species. m-tyrosine itself is an oxidation product of phenylalanine which has been shown to induce ASD type symptoms in laboratory animals. HPHPA has also been found in individuals with *Clostridium difficile* infection and in extremely high levels in schizophrenic patients during acute psychotic episodes. In one case psychosis was successfully treated with vancomycin and levels of HPHPA decreased concomitantly in urine. Interestingly, vancomycin, often active against *Clostridium difficile* in oral form, has been shown to improve cognitive function and social markers in children with ASD.[97]

Currently, we know little about the microbial species or indeed metabolic processes involved in AA bioavailability and biotransformation by the gut microbiota and less about the possible implications for neurological function. We also know little about how different foods and food components interact to regulate AA uptake or metabolism by the gut microbiota. Early *in vitro* studies by Smith and Macfarlane[88] showed that AA fermentation by the human gut microbiota is inhibited both by low pH and by the presence of fermentable fiber/carbohydrate (resistant starch). However, we do not know how this process translates *in vivo* or how it changes in disease states such as ASD or upon antibiotic therapy. From studies in piglets, a model used to mimic human infant nutrition, we know that cholesterol supplementation of infant formula alters the profile of AA in the brain reducing concentrations of glutamate, serine, glutamine, threonine, β-alanine, alanine, methionine, isoleucine, leucine and γ-aminobutyrate, while increasing concentrations of glycine and lysine. The fatty acid docosahexaenoic acid (DHA) had a similar effect except for reducing taurine levels and having no effect on isoleucine and lysine. AA levels in the liver, muscle and plasma of piglets was also effected by cholesterol or DHA dietary supplementation. DHA also reduced muscle carnosine and ammonia in muscle and brain.[98] Little information exists on the effect of food processing on AA profiles in ready-to-eat or convenience foods. Hermanussen et al.[99] profiling AA patterns in 17 commonly consumed convenience foods in Germany found that not only did AA profiles not match those on food labels or packing, but concentrations of glutamate were often well above those commonly reported for the foodstuff, free-cysteine was generally low, probably as a result of thermal lysis during food processing, and some products labeled as containing meat contained no detectable carnosine. GABA was commonly found at relatively high concentrations. More recently, the same authors have reported that industrial food processing can radically modify the profile of AA in chicken-based foods, elevating concentrations of glutamate, and eliminating anserine or carnosine present in the home-cooked equivalent product. Similar results were found in vegetable soups, with glutamate sometimes making up 96% of all free AA.[100] In general, although AA metabolism appears to be intimately involved in the biology of ASD, clear understanding of the underlying mechanisms or direct link to pathological processes is currently lacking and further studies in well-documented human cohorts are necessary to explore further the links between AA metabolism, nutrition and gut microbiota in ASD.

Such observations do, however, have implications for how food choice or the prevailing household diet can impact on nutrient availability and metabolism with implications for brain development and function in early childhood.

LIPID METABOLISM AND THE BRAIN

Linoleic acid (LA, C18:2n-6, precursor of n-6 fatty acids) and α-linolenic acid (ALA, C18:3n-3, precursor of n-3 fatty acids) are essential fatty acids (EFAs) not produced by mammals which must be ingested through diet. These essential fatty acids, their derivatives, their relative proportions and individual molecular species play critical roles in a number of mammalian processes including phospholipid generation, membrane fluidity and brain development.[101] In contrast to other bodily tissues, the brain has lower LA and ALA concentrations compared to that of their derivatives, especially arachidonic acid (ARA) and DHA.[102] DHA has an important structural role in the brain and plays a role in cell signaling and cellular proliferation. ARA is involved in signal transduction and cell growth and another ALA derivative, eicosapentaenoic acid (EPA), is thought to play an important role in brain function.[102] LA and ALA derivatives may be further modified by host phospholipases and converted into eicosanoids, derived mainly from ARA such as prostaglandins, leukotrienes and thromboxanes. These eicosanoids are considered proinflammatory molecules acting as local hormones to activate immune cells, initiate platelet aggregation, and induce birth.[103] Conversely, DHA and EPA can be further converted into anti-inflammatory resolvins and protectins.[102] A number of small-scale dietary interventions with n-3 and n-6 fatty acids have shown symptom relief in ASD patients, although not all studies show an improvement.[102] Yui et al.[104] found that dietary supplementation with ARA plus DHA (ARA in excess) significantly improved Aberrant Behavior checklist-community-measured social withdrawal and Social responsiveness Scale-measured communication in ASD patients ($n = 13$). Although small, this double-blind, placebo-controlled randomized trial illustrates the possible benefits of modulating brain fatty acid profiles through diet, the same fatty acid modulations which have been achieved with probiotic microorganisms by others in animal studies. ASD and schizophrenic patients are less likely to have been breastfed than healthy controls, implicating human breast milk, rich in ARA, EPA and DHA in optimal infant diet for brain development.[31,105] Conversely, early weaning has been associated with increased risk of ASD.[102,106] Such observations not only emphasize the importance of early postnatal diet in brain development and ASD risk, but also hint at possible early roles for the gut:brain axis and gut microbiome in this developmental process.

A number of gut microorganisms, most notably certain species of *Lactobacillus* and *Bifidobacterium* possess the enzymes necessary to carry out biohydrogenation of fatty acids increasing their degree of unsaturation.[47] Wall et al.[107] showed that dietary supplementation with ALA altered fatty acid profiles in liver, adipose tissue and brain of mice, and that fatty acid profiles could be further modified upon co-administration of ALA and the probiotic *Bifidobacterium breve* NCIMB 702258. Animals fed the n3-fatty acid plus probiotic had elevated brain DHA and reduced ARA compared to control feeding or ALA alone. The same authors later showed that *B. breve* NCIMB 702258 alone, increases brain DHA and ARA in mice compared to another conjugated linoleic acid (CLA) producing *B. breve* strain and the control group, confirming the ability of probiotics to modulate brain fatty acid profiles and the apparent strain specificity of this activity.[108] However, we know little about the biohydrogenation potential of the gut microbiota as a whole, the contribution of their combined activities to fatty acid profiles in the brain or other bodily tissues for that matter, or how this fatty acid biohydrogenation service changes upon modulation of gut microbiota profiles, e.g., in response to high-fat, Western-style diets or alternatively, upon prebiotic modulation.[48]

Although considered EFAs for phospholipid formation, when LA and ALA are low in the diet, other fats, such as saturated fatty acids can sometimes be used as substitutes, with structural and possibly functional implications for the final phospholipid.[102,109,110] Irregular phospholipid metabolism, fatty acid deficiencies or dyslipidemia have now been implicated in a number of neurological and brain developmental or degenerative disorders including schizophrenia, ADHD, depression, pervasive developmental disorder, developmental coordination disorder, epilepsy, bipolar disorder, Alzheimer's, Parkinson's, Niemann–Pich's and Huntington's diseases and stroke.[102] In autism too, both aberrant cholesterol metabolism and phospholipid metabolism have been suspected to play a pathological role. ASD patients present with higher levels phospholipase A2 in the blood, reduced ARA and ADH in cell membranes and, possibly, a higher ratio of n-6 to n-3 fatty acids. Children with autism ($n = 16$) had modified lipid profiles in red blood cells compared to healthy controls, lower cholesterol and higher proportion of monosialotetrahexosylganglioside, GM1 in cellular membranes.[111] The authors suggested that this might reflect a more generalized defect in cholesterol synthesis, which in the brain, and together with alterations in GM1 expression, might contribute to ASD pathobiology.

Brain-derived neurotrophic factor (BDNF) facilitates long-term potentiating and excitatory synaptic transmission by enhancing the release of neurotransmitters.[112–114] Guirland et al.[115] showed that cholesterol-rich lipid-rafts in synapses are required for BDNF-induced synaptic modulation and nerve cone chemotrophic growth. They later showed that BDNF elicits cholesterol biosynthesis in cultured cortical and hippocampal neurons but not in glial cells by upregulating expression of enzymes involved in *de novo* cholesterol synthesis.[112] Moreover, BDNF increased cholesterol specifically in neuronal lipid rafts implicating BDNF-mediated cholesterol biosynthesis in the development of synaptic vesicles. Interestingly, BDNF is reduced in ASD and moreover, genetic polymorphism in the gene responsible for BDNF production leads to production of isoforms of the protein with different proteolytic activities.[116] Probiotic feeding in animal models has been related to improvements in behavioral measures, which have been linked to changes in BDNF in the hippocampus and the amygdala.[117,118] However, not all probiotic studies have implicated BNDF in observed improved brain function in laboratory animals.[119] Similarly, prebiotic dietary fibers have recently been shown to elevate brain BDNF in mice.[120] Interestingly, BDNF, which is involved in upregulating *de novo* lipogenesis supplying neuronal lipid rafts and synaptic vesicle formation, and is reduced in ASD patients. Similarly, the ASD diet is characterized by reduced dietary fiber and likely, reduced contribution of colonic acetate to circulating acetate concentrations, a key substrate for *de novo* lipogenesis in the brain. These observations raise the intriguing possibility that upregulation of BDNF by probiotics, combined with enhanced acetate production from prebiotics in the colon could impact on *de novo* lipogenesis in the brain, elevated lipid raft formation in neurons and facilitated synaptic vesicle formation leading to improved brain function in children afflicted by ASD and ADHD, and other neurodevelopmental or functional disorders, including depression. However, fundamental studies both in human subjects and relevant laboratory models are necessary to test the validity of these suppositions and to link mechanistically fermentation and gut bacteria with these debilitating conditions.

SHORT-CHAIN FATTY ACIDS (SCFA) AND THE BRAIN

The brain represents about 2% of the body mass but about 20% of its cholesterol. The BBB is impervious to lipoproteins which means that cholesterol for the brain is endogenously formed, with astrocytes and neurons, net cholesterol producers and users, respectively, representing a unique compartmentalization of cholesterol biosynthetic machinery.[121] Neurons have a massive demand for cholesterol to maintain their extensive membranal surface area and to supply presynaptic vesicle formation. They also have slightly different enzymatic pathways for the conversion of squalene into cholesterol. Postnatal cholesterol is mainly supplied by astrocytes and derives preferentially from acetate.[122,123] Astrocytes are cells which embrace neuronal cells and which are responsible for supplying extracellular potassium, glutamate, energy and antioxidants, and mediate activity-dependent regulation of blood flow in the brain and possibly influence synaptic activity.[121] The early postnatal period appears to be particularly important for brain cholesterol formation with increased cholesterol synthesis rate: in adulthood, cholesterol biosynthesis falls to one-tenth of postnatal peak irrespective of blood cholesterol concentrations.[121,124] Interestingly, this period is also when acetate production and its relative molar concentrations to other SCFA in the gut is at its highest, during natural breastfeeding.[125,126]

Astrocytes produce apolipoproteins, including ApoE, responsible for cholesterol transport in the brain.[127] Rodent models deficient for ApoE display various behavioral and neurological symptoms and deficits in sensory systems linked to age-dependent loss of synapses and dendrites[128,129] and modified synaptic membrane cholesterol distribution.[130] Cholesterol turnover is completed by neurons through CYP46 by excretion of 24 S-hydroycholesterol which passes through the BBB into the blood and is excreted in bile.[121] Concentrations of 24 S-hydroycholesterol are also thought to control cholesterol homeostasis in the brain by communication with astrocytes through liver X receptors and can impact on HDL synthesis and remodeling at the BBB.[131,132]

Building blocks for cholesterol include acetate, acetyl-CoA and acetoacetyl CoA. There are many sources of these metabolites in the body including the tricarboxylic acid cycle (TCA) but also microbial-derived acetate, the main end-product of carbohydrate fermentation in the colon which is converted into acetyl-CoA by acetyl coenzyme A synthetase 1 (AceCS1) in the cytosol. Acetate may also be taken up by mitochondria and converted into acetyl-coA by AceCS2 for respiration through the TCA, especially under ketogenic or fasting conditions. In fact both the liver and gut release acetate into circulation when plasma acetate levels are low suggesting a special role for acetate systemically and possibly in the brain.[133,134] Indeed, acetate may be viewed as a special metabolite and the cusp of energy production (through respiration) and storage (through fatty acid biosynthesis) and one which is tightly controlled by glucose availability. Ariyannur et al.[134] have recently shown that AceCS1 is predominantly expressed in brain cell nuclei rather than cytoplasm alone, indicating an additional role in providing acetate for histone acetyltransferase and epigenetic control of gene expression. Interestingly, nuclear AceCS1 expression was highest during the postnatal period compared to adulthood, implicating acetate in postnatal neurological development, cellular differentiation and myelination.[134] The ratio of acetate to other SCFA produced in the colon is at its highest in the neonate during breastfeeding, especially when compared to formula-fed infants.[135–137] AceCS1 was also found in the cytoplasm of basal forebrain neurons, on the surface of cortical and hippocampal neurons, in the cell bodies of some brainstem neurons and ganglion cells, and in axons in many brainstem fiber pathways which may indicate a role in fatty acid/cholesterol synthesis in these cells or other cytoplasmic Acetyl-coA utilizing reactions.[134]

N-acetylaspartate (NAA) is found in high levels (8–11 mM) representing one of the most concentrated sources of acetate in the brain and may serve as an acetate storage molecule serving lipogenesis, protein acetylation or in the nuclei, histone acetylation and regulation of gene expression.[138] It is formed in neuronal cytoplasm and mitochondria by aspartate *N*-acetyltransferase, presumably when acetyl-CoA is in excess.[139] A recent metanalysis showed that children with autism had significantly lower NAA concentrations in all the examined brain regions but cerebellum compared with healthy children, raising the intriguing possibility that altered brain metabolism in ASD may be linked to colonic fermentation. However, the quantitative contribution of colonically derived acetate to either plasma levels or brain supply of acetate remain poorly studied.[140] Interestingly, another SCFA, propionate, appears to have the opposite effect in the brain when introduced in high concentrations. Direct injection with propionic acid in the mM range induces neuroinflammation, developmental delay and cognitive impairments.[141] Although propionate, in common with other SCFA can activate cell signaling processes and gene expression through G-coupled proteins GPR41 and GPR-43, including modulation of immune function, it has also been shown to inhibit the enzymes hydroxyl-methylglutaryl-CoA (HMG-CoA) reductase or fatty acid synthase (FAS), two key enzymes in mammalian cholesterol biosynthesis. Indeed, Frye et al.[142] have shown that propionate intracerebroventricular infusion impacts on fatty acid metabolism resulting in alterations in acylcarnitine and other mitochondrial metabolites in animals models, resulting in similar metabolite profiles observed in ASD children. However, we know little about how diet impacts on SCFA availability or utilization in the brain. In the gut, the relative proportions of saturated fat, protein and fermentable fiber/prebiotics affects the quantity and ratio of SCFA produced by the gut microbiota and also the amount of acetate in particular, entering the peripheral blood, as much butyrate is used by the gut wall and propionate is effectively cleared by the liver. However, blood acetate concentrations do respond to quantities of fermentable fiber in particular, and it is likely that high-fiber diets can contribute significantly to circulating acetate concentrations with knock-on implications for cholesterol biosynthesis and/or respiration in the brain. This is of particular interest in autism where reactive oxygen species produced during respiration and resultant oxidative stress and inflammation have been linked to pathophysiology, and where in some individuals at least, mutations in genes encoding inborn errors in metabolism and mitochondrial function appear to be more common.[143,144] Similarly, alterations in cholesterol metabolism are also recognized in ASD.[111,145]

Epigenetic mechanisms have also been investigated in ASD. Histone deacetylase (HDAC) inhibitors have recently been shown to regulate both GAD2 (encoding GAD65 glutamate decarboxylase, one of two isoforms of the enzyme responsible for GABA production in mammalian cells) and GABA transaminase (responsible for GABA catabolism) in the rat brain.[146–149] Indeed high doses of the HDAC inhibitor, valproic acid, either prenatally or in the postnatal period are used to induce ASD-type disease in laboratory animals.[150] Fukuchi et al.[146] showed that valproate modulates neuronal excitation and inhibition in rat cortical cells via HDAC regulation of gene expression including expression of GABA receptors. Such observations suggest that epigenetic control of neurological function through HDAC inhibitors plays a role in neurological development and might regulate postnatal switching of GABAergic responses from excitatory to inhibitory (a process mediated by GABA concentrations).[59] The subtleties of HDAC inhibitor epigenetic regulation of gene expression in the brain may therefore represent a potentially valuable target for novel ASD therapies considering we still do not have a full understanding of how HDAC inhibitor dose, molecular structure or time of gene expression modulation determines optimal early neurological programming or the initiation pathological processes which present in later life as altered behavior.[151–154] Interestingly, the SCFA n-butyrate, produced by the human gut microbiota mainly upon fiber/prebiotic fermentation in the colon, is a recognized epigenetic effector working through HDAC inhibition as well as other mechanisms.[155,156] Colonic butyrate production is low in infants during suckling and increases alongside the weaning process with the introduction of solid foods and the development of a more diverse, adult-type microbiota. In people on the low fermentable fiber "Western" style diet, butyrate usually does not reach the blood in high levels due to low production and utilization as an energy source by the gut mucosa. However, recent studies with high-fiber feeding suggest that plasma and, therefore, possibly brain levels of butyrate can be modulated through diet, for example through dietary supplementation with cereal fermentable fiber or prebiotics.[157] This raises the intriguing possibility that diets enriched for fermentable fiber/prebiotics or more traditional, whole plant foods, may generate enough butyrate through colonic fermentation to raise blood and possibly brain butyrate concentrations to levels sufficient to mediate subtle epigenetic regulation of key processes in brain development. Indeed this process might be particularly relevant during and post-weaning, a stage where the protective effects of mothers milk are diminishing and when the molar ratio of SCFA in the colon changes. However, introduction of new foods nowadays usually comprises of processed foods with high protein, high fat, high refined carbohydrate and low fermentable fiber/prebiotics and reduced exposure to safe food microorganisms, all of which can impact on the metabolic output of the gut microbiota.

GUT MICROBIOTA AND DIGESTIVE FUNCTION

Intestinal maturation and digestive function changes in the postnatal period with host-derived factors such as hormones, and environmental factors like diet and the successional development of the gut microbiota all playing a role.[158] In piglets, changes occur in the expression and kinetics of brush-border enzymes, and this process appears to be impacted by the gut microbiota.[159,160] Each brush-border enzyme (BBE) changes in its own developmental pattern associated with enterocyte maturation. For example, during suckling, piglets have high lactase phlorizin hydrolase (LPH) but low sucrase activities and this pattern reverses upon weaning.[160,161] Expression of BBE also changes as enterocytes mature and migrate from crypt to villus tip, for example aminopeptidase N activity is low in crypt and high at villus tip. Various bacteria have been shown to effect BBE activity during the early postnatal period including *E. coli*,[162] *Bifidobacterium bifidum*,[163] *Lactobacillus fermentum*[164] and *Bacteroides thetaiotaomicron*[165] but the underlying mechanisms and role of specific gut bacteria has yet to be determined. Further, intestinal bacteria have been shown to deactivate intestinal enzymes, including BBE, *in vivo* and *in vitro*, and these activities appear to be different for different bacteria and different BBE differ in their recalcitrance to microbial deactivation. For examples Willing and Van Kessel[160] found that *E. coli* can deactivate aminopeptidase N, while *L. fermentum* does not appear to deactivate the same enzyme, with both species displaying different effects on BBE gene expression.

The role of the gut microbiota in maturation of the intestinal mucosa, including its digestive functions is only now being fully appreciated. Kozakova et al.[162] showed that germ-free (GF) piglets have a very different pattern of digestive enzyme development compared to conventional animals, with conventionalization speeding up mucosal and digestive enzyme maturation. Mono-association of GF piglets with either commensal/probiotic or pathogenic strains of *E. coli* also affected mucosal maturation, localization and expression, of BBE, specifically speeding up the maturation-associated decreases in lactase, peptidases (aminopeptidase N which is distributed evenly along the length of the gut) and alkaline phosphatase activities and increased expression of sucrase in the jejunum and glucoamylase in the duodenum and jejunum.

More recently, examining the expression of disaccharidases and hexose transporters in ileal biopsies taken from children with gastrointestinal disease with or without concomitant autism, found that in autistic children expression of both disaccharidases (a significant reduction in at least one of the three common gut disaccharidases, SI, MGAM or LCT) and hexose transporters (SGLT1 and GLUT2) was significantly reduced and that these changes correlated with changes within the make-up of the gut microbiota. Changes were associated with higher relative abundance of Firmicutes and reduced Bacteroidetes and elevated β-proteobacteria.[164] Interestingly, animal studies have shown that carbohydrate digestive enzymes and sugar transporter expression is reduced in animals fed high-fat diets compared to high-starch diets,[166] and that the glucose sensor GLP-2 regulates expression of SGLT1 through activation of the GLP-2 receptor present in enteric-neuronal cells rather than endothelial cells[167] indicating that both expression of carbohydrate degrading enzymes and sugar transporters are both induced by diet and regulated via the enteric-nervous system.[168] Additionally, gut bacteria themselves,[165] or through the activity of butyrate produced upon carbohydrate fermentation regulating the transcription factor CDX2, may also positively control expression of intestinal disaccharidases and sugar transporters.[164]. The role of butyrate in maturation of the intestinal mucosa may be important given its known role as preferred energy source for colonocytes, mediator of mucosal turnover and differentiation, and the fact that production rates appear to change upon successional development of the gut microbiota in early life.[57,169]

Uptake of SCFA from the intestine has recently been shown to involve an active component mediated through monocarboxylic acid transporters (MCT) and sodium-dependent MCT (SMCT or slc5a8).[170] Indeed, MCTs are key transporter molecules of SCFA throughout the gut and in other tissues of the body including the brain.[171] In the intestine, MCT1 is the main transporter of n-butyrate. Butyrate induces not only expression of MCT1 but also induces its presentation in "real-time" at the apical surface of colonocytes through a newly recognized rapid nutrient sensing mechanism via GPR109A which allows the mammalian mucosa to effectively compete for luminal butyrate produced by the gut microbiota.[172] High-fiber diets and supplementation with inulin, pectin and β-glucan have been shown to increase MCT1 expression on the intestinal wall thereby increasing SCFA uptake and entry into systemic circulation.[172–175] Pectin has also been shown to upregulate MCT1 in the rat adrenal gland, hinting at homeostatic processes linking cholesterol metabolism, acetate substrate availability and adrenal hormone production.[176] Interestingly, the primary bile salt chenodeoxycholic acid, which is induced in the gut upon ingestion of fat, has recently been shown to inhibit the uptake of SCFA by MCT through competitive exclusion.[177] Raising the possibility that high-fat diets may inhibit SCFA uptake from the colon, important given the role of SCFA in processes linked to metabolic health, including glucose homeostasis, lipid metabolism, adipocyte function (fat storage and adipokine production) and inflammation.[41] Similarly, MCT-1 expression is downregulated by intestinal inflammation and oxidative stress.[177,178] Such mechanisms might explain why excretion of SCFA in feces of obese individuals or laboratory animals appears to be elevated compared to fecal SCFA excretion in lean individuals or animals. Conversely, leptin, the obesity hormone, has been shown to increase uptake of butyrate by intestinal cells via upregulation of MCT1 gene expression and increased translocation of CD147/MCT-1 to the apical surface in Caco2-BBE cell monolayers.[179] However, it is not known how these processes operate under a leptin resistant host environment. Also the gastrointestinal neuropeptide somatostatin (SST) induces butyrate uptake due to increased MCT-1/CD147 and apical presentation in a p38-MAP-kinase-dependent manner.[180] Butyrate has a central role as mucosal fuel and plays important roles in cellular differentiation, thermogenesis, immune tolerance and epigenetics, and may be an important player both within the gut and systemically in the postnatal period given sufficient dietary fiber/prebiotic as substrate for saccharolytic fermentation by the emerging adult type microbiota at this stage in development.[41,157,181] Environmental factors, therefore, such as microbiota composition and diet will have considerable consequences for optimal butyrate production, uptake and utilization for postnatal development of mucosal architecture and possibly elsewhere in the body through HDAC epigenetic activities.

Pierre et al.[182] showed that the same MCT transporters (MCT1, MCT2 and MCT4) appear to be upregulated in certain regions and cell types of the brain in obese animals. Obese mice, whether by virtue of diet (high-fat diet) or genotype (ob/ob and db/db mice) all appear to have increased expression of MCT1-4 specifically in neurons and neuronal somata. Previously, MCT expression has been shown to increase in response to ketogenic diets[183] and during suckling,[184] especially in endothelial cells of the brain capillaries and parenchymal cells (i.e., astrocytes and neurons) with the suggestion that increased density of MCT transporters served to allow increased energy supply to the brain from circulating ketone bodies generated from fatty acids in the mother's breast milk or colonic acetate. The significance of altered MCT expression in the brain with obesity may also indicate aberrant energy metabolism within different brain regions but the fact that upregulation of MCT expression was observed in both diet-induced and genetic animal models of obesity suggests that diet per se might not be directly responsible but rather altered MCT expression may result indirectly as a result of hormonal changes induced by the obese state itself.[182,185,186] Peirre et al.[182] suggested that upregulation of MCT in the obese brain could be a response to increased insulin due to insulin resistance systemically and that these alterations in MCT expression could contribute to the development of neurological disorders associated with obesity, insulin resistance and the metabolic syndrome. Considering the important role SCFA appear to play in brain energy metabolism, cholesterol biosynthesis and epigenetic processes, aberrant expression of the main SCFA transporters within the brain induced by environmental factors such as obesity is likely to impact considerably on brain function, and when these alterations occur early in life could play an important role in altered developmental processes linked to the emergence of neurological disorders as well as the establishment of metabolic "set-points" thought to be involved in individual susceptibility to obesity in later life. This raises the intriguing possibility that deviation from postnatal developmental programming regulated by major dietary shifts from human breast milk to solid food and the prevailing household diet post weaning (between 6 months and 3 years) has considerable consequences for physiological development, the emergence of metabolic pathways including those involving the gut microbiota and key transporters regulating the flow of nutrients from the gut and within other organs including the brain. Aberrant diet, especially the adoption of modern Western-style diets upon weaning may in fact be downregulating the flow of essential nutrients (amino acids, fats and SCFA) from the gut by effectively turning off the tap as represented by nutrient transporters and redirecting the flow of nutrients through altered microbiota metabolic turnpikes.

Dietary Patterns, Gut Microbiota and Brain Development

Comparisons of the modern Western-style diet (WSD) with more ancient and traditional diets are helping to redefine the paradigm of malnutrition. Malnutrition is no longer restricted to the lack of certain essential nutrients, but also encompasses over nutrition and aberrant nutrient ratios and profiles. The WSD is typified by altered fat profiles with elevated saturated fats and synthetic trans-fatty acids compared to essential fatty acids and unsaturated fatty acids (mono- and polyunsaturated) and low availability of saturated fats in more traditional diets. The WSD is also defined by high levels of refined carbohydrates or sugars as opposed to traditional diets where foods enriched with complex, lowly digestible highly fermentable carbohydrates (e.g., fiber and prebiotics) form staples and support a saccharolytic, SCFA-producing gut microbiota. For proteins too, major differences occur between WSD and more traditional diets and foods, with amino acid composition and profile of foods being altered by industrial food-processing technologies. The WSD has largely replaced plant polyphenols as preservatives and to a lesser extent flavorings with chemical substitutes and finally, modern foods have substituted the phylogenetically diverse and numerically dense microbial food passengers found on traditional fermented and raw foods with monocultures of strains selected for technological purposes or more commonly, with sterility. Within the human "super-organism," nowhere are the metabolic consequences of this altered nutritional environment more obvious than in the interactions between the gut microbiome and host energy metabolism and brain function. We have known for a while now that the gut microbiome and its interactions with diet plays a critical role in energy homeostasis and immune tolerance.[187–189] We have linked aberrant gut microbiota profiles with diseases of immune function or autoimmune diseases and metabolic diseases like obesity, diabetes and non-alcoholic fatty liver disease.[188,189] However, only very recently has altered brain development been considered a consequence of our closely co-evolved gut microbiome being out of step with our modern diet. Within a given genetic predisposition to ASD, environment appears to account for an appreciable contribution (55%) of disease risk.[30] Since ASD symptomology presents within the first 3 years of live, infancy may be a critical period for aberrant neurological development underpinning the pathobiology. Infancy is a key transition period in human development, with major nutritional changes from mothers' milk to solid foods and subsequent adoption of adult foods. Diet in infancy drives the successional development of the gut microbiota from suckling regime (type and quality of milk, mothers or formula) through to the foods introduced during the weaning process. This nutritional transition too is critical in shaping the structure of our gut microbiota and it may be that it also shapes the co-metabolic processes which emerge upon gut colonization by an adult type microbiota which are recognized as playing critical roles within the human metabonome – the enterohepatic circulation of bile acids and the co-metabolism of polyphenols, polysaccharides, fats, phosphatidylcholine, amino acids and biogenic amines.[157,181] Indeed, when you consider the epigenetic potential of molecules such as butyrate, acetate and small phenolic microbial flavonoid catabolites,[41,133,190] the intriguing possibility emerges that infancy is a time where host metabolic pathways and indeed postnatal brain development, are shaped by metabolite efflux emerging from the colon and derived from microbial digestion of food.

Dietary Modulation of the Gut Microbiota for Improved Brain Function

Obesity and poor diet have been associated with increased incidence of depression and ASD.[191–193] Similarly, poor maternal diet, especially high fat, and health status, especially obesity/metabolic syndrome, impact on fetal and neonatal brain developmental processes increasing the risk of neurological disorders like anxiety, depression, ADHD and ASD.[194] Recent studies in laboratory animals have shown that high-fat diets lead to an upregulation of mechanisms involved in glutamate transport and a downregulation in enzyme activities involved in glutamate catabolism with concomitant impaired neurotransmission and impaired basal synaptic transmission and hindered NMDA-induced long-term depression.[195] These high-fat-diet-induced changes, linked to the development of leptin resistance in the hippocampus.[195] Diets high in fat and low in fermentable fiber also lead to metabolic endotoxemia, insulin resistance and obesity in man and animals which has been linked to increased BBB permeability and cerebral inflammation.[65,188,196] This diet in humans closely correlates with the Western-style diet, low in fiber and whole plant foods and enriched in fat, animal protein/meat and refined carbohydrates.

Interestingly, we have shown using 16 S rRNA amplicon 454-pyrosequencing that the gut microbiota of children living in rural Africa differs considerably from those living in urban, westernized, Florence.[197] The Europeans had a microbiota enriched in *Firmicutes* and *γ-Proteobacteria* (the enterobacteria in particular), while the African children had higher relative abundance of *Bacteroidetes*, particularly *Provotella*, and *Xylanibacter*, which are important fiber degraders. Moreover, the African children had about three-fold higher concentrations of SCFAs in their feces. Recently, dietary supplementation with prebiotic fermentable fibers, which selectively stimulate beneficial gut bacteria like the bifidobacteria, has been shown to mediate important changes in the brain. Laboratory animals fed frutocoligosaccharides (FOS) or galactooligosaccharides (GOS), showed increased hippocampal BDNF and the *N*-methyl-D-aspartate receptor (NMDARs) subunit NR1 expression, and GOS appeared to mediate this process though induction of the gut incretin PYY.[120] Both GOS and FOS have been shown to upregulate SCFA production, especially acetate and butyrate, by the human gut microbiota[198–200] as well as increasing the relative abundance of intestinal bifidobacteria.[53]

Despite the fact that an estimated 95% of dietary plant polyphenols escape digestion and absorption in the upper gut and reach the gut microbiota in the colon,[190] microbial metabolism of plant polyphenols has scarcely been studied and few studies have examined the impact of polyphenol-derived metabolites on brain development, function or degeneration.[201] However, those studies which do exist, confirm epidemiological data suggesting that polyphenols and their metabolites may be capable of promoting brain health.[201,202] Proposed mechanisms of action include, antioxidant activities, improving vascular function and blood flow to the brain, direct effects on signaling to enhance neuronal communication, ability to buffer against calcium, enhancement of neuroprotective stress shock proteins and reduction of stress signals.[203] Recently also mitochondrial dysfunction has gained the attention in the pathogenesis of ASD, neurodegenerative diseases and general brain aging.[144] Mitochondria are often considered as both initiator and target of oxidative stress in which polyphenolic metabolites may prove protective.[204] However it is not clear which polyphenol metabolites reach the brain in sufficient concentrations and there is little information about the interaction of polyphenols or their metabolites with the BBB.[205] In addition, reliable data on uptake into the brain are limited and there is little agreement currently in the literature.[206] The majority of studies reporting biological activities and targets of polyphenols and their metabolites in the brain have used *in vitro* models of the BBB which are prone to artefacts or often do not reproduce accurately the real complexity of BBB transport kinetics.[204] Nevertheless, *in vitro* the bioactivity of selected polyphenol metabolites have been tested at physiologically relevant doses against advanced glycation end-product formation and their ability to counteract mild oxidative stress in human neuronal cells.[207] For example, urolithin A and B, from the microbial catabolism of ellagitannins, significantly reduced protein glycation at 1 μmol/l. Using PC12 cells, protocatechuic acid, a microbial catabolite of the anthocyanins, has been shown to reduce mitochondrial dysfunction.[208] Ferruzzi et al.[209] reported the presence of gallic acid in the brain of rats following the gastric administration of proanthocyanidins in grape polyphenol extract. Gallic acid and some further methylated metabolites were observed to be present in brain, albeit in trace amounts. A few studies have examined the impact of polyphenolic extracts from medicinal plants in both animal models of ASD. Banji et al.[210] found that a high dose of green tea extract (300 mg/kg) was able to reduce oxidative damage and improve brain histopathological scores and animal behavior following postnatal challenge with valproate (400 mg/kg). Similar effects were observed with *Bacopa monniera*, a plant used in ayurvedic medicine.[211] Kim et al.[212] showed that Korean red ginseng dose-dependently improved social interactions in valproate-treated animals. However, it remains to be determined whether the antioxidant activities of these high-dose plant extracts are relevant only in alleviating the oxidative damage caused by valproate or are more broadly relevant to neuropathological conditions such as ASD where oxidative damage may be only one contributing factor. Targeting neurosignaling liquorice (*Glycyrrhiza glabra*) extract, at antiamnestic doses (150 mg/kg for 7 days), has been shown to disrupt startle response prepulse inhibition via augmentation of monoaminergic transmission in the cortex, hippocampus and striatum of the acoustic startle response mouse model. Hasanzadeh et al.[213] in a double-blind, parallel study tested the effect of *Ginkgo biloba* extract with risperidone (an atypical antipsychotic drug sometimes used to treat irritability in ASD patients) compared to risperidone alone, showed no significant impact on Aberrant Behavior Checklist-Community (ABC-C) rating in 47 ASD patients. Conversely, Taliou et al.[214] have recently shown that a mixture of two flavonoids, luteolin and quercetin, carried in olive kernel oil improved adaptive functioning as measured using the Vineland Adaptive Behavior Scale age-equivalent score and behavior as measured using the Aberrant Behavior checklist subscale scores in 40 children with autism. Although promising in terms of proving an improvement in

ASD patient symptoms upon polyphenol consumption, this study did not have a control or placebo treatment, a significant limitation and necessitates confirmation in further human studies.

PROBIOTICS, GUT MICROBIOTA SUCCESSIONAL DEVELOPMENT AND BRAIN FUNCTION

The neuroendocrine system, specifically the hypothalamic–pituitary–adrenal (HPA) axis represents the key communication highway between the gut environment and the central nervous system (CNS) or the gut:brian axis.[57] Recent studies have proven that the gut microbiota and its interaction with dietary components and digestion is a key player in development and regulation of the gut–brain axis. Sudo et al.[215] showed that the stress response of GF mice differed to mice carrying specific pathogen free (SPF) microbiota, in that under stress, they had exaggerated release of corticosterone and adrenocorticotrophin hormone. GF colonization with the probiotic *Bifidobacterium infantis* in the postnatal period completely reversed this situation in a time-dependent manner but colonization with *E. coli* exaggerated the stress response of the ex-GF mice. The authors concluded that exposure to microorganisms, and the "right" microorganisms, early in life was required for the HPA system to become fully susceptible to inhibitory neuronal activity. Neufeld et al.[216] reported that GF Swiss Webster female mice, compared to conventional counterparts, had altered anxiolytic behavior to elevated plus maze stress together with altered brain mRNA expression [central amygdala decreased *N*-methyl-D-aspartate receptor subunit NR2B expression, increased BDNF expression and decreased serotonin receptor 1A (5HT1A) expression in the dentate granule layer of the hippocampus]. Diaz-Heijtz et al.[217] reported that GF mice showed increased motor activity and reduced anxiety compared to SPF mice together with altered expression of secondary messenger pathway and synaptic long-term potentiation genes in brain regions involved in motor control and anxiety-like behavior. These effects could be reversed by early colonization with SPF microbiota in the neonatal period. Intervention with various probiotic microorganisms, both lactobacilli and bifidobacteria, has been shown in animal models to alleviate anxiety-like behavior to standard animal stress challenges. Desbonnet et al.[218] showed that rats fed *Bifidobacterium infantis* chronically for 14 days had a significant reduction in inflammatory cytokines systemically (INF-γ, TNF-α, and IL-6 following mitogen activation) and elevated concentrations of tryptophan, the precursor of serotonin and kynurenic acid and 5-hydroxyindole acetic acid (the main serotonin metabolite) in the frontal cortex and 3,4-dihydroxyphenylacetic acid (the main dopamine metabolite) in the amygdaloid cortex compared to control animals. The same authors later showed that this probiotic strain could counteract the stress effects of forced swim stress test (reduced swim behavior, increased immobility, decreased noradrenalin in the brain, elevated systemic IL-6 and amygdala corticotrophin-releasing factor mRNA) in mice who had been subjected to maternal separation, a recognized model of depression.[219] In rats fed a probiotic mix (*L. helveticus* R0052 and *B. longue* R0175) daily for two weeks, reduced anxiety-like behavior (defensive burying test) was observed and the same probiotic mix in humans alleviated psychological distress measured by the Hopkins Symptom Checklist (HSCL-90).[220] Bravo et al.[221] showed that the probiotic *L. rhamnosus* JB-1 reduced stress induced corticosterone and anxiety- and depression-related behavior, and concomitantly induced changes in GABA receptor expression in different regions of the brain. In the probiotic fed animals, $GABA_{B1b}$ was upregulated in cingulated and relimbic cortical regions and downregulated in the hippocampus, amygdala and locus coeruleus compared to control animals. Similarly, $GABA_{A\alpha2}$ expression in the prefrontal cortex and amygdala were reduced and increased in the hippocampus. Importantly, both the behavioral changes and changes in GABA-receptor expression induced by the probiotic did not occur in vagotomized mice indicating the central role of the vagus nerve in the gut:brain communication highway. Conversely, infection with pathogenic bacteria, for example *Citrobacter rodentium*, a γ-Protobacterium used as a model of human enteropathogenic *E. coli* in the mouse model, has been shown recently to induce memory loss in stressed mice.[222] The same group showed earlier, that *C. rodentium* infection induced stress in mice, as measured by increased c-Fos protein-positive neurons in vagal sensory ganglia and increased risk behaviors,[223] and conversely, stress-induced changes in the composition of the mouse gut microbiota reduces colonization resistance to *C. rodentium*.[224]

Recent studies in an animal model of ASD have highlighted the potential of probiotics or commensal gut bacteria to alleviate not only gastrointrestinal symptoms associated with ASD, but also overt ASD pathological traits. Hsiao et al.[225] used a mouse model of inflammation, the maternal immune activation (MIA) model which leads to pups which exhibit the core communicative, social and stereotyped impairments of ASD. The offspring of MIA dams also exhibit localized deficiency in GABAergic cerebellar Purkinje cells, a pathological feature of

ASD, a "leaky gut" with associated aberrant microbiota profiles (especially altered species composition within the clostridia and Bacteroides compared to control animals), and modified intestinal cytokine profiles. Treating MIA mice at weaning with the human commensal *Bacteroides fragilis* NCTC 9343, restores intestinal barrier function through normalization of colonic tight junction proteins, and resolution of IL-6-mediated inflammation. Supplementation with *B. fragilis* reduced anxiety-like behavior in the mouse open-field test, improved sensorimotor gating in the prepulse inhibition test, descreased levels of stereotyped marble burying and improved mouse communication as measured by enrichment of vocal calls of longer syllable, changes in behavior tests considered to indicate reduced ASD like behavior in rodents. However, *B. fragilis* did not improve tests measuring social behavior. The authors also reported that although similar results were obtained with *B. thetaiotaomicron, Enterococcus faecalis,* another mammalian commensal microorganisms had no effect on anxiety-like and repetitive behavior in MIA offspring, indicating that beneficial modulation of the gut:brain axis in ASD models is not a universal activity associated with bacterial challenge but rather confined to certain bacteria or groups of bacteria. Surprisingly, in this study, *B. fragilis* supplementation did not appear to mediate an impact on systemic inflammatory markers, a suspected pathophysiological processes linking "leaky gut" and inflammatory processes at extra-intestinal body sites. However, both MIA and *B. fragilis* induced significant changes within the mouse serum metabolome, including, amongst other metabolic changes, restoration of MIA induced elevation of 4-ethyphenylsulfate (4EPS) and indolpyruvate by *B. fragilis* intervention. These metabolites are linked metabolically with other metabolites and pathways already implicated in ASD, specifically p-cresol and tryptophan metabolism, respectively, which is biochemically related to p-cresol, has previously been shown to impact on mouse communication.[226] Morever, systemic challenge of naive mice with 4EPS between 3 weeks and 6 weeks of age, the period when MIA "leaky gut" develops, induced anxiety-like behaviors, linking directly, this microbial metabolite with common ASD-type symptoms at least in mice.

CONCLUSIONS

In this chapter we have explored the links between diet and the gut:brain axis, presenting evidence that diet:microbe interactions within the gut can impact on systemic processes linked both to brain development and function. Disparate evidence from diverse models systems and small human observational, and rare intervention studies, show that many nutrients, their availability and metabolism/catabolism, linked to gut microbiota activities appear altered in ASD compared to healthy controls, highlighting the putative role of the gut microbiome in disease onset or maintenance. Early childhood appears to be a critical period, both in terms of successional development within the brain and the gut, and a process which may be dramatically influenced by the host's diet. Thus it may therefore be that the successional development of the gut microbiota, relative exposure to probiotic or pathogenic microorganisms, the emergence of host:microbe co-metabolic processes and the epigenetic development of gut digestive and/or absorptive functions during the transition, suckling/ milk formula → weaning → adult "house-hold" diet, may all play critical roles in brain development at a crucial juncture in neurological development that can have far reaching consequences for behavior and disease risk throughout life, including the onset or severity of ASD. However, there is a clear lack of human data describing the contribution of gut microbiota or their metabolic activities in ASD and moreover, controlled human feeding studies showing the efficacy of microbiota modulation, with probiotics and/or prebiotics in ASD are lacking. However, it is probably fair to say that dietary modulation of the gut microbiota and subsequent modulation in hormonal, immunological or metabolic communication along the information highway comprising the gut: brain axis offers considerable potential for both preventing and treating neurological disorders like ASD.

References

1. American Psychiatric Association. *Diagnostic and Statistical Manual of Mental Disorders*. 4th ed. Washington DC: American Psychiatric Association; 1994.
2. Hughes V. Epidemiology: complex disorder. *Nature*. 2012;491:S2–S3.
3. Baron-Cohen S, Scott FJ, Allison C, et al. Prevalence of autism-spectrum conditions: UK school-based population study. *Br J Psychiatry*. 2009;194:500–509.
4. Brugha TS, McManus S, Bankart J, et al. Epidemiology of autism spectrum disorders in adults in the community in England. *Arch Gen Psychiatry*. 2011;68:459–466.
5. Abrahams BS, Geschwind DH. Advances in autism genetics: On the threshold of a new neurobiology. *Nat Rev Genet*. 2008;9:341–355.

6. Ozonoff S, Williams BJ, Landa R. Parental report of the early development of children with regressive autism: The delays-plus-regression phenotype. *Autism*. 2005;9:461–486.
7. Ozonoff S, Young GS, Steinfeld MB, et al. How early do parent concerns predict later autism diagnosis? *J Dev Behav Pediatr*. 2009;30:367–375.
8. Esposito G, Venuti P. Symmetry in infancy: Analysis of motor development in autism spectrum disorders. *Symmetry*. 2009;1:215–225.
9. Esposito G, Venuti P, Apicella F, Muratori F. Analysis of unsupported gait in toddlers with autism. *Brain and Development*. 2011;33:367–373.
10. Happé F, Ronald A, Plomin R. Time to give up on a single explanation for autism. *Nat Neurosci*. 2006;9:1218–1220.
11. Charman T, Jones CRG, Pickles A, Simonoff E, Baird G, Happé F. Defining the cognitive phenotype of autism. *Brain Res*. 2011;1380:10–21.
12. Krakowiak P, Walker CK, Bremer AA, et al. Maternal metabolic conditions and risk for autism and other neurodevelopmental disorders. *Pediatrics*. 2012;129:e1121–e1128.
13. Hollway JA, Aman MG. Pharmacological treatment of sleep disturbance in developmental disabilities: a review of the literature. *Res Dev Disabil*. 2011;32:939–962.
14. Molloy CA, Manning-Courtney P. Prevalence of chronic gastrointestinal symptoms in children with autism and autistic spectrum disorders. *Autism*. 2003;7:165–171.
15. Keen DV. Childhood autism, feeding problems and failure to thrive in early infancy: Seven case studies. *Eur Child Adolesc Psychiatr*. 2008;17:209–216.
16. Bandini LG, Anderson SE, Curtin C, et al. Food selectivity in children with autism spectrum disorders and typically developing children. *J Pediatr*. 2010;157:259–264.
17. Cermak SA, Curtin C, Bandini LG. Food selectivity and sensory sensitivity in children with autism spectrum disorders. *J Am Diet Assoc*. 2010;110:238–246.
18. Martins Y, Young RL, Robson DC. Feeding and eating behaviors in children with autism and typically developing children. *J Autism Dev Disord*. 2008;38:1878–1887.
19. Cornish E. A balanced approach towards healthy eating in autism. *J Hum Nutr Diet*. 1998;11:501–509.
20. Whiteley P, Rodgers J, Savery D, Shattock P. A gluten-free diet as an intervention for autism and associated spectrum disorders: Preliminary findings. *Autism*. 1999;3:45–65.
21. Fitzgerald K, Hyman M, Swift K. Autism spectrum disorders. *Glob Adv Health Med*. 2012;1:62–74.
22. Courchesne E, Campbell K, Solso S. Brain growth across the life span in autism: age-specific changes in anatomical pathology. *Brain Res*. 2011;1380:138–145.
23. Rutter M. Aetiology of autism: findings and questions. *J Intell Disabil Res*. 2005;49:231–238.
24. Francis A, Msall M, Obringer E, Kelley K. Children with autism spectrum disorder and epilepsy. *Pediatr Ann*. 2013;42:255–260.
25. Moss J, Oliver C, Nelson L, Richards C, Hall S. Delineating the profile of autism spectrum disorder characteristics in cornelia de lange and fragile x syndromes. *Am J Intellect Dev Disabil*. 2013;118:55–73.
26. Michaelson JJ, Shi Y, Gujral M, et al. Whole-genome sequencing in autism identifies hot spots for de novo germline mutation. *Cell*. 2012;151:1431–1442.
27. Vardarajan BN, Eran A, Jung JY, Kunkel LM, Wall DP. Haplotype structure enables prioritization of common markers and candidate genes in autism spectrum disorder. *Transl Psychiatry*. 2013;3:e262.
28. Veenstra-Vanderweele J, Blakely RD. Networking in autism: Leveraging genetic, biomarker and model system findings in the search for new treatments. *Neuropsychopharmacol*. 2012;37:196–212.
29. Kleijer KTE, Schmeisser MJ, Krueger DD, et al. Neurobiology of autism gene products: towards pathogenesis and drug targets. *Psychopharmacology (Berl)*. 2014;231(6):1037–1062.
30. Hallmayer J, Cleveland S, Torres A, et al. Genetic heritability and shared environmental factors among twin pairs with autism. *Arch Gen Psychiatry*. 2011;68:1095–1102.
31. Das UN. Autism as a disorder of deficiency of brain-derived neurotrophic factor and altered metabolism of polyunsaturated fatty acids. *Nutrition*. 2013;29:1275–1285.
32. Emond P, Mavel S, Aïdoud N, et al. GC-MS-based urine metabolic profiling of autism spectrum disorders. *Anal Bioanal Chem*. 2013;405:5291–5300.
33. Frye RE, Melnyk S, Fuchs G, et al. Effectiveness of methylcobalamin and folinic acid treatment on adaptive behavior in children with autistic disorder is related to glutathione redox status. *Autism Res Treat*. 2013;2013:Article ID 609705.
34. Yap IKS, Angley M, Veselkov KA, Holmes E, Lindon JC, Nicholson JK. Urinary metabolic phenotyping differentiates children with autism from their unaffected siblings and age-matched controls. *J Proteome Res*. 2010;9:2996–3004.
35. Wang Y, Kasper LH. The role of microbiome in central nervous system disorders. *Brain Behav Immun*. 2014;38:1–12.
36. Hsiao E, McBride S, Hsien S, et al. Microbiota modulate behavioral and physiological abnormalities associated with neurodevelopmental disorders. *Cell*. 2013;155:1451–1463.
37. Louis P. Does the human gut microbiota contribute to the etiology of autism spectrum disorders? *Dig Dis Sci*. 2012;57:1987–1989.
38. Forsythe P, Kunze WA, Bienenstock J. On communication between gut microbes and the brain. *Curr Opin Gastroenterol*. 2012;28:557–562.
39. Stilling RM, Dinan TG, Cryan JF. Microbial genes, brain & behaviour – epigenetic regulation of the gut–brain axis. *Genes Brain Behav*. 2014;13:69–86.
40. Samala R, Klein J, Borges K. The ketogenic diet changes metabolite levels in hippocampal extracellular fluid. *Neurochem Int*. 2011;58:5–8.
41. Conterno L, Fava F, Viola R, Tuohy KM. Obesity and the gut microbiota: Does up-regulating colonic fermentation protect against obesity and metabolic disease? *Genes Nutr*. 2011;6:241–260.
42. Joyce SA, Gahan CGM. The gut microbiota and the metabolic health of the host. *Curr Opin Gastroenterol*. 2014;30:120–127.
43. Gibson GR, Roberfroid MB. Dietary modulation of the human colonic microbiota: Introducing the concept of prebiotics. *J Nutr*. 1995;125:1401–1412.
44. Tuohy KM, Probert HM, Smejkal CW, Gibson GR. Using probiotics and prebiotics to improve gut health. *Drug Discov Today*. 2003;8:692–700.

45. Dongarrà ML, Rizzello V, Muccio L, et al. Mucosal immunology and probiotics. *Curr Allergy Asthma Rep*. 2013;13:19–26.
46. Kleessen B, Blaut M. Modulation of gut mucosal biofilms. *Br J Nutr*. 2005;93:S35–S40.
47. O'Shea EF, Cotter PD, Stanton C, Ross RP, Hill C. Production of bioactive substances by intestinal bacteria as a basis for explaining probiotic mechanisms: Bacteriocins and conjugated linoleic acid. *Int J Food Microbiol*. 2012;152:189–205.
48. Druart C, Dewulf EM, Cani PD, Neyrinck AM, Thissen JP, Delzenne NM. Gut microbial metabolites of polyunsaturated fatty acids correlate with specific fecal bacteria and serum markers of metabolic syndrome in obese women. *Lipids*. 2014;49(4):394–402.
49. D'Aimmo MR, Mattarelli P, Biavati B, Carlsson NG, Andlid T. The potential of bifidobacteria as a source of natural folate. *J Appl Microbiol*. 2012;112:975–984.
50. Barrett E, Ross RP, O'Toole PW, Fitzgerald GF, Stanton C. γ-Aminobutyric acid production by culturable bacteria from the human intestine. *J Appl Microbiol*. 2012;113:411–417.
51. Felis GE, Dellaglio F. Taxonomy of lactobacilli and bifidobacteria. *Curr Issues Intest Microbiol*. 2007;8:44–61.
52. Spencer JPE. The impact of flavonoids on memory: physiological and molecular considerations. *Chem Soc Rev*. 2009;38:1152–1161.
53. Tuohy KM, Conterno L, Gasperotti M, Viola R. Up-regulating the human intestinal microbiome using whole plant foods, polyphenols, and/or fiber. *J Agric Food Chem*. 2012;60:8776–8782.
54. De Vadder F, Kovatcheva-Datchary P, Goncalves D, et al. Microbiota-generated metabolites promote metabolic benefits via gut–brain neural circuits. *Cell*. 2014;156:84–96.
55. Ray K. Gut microbiota: Microbial metabolites feed into the gut–brain–gut circuit during host metabolism. *Nat Rev Gastroenterol Hepatol*. 2014;11:76.
56. Mendoza C, Matheus N, Iceta R, Mesonero JE, Alcalde AI. Lipopolysaccharide induces alteration of serotonin transporter in human intestinal epithelial cells. *Innate Immun*. 2009;15:243–250.
57. Rosenblat JD, Cha DS, Mansur RB, McIntyre RS. Inflamed moods: A review of the interactions between inflammation and mood disorders. *Prog Neuropsychopharmacol Biol Psychiatry*. 2014;53C:23–24.
58. Gaetz W, Bloy L, Wang DJ, et al. GABA estimation in the brains of children on the autism spectrum: Measurement precision and regional cortical variation. *NeuroImage*. 2014;86(1–9).
59. Kilb W. Development of the GABAergic system from birth to adolescence. *Neuroscientist*. 2012;18:613–630.
60. Kinast K, Peeters D, Kolk SM, Schubert D, Homberg JR. Genetic and pharmacological manipulations of the serotonergic system in early life: neurodevelopmental underpinnings of autism-related behaviour. *Front Cell Neurosci*. 2013;7:72:article.
61. Rolf LH, Haarmann FY, Grotemeyer KH, Kehrer H. Serotonin and amino acid content in platelets of autistic children. *Acta Psychiatr Scand*. 1993;87:312–316.
62. Fisher MA, Hagen DQ, Colvin RB. Aminooxyacetic acid: Interactions with gamma-aminobutyric acid and the blood–brain barrier. *Science*. 1966;153:1668–1670.
63. Deelchand DK, Shestov AA, Koski DM, Uğurbil K, Henry PG. Acetate transport and utilization in the rat brain. *J Neurochem*. 2009;109:46–54.
64. Onore CE, Nordahl CW, Young GS, Van De Water JA, Rogers SJ, Ashwood P. Levels of soluble platelet endothelial cell adhesion molecule-1 and P-selectin are decreased in children with autism spectrum disorder. *Biol Psychiatry*. 2012;72:1020–1025.
65. Cani PD, Amar J, Iglesias MA, et al. Metabolic endotoxemia initiates obesity and insulin resistance. *Diabetes*. 2007;56:1761–1772.
66. Li J, Ye L, Wang X, et al. (−)-Epigallocatechin gallate inhibits endotoxin-induced expression of inflammatory cytokines in human cerebral microvascular endothelial cells. *J Neuroinflammation*. 2012;9:161.
67. Stolp HB, Dziegielewska KM, Ek CJ, Potter AM, Saunders NR. Long-term changes in blood–brain barrier permeability and white matter following prolonged systemic inflammation in early development in the rat. *Eur J Neurosci*. 2005;22:2805–2816.
68. Stolp HB, Johansson PA, Habgood MD, Dziegielewska KM, Saunders NR, Ek CJ. Effects of neonatal systemic inflammation on blood–brain barrier permeability and behaviour in juvenile and adult rats. *Cardiovasc Psychiatry Neurol*. 2011;2011:469046.
69. Lin YL, Wang S. Prenatal lipopolysaccharide exposure increases depression-like behaviors and reduces hippocampal neurogenesis in adult rats. *Behav Brain Res*. 2014;259:24–34.
70. Angelidou A, Asadi S, Alysandratos KD, Karagkouni A, Kourembanas S, Theoharides TC. Perinatal stress, brain inflammation and risk of autism—Review and proposal. *BMC Pediatr*. 2012;12:89.
71. Frye RE, Sequeira JM, Quadros EV, James SJ, Rossignol DA. Cerebral folate receptor autoantibodies in autism spectrum disorder. *Mol Psychiatry*. 2013;18:369–381.
72. Mostafa GA, Al-Ayadhi LY. A lack of association between hyperserotonemia and the increased frequency of serum anti-myelin basic protein auto-antibodies in autistic children. *J Neuroinflammation*. 2011;8:71.
73. Wills S, Rossi CC, Bennett J, et al. Further characterization of autoantibodies to GABAergic neurons in the central nervous system produced by a subset of children with autism. *Mol Autism*. 2011;2:5.
74. Fetissov SO, Hamze Sinno M, Coëffier M, et al. Autoantibodies against appetite-regulating peptide hormones and neuropeptides: Putative modulation by gut microflora. *Nutrition*. 2008;24:348–359.
75. Sysi-Aho M, Ermolov A, Gopalacharyulu PV, et al. Metabolic regulation in progression to autoimmune diabetes. *PLoS Comput Biol*. 2011;7: e1002257.
76. Nanau RM, Neuman MG. Metabolome and inflammasome in inflammatory bowel disease. *Transl Res*. 2012;160:1–28.
77. Perry TL, Hansen S, Christie RG. Amino compounds and organic acids in CSF, plasma, and urine of autistic children. *Biol Psychiatry*. 1978;13:575–586.
78. Aldred S, Moore KM, Fitzgerald M, Waring RH. Plasma amino acid levels in children with autism and their families. *J Autism Dev Disord*. 2003;33:93–97.
79. Shimmura C, Suda S, Tsuchiya KJ, et al. Alteration of plasma glutamate and glutamine levels in children with high-functioning autism. *PLoS ONE*. 2011;6:e25340.
80. Fernell E, Karagiannakis A, Edman G, Bjerkenstedt L, Wiesel FA, Venizelos N. Aberrant amino acid transport in fibroblasts from children with autism. *Neurosci Lett*. 2007;418:82–86.

81. Flyckt L, Venizelos N, Edman G, Bjerkenstedt L, Hagenfeldt L, Wiesel FA. Aberrant tyrosine transport across the cell membrane in patients with schizophrenia. *Arch Gen Psychiatry*. 2001;58:953–958.
82. Persson ML, Johansson J, Vumma R, et al. Aberrant amino acid transport in fibroblasts from patients with bipolar disorder. *Neurosci Lett*. 2009;457:49–52.
83. Johansson J, Landgren M, Fernell E, et al. Altered tryptophan and alanine transport in fibroblasts from boys with attention-deficit/hyperactivity disorder (ADHD): An in vitro study. *Behav Brain Funct*. 2011;7:40.
84. Ming X, Stein TP, Barnes V, Rhodes N, Guo L. Metabolic perturbance in autism spectrum disorders: A metabolomics study. *J Proteome Res*. 2012;11:5856–5862.
85. Suárez I, Bodega G, Fernández B. Glutamine synthetase in brain: effect of ammonia. *Neurochem Int*. 2002;41:123–142.
86. Abu Shmais GA, Al-Ayadhi LY, Al-Dbass AM, El-Ansary AK. Mechanism of nitrogen metabolism-related parameters and enzyme activities in the pathophysiology of autism. *J Neurodevl Disord*. 2012;4:1–11.
87. Mortensen PB, Holtug K, Bonnen H, Clausen MR. The degradation of amino acids, proteins, and blood to short-chain fatty acids in colon is prevented by lactulose. *Gastroenterology*. 1990;98:353–360.
88. Smith EA, Macfarlane GT. Dissimilatory amino acid metabolism in human colonic bacteria. *Anaerobe*. 1997;3:327–337.
89. Sharma BC, Sharma P, Lunia MK, Srivastava S, Goyal R, Sarin SK. A randomized, double-blind, controlled trial comparing rifaximin plus lactulose with lactulose alone in treatment of overt hepatic encephalopathy. *Am J Gastroenterol*. 2013;108:1458–1463.
90. Chez MG, Buchanan CP, Aimonovitch MC, et al. Double-blind, placebo-controlled study of L-carnosine supplementation in children with autistic spectrum disorders. *J Child Neurol*. 2002;17:833–837.
91. Cummings JH, MacFarlane GT. Role of intestinal bacteria in nutrient metabolism. *Clin Nutr*. 1997;16:3–11.
92. Bergen WG, Wu G. Intestinal nitrogen recycling and utilization in health and disease. *J Nutr*. 2009;139:821–825.
93. Dai ZL, Li XL, Xi PB, Zhang J, Wu G, Zhu WY. Regulatory role for l-arginine in the utilization of amino acids by pig small-intestinal bacteria. *Amino Acids*. 2012;43:233–244.
94. Nielsen CU, Carstensen M, Brodin B. Carrier-mediated γ-aminobutyric acid transport across the basolateral membrane of human intestinal Caco-2 cell monolayers. *Eur J Pharm Biopharm*. 2012;81:458–462.
95. Burrus CJ. A biochemical rationale for the interaction between gastrointestinal yeast and autism. *Med Hypotheses*. 2012;79:784–785.
96. Shaw W. Increased urinary excretion of a 3-(3-hydroxyphenyl)- 3-hydroxypropionic acid (HPHPA),an abnormal phenylalanine metabolite of *Clostridia* spp. in the gastrointestinal tract, in urine samples from patients with autism and schizophrenia. *Nutr Neurosci*. 2010;13:135–143.
97. Sandler RH, Finegold SM, Bolte ER, et al. Short-term benefit from oral vancomycin treatment of regressive-onset autism. *J Child Neurol*. 2000;15:429–435.
98. Li P, Kim SW, Li X, Datta S, Pond WG, Wu G. Dietary supplementation with cholesterol and docosahexaenoic acid affects concentrations of amino acids in tissues of young pigs. *Amino Acids*. 2009;37:709–716.
99. Hermanussen M, Gonder U, Jakobs C, Stegemann D, Hoffmann G. Patterns of free amino acids in German convenience food products: Marked mismatch between label information and composition. *Eur J Clin Nutr*. 2010;64:88–98.
100. Hermanussen M, Gonder U, Stegemann D, et al. How much chicken is food? Questioning the definition of food by analyzing amino acid composition of modern convenience products. *Anthropol Anz*. 2012;69:57–69.
101. Calder PC, Dangour AD, Diekman C, et al. Essential fats for future health. Proceedings of the 9th unilever nutrition symposium, 26–27 May 2010. *Eur J Clin Nutr*. 2010;64:S1–S13.
102. Brown CM, Austin DW. Autistic disorder and phospholipids: a review. *Prostaglandins Leukot Essent Fatty Acids*. 2011;84:25–30.
103. Lauritzen L, Hansen HS, Jorgensen MH, Michaelsen KF. The essentiality of long chain n-3 fatty acids in relation to development and function of the brain and retina. *Prog Lipid Res*. 2001;40:1–94.
104. Yui K, Koshiba M, Nakamura S, Kobayashi Y. Effects of large doses of arachidonic acid added to docosahexaenoic acid on social impairment in individuals with autism spectrum disorders: A double-blind, placebo-controlled, randomized trial. *J Clin Psychopharmacol*. 2012;32:200–206.
105. Selim ME, Al-Ayadhi LY. Possible ameliorative effect of breastfeeding and the uptake of human colostrum against coeliac disease in autistic rats. *World J Gastroenterol*. 2013;19:3281–3290.
106. Tanoue Y, Oda S. Weaning time of children with infantile autism. *J Autism Dev Disord*. 1989;19:425–434.
107. Wall R, Ross RP, Shanahan F, et al. Impact of administered bifidobacterium on murine host fatty acid composition. *Lipids*. 2010;45:429–436.
108. Wall R, Marques TM, O'Sullivan O, et al. Contrasting effects of Bifidobacterium breve NCIMB 702258 and Bifidobacterium breve DPC 6330 on the composition of murine brain fatty acids and gut microbiota. *Am J Clin Nutr*. 2012;95:1278–1287.
109. Horrobin DF. The membrane phospholipid hypothesis as a biochemical basis for the neurodevelopmental concept of schizophrenia. *Schizophr Res*. 1998;30:193–208.
110. Simopoulos AP. Omega-3 fatty acids in health and disease and in growth and development. *Am J Clin Nutr*. 1991;54:438–463.
111. Schengrund CL, Ali-Rahmani F, Ramer JC. Cholesterol, GM1, and autism. *Neurochem Res*. 2012;37:1201–1207.
112. Suzuki S, Kiyosue K, Hazama S, et al. Brain-derived neurotrophic factor regulates cholesterol metabolism for synapse development. *J Neurosci*. 2007;27:6417–6427.
113. Figurov A, Pozzo-Miller LD, Olafsson P, Wang T, Lu B. Regulation of synaptic responses to high-frequency stimulation and LTP by neurotrophins in the hippocampus. *Nature*. 1996;381:706–709.
114. Gottmann K, Mittmann T, Lessmann V. BDNF signaling in the formation, maturation and plasticity of glutamatergic and GABAergic synapses. *Experimental Brain Res*. 2009;199:203–234.
115. Guirland C, Suzuki S, Kojima M, Lu B, Zheng JQ. Lipid rafts mediate chemotropic guidance of nerve growth cones. *Neuron*. 2004;42:51–62.
116. Garcia KLP, Yu G, Nicolini C, et al. Altered balance of proteolytic isoforms of pro-brain-derived neurotrophic factor in autism. *J Neuropathol Exp Neurol*. 2012;71:289–297.

117. Bercik P, Verdu EF, Foster JA, et al. Chronic gastrointestinal inflammation induces anxiety-like behavior and alters central nervous system biochemistry in mice. *Gastroenterology*. 2010;139:2102–2112.e2101.
118. O'Sullivan E, Barrett E, Grenham S, et al. BDNF expression in the hippocampus of maternally separated rats: Does Bifidobacterium breve 6330 alter BDNF levels? *Benef Microb*. 2011;2:199–207.
119. Bercik P, Park AJ, Sinclair D, et al. The anxiolytic effect of *Bifidobacterium longum* NCC3001 involves vagal pathways for gut–brain communication. *Neurogastroenterol Motil*. 2011;23:1132–1139.
120. Savignac HM, Corona G, Mills H, et al. Prebiotic feeding elevates central brain derived neurotrophic factor, *N*-methyl-D-aspartate receptor subunits and d-serine. *Neurochem Int*. 2013;63:756–764.
121. Pfrieger FW, Ungerer N. Cholesterol metabolism in neurons and astrocytes. *Prog Lipid Res*. 2011;50:357–371.
122. Lopes-Cardozo M, Larsson OM, Schousboe A. Acetoacetate and glucose as lipid precursors and energy substrates in primary cultures of astrocytes and neurons from mouse cerebral cortex. *J Neurochem*. 1986;46:773–778.
123. Waniewski RA, Martin DL. Preferential utilization of acetate by astrocytes is attributable to transport. *J Neurosci*. 1998;18:5225–5233.
124. Quan G, Xie C, Dietschy JM, Turley SD. Ontogenesis and regulation of cholesterol metabolism in the central nervous system of the mouse. *Dev Brain Res*. 2003;146:87–98.
125. Bakker-Zierikzee AM, Alles MS, Knol J, Kok FJ, Tolboom JJM, Bindels JG. Effects of infant formula containing a mixture of galacto- and fructo-oligosaccharides or viable *Bifidobacterium animalis* on the intestinal microflora during the first 4 months of life. *Br J Nutr*. 2005;94:783–790.
126. Mischke M, Plösch T. More than just a gut instinct—the potential interplay between a baby's nutrition, its gut microbiome, and the epigenome. *Am J Physiol Regul Integr Comp Physiol*. 2013;304:R1065–R1069.
127. Chen J, Zhang X, Kusumo H, Costa LG, Guizzetti M. Cholesterol efflux is differentially regulated in neurons and astrocytes: Implications for brain cholesterol homeostasis. *Biochim Biophys Acta*. 2013;1831:263–275.
128. Masliah E, Mallory M, Ge N, Alford M, Veinbergs I, Roses AD. Neurodegeneration in the central nervous system of apoE-deficient mice. *Exp Neurol*. 1995;136:107–122.
129. Buttini M, Orth M, Bellosta S, et al. Expression of human apolipoprotein E3 or E4 in the brains of Apoe(−/−) mice: Isoform-specific effects on neurodegeneration. *J Neurosci*. 1999;19:4867–4880.
130. Igbavboa U, Avdulov NA, Chochina SV, Wood WG. Transbilayer distribution of cholesterol is modified in brain synaptic plasma membranes of knockout mice deficient in the low-density lipoprotein receptor, apolipoprotein E, or both proteins. *J Neurochem*. 1997;69:1661–1667.
131. Wang Y, Muneton S, Sjövall J, Jovanovic JN, Griffiths WJ. The effect of 24 S-hydroxycholesterol on cholesterol Homeostasis in neurons: Quantitative changes to the cortical neuron proteome. *J Proteome Res*. 2008;7:1606–1614.
132. Chirackal Manavalan AP, Kober A, Metso J, et al. Phospholipid transfer protein is expressed in cerebrovascular endothelial cells and involved in high density lipoprotein biogenesis and remodeling at the blood–brain barrier. *J Biol Chem*. 2014;289:4683–4698.
133. Skutches CL, Holroyde CP, Myers RN, Paul P, Reichard GA. Plasma acetate turnover and oxidation. *J Clin Invest*. 1979;64:708–713.
134. Ariyannur PS, Moffett JR, Madhavarao CN, et al. Nuclear-cytoplasmic localization of acetyl coenzyme a synthetase-1 in the rat brain. *J Comp Neurol*. 2010;518:2952–2977.
135. Siigur U, Ormisson A, Tamm A. Faecal short-chain fatty acids in breast-fed and bottle-fed infants. *Acta Paediatr*. 1993;82:536–538.
136. Edwards CA, Parrett AM, Balmer SE, Wharton BA. Faecal short chain fatty acids in breast-fed and formula-fed babies. *Acta Paediatr*. 1994;83:459–462.
137. Christian MT, Edwards CA, Preston T, Johnston L, Varley R, Weaver LT. Starch fermentation by faecal bacteria of infants, toddlers and adults: Importance for energy salvage. *Eur J Clin Nutr*. 2003;57:1486–1491.
138. Arun P, Madhavarao CN, Moffett JR, et al. Metabolic acetate therapy improves phenotype in the tremor rat model of Canavan disease. *J Inherit Metab Dis*. 2010;33:195–210.
139. Wiame E, Tyteca D, Pierrot N, et al. Molecular identification of aspartate *N*-acetyltransferase and its mutation in hypoacetylaspartia. *Biochem J*. 2010;425:127–136.
140. Aoki Y, Kasai K, Yamasue H. Age-related change in brain metabolite abnormalities in autism: A meta-analysis of proton magnetic resonance spectroscopy studies. *Transl Psychiatry*. 2012;2:e69.
141. MacFabe DF, Cain NE, Boon F, Ossenkopp KP, Cain DP. Effects of the enteric bacterial metabolic product propionic acid on object-directed behavior, social behavior, cognition, and neuroinflammation in adolescent rats: Relevance to autism spectrum disorder. *Behav Brain Res*. 2011;217:47–54.
142. Frye RE, Melnyk S, Macfabe DF. Unique acyl-carnitine profiles are potential biomarkers for acquired mitochondrial disease in autism spectrum disorder. *Transl Psychiatry*. 2013;3:e220.
143. Ghanizadeh A, Akhondzadeh S, Hormozi M, Makarem A, Abotorabi-Zarchi M, Firoozabadi A. Glutathione-related factors and oxidative stress in autism, a review. *Curr Med Chem*. 2012;19:4000–4005.
144. Frye RE, Delatorre R, Taylor H, et al. Redox metabolism abnormalities in autistic children associated with mitochondrial disease. *Transl Psychiatry*. 2013;3:e273.
145. Paul SM, Doherty JJ, Robichaud AJ, et al. The major brain cholesterol metabolite 24(S)-hydroxycholesterol is a potent allosteric modulator of *N*-methyl-D-aspartate receptors. *J Neurosci*. 2013;33:17290–17300.
146. Fukuchi M, Nii T, Ishimaru N, et al. Valproic acid induces up- or down-regulation of gene expression responsible for the neuronal excitation and inhibition in rat cortical neurons through its epigenetic actions. *Neurosci Res*. 2009;65:35–43.
147. Suzuki Y, Takahashi H, Fukuda M, et al. β-hydroxybutyrate alters GABA-transaminase activity in cultured astrocytes. *Brain Res*. 2009;1268:17–23.
148. Zhang Z, Cai YQ, Zou F, Bie B, Pan ZZ. Epigenetic suppression of GAD65 expression mediates persistent pain. *Nat Med*. 2011;17:1448–1455.
149. Gräff J, Tsai LH. Histone acetylation: Molecular mnemonics on the chromatin. *Nat Rev Neurosci*. 2013;14:97–111.
150. Chomiak T, Turner N, Hu B. What we have learned about autism spectrum disorder from valproic acid. *Pathol Res Int*. 2013;2013:712758.

151. Fagiolini M, Leblanc JJ. Autism: a critical period disorder? *Neural Plast.* 2011;2011:921680.
152. Foti SB, Chou A, Moll AD, Roskams AJ. HDAC inhibitors dysregulate neural stem cell activity in the postnatal mouse brain. *Int J Dev Neurosci.* 2013;31:434–447.
153. Kim BW, Yang S, Lee CH, Son H. A critical time window for the survival of neural progenitor cells by HDAC inhibitors in the hippocampus. *Mol Cells.* 2011;31:159–164.
154. Fass DM, Shah R, Ghosh B, et al. Effect of inhibiting histone deacetylase with short-chain carboxylic acids and their hydroxamic acid analogs on vertebrate development and neuronal chromatin. *ACS Med Chem Lett.* 2010;2:39–42.
155. Intlekofer KA, Berchtold NC, Malvaez M, et al. Exercise and sodium butyrate transform a subthreshold learning event into long-term memory via a brain-derived neurotrophic factor-dependent mechanism. *Neuropsychopharmacol.* 2013;38:2027–2034.
156. Steliou K, Boosalis MS, Perrine SP, Sangerman J, Faller DV. Butyrate histone deacetylase inhibitors. *Biores Open Access.* 2012;1:192–198.
157. Hamer HM, Jonkers D, Venema K, Vanhoutvin S, Troost FJ, Brummer RJ. Review article: The role of butyrate on colonic function. *Aliment Pharmacol Ther.* 2008;27:104–119.
158. Nauta AJ, Amor KB, Knol J, Garssen J, Van Der Beek EM. Relevance of pre- and postnatal nutrition to development and interplay between the microbiota and metabolic and immune systems. *Am J Clin Nutr.* 2013;98:586S–593S.
159. Guilloteau P, Zabielski R, Hammon HM, Metges CC. Nutritional programming of gastrointestinal tract development. Is the pig a good model for man? *Nutr Res Rev.* 2010;23:4–22.
160. Willing BP, Van Kessel AG. Intestinal microbiota differentially affect brush border enzyme activity and gene expression in the neonatal gnotobiotic pig. *J Anim Physiol Anim Nutr (Berl).* 2009;93:586–595.
161. Cummins AG, Steele TW, LaBrooy JT, Shearman DJC. Maturation of the rat small intestine at weaning: Changes in epithelial cell kinetics, bacterial flora, and mucosal immune activity. *Gut.* 1988;29:1672–1679.
162. Kozakova H, Kolinska J, Lojda Z, et al. Effect of bacterial monoassociation on brush-border enzyme activities in ex-germ-free piglets: comparison of commensal and pathogenic Escherichia coli strains. *Microbes Infect.* 2006;8:2629–2639.
163. Kozáková H, Řeháková Z, Kolínská J. *Bifidobacterium bifidum* monoassociation of gnotobiotic mice: effect on enterocyte brush-border enzymes. *Folia Microbiol (Praha).* 2001;46:573–576.
164. Williams BL, Hornig M, Buie T, et al. Impaired carbohydrate digestion and transport and mucosal dysbiosis in the intestines of children with autism and gastrointestinal disturbances. *PLoS One.* 2011;6:e24585.
165. Hooper LV, Wong MH, Thelin A, Hansson L, Falk PG, Gordon JI. Molecular analysis of commensal host–microbial relationships in the intestine. *Science.* 2001;291:881–884.
166. Inoue S, Mochizuki K, Goda T. Jejunal induction of SI and SGLT1 genes in rats by high-starch/low-fat diet is associated with histone acetylation and binding of GCN5 on the genes. *J Nutr Sci Vitaminol (Tokyo).* 2011;57:162–169.
167. Shirazi-Beechey SP, Moran AW, Batchelor DJ, Daly K, Al-Rammahi M. Glucose sensing and signalling; regulation of intestinal glucose transport. *Proc Nutr Soc.* 2011;70:185–193.
168. Mourad FH, Saadé NE. Neural regulation of intestinal nutrient absorption. *Prog Neurobiol.* 2011;95:149–162.
169. Rasmussen HS, Holtug K, Ynggard C, Mortensen PB. Faecal concentrations and production rates of short chain fatty acids in normal neonates. *Acta Paediatr Scand.* 1988;77:365–368.
170. Iwanaga T, Takebe K, Kato I, Karaki SI, Kuwahara A. Cellular expression of monocarboxylate transporters (MCT) in the digestive tract of the mouse, rat, and humans, with special reference to slc5a8. *Biomed Res.* 2006;27:243–254.
171. Moschen I, Bröer A, Galic S, Lang F, Bröer S. Significance of short chain fatty acid transport by members of the monocarboxylate transporter family (MCT). *Neurochem Res.* 2012;37:2562–2568.
172. Borthakur A, Priyamvada S, Kumar A, et al. A novel nutrient sensing mechanism underlies substrate-induced regulation of monocarboxylate transporter-1. *Am J Physiol Gastrointest Liver Physiol.* 2012;303:G1126–G1133.
173. Kirat D, Kondo K, Shimada R, Kato S. Dietary pectin up-regulates monocaboxylate transporter 1 in the rat gastrointestinal tract. *Exp Physiol.* 2009;94:422–433.
174. Lacorn M, Goerke M, Claus R. Inulin-coated butyrate increases ileal MCT1 expression and affects mucosal morphology in the porcine ileum by reduced apoptosis. *J Anim Physiol Anim Nutr (Berl).* 2010;94:670–676.
175. Metzler-Zebeli BU, Nzle MGG, Mosenthin R, Zijlstra RT. Oat β-glucan and dietary calcium and phosphorus differentially modify intestinal expression of proinflammatory cytokines and monocarboxylate transporter 1 and cecal morphology in weaned pigs. *J Nutr.* 2012;142:668–674.
176. Kirat D. Effect of pectin feeding on monocarboxylate transporters in rat adrenal gland. *J Comp Physiol [B].* 2010;180:57–65.
177. Gonçalves P, Catarino T, Gregõrio I, Martel F. Inhibition of butyrate uptake by the primary bile salt chenodeoxycholic acid in intestinal epithelial cells. *J Cell Biochem.* 2012;113:2937–2947.
178. Thibault R, De Coppet P, Daly K, et al. Down-regulation of the monocarboxylate transporter 1 is involved in butyrate deficiency during intestinal inflammation. *Gastroenterology.* 2007;133:1916–1927.
179. Buyse M, Sitaraman SV, Liu X, Bado A, Merlin D. Luminal leptin enhances CD147/MCT-1-mediated uptake of butyrate in the human intestinal cell line Caco2-BBE. *J Biol Chem.* 2002;277:28182–28190.
180. Saksena S, Theegala S, Bansal N, et al. Mechanisms underlying modulation of monocarboxylate transporter 1 (MCT1) by somatostatin in human intestinal epithelial cells. *Am J Physiol Gastrointest Liver Physiol.* 2009;297:G878–G885.
181. Wang M, Radlowski EC, Monaco MH, Fahey Jr GC, Gaskins HR, Donovan SM. Mode of delivery and early nutrition modulate microbial colonization and fermentation products in neonatal piglets. *J Nutr.* 2013;143:795–803.
182. Pierre K, Parent A, Jayet PY, Halestrap AP, Scherrer U, Pellerin L. Enhanced expression of three monocarboxylate transporter isoforms in the brain of obese mice. *J Physiol.* 2007;583:469–486.
183. Leino RL, Gerhart DZ, Duelli R, Enerson BE, Drewes LR. Diet-induced ketosis increases monocarboxylate transporter (MCT1) levels in rat brain. *Neurochem Int.* 2001;38:519–527.

184. Vannucci SJ, Simpson IA. Developmental switch in brain nutrient transporter expression in the rat. *Am J Physiol Endocrinol Metab.* 2003;285:E1127–E1134.
185. Lizarraga-Mollinedo E, Fernández-Millán E, de Toro Martín J, Martínez-Honduvilla C, Escrivá F, Álvarez C. Early undernutrition induces glucagon resistance and insulin hypersensitivity in the liver of suckling rats. *Am J Physiol Endocrinol Metab.* 2012;302: E1070–E1077.
186. Pierre K, Pellerin L. Monocarboxylate transporters in the central nervous system: Distribution, regulation and function. *J Neurochem.* 2005;94:1–14.
187. Geurts L, Neyrinck AM, Delzenne NM, Knauf C, Cani PD. Gut microbiota controls adipose tissue expansion, gut barrier and glucose metabolism: novel insights into molecular targets and interventions using prebiotics. *Benef Microbes.* 2014;5:3–17.
188. Tuohy KM, Fava F, Viola R. 'The way to a man's heart is through his gut microbiota' – dietary pro- and prebiotics for the management of cardiovascular risk. *Proc Nutr Soc.* 2014;73:172–185.
189. Fava F, Danese S. Intestinal microbiota in inflammatory bowel disease: friend of foe? *World J Gastroenterol.* 2011;17:557–566.
190. Del Rio D, Rodriguez-Mateos A, Spencer JP, Tognolini M, Borges G, Crozier A. Dietary (poly)phenolics in human health: structures, bioavailability, and evidence of protective effects against chronic diseases. *Antioxid Redox Signal.* 2013;18:1818–1892.
191. Zhao G, Ford ES, Li C, Tsai J, Dhingra S, Balluz LS. Waist circumference, abdominal obesity, and depression among overweight and obese U.S. adults: National health and nutrition examination survey 2005–2006. *BMC Psychiatry.* 2011;11:130.
192. Yu ZM, Parker L, Dummer TJB. *Depressive symptoms, diet quality, physical activity, and body composition among populations in Nova Scotia. Preventive Medicine.* Canada: Report from the Atlantic Partnership for Tomorrow's Health; 2014;61:106–113.
193. Zuckerman KE, Hill AP, Guion K, Voltolina L, Fombonne E. Overweight and obesity: prevalence and correlates in a large clinical sample of children with autism spectrum disorder. *J Autism Dev Disord.* 2014 [Epub ahead of print].
194. Sullivan EL, Nousen EK, Chamlou KA. Maternal high fat diet consumption during the perinatal period programs offspring behavior. *Physiol Behav.* 2014;123:236–242.
195. Valladolid-Acebes I, Merino B, Principato A, et al. High-fat diets induce changes in hippocampal glutamate metabolism and neurotransmission. *Am J Physiol Endocrinol Metab.* 2012;302:E396–E402.
196. Ouyang S, Hsuchou H, Kastin AJ, Wang Y, Yu C, Pan W. Diet-induced obesity suppresses expression of many proteins at the blood–brain barrier. *J Cereb Blood Flow Metab.* 2014;34:43–51.
197. De Filippo C, Cavalieri D, Di Paola M, et al. Impact of diet in shaping gut microbiota revealed by a comparative study in children from Europe and rural Africa. *Proc Natl Acad Sci USA.* 2010;107:14691–14696.
198. Maathuis AJ, van den Heuvel EG, Schoterman MH, Venema K. Galacto-oligosaccharides have prebiotic activity in a dynamic in vitro colon model using a (13)C-labeling technique. *J Nutr.* 2012;142:1205–1212.
199. Morrison DJ, Mackay WG, Edwards CA, Preston T, Dodson B, Weaver LT. Butyrate production from oligofructose fermentation by the human faecal flora: What is the contribution of extracellular acetate and lactate? *Br J Nutr.* 2006;96:570–577.
200. Van Dokkum W, Wezendonk B, Srikumar TS, Van Den Heuvel EGHM. Effect of nondigestible oligosaccharides on large-bowel functions, blood lipid concentrations and glucose absorption in young healthy male subjects. *Eur J Clin Nutr.* 1999;53:1–7.
201. Spencer JPE. The impact of fruit flavonoids on memory and cognition. *Br J Nutr.* 2010;104:S40–S47.
202. Bonaccio M, Di Castelnuovo A, Bonanni A, et al. Adherence to a Mediterranean diet is associated with a better health-related quality of life: A possible role of high dietary antioxidant content. *BMJ Open.* 2013;3:e003003.
203. Joseph J, Cole G, Head E, Ingram D. Nutrition, brain aging, and neurodegeneration. *J Neurosci.* 2009;29:12795–12801.
204. Schaffer S, Asseburg H, Kuntz S, Muller WE, Eckert GP. Effects of polyphenols on brain ageing and Alzheimer's disease: Focus on mitochondria. *Mol Neurobiol.* 2012;46:161–178.
205. Singh M, Arseneault M, Sanderson T, Murthy V, Ramassamy C. Challenges for research on polyphenols from foods in Alzheimer's disease: bioavailability, metabolism, and cellular and molecular mechanisms. *J Agric Food Chem.* 2008;56:4855–4873.
206. Schaffer S, Halliwell B. Do polyphenols enter the brain and does it matter? Some theoretical and practical considerations. *Genes Nutr.* 2012;7:99–109.
207. Verzelloni E, Pellacani C, Tagliazucchi D, et al. Antiglycative and neuroprotective activity of colon-derived polyphenol catabolites. *Mol Nutr Food Res.* 2011;55:S35–S43.
208. An LJ, Guan S, Shi GF, Bao YM, Duan YL, Jiang B. Protocatechuic acid from Alpinia oxyphylla against MPP+ - induced neurotoxicity in PC12 cells. *Food Chem Toxicol.* 2006;44:436–443.
209. Ferruzzi MG, Lobo JK, Janle EM, et al. Bioavailability of gallic acid and catechins from grape seed polyphenol extract is improved by repeated dosing in rats: Implications for treatment in Alzheimer's disease. *J Alzheimers Dis.* 2009;18:113–124.
210. Banji D, Banji OJF, Abbagoni S, Hayath MS, Kambam S, Chiluka VL. Amelioration of behavioral aberrations and oxidative markers by green tea extract in valproate induced autism in animals. *Brain Res.* 2011;1410:141–151.
211. Sandhya T, Sowjanya J, Veeresh B. *Bacopa monniera* (L.) *Wettst* ameliorates behavioral alterations and oxidative markers in sodium valproate induced autism in rats. *Neurochem Res.* 2012;37:1121–1131.
212. Kim P, Park JH, Kwon KJ, et al. Effects of Korean red ginseng extracts on neural tube defects and impairment of social interaction induced by prenatal exposure to valproic acid. *Food Chem Toxicol.* 2013;51:288–296.
213. Hasanzadeh E, Mohammadi MR, Ghanizadeh A, et al. A double-blind placebo controlled trial of ginkgo biloba added to risperidone in patients with autistic disorders. *Child Psychiatry Hum Dev.* 2012;43:674–682.
214. Taliou A, Zintzaras E, Lykouras L, Francis K. An open-label pilot study of a formulation containing the anti-inflammatory flavonoid luteolin and its effects on behavior in children with autism spectrum disorders. *Clin Ther.* 2013;35:592–602.
215. Sudo N, Chida Y, Aiba Y, et al. Postnatal microbial colonization programs the hypothalamic–pituitary–adrenal system for stress response in mice. *J Physiol.* 2004;558:263–275.
216. Neufeld KM, Kang N, Bienenstock J, Foster JA. Reduced anxiety-like behavior and central neurochemical change in germ-free mice. *Neurogastroenterol Motil.* 2011;23:255–264. e119.

217. Heijtz RD, Wang S, Anuar F, et al. Normal gut microbiota modulates brain development and behavior. *Proc Natl Acad Sci USA*. 2011;108:3047–3052.
218. Desbonnet L, Garrett L, Clarke G, Bienenstock J, Dinan TG. The probiotic *Bifidobacteria infantis*: An assessment of potential antidepressant properties in the rat. *J Psychiatr Res*. 2008;43:164–174.
219. Desbonnet L, Garrett L, Clarke G, Kiely B, Cryan JF, Dinan TG. Effects of the probiotic *Bifidobacterium infantis* in the maternal separation model of depression. *Neuroscience*. 2010;170:1179–1188.
220. Messaoudi M, Violle N, Bisson J, Desor D, Javelot H, Rougeot C. Beneficial psychological effects of a probiotic formulation (*Lactobacillus helveticus* R0052 and *Bifidobacterium longum* R0175) in healthy human volunteers. *Gut Microbes*. 2011;2:256–261.
221. Bravo JA, Forsythe P, Chew MV, et al. Ingestion of *Lactobacillus* strain regulates emotional behavior and central GABA receptor expression in a mouse via the vagus nerve. *Proc Natl Acad Sci USA*. 2011;108:16050–16055.
222. Gareau MG, Wine E, Rodrigues DM, et al. Bacterial infection causes stress-induced memory dysfunction in mice. *Gut*. 2011;60:307–317.
223. Lyte M, Li W, Opitz N, Gaykema RPA, Goehler LE. Induction of anxiety-like behavior in mice during the initial stages of infection with the agent of murine colonic hyperplasia Citrobacter rodentium. *Physiol Behav*. 2006;89:350–357.
224. Bailey MT, Dowd SE, Parry NMA, Galley JD, Schauer DB, Lyte M. Stressor exposure disrupts commensal microbial populations in the intestines and leads to increased colonization by *Citrobacter rodentium*. *Infect Immun*. 2010;78:1509–1519.
225. Hsiao EY, McBride SW, Hsien S, et al. Microbiota modulate behavioral and physiological abnormalities associated with neurodevelopmental disorders. *Cell*. 2013;155(7):1451–1463.
226. Lafaye A, Junot C, Ramounet-Le Gall B, Fritsch P, Ezan E, Tabet JC. Profiling of sulfoconjugates in urine by using precursor ion and neutral loss scans in tandem mass spectrometry. Application to the investigation of heavy metal toxicity in rats. *J Mass Spectrom*. 2004;39(6):655–664.

Index

Note: Page numbers followed by "*f*" and "*t*" refer to figures and tables, respectively.

E

F

G

Printed in the United States
By Bookmasters

Printed in the United States
By Bookmasters